shape optimization and optimal design

proceedings of the IFIP conference

edited by

John Cagnol
Pôle Universitaire Léonard de Vinci
Paris, France

Michael P. Polis
Oakland University
Rochester, Michigan

Jean-Paul Zolésio
Ecole des Mines de Paris
Sophia Antipolis, France

MARCEL DEKKER

MARCEL DEKKER, INC. NEW YORK • BASEL

ISBN: 0-8247-0556-4

This book is printed on acid-free paper.

Headquarters
Marcel Dekker, Inc.
270 Madison Avenue, New York, NY 10016
tel: 212-696-9000; fax: 212-685-4540

Eastern Hemisphere Distribution
Marcel Dekker AG
Hutgasse 4, Postfach 812, CH-4001 Basel, Switzerland
tel: 41-61-261-8482; fax: 41-61-261-8896

World Wide Web
http://www.dekker.com

The publisher offers discounts on this book when ordered in bulk quantities. For more information, write to Special Sales/Professional Marketing at the headquarters address above.

Current printing (last digit):
10 9 8 7 6 5 4 3 2 1

PRINTED IN THE UNITED STATES OF AMERICA

shape optimization and optimal design

LECTURE NOTES IN PURE AND APPLIED MATHEMATICS

1. *N. Jacobson,* Exceptional Lie Algebras
2. *L.-Å. Lindahl and F. Poulsen,* Thin Sets in Harmonic Analysis
3. *I. Satake,* Classification Theory of Semi-Simple Algebraic Groups
4. *F. Hirzebruch et al., Differentiable Manifolds and Quadratic Forms*
5. *I. Chavel,* Riemannian Symmetric Spaces of Rank One
6. *R. B. Burckel,* Characterization of C(X) Among Its Subalgebras
7. *B. R. McDonald et al.,* Ring Theory
8. *Y.-T. Siu,* Techniques of Extension on Analytic Objects
9. *S. R. Caradus et al.,* Calkin Algebras and Algebras of Operators on Banach Spaces
10. *E. O. Roxin et al.,* Differential Games and Control Theory
11. *M. Orzech and C. Small,* The Brauer Group of Commutative Rings
12. *S. Thomier,* Topology and Its Applications
13. *J. M. Lopez and K. A. Ross,* Sidon Sets
14. *W. W. Comfort and S. Negrepontis,* Continuous Pseudometrics
15. *K. McKennon and J. M. Robertson,* Locally Convex Spaces
16. *M. Carmeli and S. Malin,* Representations of the Rotation and Lorentz Groups
17. *G. B. Seligman,* Rational Methods in Lie Algebras
18. *D. G. de Figueiredo,* Functional Analysis
19. *L. Cesari et al.,* Nonlinear Functional Analysis and Differential Equations
20. *J. J. Schäffer,* Geometry of Spheres in Normed Spaces
21. *K. Yano and M. Kon,* Anti-Invariant Submanifolds
22. *W. V. Vasconcelos,* The Rings of Dimension Two
23. *R. E. Chandler,* Hausdorff Compactifications
24. *S. P. Franklin and B. V. S. Thomas,* Topology
25. *S. K. Jain,* Ring Theory
26. *B. R. McDonald and R. A. Morris,* Ring Theory II
27. *R. B. Mura and A. Rhemtulla,* Orderable Groups
28. *J. R. Graef,* Stability of Dynamical Systems
29. *H.-C. Wang,* Homogeneous Branch Algebras
30. *E. O. Roxin et al.,* Differential Games and Control Theory II
31. *R. D. Porter,* Introduction to Fibre Bundles
32. *M. Altman,* Contractors and Contractor Directions Theory and Applications
33. *J. S. Golan,* Decomposition and Dimension in Module Categories
34. *G. Fairweather,* Finite Element Galerkin Methods for Differential Equations
35. *J. D. Sally,* Numbers of Generators of Ideals in Local Rings
36. *S. S. Miller,* Complex Analysis
37. *R. Gordon,* Representation Theory of Algebras
38. *M. Goto and F. D. Grosshans,* Semisimple Lie Algebras
39. *A. I. Arruda et al.,* Mathematical Logic
40. *F. Van Oystaeyen,* Ring Theory
41. *F. Van Oystaeyen and A. Verschoren,* Reflectors and Localization
42. *M. Satyanarayana,* Positively Ordered Semigroups
43. *D. L Russell,* Mathematics of Finite-Dimensional Control Systems
44. *P.-T. Liu and E. Roxin,* Differential Games and Control Theory III
45. *A. Geramita and J. Seberry,* Orthogonal Designs
46. *J. Cigler, V. Losert, and P. Michor,* Banach Modules and Functors on Categories of Banach Spaces
47. *P.-T. Liu and J. G. Sutinen,* Control Theory in Mathematical Economics
48. *C. Byrnes,* Partial Differential Equations and Geometry
49. *G. Klambauer,* Problems and Propositions in Analysis
50. *J. Knopfmacher,* Analytic Arithmetic of Algebraic Function Fields
51. *F. Van Oystaeyen,* Ring Theory
52. *B. Kadem,* Binary Time Series
53. *J. Barros-Neto and R. A. Artino,* Hypoelliptic Boundary-Value Problems
54. *R. L. Sternberg et al.,* Nonlinear Partial Differential Equations in Engineering and Applied Science
55. *B. R. McDonald,* Ring Theory and Algebra III
56. *J. S. Golan,* Structure Sheaves Over a Noncommutative Ring
57. *T. V. Narayana et al.,* Combinatorics, Representation Theory and Statistical Methods in Groups
58. *T. A. Burton,* Modeling and Differential Equations in Biology
59. *K. H. Kim and F. W. Roush,* Introduction to Mathematical Consensus Theory

60. *J. Banas and K. Goebel,* Measures of Noncompactness in Banach Spaces
61. *O. A. Nielson,* Direct Integral Theory
62. *J. E. Smith et al.,* Ordered Groups
63. *J. Cronin,* Mathematics of Cell Electrophysiology
64. *J. W. Brewer,* Power Series Over Commutative Rings
65. *P. K. Kamthan and M. Gupta,* Sequence Spaces and Series
66. *T. G. McLaughlin,* Regressive Sets and the Theory of Isols
67. *T. L. Herdman et al.,* Integral and Functional Differential Equations
68. *R. Draper,* Commutative Algebra
69. *W. G. McKay and J. Patera,* Tables of Dimensions, Indices, and Branching Rules for Representations of Simple Lie Algebras
70. *R. L. Devaney and Z. H. Nitecki,* Classical Mechanics and Dynamical Systems
71. *J. Van Geel,* Places and Valuations in Noncommutative Ring Theory
72. *C. Faith,* Injective Modules and Injective Quotient Rings
73. *A. Fiacco,* Mathematical Programming with Data Perturbations I
74. *P. Schultz et al.,* Algebraic Structures and Applications
75. *L Bican et al.,* Rings, Modules, and Preradicals
76. *D. C. Kay and M. Breen,* Convexity and Related Combinatorial Geometry
77. *P. Fletcher and W. F. Lindgren,* Quasi-Uniform Spaces
78. *C.-C. Yang,* Factorization Theory of Meromorphic Functions
79. *O. Taussky,* Ternary Quadratic Forms and Norms
80. *S. P. Singh and J. H. Burry,* Nonlinear Analysis and Applications
81. *K. B. Hannsgen et al.,* Volterra and Functional Differential Equations
82. *N. L. Johnson et al.,* Finite Geometries
83. *G. I. Zapata,* Functional Analysis, Holomorphy, and Approximation Theory
84. *S. Greco and G. Valla,* Commutative Algebra
85. *A. V. Fiacco,* Mathematical Programming with Data Perturbations II
86. *J.-B. Hiriart-Urruty et al.,* Optimization
87. *A. Figa Talamanca and M. A. Picardello,* Harmonic Analysis on Free Groups
88. *M. Harada,* Factor Categories with Applications to Direct Decomposition of Modules
89. *V. I. Istrătescu,* Strict Convexity and Complex Strict Convexity
90. *V. Lakshmikantham,* Trends in Theory and Practice of Nonlinear Differential Equations
91. *H. L. Manocha and J. B. Srivastava,* Algebra and Its Applications
92. *D. V. Chudnovsky and G. V. Chudnovsky,* Classical and Quantum Models and Arithmetic Problems
93. *J. W. Longley,* Least Squares Computations Using Orthogonalization Methods
94. *L. P. de Alcantara,* Mathematical Logic and Formal Systems
95. *C. E. Aull,* Rings of Continuous Functions
96. *R. Chuaqui,* Analysis, Geometry, and Probability
97. *L. Fuchs and L. Salce,* Modules Over Valuation Domains
98. *P. Fischer and W. R. Smith,* Chaos, Fractals, and Dynamics
99. *W. B. Powell and C. Tsinakis,* Ordered Algebraic Structures
100. *G. M. Rassias and T. M. Rassias,* Differential Geometry, Calculus of Variations, and Their Applications
101. *R.-E. Hoffmann and K. H. Hofmann,* Continuous Lattices and Their Applications
102. *J. H. Lightbourne III and S. M. Rankin III,* Physical Mathematics and Nonlinear Partial Differential Equations
103. *C. A. Baker and L. M. Batten,* Finite Geometrics
104. *J. W. Brewer et al.,* Linear Systems Over Commutative Rings
105. *C. McCrory and T. Shifrin,* Geometry and Topology
106. *D. W. Kueke et al.,* Mathematical Logic and Theoretical Computer Science
107. *B.-L. Lin and S. Simons,* Nonlinear and Convex Analysis
108. *S. J. Lee,* Operator Methods for Optimal Control Problems
109. *V. Lakshmikantham,* Nonlinear Analysis and Applications
110. *S. F. McCormick,* Multigrid Methods
111. *M. C. Tangora,* Computers in Algebra
112. *D. V. Chudnovsky and G. V. Chudnovsky,* Search Theory
113. *D. V. Chudnovsky and R. D. Jenks,* Computer Algebra
114. *M. C. Tangora,* Computers in Geometry and Topology
115. *P. Nelson et al.,* Transport Theory, Invariant Imbedding, and Integral Equations
116. *P. Clément et al.,* Semigroup Theory and Applications
117. *J. Vinuesa,* Orthogonal Polynomials and Their Applications
118. *C. M. Dafermos et al.,* Differential Equations
119. *E. O. Roxin,* Modern Optimal Control
120. *J. C. Díaz,* Mathematics for Large Scale Computing

121. *P. S. Milojević* Nonlinear Functional Analysis
122. *C. Sadosky,* Analysis and Partial Differential Equations
123. *R. M. Shortt,* General Topology and Applications
124. *R. Wong,* Asymptotic and Computational Analysis
125. *D. V. Chudnovsky and R. D. Jenks,* Computers in Mathematics
126. *W. D. Wallis et al.,* Combinatorial Designs and Applications
127. *S. Elaydi,* Differential Equations
128. *G. Chen et al.,* Distributed Parameter Control Systems
129. *W. N. Everitt,* Inequalities
130. *H. G. Kaper and M. Garbey,* Asymptotic Analysis and the Numerical Solution of Partial Differential Equations
131. *O. Arino et al.,* Mathematical Population Dynamics
132. *S. Coen,* Geometry and Complex Variables
133. *J. A. Goldstein et al.,* Differential Equations with Applications in Biology, Physics, and Engineering
134. *S. J. Andima et al.,* General Topology and Applications
135. *P Clément et al.,* Semigroup Theory and Evolution Equations
136. *K. Jarosz,* Function Spaces
137. *J. M. Bayod et al.,* *p*-adic Functional Analysis
138. *G. A. Anastassiou,* Approximation Theory
139. *R. S. Rees,* Graphs, Matrices, and Designs
140. *G. Abrams et al.,* Methods in Module Theory
141. *G. L. Mullen and P. J.-S. Shiue,* Finite Fields, Coding Theory, and Advances in Communications and Computing
142. *M. C. Joshi and A. V. Balakrishnan,* Mathematical Theory of Control
143. *G. Komatsu and Y. Sakane,* Complex Geometry
144. *I. J. Bakelman,* Geometric Analysis and Nonlinear Partial Differential Equations
145. *T. Mabuchi and S. Mukai,* Einstein Metrics and Yang–Mills Connections
146. *L. Fuchs and R. Göbel,* Abelian Groups
147. *A. D. Pollington and W. Moran,* Number Theory with an Emphasis on the Markoff Spectrum
148. *G. Dore et al.,* Differential Equations in Banach Spaces
149. *T. West,* Continuum Theory and Dynamical Systems
150. *K. D. Bierstedt et al.,* Functional Analysis
151. *K. G. Fischer et al.,* Computational Algebra
152. *K. D. Elworthy et al.,* Differential Equations, Dynamical Systems, and Control Science
153. *P.-J. Cahen, et al.,* Commutative Ring Theory
154. *S. C. Cooper and W. J. Thron,* Continued Fractions and Orthogonal Functions
155. *P. Clément and G. Lumer,* Evolution Equations, Control Theory, and Biomathematics
156. *M. Gyllenberg and L. Persson,* Analysis, Algebra, and Computers in Mathematical Research
157. *W. O. Bray et al.,* Fourier Analysis
158. *J. Bergen and S. Montgomery,* Advances in Hopf Algebras
159. *A. R. Magid,* Rings, Extensions, and Cohomology
160. *N. H. Pavel,* Optimal Control of Differential Equations
161. *M. Ikawa,* Spectral and Scattering Theory
162. *X. Liu and D. Siegel,* Comparison Methods and Stability Theory
163. *J.-P. Zolésio,* Boundary Control and Variation
164. *M. Křížek et al.,* Finite Element Methods
165. *G. Da Prato and L. Tubaro,* Control of Partial Differential Equations
166. *E. Ballico,* Projective Geometry with Applications
167. *M. Costabel et al.,* Boundary Value Problems and Integral Equations in Nonsmooth Domains
168. *G. Ferreyra, G. R. Goldstein, and F. Neubrander,* Evolution Equations
169. *S. Huggett,* Twistor Theory
170. *H. Cook et al.,* Continua
171. *D. F. Anderson and D. E. Dobbs,* Zero-Dimensional Commutative Rings
172. *K. Jarosz,* Function Spaces
173. *V. Ancona et al.,* Complex Analysis and Geometry
174. *E. Casas,* Control of Partial Differential Equations and Applications
175. *N. Kalton et al.,* Interaction Between Functional Analysis, Harmonic Analysis, and Probability
176. *Z. Deng et al.,* Differential Equations and Control Theory
177. *P. Marcellini et al.* Partial Differential Equations and Applications
178. *A. Kartsatos,* Theory and Applications of Nonlinear Operators of Accretive and Monotone Type
179. *M. Maruyama,* Moduli of Vector Bundles
180. *A. Ursini and P. Agliànò,* Logic and Algebra
181. *X. H. Cao et al.,* Rings, Groups, and Algebras
182. *D. Arnold and R. M. Rangaswamy,* Abelian Groups and Modules
183. *S. R. Chakravarthy and A. S. Alfa,* Matrix-Analytic Methods in Stochastic Models

184. *J. E. Andersen et al.*, Geometry and Physics
185. *P.-J. Cahen et al.*, Commutative Ring Theory
186. *J. A. Goldstein et al.*, Stochastic Processes and Functional Analysis
187. *A. Sorbi*, Complexity, Logic, and Recursion Theory
188. *G. Da Prato and J.-P. Zolésio*, Partial Differential Equation Methods in Control and Shape Analysis
189. *D. D. Anderson*, Factorization in Integral Domains
190. *N. L. Johnson*, Mostly Finite Geometries
191. *D. Hinton and P. W. Schaefer*, Spectral Theory and Computational Methods of Sturm–Liouville Problems
192. *W. H. Schikhof et al.*, *p*-adic Functional Analysis
193. *S. Sertöz*, Algebraic Geometry
194. *G. Caristi and E. Mitidieri*, Reaction Diffusion Systems
195. *A. V. Fiacco*, Mathematical Programming with Data Perturbations
196. *M. Křížek et al.*, Finite Element Methods: Superconvergence, Post-Processing, and A Posteriori Estimates
197. *S. Caenepeel and A. Verschoren*, Rings, Hopf Algebras, and Brauer Groups
198. *V. Drensky et al.*, Methods in Ring Theory
199. *W. B. Jones and A. Sri Ranga*, Orthogonal Functions, Moment Theory, and Continued Fractions
200. *P. E. Newstead*, Algebraic Geometry
201. *D. Dikranjan and L. Salce*, Abelian Groups, Module Theory, and Topology
202. *Z. Chen et al.*, Advances in Computational Mathematics
203. *X. Caicedo and C. H. Montenegro*, Models, Algebras, and Proofs
204. *C. Y. Yıldırım and S. A. Stepanov*, Number Theory and Its Applications
205. *D. E. Dobbs et al.*, Advances in Commutative Ring Theory
206. *F. Van Oystaeyen*, Commutative Algebra and Algebraic Geometry
207. *J. Kakol et al.*, *p*-adic Functional Analysis
208. *M. Boulagouaz and J.-P. Tignol*, Algebra and Number Theory
209. *S. Caenepeel and F. Van Oystaeyen*, Hopf Algebras and Quantum Groups
210. *F. Van Oystaeyen and M. Saorin*, Interactions Between Ring Theory and Representations of Algebras
211. *R. Costa et al.*, Nonassociative Algebra and Its Applications
212. *T.-X. He*, Wavelet Analysis and Multiresolution Methods
213. *H. Hudzik and L. Skrzypczak*, Function Spaces: The Fifth Conference
214. *J. Kajiwara et al.*, Finite or Infinite Dimensional Complex Analysis
215. *G. Lumer and L. Weis*, Evolution Equations and Their Applications in Physical and Life Sciences
216. *J. Cagnol et al.*, Shape Optimization and Optimal Design

Additional Volumes in Preparation

Preface

This volume comprises selected papers from the sessions "Distributed Parameter Systems" and "Optimization Methods and Engineering Design" held within the 19th conference System Modeling and Optimization in Cambridge, England.

Those sessions were organized by the Working Groups 7.2 (Computational Techniques in Distributed Systems) and 7.4 (Discrete Optimization) of the Technical Committee 7 (Modeling and Optimization Techniques) of the International Federation for Information Processing (IFIP).

The aim of these sessions was to present the latest developments and major advances in the fields of passive and active control for systems governed by partial differential equations. Shape analysis and optimal shape design were particularly emphasized during the talks. The active control portion includes exact controllability and nonlinear boundary control/ stabilization.

We would like to acknowledge the contribution of M. J. D. Powell, who was the main organizer of the conference, and of M. Kocvara and K. Zowe, who were the organizers of the session "Optimization Methods and Engineering Design."

John Cagnol
Michael P. Polis
Jean-Paul Zolésio

Contents

Contributors

Konstantin Afanasiev Technische Universität, Berlin, Berlin, Germany

Sébastien Boisgérault Centre de Mathématiques Appliquées, Ecole des Mines de Paris, Sophia Antipolis, France

John Cagnol Pôle Universitaire Léonard de Vinci, Paris La Defense, France

Michel C. Delfour Université de Montréal, Montréal, Quebec, Canada

Jean Deteix Université de Montréal, Montréal, Quebec, Canada

Zdeněk Dostál VSB-Technical University Ostrava, Ostrava, Czech Republic

Raja Dziri Université de Tunis, Tunis, Tunisia

Matthias Eller Georgetown University, Washington, DC

Jamel Ferchichi Centre de Mathématiques Appliquées, Ecole des Mines de Paris, Sophia Antipolis, France, and LAMSIN, Université de Tunis, Tunis, Tunisia

Gilles Fremiot Institut Eli Cartan, Université Henri Poincaré Nancy I, Vandoeuvre lès Nancy, France

Leszek Gasiński Institute of Computer Science, Jagiellonian University, Kraków, Poland

Francisco A. M. Gomes University of Campinas, Campinas SP, Brazil

Nicolas Gomez School of Engineering and Computer Science, Oakland University, Rochester, Michigan

Jacques Henry INRIA, Le Chesnay, France

Michael Hinze Technische Universität, Berlin, Berlin, Germany

Irena Lasiecka University of Virginia, Charlottesville, Virginia

Jean-Paul Marmorat Centre de Mathématiques Appliquées, Ecole des Mines de Paris, Sophia Antipolis, France

Andrzej Myśliński Institute of Transport, Warsaw University of Technology, Warsaw, Poland

Cătălin Popa Universitatea "Al.I. Cuza," Iaşi, Romania

Edward Rydygier The Andrzej Soltan Institute for Nuclear Studies, Otwock-Swierk, Poland

Sandra A. Santos University of Campinas, Campinas SP, Brazil

Marianna A. Shubov Texas Tech University, Lubbock, Texas

Jan Sokolowski Institut Eli Cartan, Université Henri Poincaré Nancy I, Vandoeuvre lès Nancy, France

Roberto Triggiani University of Virginia, Charlottesville, Virginia

Zdzislaw W. Trzaska Department of Electrical Engineering, Warsaw University of Technology, Warsaw, Poland

Antoine Viel UTC, Compiegne, France

Jean-Pierre Yvon INSA-Rennes, Rennes, France

Jean-Paul Zolésio CNRS and Ecole des Mines de Paris, Sophia Antipolis, France

Boundary Variations in the Navier-Stokes Equations and Lagrangian Functionals

Sébastien Boisgérault and Jean-Paul Zolésio

Abstract. We study the shape sensitivity of the stationary Navier-Stokes Equations in the general case of non-homogeneous and shape-dependent forces and boundary conditions. Under an assumption of non-singularity of the equations, the shape differentiability of the velocity and the pressure are obtained in some Sobolev spaces. The influence of the regularity of the geometrical and functional data on the best space for which the result holds is stressed. We apply these results on a class of shape functionals where a high regularity is required: the Lagrangian functionals. Their main characteristic is to take into account the paths of the fluid particles. The usual shape calculus is extended to take into account such features. We determine the shape derivative of a shape-dependent flow and develop the methods to achieve the explicit calculation of the shape gradient.

1 Introduction

1.1 The Navier-Stokes Equations

We consider the stationary incompressible Navier-Stokes Equations (NSE) in some smooth enough open and bounded sets $\Omega \subset \mathbb{R}^3$ of boundary Γ

$$\begin{aligned} -\nu\Delta u + (u\cdot\nabla)u + \nabla p &= f \quad \text{in } \Omega \\ \operatorname{div} u &= 0 \quad \text{in } \Omega \\ u &= g \quad \text{on } \Gamma \end{aligned} \tag{1.1}$$

The shape analysis will therefore include as special cases the situation where the flow is uniquely driven by the force field (with homogeneous Dirichlet boundary conditions, cf. [3]), as well as the one where the flow is induced by a body moving at a constant velocity (cf. [1]).

Moreover, for a greater generality, we assume that the data (f, g) may explicitly depend on the shape Ω. For a given set Ω, the corresponding value of f, denoted f_Ω, is *a priori* defined only on Ω (and g_Γ only on Γ). We assume that $f_\Omega \in H^{-1}(\Omega; \mathbb{R}^3)$, $g_\Gamma \in H^{1/2}(\Gamma; \mathbb{R}^3)$ and moreover that for any admissible set Ω, and for any connected component Λ of Γ, we have

$$\int_\Lambda \langle g_\Gamma, n \rangle_{\mathbb{R}^3} \, d\mathcal{H}^2 = 0 \tag{1.2}$$

(n_Ω, or simply n when no doubt is possible, is the unit outer normal to Ω).

We recall briefly the corresponding abstract setting. We denote by $V^1(\Omega)$ the space $\{u \in H^1(\Omega; \mathbb{R}^3),\ \operatorname{div} u = 0\}$, we set $V_0^1(\Omega) = V^1(\Omega) \cap H_0^1(\Omega; \mathbb{R}^3)$ and $V^{1/2}(\Gamma) = \{g \in H^{1/2}(\Gamma; \mathbb{R}^3),\ \int_\Gamma \langle g, n \rangle \, d\mathcal{H}^2 = 0\}$. The linear operator $\pi : H^{-1}(\Omega; \mathbb{R}^3) \to (V_0^1(\Omega))'$ is Leray's projector: $\pi(f)$ is the restriction of the linear form f on $H_0^1(\Omega; \mathbb{R}^3)$ to $V_0^1(\Omega)$. For any u and $v \in V^1(\Omega)$, we set $Au = -\pi(\Delta u)$ and $B(u, v) = \pi((u \cdot \nabla) v)$. The (nonlinear) Navier-Stokes operator is the mapping

$$F : \begin{array}{ccc} V^1(\Omega) & \to & V_0^1(\Omega)' \times V^{1/2}(\Gamma) \\ u & \mapsto & (\nu A u + B(u, u), u|_\Gamma) \end{array} \tag{1.3}$$

1.2 Shape Analysis Framework

The shape sensitivity of this equation is studied in the framework of the *Speed Method.* The first step is to generate some of the geometries around the reference set Ω while staying in a given design region. In that purpose, we associate to a smooth, open and bounded set D (designed later on as the *hold-all*), a velocity space $\mathcal{V}$, chosen among the $\mathcal{V}_k$ for a $k \geq 1$

$$\mathcal{V}_k = \{V \in C^0([-T; T]; C^k(\overline{D}; \mathbb{R}^3)),\ \langle V, n_D \rangle_{\mathbb{R}^3} = 0 \text{ on } \partial D\} \tag{1.4}$$

Then for any $V \in \mathcal{V}$, a one-parameter family of deformations $T_s : \overline{D} \to \overline{D}$ is given by the following initial-value problem

$$\begin{aligned} \partial_s T_s &= V(s) \circ T_s \\ T_0 &= I \end{aligned} \tag{1.5}$$

We denote $\Omega_s := T_s(\Omega)$ the corresponding transported sets generated by Ω and V. We also set $\Gamma_s := T_s(\Gamma)$.

The regularity of shape-dependent mappings such as f and g are defined in the following way. Let $W(\Omega)$ be either one of the Sobolev spaces $W^{m,p}(\Omega; \mathbb{R}^m)$ ($m \geq 0$, $p \geq 1$) or one of the spaces $C^k(\overline{\Omega}; \mathbb{R}^m)$ ($k \geq 0$).

DEFINITION 1.1. *We say that the mapping f is C^k with respect to the shape in $W(\Omega)$ if for any $V \in \mathcal{V}$, the mapping $s \mapsto f_{\Omega_s} \circ T_s$ is in $C^k(I; W(\Omega))$ on a neighborhood I of* 0.

The shape sensitivity results of the section 2 are expressed in Sobolev spaces whereas the C^k-spaces are needed for the study of the Lagrangian functionals (section 3).

Of course, an analogous definition holds for the shape-dependent mappings g defined on the boundary Γ. Two different types of derivatives with respect to the geometry may be used to describe the variations of these fields: the *material derivative* of f and g, given by

$$\dot{f}_\Omega = \frac{d}{ds} f_{\Omega_s} \circ T_s|_{s=0} \text{ and } \dot{g}_\Gamma = \frac{d}{ds} g_{\Gamma_s} \circ T_s|_{s=0} \tag{1.6}$$

and the *shape derivative* and *boundary shape derivative*

$$f'_\Omega = \dot{f}_\Omega - Df_\Omega \cdot V(0) \text{ and } g'_\Gamma = \dot{g}_\Gamma - D_\tau g_\Gamma \cdot V(0) \tag{1.7}$$

($D_\tau g$ is the tangential Jacobian matrix of g). When the choice of Ω is clear, the corresponding subscript may be dropped.

In order to characterize the regularity of shape-dependent forces f that belong to $H^{-1}(\Omega; \mathbb{R}^3)$, we define $\phi_s : H_0^1(\Omega; \mathbb{R}^3) \to H_0^1(\Omega_s; \mathbb{R}^3)$ by $\phi_s(f) = f \circ T_s^{-1}$ and set $\gamma_s = \det DT_s$ (see also section 2.2). Then for $W(\Omega) = H^{-1}(\Omega; \mathbb{R}^3)$, $s \mapsto f_{\Omega_s} \circ T_s$ has to be replaced by $s \mapsto \gamma_s^{-1}\phi_s^*(f_{\Omega_s})$ in the definition 1.1 as well as in the definition of the material derivative (formula (1.6)). This extension of the definition is consistent : these two expressions are equal when f_Ω belongs to $L^2(\Omega; \mathbb{R}^3)$. The shape derivative (when it exist) is still given by the equation (1.7).

2 Sensitivity Results

In this section, we derive a result on the regularity of the solution (u, p) of the Navier-Stokes Equations with respect to a perturbation of the boundary. The first part is dedicated to the description of the set of assumptions and of the statement of the result. The corresponding proof is developed in the sections (2.2) to (2.4).

2.1 Assumptions and Main Statement

First, we describe the regularity needed on the geometrical and functional data to derive our sensitivity result. The degree of regularity is given, in this assumption as well as in the proposition by an integer $k \geq 1$.

($\mathcal{A}_1$: Regularity of the data)

• Ω and D are C^{k+1} and the velocity space $\mathcal{V}$ is $\mathcal{V}_{k+2}$.

• the mappings f and g are continuous with respect to the shape in $H^{k-1}(\Omega;\mathbb{R}^3)$ and $H^{k+1/2}(\Gamma;\mathbb{R}^3)$ respectively.

• the mappings f and g are continuously differentiable with respect to the shape in $H^{k-2}(\Omega;\mathbb{R}^3)$ and $H^{k-1/2}(\Gamma;\mathbb{R}^3)$ respectively.

REMARK 2.1. *This set of assumptions is rather designed to handle the high regularity case, even if the methods developed in the following sections could be used as well to study the situations where such a regularity is not available.*

The smoothness of the geometry is determined so that the regularity of the solutions of the Navier-Stokes Equations is maximal with respect to the known regularity of f and g.

The assumptions made for these mappings may be surprising at first: the spatial regularity of f and $\dot{f}$ on one hand, of g and $\dot{g}$ on the other hand, are not the same. However this is the usual situation; when f does not depend on Ω, that is when there is for any $V \in \mathcal{V}$ a $F \in H^k(D;\mathbb{R}^3)$ such that $f_{\Omega_s} = F|_{\Omega_s}$, the existence of $\dot{f}$ takes place a priori *only in H^{k-1}, the material derivative being given by $\dot{f} = \langle D_x f, V(0)\rangle$. An analogous property holds for g.*

An interesting consequence of that gap is that the material and shape derivatives of the data have the same spatial regularity as

$$f'_\Omega = \dot{f}_\Omega - Df_\Omega \cdot V(0) \quad \text{and} \quad g'_\Gamma = \dot{g}_\Gamma - D_\tau g_\Gamma \cdot V(0)$$

The theorem 2.1 shows that the solutions u and p of the Navier-Stokes Equations exhibit the same kind of regularity. Therefore, the same property holds for their derivatives with respect to the shape.

The second assumption, less common, results from the non-linearity of the Navier-Stokes Equations: it has no equivalent for the linear shape optimization problems as the linearized problem is automatically well-posed when the initial one is.

($\mathcal{A}_2$: Non-Singularity of the NSE) The couple (f_Ω, g_Γ) is a regular value of the Navier-Stokes operator (1.3): for any solution u of the system (1.1), the operator $DF(u)$ is an isomorphism.

REMARK 2.2. *Explicitly, the assumption $\mathcal{A}_2$ states that for any $k \in H^{-1}(\Omega;\mathbb{R}^3)$, $l \in H^{1/2}(\Gamma;\mathbb{R}^3)$ such that $\int_\Gamma \langle k, n\rangle \, d\mathcal{H}^2 = 0$ and any solution of the Navier-Stokes Equations, the system*

$$\begin{aligned} -\nu\Delta v + (u\cdot\nabla)v + (v\cdot\nabla)u + \nabla q = h \quad & \text{in } \Omega \\ \operatorname{div} v = 0 \quad & \text{in } \Omega \\ v = k \quad & \text{on } \Gamma \end{aligned} \tag{2.1}$$

has a unique solution $(v, q) \in H^1(\Omega; \mathbb{R}^3) \times L^2(\Omega; \mathbb{R})/\mathbb{R}$.

This assumption is naturally fulfilled in the high viscosity (or small data) case: when ν *large enough with respect to the data* (f, g) *or conversely when* f *and* g *are small enough in* $H^1(\Omega; \mathbb{R}^3)$ *and* $H^{1/2}(\Gamma; \mathbb{R}^3)$ *for a given viscosity* ν, *then the linearized Navier-Stokes Equations at the considered solution are well-posed.*

In general, without such an assumption on the viscosity, the Navier-Stokes operator is still generically *non-singular with respect to* (f, g) *or even with respect to* g *for a fixed value of* f *(see [6], [9]). However, the well-posedness of the linearized problem is ensured only in more regular (Hölder or Sobolev) spaces than those considered in the section 1.1.*

We may now state the main result of this section:

THEOREM 2.1. *Assume that* $\mathcal{A}_1$ *and* $\mathcal{A}_2$ *are satisfied. Let* (u, p) *be a solution of the Navier-Stokes Equations in* Ω. *Then, for any* $V \in \mathcal{V}$, *there is a neighborhood* I *of* 0 *in* $\mathbb{R}$ *and a locally unique family* $s \in I \mapsto (u_s, p_s)$ *of solutions of the Navier-Stokes Equations in* Ω_s *such that:*

(i) Initial Condition: $(u_0, p_0) = (u, p)$

(ii) Regularity: $(\mathcal{U}, \mathcal{P}) : [s \mapsto (u_s \circ T_s, p_s \circ T_s)]$ *is such that*
- $\mathcal{U} \in C^0(I; H^{k+1}(\Omega; \mathbb{R}^3)) \cap C^1(I; H^k(\Omega; \mathbb{R}^3))$
- $\mathcal{P} \in C^0(I; H^k(\Omega; \mathbb{R}^3)/\mathbb{R}) \cap C^1(I; H^{k-1}(\Omega; \mathbb{R}^3)/\mathbb{R})$

(iii) Shape Derivatives: u' *and* p' *are the solutions of*

$$\begin{aligned} -\nu\Delta u' + (u \cdot \nabla)u' + (u' \cdot \nabla)u + \nabla p' &= f' && \text{in } \Omega \\ \operatorname{div} u' &= 0 && \text{in } \Omega \\ u' &= -\frac{\partial u}{\partial n}\langle V(0), n\rangle + g'_\Gamma && \text{on } \Gamma \end{aligned} \tag{2.2}$$

2.2 Transport

We develop a modification of the process of *transported equation* which is used to characterize the regularity with respect to the shape of the solutions of a PDE problem. Usually, we associate to any $V \in \mathcal{V}$ the family of one-to-one correspondences $(\phi_s)_{s\in\mathbb{R}}$

$$\phi_s \begin{pmatrix} W(\Omega) & \to & W(\Omega_s) \\ u & \mapsto & u \circ T_s^{-1} \end{pmatrix} \tag{2.3}$$

where $W(\Omega)$ and $W(\Omega_s)$ are the adequate functional spaces for the considered problem) and we study the properties of the problem satisfied by $\phi_s^{-1}(u_{\Omega_s})$ where u_{Ω_s} is the solution of the initial problem in Ω_s; this transported problem is expressed in the fixed space Ω. However, such an approach is not straightforward here *because the transport of the compatibility condition (1.2) would not* in general *lead to fixed functional spaces*: this property is directly true for homogeneous Dirichlet boundary conditions or with a slight adaptation for the non-homogeneous case $u|_\Gamma = g$ *when* g *does not depend on the shape* Γ (cf. [1]). However, this does not include the general case.

In order to circumvent that problem, we introduce the Piola transformations ψ_s : $L^2(\Omega;\mathbb{R}^3) \to L^2(\Omega_s;\mathbb{R}^3)$, defined by :

$$\psi_s(u) = \phi_s(\gamma_s^{-1} J_s \cdot u) \tag{2.4}$$

with $J_s = DT_s$ and $\gamma_s = \det(J_s)$. These coefficients satisfy the following lemma.

LEMMA 2.2. *For any $V \in \mathcal{V}_k$, and any $s \in \mathbb{R}$, J_s and γ_s are invertible. Moreover, J_s and J_s^{-1} are in $C^1(\mathbb{R}; C^{k-1}(\overline{D};\mathbb{R}^{3\times 3}))$ and γ_s and γ_s^{-1} are in $C^1(\mathbb{R}; C^{k-1}(\overline{D};\mathbb{R}))$.*

The proof is a direct consequence of the regularity of $s \mapsto T_s$ described in [10]. Most of the classical properties of the ϕ_s are also satisfied by the ψ_s. For any $k \geq 1$, $V \in \mathcal{V}_{k+1}$ and $s \in \mathbb{R}$,

- ϕ_s and ψ_s are isomorphisms between $L^2(\Omega;\mathbb{R}^3)$ and $L^2(\Omega_s;\mathbb{R}^3)$; their inverse are given by:

$$\phi_s^{-1}(u) = u \circ T_s \quad \text{and} \quad \psi_s^{-1}(u) = \gamma_s J_s^{-1} \phi_s^{-1}(u)$$

- The regularity of the mappings are preserved through the transport: for any integer $m \leq k$, ϕ_s and ψ_s induce isomorphisms between $H^m(\Omega;\mathbb{R}^3)$ and $H^m(\Omega_s;\mathbb{R}^3)$. As $\mathcal{D}^m(\Omega)$ is also mapped into $\mathcal{D}^m(\Omega_s)$, the adjoint operators $\phi_s^\star$ and $\psi_s^\star$ are isomorphisms from $H^{-m}(\Omega_s;\mathbb{R}^3)$ to $H^{-m}(\Omega;\mathbb{R}^3)$.
- The same property holds for boundary Sobolev spaces: for any integer m such that $m + 1/2 \leq k$, ϕ_s and ψ_s induce isomorphisms between the spaces $H^{m+1/2}(\Gamma;\mathbb{R}^3)$ and $H^{m+1/2}(\Gamma_s;\mathbb{R}^3)$.

Moreover, ψ_s satisfies the two extra following properties:

PROPOSITION 2.3. *Let $V \in \mathcal{V}_2$ and $s \in \mathbb{R}$. For any $u \in L^2(\Omega;\mathbb{R}^3)$ and $\varphi \in H_0^1(\Omega;\mathbb{R}^3)$, we have*

$$\langle \operatorname{div} u, \varphi \rangle_{H^{-1}\times H_0^1} = \langle \operatorname{div}(\psi_s u), \phi_s \varphi \rangle_{H^{-1}\times H_0^1} \tag{2.5}$$

Therefore ψ_s also induce an isomorphism between $V^1(\Omega)$ and $V^1(\Omega_s)$. Moreover, for any $u \in H^1(\Omega;\mathbb{R}^3)$ and $\varphi \in H^1(\Omega;\mathbb{R}^3)$,

$$\int_{\Gamma_s} \langle \psi_s(u), n_{\Omega_s} \rangle_{\mathbb{R}^3}\, \phi_s(\varphi)\, d\mathcal{H}^2 = \int_\Gamma \langle u, n_\Omega \rangle_{\mathbb{R}^3}\, \varphi\, d\mathcal{H}^2 \tag{2.6}$$

Proof. By a change of variable, we have the equality

$$\begin{aligned}\langle \operatorname{div} \psi_s(u), \phi_s(\varphi)\rangle_{H^{-1}\times H_0^1} &= -\int_{\Omega_s} \langle \psi_s(u), \nabla(\phi_s(\varphi))\rangle \, dx \\ &= -\int_{\Omega} \langle \psi_s(u)\circ T_s, \nabla(\phi_s(\varphi))\circ T_s\rangle \, \gamma_s \, dx\end{aligned}$$

As $\nabla(\phi_s(\varphi))\circ T_s = \nabla(\varphi\circ T_s^{-1})\circ T_s = [DT_s^*]^{-1}\nabla\varphi$, we obtain as desired

$$\begin{aligned}\langle \operatorname{div} \psi_s(u), \phi_s(\varphi)\rangle_{H^{-1}\times H_0^1} &= -\int_{\Omega} \langle \gamma_s J_s^{-1}\psi_s(u)\circ T_s, \nabla\varphi\rangle \, dx \\ &= -\int_{\Omega} \langle u, \nabla\varphi\rangle \, dx \\ &= \langle \operatorname{div} u, \varphi\rangle_{H^{-1}\times H_0^1}\end{aligned}$$

To establish the equality 2.6, we make the same change of variable. It yields

$$\int_{\Gamma_s} \langle \psi_s(u), n_{\Omega_s}\rangle \, \phi_s(\varphi) \, d\mathcal{H}^2 = \int_{\Gamma} \langle \psi_s(u)\circ T_s, n_{\Omega_s}\circ T_s\rangle \, (\phi_s(\varphi)\circ T_s) \, \omega_s \, d\mathcal{H}^2$$

with $\omega_s = \gamma_s \|(J_s^*)^{-1} n_\Omega\|$. As $n_{\Omega_s}\circ T_s = \dfrac{(J_s^*)^{-1} n_\Omega}{\|(J_s^*)^{-1} n_\Omega\|}$, we finally obtain

$$\int_{\Gamma_s} \langle \psi_s(u), n_{\Omega_s}\rangle \, \phi_s(\varphi) \, d\mathcal{H}^2 = \int_{\Gamma} \langle u, n_\Omega\rangle \, \varphi \, d\mathcal{H}^2$$

□

REMARK 2.3. *In the process of definition if the material derivative, described in the section 1.1, we could replace the mappings* ϕ_s *by the* ψ_s *and therefore defined the* Piola material derivative *of* f, $\dot{f}^P$, *by*

$$\dot{f}_\Omega^P = \frac{d}{ds}\psi_s^{-1}(f_{\Omega_s})|_{s=0}$$

In the case of a volume-preserving transformation, that is when $\operatorname{div} V = 0$ *and* $\gamma_s = 1$, *the mapping* ψ_s^{-1} *reduces to an intrinsic transformation widely used in differential geometry (see for example [2]). The shape derivative of* f *is then simply deduced from the Piola derivative by the formula*

$$f'_\Omega = \dot{f}_\Omega^P - [V(0), f_\Omega]$$

where $[\cdot,\cdot]$ *are the Lie brackets.*

2.3 Transported Equation

We prove in this section the part of the theorem 2.1 dedicated to the existence and uniqueness of $s \mapsto (u_s, p_s)$. We introduce the operators

$$A^s = \psi_s^\star \circ A \circ \psi_s \;\text{ and }\; B^s = \psi_s^\star \circ B \circ (\psi_s, \psi_s) \tag{2.7}$$

and we call *Transported Navier-Stokes Equations* the system

$$\begin{aligned} \nu A^s u + B^s(u,u) &= \psi_s^\star(\pi(f_{\Omega_s})) \\ u|_\Gamma &= \psi_s^\star(g_{\Gamma_s}) \end{aligned} \tag{2.8}$$

From the very definition of the operators, a field $u \in H^1(\Omega; \mathbb{R}^3)$ is a solution of the transported Navier-Stokes Equations if and only if $\psi_s(u)$ is a solution of the Navier-Stokes Equations in Ω_s. This is the key property for the analysis of $s \mapsto u_s$ which is made first. The regularity with respect of the shape of the pressure is deduced as a consequence in the following section.

Regularity of the Velocity with Respect to the Shape

The results concerning $s \mapsto u_s$ are consequences of the implicit function theorem applied to the mapping Ψ defined by

$$\Psi(s,u) = (\nu A^s u + B^s(u,u) - \psi_s^\star(\pi(f_{\Omega_s}))\,,\, u - \psi_s^\star(g_{\Gamma_s})) \tag{2.9}$$

The implicit function theorem is in fact applied twice: a first time to obtain the continuity in H^{k+1} and then to get the continuous differentiability in H^k only. The functional spaces involved in the theorem are

$$\mathcal{F}_k = \left\{u \in H^k(\Omega; \mathbb{R}^3),\, \operatorname{div} u = 0\right\},\; \mathcal{G}_k = \left\{f \in \pi\left(H^{k-2}(\Omega; \mathbb{R}^3)\right)\right\} \tag{2.10}$$

this latter space being endowed by the norm induced by $H^{k-2}(\Omega; \mathbb{R}^3)$, and

$$\mathcal{H}_k = \left\{g \in H^{k-1/2}(\Gamma; \mathbb{R}^3),\, \int_\Gamma \langle g, n\rangle \, d\mathcal{H}^2 = 0\right\} \tag{2.11}$$

Let (u_0, p_0) be the initial solution chosen of the Navier-Stokes Equations in the initial geometry. We have to check the following properties for Ψ:

(i) The continuity of $\Psi : \mathbb{R} \times \mathcal{F}_{k+1} \to \mathcal{G}_{k+1} \times \mathcal{H}_{k+1}$, the existence of $\partial_u \Psi(s,u)$ anywhere and its continuity in in $(0, u_0)$. The continuous differentiability of Ψ as a mapping from $\mathbb{R} \times \mathcal{F}_k$ to $\mathcal{G}_k \times \mathcal{H}_k$.

(ii) The initial solution u_0 belongs to $\mathcal{F}_{k+1}$ and the mapping $\partial_u \Psi(0, u_0)$ is an isomorphism from $\mathcal{F}_k$ to $\mathcal{G}_k \times \mathcal{H}_k$ as well as from $\mathcal{F}_{k+1}$ to $\mathcal{G}_{k+1} \times \mathcal{H}_{k+1}$.

The properties in (i) are consequences of the explicit expression of the transported operators. Simple calculations show that

$$\psi_s^\star \Delta \psi_s(u) = \gamma_s^{-1} J_s^* \cdot \text{div}(D(\gamma_s^{-1} J_s u) \gamma_s J_s^{-1} (J_s^{-1})^*) \tag{2.12}$$

$$\psi_s^\star((\psi_s(u) \cdot \nabla)\psi_s(v)) = \gamma_s^{-1} J_s^*(u \cdot \nabla)(\gamma_s^{-1} J_s v) \tag{2.13}$$

The desired regularity is obtained from the properties of the trilinear form $(u, v, w) \mapsto \int_\Omega uvw\, dx$ (see [6], [3]). The same properties also yield with a bootstrap method the properties (ii) as the existence is already known for the solution of the Navier-Stokes Equations and also for the linearized equation from the assumption $\mathcal{A}_2$. Notice that the only thing that prevents Ψ to be continuously differentiable from $\mathbb{R} \times \mathcal{F}_{k+1}$ to $\mathcal{G}_{k+1} \times \mathcal{H}_{k+1}$ is the fact that the data f and g are not regular enough.

As a consequence of (i) and (ii), the implicit function theorem asserts that there is a neighborhood $I \subset \mathbb{R}$ of 0, a unique family $(u_s)_{s \in I}$ of solutions of the Navier-Stokes Equations in Ω_s such that $s \mapsto \psi_s^{-1}(u_s) \in C^0(I; H^{k+1}(\Omega; \mathbb{R}^3))$ and $s \mapsto \psi_s^{-1}(u_s) \in C^1(I; H^k(\Omega; \mathbb{R}^3))$. From the definition of ψ_s we deduce that the same regularity is obtained for the mapping $s \mapsto \phi_s^{-1}(u_s)$.

Regularity of the Pressure

The result of the theorem 2.1 concerning the pressure is obtained by the same transport methods but applied on the classical form of the Navier-Stokes Equations, without using the projector π. We notice that $\psi_s^\star(\nabla p_s) = \nabla \phi_s^{-1}(p_s)$ and therefore that any solution p_s of the Navier-Stokes Equations in Ω_s is subject to

$$\nabla \phi_s^{-1}(p_s) = \nu \cdot \psi_s^\star \Delta u_s - \psi_s^\star((u_s \cdot \nabla)u_s) + \psi_s^\star(f_{\Omega_s})$$

That equation and the results of the previous section yield the desired regularity of $s \mapsto \phi_s^{-1}(p_s)$.

2.4 Shape Derivative and Linearized Equation

In order to establish the equation satisfied by the shape derivatives of u and p, we may use extensions of these shape-dependent mappings. Thanks to the regularity of $s \mapsto u_s$ and $s \mapsto p_s$, there exist two mappings $U \in C^1(\mathbb{R}; H^1(D; \mathbb{R}^3))$ and $P \in C^1(\mathbb{R}; L^2(D; \mathbb{R}))$ such that for s small enough, we have $u_{\Omega_s} = U(s)|_{\Omega_s}$ and $p_{\Omega_s} = P(s)|_{\Omega_s}$. Then, the shape derivatives are given by $u'_\Omega = \partial_s U(0)$ and $p'_\Omega = \partial_s P(0)$.

However, this method cannot be applied directly to the right-hand side f for the minimal regularity ($k = 1$). But the shape derivative of f may still be characterized weakly by the following lemma.

LEMMA 2.4. *Let f be a shape-dependent mapping which is C^0 w.r. to the shape in $L^2(\Omega;\mathbb{R}^3)$ and C^1 w.r. to the shape in $H^{-1}(\Omega;\mathbb{R}^3)$. Then for any $\varphi \in \mathcal{D}(\Omega;\mathbb{R}^3)$, the function*

$$s \mapsto \int_{\mathbb{R}^3} \langle f_{\Omega_s}, \varphi\rangle_{\mathbb{R}^3}\, dx$$

is differentiable at $s = 0$ and its derivative is given by

$$\frac{\partial}{\partial s}\left(\int_{\mathbb{R}^3} \langle f_{\Omega_s}, \varphi\rangle_{\mathbb{R}^3}\, dx\right)|_{s=0} = \langle f'_\Omega, \varphi\rangle_{H^{-1}\times H^1_0} \tag{2.14}$$

Proof. From the definition of the mappings ϕ_s, we deduce that

$$\int_{\mathbb{R}^3} \langle f_{\Omega_s}, \varphi\rangle_{\mathbb{R}^3}\, dx = \langle \gamma_s^{-1}\phi_s^\star(f_{\Omega_s}), \gamma_s\varphi\circ T_s\rangle_{H^{-1}\times H^1_0}$$

Both arguments of the duality brackets are strongly differentiable. That proves the existence of the derivative. By definition of the material derivative, we have $\dot{f}_\Omega = \frac{\partial}{\partial s}\gamma_s^{-1}\phi_s^\star(f_{\Omega_s})|_{s=0}$ and clearly, $\frac{\partial}{\partial s}\gamma_s\varphi\circ T_s|_{s=0} = \operatorname{div} V(0)\cdot\varphi + D\varphi\cdot V(0) = \operatorname{div}(\varphi\otimes V(0))$. Consequently, we have

$$\begin{aligned}\frac{\partial}{\partial s}\left(\int_{\mathbb{R}^3} \langle f_{\Omega_s}, \varphi\rangle_{\mathbb{R}^3}\, dx\right)|_{s=0} &= \left\langle \dot{f}_\Omega, \varphi\right\rangle_{H^{-1}\times H^1_0} \\ &\quad + \int_{\mathbb{R}^3} \langle f_\Omega, \operatorname{div}(\varphi\otimes V(0))\rangle_{\mathbb{R}^3}\, dx\end{aligned}$$

On the other hand, as

$$\begin{aligned}\int_{\mathbb{R}^3} \langle f_\Omega, \operatorname{div}(\varphi\otimes V(0))\rangle_{\mathbb{R}^3}\, dx &= -\langle Df_\Omega, \varphi\otimes V(0)\rangle_{H^{-1}\times H^1_0} \\ &= -\langle Df_\Omega\cdot V(0), \varphi\rangle_{H^{-1}\times H^1_0}\end{aligned}$$

using the definition of f'_Ω, we obtain the equality (2.14). □

Let $\varphi \in \mathcal{D}(\mathbb{R}^3;\mathbb{R}^3)$ such that $\operatorname{Supp}(\varphi) \subset \Omega$. For small values of s, the support of φ is also included in Ω_s and therefore

$$\begin{aligned}&\int_D \nu\partial_x U(s)\cdot\cdot\,\partial_x\varphi + \langle\partial_x U(s)\cdot U(s), \varphi\rangle_{\mathbb{R}^3}\, dx \\ &\qquad = \int_{\mathbb{R}^3} P(s)\cdot\operatorname{div}\varphi + \langle f_{\Omega_s}, \varphi\rangle_{\mathbb{R}^3}\, dx\end{aligned}$$

The differentiation of this equation with respect to s gives

$$\begin{aligned}&\int_D \nu\partial_x\partial_s U(0)\cdot\cdot\,\partial_x\varphi + \langle[\partial_x\partial_s U(0)]\,U(0) + [\partial_x U(0)]\partial_s U(0), \varphi\rangle_{\mathbb{R}^3}\, dx \\ &\qquad = \int_D \partial_s P(0)\cdot\operatorname{div}\varphi\, dx + \langle f'_\Omega, \varphi\rangle_{H^{-1}\times H^1_0}\end{aligned}$$

which implies that u' and p' are solutions of (2.2). The equation $\operatorname{div} u' = 0$ is a direct consequence of $\operatorname{div} u_s = 0$ and of the existence of the extension $s \mapsto U(s)$.

The boundary condition is proved as follows. As $u|_\Gamma = g_\Gamma$, $u'_\Gamma = g'_\Gamma$. On the other hand

$$u'_\Gamma = u'|_\Gamma + \frac{\partial u}{\partial n}\langle V(0), n\rangle$$

therefore, we obtain the desired Dirichlet boundary condition.

3 Lagrangian Functionals

In order to investigate the sensitivity of shape functionals of Lagrangian type, we state some sensitivity results for the solutions of the ordinary differential equations. The analysis is done for time-dependent mappings with values in C^k spaces rather than in the classical framework of two-variable mappings (time and space).

Before this, we briefly analyze the shape differentiability of a simple distributed shape functional. The corresponding results are needed as tools for the analysis of the Lagrangian functionals.

3.1 Shape Differentiability of a Distributed Functional

From now on, we assume that the data considered in the theorem 2.1 satisfy the assumptions $\mathcal{A}_1$ with $k = 1$ and $\mathcal{A}_2$. Let $\rho : \mathbb{R}^3 \to \mathbb{R}^3$ be a C^1 function whose support is compactly included in the reference set $\Omega_0 \subset D$. We consider the (possibly multi-valued) shape functional J, given by

$$J(\Omega) = \int_\Omega \langle \rho, u\rangle_{\mathbb{R}^3}\, dx \tag{3.1}$$

where u is a solution of the system of equations (1.1). For any $V \in \mathcal{V}$ and any solution u_0 of the Navier-Stokes Equations in Ω_0, the theorem 2.1 provides a family $(u_s)_{s\in I}$ that determines uniquely a value of the shape functional in Ω_s, value still denoted $J(\Omega_s)$. As a consequence of the regularity of $(u_s)_{s\in I}$ given in the theorem 2.1 and by the Reynolds formula (see [10]), the Eulerian derivative of J in the direction $V \in \mathcal{V}$, defined as $dJ(\Omega; V) = \frac{d}{ds}J(\Omega_s)|_{s=0}$, exists and is given by

$$dJ(\Omega; V) = \int_\Omega \langle \rho, u'\rangle\, dx \tag{3.2}$$

where u' is the solution of (2.2). The introduction of the corresponding adjoint system allows further simplifications:

$$\begin{aligned} -\nu\Delta\eta - D\eta\cdot u_0 + [Du_0]^*\cdot\eta + \nabla\pi &= \rho \ \text{ in } \Omega_0 \\ \operatorname{div}\eta &= 0 \ \text{ in } \Omega_0 \\ \eta &= 0 \ \text{ on } \Gamma_0 \end{aligned} \tag{3.3}$$

The assumption $\mathcal{A}_2$ ensures the existence and uniqueness of the solution η. The regularity of u, deduced from the assumption $\mathcal{A}_1$ and a bootstrapping method shows that $\eta \in H^2(\Omega;\mathbb{R}^3)$. Therefore, we have

$$\begin{aligned} dJ(\Omega;V) &= \int_\Omega \langle -\nu\Delta\eta - D\eta\cdot u_0 + [Du_0]^*\cdot\eta + \nabla\pi, u'\rangle \ dx \\ &= \int_\Omega \langle \eta, f'_\Omega\rangle \ dx + \int_\Gamma \left\langle -\nu\frac{\partial\eta}{\partial n} + \pi n, u'\right\rangle \ dx \\ &= \int_\Omega \langle \eta, f'_\Omega\rangle \ dx + \int_\Gamma \left\langle -\nu\frac{\partial\eta}{\partial n} + \pi n \ , \ g'_\Gamma - \frac{\partial u}{\partial n}\langle V(0), n\rangle \right\rangle \ dx \end{aligned}$$

This expression may be rewritten in terms of *stress tensor*. We associate to $u \in H^1(\Omega;\mathbb{R}^3)$ and $p \in L^2(\Omega;\mathbb{R})$ the matrix

$$\sigma(u,p) = pI - \nu(Du + Du^*) \tag{3.4}$$

Some additional information on the structure of the normal component of $\sigma(u,p)$ on the boundary may be obtained. Let D_τ and div_τ be respectively the tangential Jacobian and the tangential divergence operators. We state the

LEMMA 3.1. *Let u be a divergence-free field of $H^2(\Omega;\mathbb{R}^3)$ such that $u|_\Gamma = h$. Then, we have*

$$Du^* n = [D_\tau h^* - (\operatorname{div}_\tau h)I]\, n \tag{3.5}$$

Proof. Using the decomposition of the operators D and div on the boundary, we obtain

$$Du = D_\tau h + \frac{\partial u}{\partial n}\otimes n \ \text{ and } \ \operatorname{div} u = \operatorname{div}_\tau h + \left\langle \frac{\partial u}{\partial n}, n\right\rangle = 0 \tag{3.6}$$

Consequently, we have

$$\begin{aligned} [Du^*]n &= \left[D_\tau h^* + n\otimes\frac{\partial u}{\partial n}\right] n \\ &= [D_\tau h^*]n + \left\langle \frac{\partial u}{\partial n}, n\right\rangle n \\ &= [D_\tau h^* - (\operatorname{div}_\tau h)I]\, n \end{aligned}$$

□

As a direct consequence of that lemma, the derivative of J is equal to

$$dJ(\Omega;V) = \int_\Omega \langle \eta, f'_\Omega \rangle \, dx + \int_\Gamma \left\langle \sigma(\eta,\pi)n \, , \, g'_\Gamma - \frac{\partial u}{\partial n} \langle V(0), n \rangle \right\rangle d\mathcal{H}^2 \tag{3.7}$$

3.2 Shape Derivation of the Velocity Flow

Sensitivity in the ODE's

In this section, K is a compact set of $\mathbb{R}^n$ such that $\overline{D} \subset K$ and $T > 0$. We define the set $\mathcal{H} = \{f \in C^0(\mathbb{R}^n;\mathbb{R}^n),\ \mathrm{Supp}(f) \subset K\}$ and denote $\|\cdot\|_k$ the usual norm on the set $C^k(\overline{A};F)$. If F is itself of the form $F = C^l(\overline{B};G)$, we also use the notation $\|\cdot\|_{k,l}$.

LEMMA 3.2. *The mappings*

$$\begin{array}{ccc} C^0([-T;T];C^0(\overline{D};\mathbb{R}))^2 & \to & C^0([-T;T];C^0(\overline{D};\mathbb{R})) \\ (f,g) & \mapsto & fg \end{array}$$

and

$$\begin{array}{ccc} C^0([-T;T];\mathcal{H}) \times C^0([-T;T];C^0(\overline{D};\mathbb{R}^n)) & \to & C^0([-T;T];C^0(\overline{D};\mathbb{R}^n)) \\ (f,g) & \mapsto & [t \mapsto f(t)\circ g(t)] \end{array}$$

are continuous.

Proof. i) clear (remember that $C^0([-T;T];C^0(\overline{D};\mathbb{R})) \simeq C^0([-T;T]\times\overline{D};\mathbb{R})$).

ii) for any f, f' in $C^0([-T;T];\mathcal{H})$, and g, g' in $C^0([-T;T];C^0(\overline{D};\mathbb{R}^n))$, we have

$$\|f'\circ g' - f\circ g\|_{0,0} \le \|f' - f\|_{0,0} + \|f\circ g' - f\circ g\|_{0,0}$$

It is therefore enough to show the uniform equicontinuity of the family $(f(t))_{t\in[-T;T]}$. It is proved as follows: For any $t \in [-T;T]$, and any $\varepsilon > 0$, there is a $\eta > 0$ such that $\|x-y\| < \eta$ implies $\|f(t)(x) - f(t)(y)\| < \varepsilon/3$ (by uniform continuity of $f(t)$). Let $\tau > 0$ be such that for any $t' \in [-T;T]\cap]t-\tau;t+\tau[$, $\|f(t') - f(t)\|_0 < \varepsilon/3$. Then, for any x and y such that $\|x-y\| < \eta$, $\|f(t')(x) - f(t')(y)\| < \varepsilon$. By compactness of $[-T;T]$, the result is proved. □

Assume that D is of class C^2. Using an extension operator, we may identify the velocity space $\mathcal{V}_2$ with a linear subspace of the set of continuous mappings from $[-T;T]$ to $\{f \in C^2(\mathbb{R}^n;\mathbb{R}^n),\ \mathrm{Supp}(f) \subset K\}$. For any mapping $\xi \in C^1([-T;T];C^1(\overline{D};\mathbb{R}^n))$ and $v \in \mathcal{V}_2$, we set

$$H(\xi, v) = [t \mapsto v(t)\circ\xi(t)] \tag{3.8}$$

and we state the

LEMMA 3.3. *The expression (3.8) defines a C^1 mapping*

$$H : C^0([-T;T];C^1(\overline{D};\mathbb{R}^n)) \times \mathcal{V}_2 \to C^0([-T;T];C^1(\overline{D};\mathbb{R}^n))$$

whose differential is given by:

$$DH(\xi,v)\cdot(\zeta,W) = \partial_\xi H(\xi,v)\cdot\zeta + \partial_v H(\xi,v)\cdot W \quad (3.9)$$

$$= [t \mapsto [D_x v(t)\circ\xi(t)]\,\zeta(t) + W(t)\circ\xi(t)] \quad (3.10)$$

Proof. The mapping H is well-defined: indeed for any $t \in [-T;T]$, ξ belongs to $C^0([-T;T];C^1(\overline{D};\mathbb{R}^n))$ and $v \in \mathcal{V}_2$, the function $H(\xi,v)(t)$ is differentiable and

$$D_x[H(\xi,v)(t)] = [D_x v(t)]\circ\xi(t)\cdot D_x\xi(t)$$

. The lemma 3.2 yields that $D_x H(\xi,v)$ belongs to $C^0([-T;T];C^0(\overline{D};\mathbb{R}^{n\times n}))$ and therefore $H(\xi,v) \in C^0([-T;T];C^1(\overline{D};\mathbb{R}^{n\times n}))$.

• Existence and continuity of $\partial_\xi H$: let $\mathcal{A}_{\xi,v}$ be the linear operator

$$\zeta \mapsto [t \mapsto A_{\xi,v}(t)\zeta(t)]$$

with $A_{\xi,v}(t) = D_x v(t)\circ\xi(t)$. The function $A_{\xi,v}(t)$ is differentiable for any $t \in [-T;T]$ and $D_x A_{\xi,v}(t) = \left[D_x^2 v(t)\circ\xi(t)\right][D_x\xi(t)]$. Therefore $A_{\xi,v}$ belongs to $C^0([-T;T];C^1(\overline{D};\mathbb{R}^n))$ and depends continuously on (ξ,v) (lemma 3.2). Consequently, $\mathcal{A}_{\xi,v}$ is a well-defined and continuous mapping from $C^0([-T;T];C^1(\overline{D};\mathbb{R}^n))$ to $C^0([-T;T];C^1(\overline{D};\mathbb{R}^n))$ and the mapping $(\xi,v)\mapsto\mathcal{A}_{\xi,v}$ is continuous.

Let us prove the existence of $\partial_\xi H$ and the equality $\partial_\xi H(\xi,v) = \mathcal{A}_{\xi,v}$. We set

$$\Sigma = v\circ(\xi+\zeta) - v\circ\xi - D_x v\circ\xi\cdot\zeta \quad (3.11)$$

Straightforward calculations show that for any $i \in \{1,...,n\}$, $D_x\Sigma_i = D_x v_i\circ(\xi+\zeta)\cdot D_x(\xi+\zeta) - D_x v_i\circ\xi\cdot D_x\xi - \zeta^*\cdot[D_x^2 v_i\circ\xi]\cdot D_x\xi - D_x v_i\circ\xi\cdot D_x\zeta$ and therefore,

$$D_x\Sigma_i = K_i^*\cdot D_x\xi + L_i^*\cdot D_x\zeta \quad (3.12)$$

$$\text{with } \begin{cases} K_i = L_i - D_x^2 v_i\circ\xi\cdot\zeta \\ L_i = \nabla_x v_i\circ(\xi+\zeta) - \nabla_x v_i\circ\xi \end{cases}$$

We use the following intermediate result: we set $\varphi = v$ or $\varphi = \nabla_x v_i$ for a $i \in \{1,...,n\}$. Then, there is a function $\varepsilon : \mathbb{R}_+ \to \mathbb{R}_+$, independent of (t,x), with

$$\lim_{h\to 0}\varepsilon(h) = 0$$

and such that for any $t \in [-T;T]$ and $x \in \mathbb{R}^n$, the inequality

$$\|\varphi(t)(x) - \varphi(t)(x+h) - D_x\varphi(t)(x)\cdot h\| = \varepsilon(h)\cdot\|h\| \quad (3.13)$$

holds. To prove this, we define $\psi(t,x,h) = \varphi(t)(x+h) - D_x\varphi(t)(x)\cdot(x+h)$ and notice that $D_h\psi(t,x,h) = D_x\varphi(t)(x+h) - D_x\varphi(t)(x)$. From the uniform equicontinuity of $(D_x\varphi(t))_{t\in[-T;T]}$ (see again the point *ii)* in the proof of the lemma 3.2), we deduce that there is a function $\lambda : \mathbb{R}_+ \to \mathbb{R}_+$ such that $\lim_{h\to 0}\lambda(h) = 0$ and $\|D_h\psi(t,x,h)\| \le \lambda(h)$. Therefore

$$\begin{aligned}\varphi(t)(x) - \varphi(t)(x+h) - D_x\varphi(t)(x)\cdot h &= \psi(t,x,0) - \psi(t,x,h) \\ &= \int_0^1 [D_h\psi(t,x,\theta h)]\cdot h\, d\theta\end{aligned}$$

and consequently, the equation (3.13) holds with $\varepsilon(h) = \sup_{\delta\in[0,h]}\lambda(\delta)$.

From this result, we deduce the inequalities $\|\Sigma\|_{0,0} \le \varepsilon(\|\zeta\|_{0,1})\cdot\|\zeta\|_{0,1}$, $\|K_i\|_{0,0} \le \varepsilon(\|\zeta\|_{0,1})\cdot\|\zeta\|_{0,1}$ and $\|L_i\|_{0,0} \le (\varepsilon(\|\zeta\|_{0,1}) + \|v\|_{\mathcal{V}_2})\cdot\|\zeta\|_{0,1}$. Finally, using the expressions (3.11) and (3.12), we end up with

$$\|\Sigma\|_{0,0} + \|D_x\Sigma\|_{0,0} \le \eta(\|\zeta\|_{0,1})\cdot\|\zeta\|_{0,1} \quad \text{with} \quad \lim_{h\to 0}\eta(h) = 0 \tag{3.14}$$

which proves the desired differentiability result.

• Existence and continuity of $\partial_v H$: the existence and expression of $\partial_v H(\xi,v)$ is clear: the mapping $[v \mapsto H(\xi,v)]$ is linear continuous and therefore $\partial_v H(\xi,v)\cdot W = H(\xi,W)$. The continuity of this partial derivative is a direct consequence of the lemma 3.2. □

Differentiability of the Flow

Let v be a vector field of $\mathcal{V}_1$. The corresponding flow is hereafter denoted T_v or $T(v)$ to emphasize its dependence on v; it is the solution of the initial-value problem

$$\begin{aligned}&\partial_t T_v(t) = v(t)\circ T_v(t)\\ &T_v(0) = I\end{aligned}$$

We know that $T(v)$ belongs to $C^1([-T;T];C^1(\overline{D};\mathbb{R}^n))$ (see [10]). The following proposition characterizes the regularity of the correspondence $v \mapsto T(v)$ under a stronger assumption on the regularity of the variable v.

PROPOSITION 3.4. *The mapping*

$$\begin{aligned}\mathcal{V}_2 &\to C^1([-T;T];C^1(\overline{D};\mathbb{R}^n))\\ v &\mapsto T(v)\end{aligned}$$

is continuously differentiable. For any $W \in \mathcal{V}_2$, $\zeta = \partial_v T(v)\cdot W$ is the solution of

$$\begin{aligned}&\partial_t\zeta(t) = [D_x v(t)\circ T_v(t)]\cdot\zeta(t) + W(t)\circ T_v(t)\\ &\zeta(0) = 0\end{aligned} \tag{3.15}$$

Proof. For any $\xi \in C^1([-T;T];C^1(\overline{D};\mathbb{R}^n))$ and $v \in \mathcal{V}_2$, we set

$$F(\xi, v) = \xi - \left[t \mapsto I + \int_0^t v(\tau) \circ \xi(\tau)\, d\tau\right] \tag{3.16}$$

The unique solution of $F(\xi, v) = 0$ is $\xi = T(v)$. Moreover, from the lemma 3.3 we deduce that $(\xi, v) \mapsto F(\xi, v)$ is a C^1 mapping from $C^1([-T;T];C^1(\overline{D};\mathbb{R}^n)) \times \mathcal{V}_2$ to $C^1([-T;T];C^1(\overline{D};\mathbb{R}^n))$. Its differential is given by:

$$DF(\xi, v) \cdot (\zeta, W) = \zeta - \left[t \mapsto \int_0^t ([D_x v(\tau) \circ \xi(\tau)] \cdot \zeta(\tau) + W(\tau) \circ \xi(\tau))\, d\tau\right]$$

Therefore, $\partial_\xi F(\xi, v)$ is an isomorphism: for any $\phi \in C^1([-T;T];C^1(\overline{D};\mathbb{R}^n))$, we set $\psi = \partial_t \phi$. The mapping ϕ satisfies $\partial_\xi F(\xi, v) \cdot \zeta = \phi$ iff it is the solution of

$$\begin{aligned} &\partial_t \zeta(t) = [D_x v(\tau) \circ \xi(t)] \cdot \zeta(t) + \psi(t) \\ &\zeta(0) = \phi(0) \end{aligned}$$

which is given by

$$\zeta(t) = U(0, t)\phi(0) + \int_0^t U(t, \tau)\psi(\tau)\, d\tau$$

with $U(s, t) = \exp\left(\int_s^t D_x v(\tau) \circ \xi(\tau)\, d\tau\right)$.

The regularity of $v \mapsto T(v)$ is a consequence of the implicit function theorem. The expression $\zeta = \partial_v T(v) \cdot W$ is solution of $\partial_\xi F(\xi, v) \cdot \zeta = -\partial_v F(\xi, v) \cdot W$ or equivalently, of the system (3.15). □

Now, we investigate the regularity with respect to the shape of a flow based on a shape-dependent mapping v. We assume that this mapping satisfies the following properties:

$$\begin{aligned} &v \text{ is } C^1 \text{ w.r. to the shape in } C^2(\overline{\Omega};\mathbb{R}^3) \text{ and } C^0 \text{ w.r. to the shape in} \\ &C^3(\overline{\Omega};\mathbb{R}^3). \text{ Moreover, for any admissible } \Omega,\ \langle v_\Omega, n_\Omega \rangle = 0 \end{aligned} \tag{3.17}$$

We describe the regularity of the corresponding shape-dependent flow $T(v)$ by the regularity of its extensions to $\overline{D}$.

PROPOSITION 3.5. *Let $\Omega \subset D$ be an open bounded set of class C^3 such that either $\Omega \subset\subset D$ or $\overline{D} - \overline{\Omega}$ is compactly included in D. Then, for any $V \in \mathcal{V} \subset \mathcal{V}_3$*

(i) *There is a continuously differentiable mapping $s \mapsto R_s$ with values in*

$$C^1([-T;T];C^1(\overline{D};\mathbb{R}^3))$$

such that for any $s \in \mathbb{R}$ and $t \in [-T;T]$, $R_s(t)$ is an extension of $T(v_{\Omega_s})(t)$ to $\overline{D}$.

(ii) *The restriction S of $\partial_s R_s|_{s=0}$ to Ω is the solution of*

$$\begin{aligned} \dot{S}(t) &= v' \circ T_v(t) + [D_x v \circ T_v(t)] S(t) \\ S(0) &= 0 \end{aligned} \tag{3.18}$$

Proof. The assumption (3.17) yields the existence of a mapping w that belongs to $C^1(\mathbb{R}, C^2(\overline{D}; \mathbb{R}^3))$ such that for any $s \in \mathbb{R}$, $w(s)|_{\Omega_s} = v_{\Omega_s}$. The shape derivative v'_Ω is equal to $\partial_s w(0)|_\Omega$. We may moreover assume that $w(s) \in \mathcal{V}_2$. Let R_s be the flow associated to $w(s)$, solution of

$$\begin{aligned} \partial_t R_s(t) &= w(s) \circ R_s(t) \\ R_s(0) &= I \end{aligned} \tag{3.19}$$

Obviously, we have $R_s(t)|_{\Omega_s} = T_{v_{\Omega_s}}(t)$ and the proposition 3.4 yields $s \mapsto R_s \in C^1(\mathbb{R}; C^1([-T;T]; C^1(\overline{D}; \mathbb{R}^3)))$ and $\bar{S} = \partial_s R_s|_{s=0}$ satisfies

$$\begin{aligned} &\partial_t \bar{S}(t) = \partial_s w(0)(t) \circ T_{w(0)}(t) + [D_x w(0)(t) \circ T_{w(0)}(t)] \cdot \bar{S}(t) \\ &\bar{S}(0) = 0 \end{aligned} \tag{3.20}$$

which yields (3.18). □

3.3 Shape Gradient of a Lagrangian Functional

Let $\rho : [0,T] \times \mathbb{R}^3 \to \mathbb{R}^3$ be a C^∞ function such that $\forall t \in [0,T]$, $\mathrm{Supp}(\rho(t,\cdot)) \subset \Omega$. We consider the shape functional J, defined by

$$J(\Omega) = \iint_{[0,T]\times D} \langle \rho(t,x), T_u(t)(x) \rangle \, dt dx \tag{3.21}$$

where u is solution of the Navier-Stokes Equations in Ω. We assume that the assumptions $\mathcal{A}_1$ and $\mathcal{A}_2$ are satisfied with $k = 4$ and moreover that $\langle g, n_\Omega \rangle_{\mathbb{R}^3} = 0$ and either $\Omega \subset\subset D$ or $\overline{D} - \overline{\Omega} \subset\subset D$. Under these assumptions, the theorem 2.1 and the Sobolev injections imply that u satisfies the assumptions (3.17) and therefore the proposition 3.5 holds for the flow T_u. Consequently, we have the

PROPOSITION 3.6. *The Eulerian derivative of J exists and is given by*

$$dJ(\Omega; V) = \int_\Omega \langle \eta, f'_\Omega \rangle \, dx + \int_\Gamma \left\langle \sigma(\eta,\pi) n \, , \, g'_\Gamma - \frac{\partial u}{\partial n} \langle V(0), n \rangle \right\rangle d\mathcal{H}^2 \tag{3.22}$$

where

$$\begin{aligned} &\dot{Q}(t) = -[Du^*] Q(t) - [DQ(t)] u - \rho(t,\cdot) \circ [T_u(t)]^{-1} \\ &Q(T) = 0 \end{aligned} \tag{3.23}$$

and

$$\begin{aligned} -\nu\Delta\eta - D\eta\cdot u + [Du]^*\cdot\eta + \nabla\pi &= \textstyle\int_0^T Q(t)\,dt \quad \text{in } \Omega \\ \operatorname{div}\eta &= 0 \quad \text{in } \Omega \\ \eta &= 0 \quad \text{on } \Gamma \end{aligned} \tag{3.24}$$

Proof. We introduce the adjoint equation where the adjoint state is P:

$$\begin{aligned} &\dot{P}(t) = -\rho(t,\cdot) - [Du\circ T_u(t)]^*\cdot P(t) \\ &P(T) = 0 \end{aligned} \tag{3.25}$$

Then, the Eulerian derivative of J exists and thanks to the equation (3.18) we obtain:

$$\begin{aligned} dJ(\Omega;V) &= \iint_{[0,T]\times\Omega} \langle \rho(t,\cdot), S(t)\rangle\, dtdx \\ &= \iint_{[0,T]\times\Omega} \left\langle -\dot{P}(t) - [Du\circ T_u(t)]^*\cdot P(t)\,,\, S(t)\right\rangle dtdx \\ &= \iint_{[0,T]\times\Omega} \left\langle P(t)\,,\, \dot{S}(t) - [Du\circ T_u(t)]S(t)\right\rangle dtdx \\ &= \iint_{[0,T]\times\Omega} \langle P(t)\,,\, u'\circ T_u(t)\rangle\, dtdx \end{aligned}$$

This expression may be further simplified by the introduction of $Q(t)$

$$Q(t) = P(t)\circ[T_u(t)]^{-1} \tag{3.26}$$

which is the solution of the PDE problem (3.23): the differentiation with respect to t of the equation $Q(t)\circ T_u(t) = P(t)$ gives $\dot{Q}(t)\circ T_u(t) + [DQ(t)\circ T_u(t)]\cdot\dot{T}_u(t) = \dot{P}(t)$ and therefore

$$\dot{P}(t)\circ[T_t^u]^{-1} = \dot{Q}(t) + DQ(t)\cdot(\dot{T}_t^u\circ[T_t^u]^{-1})$$

The equation (3.23) simply results from the substitution of this equality in (3.25). Then, as $\operatorname{div} u = 0$, $|\det DT_t^u| = 1$ and by a change of variable, we find that

$$\begin{aligned} dJ(\Omega;V) &= \iint_{[0,T]\times\Omega} \langle P(t)\,,\, u'\circ T_t^u\rangle\, dtdx \\ &= \iint_{[0,T]\times\Omega} \langle Q(t)\,,\, u'\rangle\, dtdx \\ &= \int_\Omega \left\langle \int_0^T Q(t)\,dt\,,\, u'\right\rangle dx \end{aligned}$$

so as a consequence of the section 3.1, we finally obtain the expression of the Eulerian derivative given in the proposition 3.6. □

Bibliography

[1] J. A. Bello, E. Fernandez-Cara, J. Lemoine, and J. Simon. The differentiability of the drag with respect to the variations of a Lipschitz domain in a Navier-stokes flow. *SIAM J. Control Optimization*, 35(2):626–640, 1997.

[2] Berger, Marcel and Gostiaux, Bernard . *Differential geometry: manifolds, curves, and surfaces.* Graduate Texts in Mathematics, 115. Springer-Verlag, 1988.

[3] S. Boisgérault and J.-P. Zolésio. Shape derivative of sharp functionals governed by Navier-stokes flow. In O. John K. Najzar W. Jäger, J. Necas and J. Stará, editors, *Partial differential equations, Theory and numerical simulation*, CRC Research Notes in Mathematics, pages 49–63. Chapman & Hall, 2000.

[4] M. Delfour, Mghazli Z., and J.-P. Zolésio. Computation of shape gradients for mixed finite element formulation. In Da Prato G. and Zolésio J.-P., editors, *partial differential equation methods in control and shape analysis*, volume 188 of *lecture notes in pure and applied mathematics*, pages 77–93. Marcel Dekker, Inc., 1997.

[5] L. C. Evans and R. F. Gariepy. *Measure Theory and Fine Properties of Functions.* Studies in Advanced Mathematics, 1992.

[6] C. Foias and R. Temam. Structure of the set of stationary solutions of the Navier-stokes equations. *Communications on Pure and Applied Mathematics*, XXX:149–164, 1977.

[7] V. Girault and P.-A. Raviart. *Finite Element Methods for Navier-Stokes Equations*, volume 5 of *Springer Series in Computational Mathematics.* Springer-Verlag, 1986.

[8] J. Nečas. *Les méthodes directes en théorie des équations elliptiques.* Masson et Cie, 1967.

[9] Saut, J.C. and Temam, R. Generic properties of Navier-Stokes equations: Genericity with respect to the boundary values. *Indiana Univ. Math. J.*, (29):427–446, 1980.

[10] J. Sokolowski and J.-P. Zolésio. *Introduction to Shape Optimization. Shape Sensitivity Analysis*, volume 16 of *Springer Series in Computational Mathematics.* Springer-Verlag, 1992.

[11] R. Temam. *Navier-Stokes Equations.* North-Holland, Elsevier Science Publishers B. V, 1984.

[12] J.-P. Zolésio. Identification de Domaines par Déformations. Thèse de Doctorat d'Etat, Nice, 1979.

Sébastien Boisgérault. Centre de Mathématiques Appliquées, Ecole des Mines de Paris, 2004 route des Lucioles, BP. 93, 06902 Sophia Antipolis Cedex, France
E-mail: Sebastien.Boisgerault@sophia.inria.fr

Jean-Paul Zolésio. Research Director at CNRS, Centre de Mathématiques Appliquées, Ecole des Mines de Paris, 2004 route des Lucioles, BP. 93, 06902 Sophia Antipolis Cedex, France
E-mail: Jean-Paul.Zolesio@sophia.inria.fr

Shape Sensitivity Analysis in the Maxwell's Equations

John Cagnol, Jean-Paul Marmorat and Jean-Paul Zolésio

Abstract. The shape sensitivity analysis for hyperbolic problems yields some specific complications due to the hyperbolic regularity, or the lack thereof. In previous works we investigated the wave equation and came up with some shape sensistivity results. In this paper we investigate sensitivity of the solutions to the Maxwell equation with respect to the shape of the domain. We explicit a derivative with respect to a deformation parameter. The transport of the free divergence property requires a specific shape different quotient that is not necessary in the scalar case.

1 Introduction

1.1 Problem Formulation

We consider a bounded domain D in $\mathbb{R}^3$ and a family $\mathcal{O}_k$ of open connected domains Ω in D whose boundary $\Gamma = \partial\Omega$ is a C^k manifold oriented by the unitary normal field n outgoing to Ω. Throughout this paper we assume $k \geq 2$.

We suppose that Ω is occupied by an electromagnetic medium of constant electric permittivity ε and constant magnetic permeability μ. We suppose the electrical charge density and the current density in Ω are zero.

Let T be a non negative real and $I = [0, T]$ be the *time interval.* We note $Q =]0; T[\times\Omega$ the *cylindrical evolution domain* and $\Sigma =]0, T[\times\Gamma$ the *lateral boundary* associated to any element Ω of the family $\mathcal{O}_k$.

Let $E(t, x)$ and $H(t, x)$ denote the electric field and magnetic fields, respectively, at a point $x \in \Omega$ and a time $t \geq 0$. They satisfy the Maxwell's equations

$$\varepsilon\partial_t E - \operatorname{curl} H = 0 \quad \text{on} \quad Q \tag{1.1}$$

$$\mu\partial_t H + \operatorname{curl} E = 0 \quad \text{on} \quad Q \tag{1.2}$$

$$\operatorname{div}(E) = \operatorname{div}(H) = 0 \quad \text{on} \quad Q \tag{1.3}$$

$$H \times n = 0 \quad \text{on} \quad \Sigma \tag{1.4}$$

$$E(0) = E_0 \quad \text{on} \quad \Omega \tag{1.5}$$

$$H(0) = H_0 \quad \text{on} \quad \Omega \tag{1.6}$$

1.2 Useful Spaces

Let $m \geq 0$ be an integer. Throughout this paper we shall note

$$\mathcal{L}^m(\Omega) = (L^m(\Omega))^3, \quad \mathcal{L}^m(\Gamma) = (L^m(\Gamma))^3$$

$$\mathcal{H}^m(\Omega) = (H^m(\Omega))^3, \quad \mathcal{H}^m(\Gamma) = (H^m(\Gamma))^3$$

with the product topology in each case. The closure in $\mathcal{L}^2(\Omega)$ of the set of functions θ in $C^\infty(\bar{\Omega})$ such that $\operatorname{div}(\theta) = 0$ will be denoted $J(\Omega)$. When the function θ is only $C^\infty(\Omega)$ we shall note $\hat{J}(\Omega)$. We also introduce the following spaces

$$J^m(\Omega) = J(\Omega) \cap \mathcal{H}^m(\Omega)$$

$$J_n^m(\Omega) = \{\theta \in J^m(\Omega),\ \langle \theta n \rangle = 0 \text{ on } \Gamma\}$$

$$J_t^m(\Omega) = \{\theta \in J^m(\Omega),\ \theta \times n = 0 \text{ on } \Gamma\}$$

in each case the topology is induced by $\mathcal{H}^m(\Omega)$. We note

$$J_n^*(\Omega) = \{\theta \in J_n^2(\Omega),\ \operatorname{curl}\theta \times n = 0 \text{ on } \Gamma\}$$

$$J_t^*(\Omega) = \{\theta \in J_n^2(\Omega),\ \langle \operatorname{curl}\theta n \rangle = 0 \text{ on } \Gamma\}$$

the topology in each space in inherited from $\mathcal{H}^2(\Omega)$. Furthermore we note

$$\mathcal{Z}^m(I, \Omega) = \cap_{i=0}^m C^i(I, J^{m-i}(\Omega))$$

$$\hat{\mathcal{Z}}^m(I, \Omega) = \cap_{i=0}^m C^i(I, \hat{J}^{m-i}(\Omega))$$

1.3 Differentiation with Respect to the Domain

Let S be a real, soon to be considered small and $\mathcal{E}_k$ be the set of $V \in C([0, S]; C^k(D, \mathbb{R}^3))$ with $\langle V, n_{\partial D} \rangle = 0$. For any $V \in \mathcal{E}_k$ we consider the flow mapping $T_s(V)$. At the point x, V has the form as follows:

$$V(s)(x) = \left(\frac{\partial}{\partial s} T_s\right) \circ T_s^{-1}(x) \tag{1.7}$$

For each $s \in [0, S[$, T_s is a one-to-one mapping from D onto D such that

i) $T_0 = I$

ii) $(s, x) \mapsto T_s(x)$ belongs to $C^1([0; S[, C^k(D; D))$ with $T_s(\partial D) = \partial D$

iii) $(s, x) \mapsto T_s^{-1}(x)$ belongs to $C([0; S[, C^k(D; D))$

Such transformations were studied in [1] and [2] where a full analysis of the situation was given.

The family $\mathcal{O}_k$ is stable under the perturbations $\Omega \mapsto \Omega_s(V) = T_s(V)(\Omega)$. We denote by Q_s the *perturbed cylinder* $]0; T[\times \Omega_s(V)$, $\Gamma_s = \partial\Omega_s$ and $\Sigma_s =]0, T[\times \Gamma_s$ the *perturbed lateral boundary.* We consider a map defined on the family $\mathcal{O}_k$

$$\mathcal{O}_k \to \bigcup_{\Omega \in \mathcal{O}_k} (\mathcal{L}^2(Q) \times \mathcal{L}^2(Q))$$

To each element $\Omega \in \mathcal{O}_k$ we associate the solution $(E, H) = (E(\Omega), H(\Omega))$ of the Maxwell's equations described above.

For any $V \in \mathcal{E}_k$ and $s \in [0; S]$ we set $E_s = E(\Omega_s) \in \mathcal{L}^2(Q_s)$ and $H_s = H(\Omega_s) \in \mathcal{L}^2(Q_s)$. Following [3], [6] the mapping $\Omega \mapsto E(\Omega)$ is said to be *shape differentiable in* $L^2(I, \mathcal{H}^m(D))$

$$\exists \bar{E} \in C^1([0; S], L^2(I, \mathcal{H}^m(D))) \tag{1.8}$$

$$\bar{E}(s, \cdot, \cdot)\big|_{Q_s} = E(\Omega_s) \tag{1.9}$$

then $\partial_s \bar{E}(0, \cdot, \cdot)\big|_Q$ which is the restriction to Q of the derivative with respect to the perturbation parameter s at $s = 0$ is independent of the choice of $\bar{E}$ verifying (1.8), (1.9) (*cf.* [6]).

DEFINITION 1.1 (SHAPE DERIVATIVE). *The* shape derivative *is that unique element*

$$E'(\Omega; V) = \left(\frac{\partial}{\partial s}\bar{E}\right)\Big|_{s=0\ (t,x)\in Q} \in L^2(I, \mathcal{L}^2(\Omega))$$

DEFINITION 1.2 (MATERIAL DERIVATIVE). *The element* $\dot{E}(\Omega; V)$ *is the* material derivative *of* E *in* $L^2(I, \mathcal{H}^m(D))$ *if it is the limit in* $L^2(I, \mathcal{H}^m(D))$ *of*

$$\frac{1}{s}\left(E(\Omega_s) \circ T_s - E(\Omega)\right)$$

when s *tends to 0*

Weak differentiability may be defined similarly. We define, as well, differentiability for H. The aim of this paper is to give a full analysis of the shape differentiability for (E, H).

Let us recall that if $\theta \in \mathcal{H}^1(D)$ then $s \mapsto \theta \circ T_s$ is differentiable in $\mathcal{L}^2(D)$ and the derivative is given by

$$\left(\frac{\partial}{\partial s}(\theta \circ T_s)\right)_{s=0} = D\theta.V(0)$$

2 Main Result

PROPOSITION 2.1. *We assume* $E_0 \in J_n^*(\Omega)$ *and* $H_0 \in J_t^*(\Omega)$. *The solution to the Maxwell's equations is weakly material differentiable in* $H(\mathrm{curl}, \Omega)$, *furthermore the material derivative is solution to*

$$\varepsilon \partial_t \dot{E} - \mathrm{curl}\, \dot{H} = -(DV(0)\nabla) \times H \quad on \quad Q \tag{2.1}$$

$$\mu \partial_t \dot{H} + \mathrm{curl}\, \dot{E} = (DV(0)\nabla) \times E \quad on \quad Q \tag{2.2}$$

$$\mathrm{div}(\dot{E}) = \mathrm{div}(DV(0).E) \quad on \quad Q \tag{2.3}$$

$$\mathrm{div}(\dot{H}) = \mathrm{div}(DV(0).H) \quad on \quad Q \tag{2.4}$$

$$\dot{H} \times n = (DV(0).H) \times n \quad on \quad \Sigma \tag{2.5}$$

$$\dot{E}(0) = DE_0.V(0) \quad on \quad \Omega \tag{2.6}$$

$$\dot{H}(0) = DH_0.V(0) \quad on \quad \Omega \tag{2.7}$$

PROPOSITION 2.2. *We assume $E_0 \in J_n^*(\Omega)$ and $H_0 \in J_t^*(\Omega)$. The solution to the Maxwell's equations is weakly shape differentiable in $H(\mathrm{curl}, \Omega)$, furthermore the shape derivative is solution to*

$$\varepsilon \partial_t E' - \mathrm{curl}\, H' = 0 \quad on \quad Q \tag{2.8}$$

$$\mu \partial_t H' + \mathrm{curl}\, E' = 0 \quad on \quad Q \tag{2.9}$$

$$\mathrm{div}(E') = 0 \quad on \quad Q \tag{2.10}$$

$$\mathrm{div}(H') = 0 \quad on \quad Q \tag{2.11}$$

$$H' \times n = [V(0), H] \times n \quad on \quad \Sigma \tag{2.12}$$

$$E'(0) = 0 \quad on \quad \Omega \tag{2.13}$$

$$H'(0) = 0 \quad on \quad \Omega \tag{2.14}$$

where $[\cdot, \cdot]$ stands for the Lie brackets.

3 Proof

3.1 Regularity of the solution

We suppose the initial conditions (E_0, H_0) satisfies $E_0 \in J_n^*(\Omega)$ and $H_0 \in J_t^*(\Omega)$. According to [5] we have

$$E \in \mathcal{Z}^2(I, \Omega)$$

$$H \in \hat{\mathcal{Z}}^2(I, \Omega)$$

3.2 Shape Difference Quotient

Following [4, section 4] We note $E^s = DT_s^{-1}(E_s \circ T_s)$ and $H^s = DT_s^{-1}(H_s \circ T_s)$ Let

$$e_s = \frac{E^s - E}{s} \quad \text{and} \quad h_s = \frac{H^s - H}{s}$$

be the shape difference quotients.

LEMMA 3.1. *The shape difference quotient associated to E and H belong to $\mathcal{Z}^2(I, \Omega)$ and $\hat{\mathcal{Z}}^2(I, \Omega)$ respectively.*

Proof. For the sake of shortness we prove the lemma when $k = \infty$. The only non-trivial point to be proven is:

$$E^s \in \mathcal{Z}^2(I, \Omega) \tag{3.1}$$

$$H^s \in \hat{\mathcal{Z}}^2(I, \Omega) \tag{3.2}$$

We shall prove (3.1), the proof of (3.2) is similar.

We have $E_s \in \mathcal{Z}^2(I,\Omega_s)$ therefore $E_s(t) \in J^2(\Omega_s)$, $\partial_t E_s(t) \in J^1(\Omega_s)$ and $\partial_{tt}E_s(t) \in J^0(\Omega_s)$. Since $E_s(t) \in J^2(\Omega_s)$, there exists a sequence $(\theta_k)_k$ of functions $C^\infty(\Omega_s)$ such that $\operatorname{div}(\theta_k) = 0$, converging toward $E_s(t)$. Let $\psi_k = DT_s^{-1}\theta_k \circ T_s$ then $\psi_k \in C^\infty(\Omega)$ moreover

$$\operatorname{div}(\psi_k) = \operatorname{div}(DT_s^{-1}\theta_k \circ T_s) = (\operatorname{div}\theta_k) \circ T_s = 0$$

and $(\psi_k)_k$ converges toward $DT_s^{-1}(E_s(t) \circ T_s)$. Hence

$$E^s(t) \in J(\Omega)$$

On the other hand $E_s(t) \in \mathcal{H}^k(\Omega_s)$ therefore

$$E^s(t) \in \mathcal{H}^k(\Omega)$$

It follows $E^s(t) \in J^2(\Omega)$.

The same method applies to prove that $\partial_t E^s(t) \in J^1(\Omega)$ and $\partial_{tt}E^s(t) \in J^0(\Omega)$, therefore $E^s \in \mathcal{Z}^2(I,\Omega)$. □

Let $\mathcal{E}$ be defined on $I \times \hat{\mathcal{Z}}^2(I,\Omega)$ by

$$\mathcal{E}(t,\theta) = \frac{1}{2}\int_\Omega |\operatorname{curl}\theta|^2 + |\partial_t\theta|^2 \, dx$$

that function will be called the energy associated to the system.

LEMMA 3.2. *The energy associated to the system is constant with respect to t*

Proof. We have

$$\partial_t\mathcal{E}(t,\theta) = \int_\Omega \operatorname{curl}\theta \operatorname{curl}\partial_t\theta + \partial_t\theta\partial_{tt}\theta \, dx$$

using $\operatorname{curl}\theta \times n = 0$ we obtain

$$\partial_t\mathcal{E}(t,\theta) = \int_\Omega (\operatorname{curl}\operatorname{curl}\theta + \partial_{tt}\theta)\partial_t\theta \, dx$$

since $\operatorname{curl}\operatorname{curl}\theta + \partial_{tt}\theta = 0$, we derive $\partial_t\mathcal{E}(t,\theta) = 0$. □

REMARK 3.1. *Lemma 3.2 means the physical system is conservative.*

Using lemma 3.1 on the shape difference quotients e_s

$$\|e_s\|_{H(\mathrm{curl},\Omega)} = \left\|\frac{DT_s^{-1}(E_0 \circ T_s) - E_0}{s}\right\|_{H(\mathrm{curl},\Omega)}$$

$$= \left\|\frac{DT_s^{-1}(E_0 \circ T_s) - E_0 \circ T_s}{s} + \frac{E_0 \circ T_s - E_0}{s}\right\|_{H(\mathrm{curl},\Omega)}$$

when s tends to 0, the latter expression tends to $\| - DV(0).E_0 + DE_0.V(0)\|_{H(\mathrm{curl},\Omega)} = \|[E_0, V(0)]\|_{H(\mathrm{curl},\Omega)}$. Therefore when s is small enough, there exists M such that

$$\|e_s\|_{H(\mathrm{curl},\Omega)} \leq M$$

LEMMA 3.3. *There exists $e \in H(\mathrm{curl}, \Omega)$ such that a minimizing subsequence of e^s tends to e weakly in $H(\mathrm{curl}, \Omega)$.*

The same method can be used to prove

LEMMA 3.4. *There exists $e \in H(\mathrm{curl}, \Omega)$ such that a minimizing subsequence of e^s tends to e weakly in $H(\mathrm{curl}, \Omega)$.*

3.3 Uniqueness of e and h

LEMMA 3.5. *The following identity holds on Q*

$$\varepsilon\partial_t e - \mathrm{curl}\, h = -\varepsilon\partial_t(DV(0).E) + \mathrm{curl}(DV(0).H) - (DV(0)\nabla) \times H$$

Proof. On Q_s we have $\varepsilon\partial_t E_s - \mathrm{curl}\, H_s = 0$ therefore, the following identity holds on Q

$$\varepsilon\partial_t(E_s \circ T_s) - (\mathrm{curl}\, H_s) \circ T_s = 0$$

since $(\mathrm{curl}\, H_s) \circ T_s = ({}^*DT_s^{-1}\nabla) \times (H_s \circ T_s)$ we get $\varepsilon(\partial_t E_s) \circ T_s - ({}^*DT_s^{-1}\nabla) \times (H_s \circ T_s) = 0$. On the other hand $\partial_t E - \mathrm{curl}\, H = 0$ on Q therefore

$$\varepsilon\partial_t\left(\frac{E_s \circ T_s - E}{s}\right) + \frac{({}^*DT_s^{-1}\nabla) \times (H_s \circ T_s) - \nabla \times H}{s} = 0 \tag{3.3}$$

The first term of (3.3) can be written

$$-\varepsilon\partial_t\left(\frac{DT_s^{-1} - I}{s}\right)(E_s \circ T_s) + \varepsilon\partial_t\left(\frac{DT_s^{-1}(E_s \circ T_s) - E}{s}\right)$$

its limit in $H(\mathrm{curl}, \Omega)$ is $\varepsilon\partial_t(DV(0).E) + \varepsilon\partial_t e$. On the other hand, the second term of (3.3) can be written

$$\begin{aligned} &\frac{({}^*DT_s^{-1}\nabla) \times (H_s \circ T_s) - (DT_s^{-1}\nabla) \times (DT_s^{-1}(H_s \circ T_s))}{s} \\ &+\frac{(DT_s^{-1}\nabla) \times (DT_s^{-1}(H_s \circ T_s)) - \nabla \times (DT_s^{-1}(H_s \circ T_s))}{s} \\ &+\frac{\nabla \times (DT_s^{-1}(H_s \circ T_s)) - \nabla \times H}{s} \end{aligned}$$

its limit in $H(\mathrm{curl}, \Omega)$ is $\mathrm{curl}(DV(0).H) - (DV(0)\nabla) \times H + \mathrm{curl}(h)$ hence the limit in $H(\mathrm{curl}, \Omega)$ of the right hand side of (3.3) as s tends to 0 is

$$\varepsilon\partial_t e - \mathrm{curl}\, h + \varepsilon\partial_t(DV(0).E) - \mathrm{curl}(DV(0).H) + (DV(0)\nabla) \times H$$

which proves the lemma □

LEMMA 3.6. *The following identity holds on Q*

$$\mu\partial_t h + \mathrm{curl}\, e = -\mu\partial_t(DV(0).H) - \mathrm{curl}(DV(0).E) + (DV(0)\nabla) \times E$$

Proof. Analogous to the proof of lemma 3.5. □

LEMMA 3.7. *The following identity holds on Q*

$$\operatorname{div}(e) = \operatorname{div}(h) = 0$$

Proof. We have $\operatorname{div}(E_s) = 0$ on Q_s therefore we have $\operatorname{div}(E_s) \circ T_s = 0$ on Q. Hence

$$\operatorname{div}(DT_s^{-1}(E_s \circ T_s)) = 0$$

on Q. On the other hand $\operatorname{div}(E) = 0$ on Q hence $\operatorname{div}(e) = 0$. The same method applies to prove $\operatorname{div}(h) = 0$. □

LEMMA 3.8. *The following identity holds on Σ*

$$h \times n = 0$$

Proof. On Σ_s one has $H_s \times n_s = 0$ thus the following identity holds on Σ

$$(H_s \circ T_s) \times (n_s \circ T_s) = 0$$

since $n_s \circ T_s$ is collinear with ${}^*DT_s^{-1}n$ we get

$$((DT_s^{-1}(H_s \circ T_s)) \times n = 0$$

on the other hand $H \times n = 0$ therefore $h \times n = 0$ on Σ □

LEMMA 3.9. *The following identity holds on Ω*

$$e(0) = [E, V(0)]$$

Proof. We have $E_s(0) = E_0$ on Ω_s hence $E_s(0) \circ T_s = E_0 \circ T_s$ on Ω. Therefore the subsequent equality holds on Ω

$$\frac{DT_s^{-1}(E(0) \circ T_s) - E(0)}{s} = \frac{DT_s^{-1}(E_0 \circ T_s) - E_0 \circ T_s}{s} + \frac{E_0 \circ T_s - E_0}{s}$$

hence $e(0) = -DV(0).E_0 + DE_0.V(0) = [E_0, V(0)]$. □

LEMMA 3.10. *The following identity holds on Ω*

$$h(0) = [H_0, V(0)]$$

Proof. Analogous to the proof of lemma 3.9. □

As a consequence of the previous lemmae we get

PROPOSITION 3.11. *The functions e and h are solution to*

$$\varepsilon\partial_t e - \operatorname{curl} h = -\varepsilon\partial_t(DV(0).E) + \operatorname{curl}(DV(0).H) - (DV(0)\nabla) \times H \quad on \quad Q \qquad (3.4)$$

$$\mu\partial_t h + \operatorname{curl} e = -\mu\partial_t(DV(0).H) - \operatorname{curl}(DV(0).E) + (DV(0)\nabla) \times E \quad on \quad Q \qquad (3.5)$$

$$\operatorname{div}(e) = \operatorname{div}(h) = 0 \quad on \quad Q \qquad (3.6)$$

$$h \times n = 0 \quad on \quad \Sigma \qquad (3.7)$$

$$e(0) = [E_0, V(0)] \quad on \quad \Omega \qquad (3.8)$$

$$h(0) = [H_0, V(0)] \quad on \quad \Omega \qquad (3.9)$$

We refer to [5] for the well-posedness of that system. Therefore e and h are unique.

3.4 Material Derivative

PROPOSITION 3.12. *The material derivatives of E and H exist on Q and*

$$\dot{E} = e + DV(0).E$$

$$\dot{H} = h + DV(0).H$$

Proof. The following identity holds on Q

$$\frac{E_s \circ T_s - E}{s} = \frac{E_s \circ T_s - DT_s^{-1}(E_s \circ T_s)}{s} + \frac{DT_s^{-1}(E_s \circ T_s) - E}{s}$$

when s tends to 0 the left hand side tends to $\dot{H}$ weakly in $H(\text{curl}, \Omega)$ and the right hand side tends to $e + DV(0).E$ weakly in $H(\text{curl}, \Omega)$. The same method proves $\dot{H} = h + DV(0).H$. □

COROLLARY 3.13. *The functions $\dot{E}$ and $\dot{H}$ are solution to (2.1)–(2.7).*

3.5 Shape Derivative

PROPOSITION 3.14. *The shape derivative of E and H exist on Q and*

$$E' = e + [V(0), E]$$

$$H' = h + [V(0), H]$$

where $[\cdot, \cdot]$ denotes the Lie brackets.

Proof. We consider an extension of E and H from $\mathcal{H}^m(\Omega)$ to $\mathcal{H}^m(D)$, those extensions will be denoted E and H, as well. The following identity holds on D

$$\frac{E_s - E}{s} = \frac{E_s - E_s \circ T_s}{s} + \frac{E_s \circ T_s - E}{s}$$

therefore, when s tends to 0, we obtain $\frac{E_s - E}{s} \rightharpoonup \dot{E} + DE.V(0)$ weakly in $H(\text{curl}, D)$. □

REMARK 3.2. ***The advantages of the shape derivative as opposed to the material derivative are noteworthy. The shape derivative is intrinsic, as we shall see the problem it is solution to, only depends on the boundary condition which is governed by the vector field $V(0)$.***

We shall now prove the shape derivatives (E', H') are solution to (2.8) – (2.14). The identities (2.12), (2.13) and (2.14) are a consequence of proposition 3.14 and (3.7), (3.8), (3.9). The identities (2.10) and (2.11) are a consequence of proposition 3.14, of identity (3.6) and $\text{div}([u, v]) = 0$ when $\text{div}(u) = \text{div}(v) = 0$. The proof of (2.8) and (2.9) could be done by a direct computation but we shall give here a theoretical argument to prove them. First, we establish the subsequent lemma.

LEMMA 3.15. *Assume $F \in C^1([0; S[, L^1(I, \mathcal{L}^1(D))$, we note $f_s = F(s, \cdot, \cdot)$ and $f = f_0$, then*

$$\frac{\partial}{\partial s}\left(\int_{Q_s} f_s(x,t)\, dx\, dt\right)_{s=0} = \int_Q \partial_s F(0, x, t)\, dx\, dt + \int_\Sigma f(x,t)\, \langle V(0) n \rangle\, d\Gamma\, dt$$

Proof. Let Θ be any of the three components of F and let θ be the corresponding component for f. Let us note $\vartheta(s) = \int_{Q_s} \Theta(s,t,x)\,dx\,dt$. The function T_s maps Ω to Ω_s hence

$$\vartheta(s) = \int_Q \Theta(s,t,T_s(x)) \det(DT_s)\,dx\,dt$$

$$\frac{\partial\vartheta}{\partial s} = \int_Q (\partial_s\Theta(s,t,x) + \langle\nabla\Theta(s,t,x)V(s)\circ T_s(t,x)\rangle \det(DT_s) + \Theta(s,t,x)\partial_s\det(DT_s)\,dx\,dt$$

Using $\partial_s\det(DT_s)|_{s=0} = \operatorname{div} V(0)$, we get

$$\frac{\partial\vartheta}{\partial s}(0) = \int_Q \partial_s\Theta(0,t,x) + \langle\nabla\Theta(0,t,x)V(0)\rangle + \Theta(0,t,x)\ \operatorname{div} V(0)\ dx\,dt$$

$$\frac{\partial\vartheta}{\partial s}(0) = \int_Q \partial_s\Theta(0,t,x)\ dx\,dt\ + \int_Q \operatorname{div}(\Theta(0,t,x)\,V(0))\ dx\,dt$$

The final result is given by the Gauss' theorem on that identity. □

The proof of (2.8) and (2.9) is now straightforward when one considers the weak form of the equalities (1.1) and (1.2) and Q_s. For instance

$$\forall\varphi \in C_0^\infty(I\times D),\ \int_{Q_s} \varepsilon(\partial_t E_s)\varphi - (\operatorname{curl} H_s)\varphi = 0$$

Lemma 3.15 yields

$$\forall\varphi \in C_0^\infty(I\times D),\ \int_Q \varepsilon(\partial_t E')\varphi - (\operatorname{curl} H')\varphi + \int_\Sigma (\varepsilon\partial_t E - \operatorname{curl} H)\varphi\,\langle V(0)n\rangle = 0$$

Since $\varepsilon\partial_t E - \operatorname{curl} H = 0$ on Q we derive $\varepsilon\partial_t E' - \operatorname{curl} H' = 0$. That proves the shape derivative is solution to (2.8)–(2.14).

Bibliography

[1] Michel C. Delfour and Jean-Paul Zolésio. Structure of shape derivatives for non smooth domains. *Journal of Functional Analysis*, 104, 1992.

[2] Michel C. Delfour and Jean-Paul Zolésio. Shape analysis via oriented distance functions. *Journal of Functional Analysis*, 123, 1994.

[3] Fabrice R. Desaint and Jean-Paul Zolésio. Manifold derivative in the laplace-beltrami equation. *Journal of Functionnal Analysis*, 151(1):234,269, 1997.

[4] Raja Dziri and Jean-Paul Zolésio. Shape existence in navier-stokes flow with heat convection. *Annali della Scuoala Normale Superiore di Pisa*, XXIV(1):165–192, 1997.

[5] John E. Lagnese. Exact boundary controllability of maxwell's equations in a general region. *Siam Journal of Control and Optimization*, 27(2):374–388, March 1989.

[6] Jean-Paul Zolésio. Introduction to shape optimization and free boundary problems. In Michel C. Delfour, editor, *Shape Optimization and Free Boundaries*, volume 380 of *NATO ASI, Series C: Mathematical and Physical Sciences*, pages 397,457, 1992.

John Cagnol. Pôle Universitaire Léonard de Vinci, Département Calcul Scientifique, Filiere Sciences et Technologie, 92916 Paris La Defense Cedex, France
E-mail: John.Cagnol@devinci.fr

Jean-Paul Marmorat. Centre de Mathématiques Appliquées, Ecole des Mines de Paris, 2004 route des Lucioles, BP. 93, 06902 Sophia Antipolis Cedex, France
E-mail: Jean-Paul.Marmorat@sophia.inria.fr

Jean-Paul Zolésio. Research Director at CNRS, Centre de Mathématiques Appliquées, Ecole des Mines de Paris, 2004 route des Lucioles, BP. 93, 06902 Sophia Antipolis Cedex, France
E-mail: Jean-Paul.Zolesio@sophia.inria.fr

Tangential Calculus and Shape Derivatives

Michel C. Delfour[1] and Jean-Paul Zolésio

Abstract. The object of this paper is to illustrate the combined strength of the *shape calculus* and the *intrinsic tangential differential calculus* (recently used in the theory of thin and asymptotic shells) to compute shape derivatives. It also makes the connection with the generic Courant metric framework of A.M. Micheletti [7].

1 Introduction

The *tangential differential calculus* naturally occurs in computations involving boundary integrals in partial differential equations. But it also occurs in the computation of *shape derivatives* of boundary integrals and in second order shape derivatives of domain integrals. An intrinsic approach to the *differential calculus on a submanifold* of co-dimension one has considerably simplified the computation and the form of the final expressions [5, 2]. It has been the basis for a completely intrinsic approach to the theory of thin and asymptotic shells. It completely avoids the use of local maps and bases and Christoffel's symbols. In this paper we combine the *shape calculus* and this *intrinsic tangential differential calculus* to compute first and second derivatives of shape functions.

We first recall basic definitions and review former results for shape derivatives. We connect the definitions with the generic Courant metric framework of A.M. Micheletti [7]. We illustrate the techniques for the first shape derivative of the norm of the normal derivative and the second order shape derivative of the domain integral. In so doing we recover earlier structural results in [1, 12] involving the half Lie bracket.

[1]The research of the first author has been supported by National Sciences and Engineering Research Council of Canada research grant A–8730 and by a FCAR grant from the Ministère de l'Éducation du Québec. This paper was completed while the first author was on sabbatical leave at INRIA-Rocquencourt (France) in the projects MACS and SOSO.

2 First Order Semiderivatives and Gradient

2.1 Equivalence Between Velocities and Transformations

Let the real number $\tau > 0$ and the map $V : [0,\tau] \times \mathbb{R}^N \to \mathbb{R}^N$ be given. The map V can be viewed as a family $\{V(t) : 0 \le t \le \tau\}$ of non-autonomous velocity fields on $\mathbb{R}^N$ defined by $x \mapsto V(t)(x) \stackrel{def}{=} V(t,x) : \mathbb{R}^N \mapsto \mathbb{R}^N$. Assume that

$$\text{(V)} \quad \begin{aligned} &\forall x \in \mathbb{R}^N, \quad V(\cdot,x) \in C([0,\tau];\mathbb{R}^N) \\ &\exists c > 0,\ \forall x,y \in \mathbb{R}^N, \quad \|V(\cdot,y) - V(\cdot,x)\|_{C([0,\tau];\mathbb{R}^N)} \le c|y-x| \end{aligned} \tag{2.1}$$

where $V(\cdot,x)$ is the function $t \mapsto V(t,x)$. Associate with V the solution $x(t;V)$ of the ordinary differential equation

$$\frac{dx}{dt}(t) = V\big(t,x(t)\big), \quad t \in [0,\tau], \quad x(0) = X \in \mathbb{R}^N, \tag{2.2}$$

and introduce the homeomorphism

$$X \mapsto T_t(V)(X) \stackrel{def}{=} x_V(t;X) : \mathbb{R}^N \to \mathbb{R}^N. \tag{2.3}$$

and the maps

$$(t,X) \mapsto T_V(t,X) \stackrel{def}{=} T_t(V)(X) : [0,\tau] \times \mathbb{R}^N \to \mathbb{R}^N, \tag{2.4}$$

$$(t,x) \mapsto T_V^{-1}(t,x) \stackrel{def}{=} T_t^{-1}(V)(x) : [0,\tau] \times \mathbb{R}^N \to \mathbb{R}^N. \tag{2.5}$$

In the sequel we shall drop the V in $T_V(t,X)$, $T_V^{-1}(t,x)$ and $T_t(V)$.

THEOREM 2.1.

(i) Under assumption (V) *the map T has the following properties*:

$$\begin{aligned} &\text{(T1)} \quad \begin{aligned} &\forall X \in \mathbb{R}^N, \quad T(\cdot,X) \in C^1([0,\tau];\mathbb{R}^N) \ \textit{and}\ \exists c > 0, \\ &\forall X,Y \in \mathbb{R}^N, \quad \|T(\cdot,Y) - T(\cdot,X)\|_{C^1([0,\tau];\mathbb{R}^N)} \le c|Y-X|, \end{aligned} \\ &\text{(T2)} \quad \forall t \in [0,\tau],\ X \mapsto T_t(X) = T(t,X) : \mathbb{R}^N \to \mathbb{R}^N \ \textit{is bijective}, \\ &\text{(T3)} \quad \begin{aligned} &\forall x \in \mathbb{R}^N, \quad T^{-1}(\cdot,x) \in C([0,\tau];\mathbb{R}^N) \ \textit{and}\ \exists c > 0, \\ &\forall x,y \in \mathbb{R}^N, \quad \|T^{-1}(\cdot,y) - T^{-1}(\cdot,x)\|_{C([0,\tau];\mathbb{R}^N)} \le c|y-x|. \end{aligned} \end{aligned} \tag{2.6}$$

(ii) Given a real number $\tau > 0$ and a map $T : [0,\tau] \times \mathbb{R}^N \to \mathbb{R}^N$ verifying assumptions (T1), (T2) *and* (T3), *then the map*

$$(t,x) \mapsto V(t,x) \stackrel{\text{def}}{=} \frac{\partial T}{\partial t}\big(t, T_t^{-1}(x)\big) : [0,\tau] \times \mathbb{R}^N \to \mathbb{R}^N \tag{2.7}$$

verifies assumption (V), *where T_t^{-1} is the inverse of $X \mapsto T_t(X) = T(t,X)$. If, in addition, $T(0,X) = X$ for all $X \in \mathbb{R}^N$, then $T(\cdot,X)$ is the solution of* (2.2).

2.2 Definitions of Semiderivatives and Derivatives

Recall that a functional $J(\Omega)$ defined on a family of subsets Ω of $\mathbb{R}^N$ is called a *shape functional* if for any transformation T of $\mathbb{R}^N$

$$T(\Omega) = \Omega \quad \Rightarrow J(T(\Omega)) = J(\Omega).$$

Under the action of a velocity field V verifying conditions (2.1), a domain Ω in $\mathbb{R}^N$ is transformed into a new domain

$$\Omega_t(V) \stackrel{def}{=} T_t(V)(\Omega) = \{T_t(V)(X) \,:\, \forall X \in \Omega\} \tag{2.8}$$

DEFINITION 2.2. *Let Θ be a topological vector subspace of* $\text{Lip}(\mathbb{R}^N, \mathbb{R}^N)$.

(i) Given a velocity field V verifying conditions (2.1), *J is said to have an* Eulerian semiderivative *at Ω in the direction V if the limit*

$$dJ(\Omega; V) \stackrel{\text{def}}{=} \lim_{t \searrow 0} \frac{J(\Omega_t(V)) - J(\Omega)}{t} \quad \textit{exists and is finite.} \tag{2.9}$$

(ii) For $\theta \in \text{Lip}(\mathbb{R}^N, \mathbb{R}^N)$ and the autonomous velocity field

$$\tilde{\theta}(t)(x) \stackrel{\text{def}}{=} \theta(x), \quad \forall t \in [0, \tau], \forall x \in D, \tag{2.10}$$

we shall use the notation $dJ(\Omega; \tilde{\theta})$ or simply $dJ(\Omega; \theta)$.

(iii) Given $\theta \in \Theta$, J is said to have a Hadamard semiderivative at Ω in the direction θ w.r.t. Θ *if for all V verifying* (2.1), *$V(t) \in \Theta$ and $V(0) = \theta$*

$$dJ(\Omega; V) \textit{ exists and only depends on } V(0) = \theta. \tag{2.11}$$

In that case the semiderivative will be denoted $d_H J(\Omega; \theta)$ and necessarily

$$d_H J(\Omega; V(0)) = dJ(\Omega; V(0)).$$

(iv) J is said to be differentiable *at Ω in Θ' if it has a Eulerian semiderivative at Ω in all directions $\theta \in \Theta$ and the map*

$$\theta \mapsto dJ(\Omega; \theta) \,:\, \Theta \to \mathbb{R} \tag{2.12}$$

is linear and continuous. The map (2.12) *is denoted $G(\Omega)$ and referred to as the* gradient *of J in the topological dual Θ' of Θ.*

The definition of a Eulerian semiderivative is quite general. It includes cases where $dJ(\Omega; V)$ is not only dependent on $V(0)$, but also on $V(t)$ in a neighborhood of $t = 0$. This will not occur under some continuity assumption on the map $V \mapsto dJ(\Omega; V)$. When $dJ(\Omega; V)$ only depends on $V(0)$, the analysis can be specialized to autonomous vector fields V and the semiderivative can be related to the gradient of J. If J has a *Hadamard semiderivative* at Ω in the direction θ, it has a Eulerian semiderivative at Ω in the direction θ and

$$d_H J(\Omega; \theta) = dJ(\Omega; \theta).$$

The following simple continuity condition can also be used to obtain the Hadamard semidifferentiability. The proofs are analogous to the ones in [10, 3, 4].

THEOREM 2.3. *Let Θ be a Banach subspace of* $\text{Lip}(\mathbb{R}^N, \mathbb{R}^N)$ *and* $\Omega \subset \mathbb{R}^N$.

(i) *Given $\theta \in \Theta$, if*

$$\forall V \in C([0,\tau];\Theta) \text{ such that } V(0) = \theta, \quad dJ(\Omega; V) \text{ exists,} \tag{2.13}$$

and if the map

$$V \mapsto dJ(\Omega; V) : C([0,\tau];\Theta) \to \mathbb{R} \tag{2.14}$$

is continuous for the subspace of V's such that $V(0) = \theta$, then J is Hadamard semidifferentiable at Ω in the direction θ w.r.t Θ and

$$\forall V \in C([0,\tau];\Theta),\ V(0) = \theta, \quad dJ(\Omega; V) = dJ(\Omega; V(0)) = d_H J(\Omega; \theta). \tag{2.15}$$

(ii) *If for all V in $C([0,\tau];\Theta)$, $dJ(\Omega; V)$ exists and the map*

$$V \mapsto dJ(\Omega; V) : C([0,\tau];\Theta) \to \mathbb{R} \tag{2.16}$$

is continuous, then J is Hadamard semidifferentiable at Ω in the direction $V(0)$ with respect to Θ and

$$\forall V \in C([0,\tau];\Theta), \quad dJ(\Omega; V) = dJ(\Omega; V(0)) = d_H J(\Omega; V(0)). \tag{2.17}$$

For simplicity the last theorem was given for a Banach space Θ. However its conclusion is not limited to Banach spaces. Consider velocity fields in

$$\vec{\mathcal{V}}^{m,k} \stackrel{def}{=} \varinjlim_{K} \left\{ V_K^{m,k} : \forall K \text{ compact in } \mathbb{R}^N \right\} \tag{2.18}$$

$$\forall m \geq 0, \quad V_K^{m,k} \stackrel{def}{=} C^m\left([0,\tau]; \mathcal{D}^k(K, \mathbb{R}^N) \cap \text{Lip}(\mathbb{R}^N, \mathbb{R}^N)\right) \tag{2.19}$$

and $\varinjlim$ denotes the inductive limit set with respect to K endowed with its natural inductive limit topology. For autonomous fields, this reduces to

$$\vec{\mathcal{V}}^{k} \stackrel{def}{=} \varinjlim_{K} \left\{ V_K^{k} : \forall K \text{ compact in } \mathbb{R}^N \right\} \tag{2.20}$$

$$V_K^k \stackrel{def}{=} \begin{cases} \mathcal{D}^0(K, \mathbb{R}^N) \cap \text{Lip}(K, \mathbb{R}^N), & k = 0 \\ \mathcal{D}^k(K, \mathbb{R}^N), & 1 \leq k \leq \infty. \end{cases} \tag{2.21}$$

For $k \geq 1$, $\vec{\mathcal{V}}^k = \mathcal{D}^k(\mathbb{R}^N, \mathbb{R}^N)$. In all cases conditions (2.1) are verified. The proof of the next theorem is analogous to the similar ones in [3, 4].

THEOREM 2.4. *Let Ω be a subset of $\mathbb{R}^N$ and $k \geq 0$ an integer.*

(i) *Given $\theta \in \vec{\mathcal{V}}^k$, assume that*

$$\forall V \in \vec{\mathcal{V}}^{0,k},\ V(0) = \theta, \quad dJ(\Omega; V) \text{ exists,} \tag{2.22}$$

and that the map $V \mapsto dJ(\Omega; V) : \vec{\mathcal{V}}^{0,k} \to \mathbb{R}$ is continuous for all V's such that $V(0) = \theta$. Then J is Hadamard semidifferentiable in Ω in the direction θ with respect to $\mathcal{V}^k$ and

$$\forall V \in \vec{\mathcal{V}}^{0,k},\ V(0) = \theta, \quad d_H J(\Omega; \theta) = dJ(\Omega; V) = dJ(\Omega; V(0)). \tag{2.23}$$

(ii) Assume that for all $V \in \vec{\mathcal{V}}^{0,k}$, $dJ(\Omega; V)$ *exists, and that the map*

$$V \mapsto dJ(\Omega; V) : \vec{\mathcal{V}}^{0,k} \to \mathbb{R} \tag{2.24}$$

is continuous. Then J *is Hadamard semidifferentiable in* Ω *in the direction* $V(0)$ *with respect to* $\mathcal{V}^k$ *and*

$$\forall V \in \vec{\mathcal{V}}^{0,k}, \quad d_H J(\Omega; V(0)) = dJ(\Omega; V) = dJ(\Omega; V(0)). \tag{2.25}$$

2.3 Perturbations of the Identity and Fréchet Derivative

Consider the generic Courant metric topologies defined in [7, 6]. Associate with a Banach subspace $\Theta \subset \text{Lip}(\mathbb{R}^N, \mathbb{R}^N)$ of transformations of $\mathbb{R}^N$ the space

$$\mathcal{F}(\Theta) \stackrel{def}{=} \{F : \mathbb{R}^N \to \mathbb{R}^N \; : \; F - I \in \Theta \text{ and } F^{-1} - I \in \Theta\}.$$

$$\forall F \in \mathcal{F}(\Theta), \quad d(I, F) \stackrel{def}{=} \inf_{(f_1,\dots,f_n)} \sum_{i=1}^{n} \|f_i\|_\Theta + \inf_{(g_1,\dots,g_m)} \sum_{i=1}^{m} \|g_i\|_\Theta, \tag{2.26}$$

where the infima are taken over all finite factorizations in $\mathcal{F}(\Theta)$ of the form

$$F = (I + f_n) \circ \cdots \circ (I + f_1) \text{ and } F^{-1} = (I + g_m) \circ \cdots \circ (I + g_1), \quad f_i, g_i \in \Theta.$$

Extend this definition to all pairs F and G in $\mathcal{F}(\Theta)$

$$d(F, G) \stackrel{def}{=} d(I, G \circ F^{-1}) \tag{2.27}$$

Define for some open or closed subset Ω_0 of $\mathbb{R}^N$ the subgroup

$$\mathcal{G}(\Omega_0) \stackrel{def}{=} \{F \in \mathcal{F}(\Theta) \; : \; F(\Omega) = \Omega_0\}$$

and the Courant metric on $\mathcal{F}(\Theta)/\mathcal{G}(\Omega_0)$

$$\rho(F, H) \stackrel{def}{=} \inf_{G, \tilde{G} \in \mathcal{G}(\Omega_0)} d(F \circ G, H \circ \tilde{G}). \tag{2.28}$$

Assume that $\mathcal{F}(\Theta)$ is complete and that $\exists$ a ball B_ε of radius $\varepsilon > 0$ in Θ and

$$\exists c > 0, \quad \forall f \in B_\varepsilon, \quad \|[I + f]^{-1} - I\|_\Theta \le c \|f\|_\Theta. \tag{2.29}$$

This is true for $k \ge 1$ and Θ equal to $C_0^k(\mathbb{R}^N, \mathbb{R}^N)$ considered in [7], $\mathcal{B}^k(\mathbb{R}^N, \mathbb{R}^N)$ in [6], and for $k \ge 0$ $C^{k,1}(\overline{\mathbb{R}^N}, \mathbb{R}^N)$ and $C^{k+1}(\overline{\mathbb{R}^N}, \mathbb{R}^N)$ in [8][2]. Hence the maps

$$f \mapsto [I + f] \mapsto [I + f] \; : \; B_\varepsilon \subset \Theta \to \mathcal{F}(\Theta) \to \mathcal{F}(\Theta)/\mathcal{G}(\Omega)$$

are well-defined and continuous in $f = 0$ since

$$\rho(I, I + f) \le d(I, I + f) \le \|f\|_\Theta + \|[I + f]^{-1} - I\|_\Theta \le (1 + c) \|f\|_\Theta.$$

[2]For Ω open $C^{k+1}(\overline{\Omega}, \mathbb{R}^N)$ is the spaces of maps f in $C^{k+1}(\Omega, \mathbb{R}^N)$ for which f and its derivatives $\partial^\alpha f$, $|\alpha| \le k + 1$, are bounded and uniformly continuous on Ω. $C^{k,1}(\overline{\Omega}, \mathbb{R}^N)$ is the space of all maps f in $C^k(\overline{\Omega}, \mathbb{R}^N)$ such that f and its derivatives $\partial^\alpha f$, $|\alpha| \le k$, are Lipschitz continuous on Ω.

For a shape functional J, the map

$$[I+f] \mapsto J_\Omega(f) \stackrel{def}{=} J([I+f](\Omega)) \, : \, \mathcal{F}(\Theta)/\mathcal{G}(\Omega) \to \mathbb{R}$$

is well-defined since J is invariant on $\mathcal{G}(\Omega)$ and the map

$$f \mapsto J_\Omega(f) \stackrel{def}{=} J([I+f](\Omega)) \, : \, \Theta \to \mathbb{R}$$

is continuous in $f = 0$ if J is continuous in Ω for the Courant metric.

The following definitions are now the standard definitions applied to the function $J_\Omega(f)$ defined on the ball B_ε in the topological vector space Θ.

DEFINITION 2.5. *Let Θ be a topological vector subspace of* Lip$(\mathbb{R}^N, \mathbb{R}^N)$ *and $f \in \Theta$ such that $[I+f] \in \mathcal{F}(\Theta)$. Denote $[I+f](\Omega)$ by Ω_f.*

(i) J_Ω is said to have a Gâteaux semiderivative *at f in the direction $\theta \in \Theta$ if the following limit exists and is finite*

$$dJ_\Omega(f;\theta) \stackrel{\text{def}}{=} \lim_{t \searrow 0} \frac{J\big([I+f+t\theta](\Omega)\big) - J([I+f](\Omega))}{t}. \tag{2.30}$$

(ii) J_Ω is said to be Gâteaux differentiable *at f if it has a* Gâteaux semiderivative *at f in all directions $\theta \in \Theta$ and the map*

$$\theta \mapsto dJ_\Omega(f;\theta) \, : \, \Theta \to \mathbb{R} \tag{2.31}$$

is linear and continuous. The map (2.31) is denoted $\nabla J_\Omega(f)$ and referred to as the gradient of J_Ω in the topological dual Θ' of Θ.

(iii) If, in addition, Θ is a normed vector space, we say that J is Fréchet differentiable *at f if J is Gâteaux differentiable at f and*

$$\lim_{\|\theta\|_\Theta \to 0} \frac{|J([I+f+\theta](\Omega)) - J([I+f](\Omega])) - < \nabla J_\Omega(f), \theta >_\Theta |}{\|\theta\|_\Theta} = 0. \tag{2.32}$$

The semiderivatives of J and J_Ω are related as follows.

THEOREM 2.6.

(i) Assume that J_Ω has a Gâteaux semiderivative at f in the direction $\theta \in \Theta$, then J has a Eulerian semiderivative at Ω_f in the direction V_θ^f and

$$dJ_\Omega(f;\theta) = dJ(\Omega_f; V_\theta^f), \quad V_\theta^f(t) \stackrel{\text{def}}{=} \theta \circ [I+f+t\theta]^{-1}. \tag{2.33}$$

(ii) If J has a Hadamard semiderivative at Ω_f in the direction $\theta \circ [I+f]^{-1}$, then J_Ω has a Gâteaux semiderivative at f in the direction θ and

$$dJ_\Omega(f;\theta) = d_H J(\Omega_f; \theta \circ [I+f]^{-1}). \tag{2.34}$$

Conversely if J_Ω has a Gâteaux semiderivative at f in the direction $\theta \circ [I+f]$, then J has a Hadamard semiderivative at Ω_f in the direction θ. If either $dJ_\Omega(f;\theta)$ or $d_H J(\Omega_f;\theta)$ is linear and continuous with respect to all θ in Θ, so does the other and

$$\forall \theta \in \Theta, \quad \begin{aligned} &< \nabla J_\Omega(f), \theta >_\Theta = < G(\Omega_f), \theta \circ [I+f]^{-1} >_\Theta \\ &< G(\Omega_f), \theta >_\Theta = < \nabla J_\Omega(f), \theta \circ [I+f] >_\Theta . \end{aligned} \tag{2.35}$$

Proof. By definition

$$\begin{aligned} dJ_\Omega(f;\theta) &= \lim_{t\searrow 0} \frac{J([I+f+t\theta](\Omega)) - J([I+f](\Omega))}{t} \\ &= \lim_{t\searrow 0} \frac{J(T_t([I+f](\Omega))) - J([I+f](\Omega))}{t} = dJ([I+f](\Omega); V_\theta^f) \end{aligned}$$

for the family of transformations

$$T_t^f \stackrel{\text{def}}{=} [I+f+t\theta] \circ [I+f]^{-1}$$

and from Theorem 2.1 in § 2.1, T_t^f corresponds to the velocity field

$$V_\theta^f(t) \stackrel{\text{def}}{=} \frac{\partial T_t^f}{\partial t} \circ (T_t^f)^{-1} = \theta \circ [I+f+t\theta]^{-1}.$$

Identity (2.34) now follows from the fact that J has a Hadamard semiderivative at Ω_f. The other properties readily follow from the definitions. □

We have standard sufficient conditions for the Fréchet differentiability.

THEOREM 2.7. *Let Ω be a subset of $\mathbb{R}^N$ and Θ be*

$$k \geq 1,\ C_0^k(\mathbb{R}^N, \mathbb{R}^N),\ \mathcal{B}^k(\mathbb{R}^N, \mathbb{R}^N), \quad k \geq 0,\ C^{k,1}(\overline{\mathbb{R}^N}, \mathbb{R}^N),\ C^{k+1}(\overline{\mathbb{R}^N}, \mathbb{R}^N).$$

If J_Ω is Gâteaux differentiable for all f in B_ε and the map $f \mapsto \nabla J_\Omega(f) : B_\varepsilon \to \Theta'$ is continuous in $f = 0$ then J_Ω is Fréchet differentiable in $f = 0$.

2.4 Shape Gradient and Structure Theorem

We now specialize to autonomous vector fields V. The choice of a *shape gradient* depends on the choice of the topological vector subspace Θ of $\text{Lip}(\mathbb{R}^N, \mathbb{R}^N)$. We choose to work in the framework of the Theory of Distributions with $\Theta = \mathcal{D}(\mathbb{R}^N, \mathbb{R}^N)$, the space of all infinitely differentiable transformations θ of $\mathbb{R}^N$ with compact support: such velocity fields V satisfy conditions (2.1).

DEFINITION 2.8. *Let Ω be a subset of $\mathbb{R}^N$.*

(i) The functional J is said to be Shape differentiable *at Ω, if it is differentiable at Ω for all θ in $\mathcal{D}(\mathbb{R}^N, \mathbb{R}^N)'$.*

(ii) The map (2.12) defines a vector distribution $G(\Omega)$ in $\mathcal{D}(\mathbb{R}^N, \mathbb{R}^N)'$ which will be referred to as the shape gradient *of J at Ω.*

(iii) When, for some finite $k \geq 0$, $G(\Omega)$ is continuous for the $\mathcal{D}^k(\mathbb{R}^N, \mathbb{R}^N)$-topology, we say that the shape gradient $G(\Omega)$ is of order k.

Associate with a subset A of $\mathbb{R}^N$ and an integer $k \geq 0$ the set

$$L_A^k \stackrel{\text{def}}{=} \left\{V \in \mathcal{D}^k(\mathbb{R}^N, \mathbb{R}^N) : \forall x \in A, V(x) \in L_A(x)\right\},$$

where $L_A(x) = \{-C_A(x)\} \cap C_A(x)$ and $C_A(x)$ is Clarke's tangent cone to A. The proof of the following theorem and its corollary can be found in [3, 4].

THEOREM 2.9 (STRUCTURE THEOREM). *Let Ω be a subset of $\mathbb{R}^N$ with boundary Γ and assume that J has a shape gradient $G(\Omega)$.*

(i) The support of the shape gradient $G(\Omega)$ is contained in Γ.

(ii) If Ω is open or closed in $\mathbb{R}^N$ and the shape gradient is of order k for some $k \geq 0$, then there exists $[G(\Omega)]$ in $(\mathcal{D}^k/L_\Omega^k)'$ such that for all V in $\mathcal{D}^k \stackrel{\text{def}}{=} \mathcal{D}^k(\mathbb{R}^N, \mathbb{R}^N)$

$$dJ(\Omega; V) = \langle [G(\Omega)], q_L V\rangle_{\mathcal{D}^k/L_\Omega^k} \tag{2.36}$$

where $q_L : \mathcal{D}^k \to \mathcal{D}^k/L_\Omega^k$ is the canonical quotient surjection. Moreover

$$G(\Omega) = (q_L)^*[G(\Omega)] \tag{2.37}$$

where $(q_L)^*$ denotes the transposed of the linear map q_L.

When the boundary Γ of Ω is compact and J is shape differentiable at Ω, the distribution $G(\Omega)$ is of finite order. Once this is known, the conclusions of Theorem 2.9 (ii) apply with k equal to the order of $G(\Omega)$. Hence $G(\Omega)$ will belong to a Hilbert space $H^{-s}(\mathbb{R}^N)$ for some $s \geq 0$. The quotient space is related to a trace on the boundary Γ and when the boundary Γ is sufficiently smooth we can indeed make that identification.

COROLLARY 2.10. *Assume that the assumptions of Theorem 2.9 are verified for an open domain Ω, that the order of $G(\Omega)$ is $k \geq 0$, and that the boundary Γ of Ω is C^{k+1}. Then for all x in Γ, $L_\Omega(x)$ is an $(N-1)$-dimensional hyperplane to Ω at x and there exists a unique outward unit normal $n(x)$ which belongs to $C^k(\Gamma; \mathbb{R}^N)$. As a result the kernel of the map*

$$V \mapsto \gamma_\Gamma(V) \cdot n : \mathcal{D}^k(\mathbb{R}^N, \mathbb{R}^N) \to C^k(\Gamma) \tag{2.38}$$

coincides with L_Ω^k where $\gamma_\Gamma : \mathcal{D}^k(\mathbb{R}^N, \mathbb{R}^N) \to C^k(\Gamma, \mathbb{R}^N)$ is the trace of V on Γ. Moreover the map $p_L(V)$

$$q_L(V) \mapsto p_L\big(q_L(V)\big) \stackrel{\text{def}}{=} \gamma_\Gamma(V) \cdot n : \mathcal{D}^k/L_\Omega^k \to C^k(\Gamma) \tag{2.39}$$

is a well-defined isomorphism. In particular there exists a scalar distribution $g(\Gamma)$ in $\mathbb{R}^N$ with support in Γ such that $g(\Gamma) \in C^k(\Gamma)'$ and for all V in $\mathcal{D}^k(\mathbb{R}^N, \mathbb{R}^N)$

$$dJ(\Omega; V) = \langle g(\Gamma), \gamma_\Gamma(V) \cdot n\rangle_{C^k(\Gamma)} \tag{2.40}$$

$$G(\Omega) = {}^*(q_L)[G(\Omega)], \quad [G(\Omega)] = {}^*(p_L)g(\Gamma). \tag{2.41}$$

Denote by γ_Γ the trace operator on Γ. When $g(\Gamma) \in L^1(\Gamma)$

$$dJ(\Omega; V) = \int_\Gamma g V \cdot n \, d\Gamma \text{ and } G = \gamma_\Gamma^*(g\,n), \tag{2.42}$$

3 Elements of Shape Calculus

We recall a number of basic formulae from [10, 11] for the derivative of *domain and boundary integrals.* We review the derivative of boundary integrals by using the intrinsic *tangential differential calculus* [5, 2].

3.1 Basic Formula for Domain Integrals

The simplest example of domain functional are is the *volume integral* over a bounded open domain Ω in $\mathbb{R}^N$. To compute its shape derivative a basic formula is used in connection with the family of transformations $\{T_t : 0 \leq t \leq \tau\}$. Assume that (2.1) is verified by the velocity field $\{V(t) : 0 \leq t \leq \tau\}$. Further assume that $V \in C^0([0,\tau]; C^1_{\text{loc}}(\mathbb{R}^N, \mathbb{R}^N))$ and that $\tau > 0$ is such that the *Jacobian* J_t is strictly positive

$$\forall t \in [0,\tau], \quad J_t(X) \stackrel{def}{=} \det DT_t(X) > 0, \quad (DT_t)_{ij} = \partial_j T_i,$$

where $DT_t(X)$ is the *Jacobian matrix* of the transformation $T_t = T_t(V)$ associated with V. Given φ in $W^{1,1}_{\text{loc}}(\mathbb{R}^N)$, consider for $0 \leq t \leq \tau$ the integral

$$J(\Omega_t(V)) \stackrel{def}{=} \int_{\Omega_t(V)} \varphi\, dx = \int_{\Omega} \varphi \circ T_t \;\, J_t \;\, dx \tag{3.1}$$

where $\Omega_t(V) \stackrel{def}{=} T_t(V)(\Omega)$. The following formulae and results are easy to check.

THEOREM 3.1. *Let φ be a function in $W^{1,1}_{\text{loc}}(\mathbb{R}^N)$. Assume that the vector field $V = \{V(t) : 0 \leq t \leq \tau\}$ satisfies condition (V).*

(i) For each $t \in [0,\tau]$ the map

$$\varphi \mapsto \varphi \circ T_t \,:\, W^{1,1}_{\text{loc}}(\mathbb{R}^N) \to W^{1,1}_{\text{loc}}(\mathbb{R}^N)$$

and its inverse are both locally Lipschitzian and

$$\nabla(\varphi \circ T_t) = {}^*DT_t\, \nabla\varphi \circ T_t.$$

(ii) If $V \in C^0([0,\tau]; C^1_{\text{loc}}(\mathbb{R}^N, \mathbb{R}^N))$, then the map

$$t \mapsto \varphi \circ T_t \,:\, [0,\tau] \to W^{1,1}_{\text{loc}}(\mathbb{R}^N)$$

is well-defined and for each t

$$\frac{d}{dt}\varphi \circ T_t = (\nabla\varphi \cdot V(t)) \circ T_t \in L^1_{\text{loc}}(\mathbb{R}^N). \tag{3.2}$$

Hence $t \mapsto \varphi \circ T_t$ belongs to $C^1([0,\tau]; L^1_{\text{loc}}(\mathbb{R}^N)) \cap C^0([0,\tau]; W^{1,1}_{\text{loc}}(\mathbb{R}^N))$.

(iii) If $V \in C^0([0,\tau]; C^1_{\text{loc}}(\mathbb{R}^N, \mathbb{R}^N))$, then the map

$$t \mapsto J_t \,:\, [0,\tau] \to C^0_{\text{loc}}(\mathbb{R}^N)$$

is differentiable and

$$\frac{dJ_t}{dt} = [\operatorname{div} V(t)] \circ T_t \; J_t \in C^0_{\text{loc}}(\mathbb{R}^N). \tag{3.3}$$

Hence the map $t \mapsto J_t$ belongs to $C^1([0,\tau]; C^0_{\text{loc}}(\mathbb{R}^N))$.

Indeed it is easy to check that

$$dJ(\Omega; V) = \int_\Omega \operatorname{div}(\varphi\, V(0))\, dx \quad \Rightarrow dJ(\Omega; V) = \int_\Gamma \varphi\, V(0) \cdot n\, d\Gamma$$

if Ω has a Lipschitzian boundary by Stokes' theorem.

THEOREM 3.2. *Assume that there exists $\tau > 0$ such that the velocity field $V(t)$ verifies conditions (V) and $V \in C^0([0,\tau]; C^1_{\mathrm{loc}}(\mathbb{R}^N, \mathbb{R}^N))$. Given a bounded measurable domain Ω with boundary Γ and a function $\varphi \in C(0,\tau; W^{1,1}_{\mathrm{loc}}(\mathbb{R}^N)) \cap C^1(0,\tau; L^1_{\mathrm{loc}}(\mathbb{R}^N))$, the semiderivative of the function*

$$J_V(t) \stackrel{\mathrm{def}}{=} \int_{\Omega_t(V)} \varphi(t)\, dx \tag{3.4}$$

at $t = 0$ is given by

$$dJ_V(0) = \int_\Omega \varphi'(0) + \operatorname{div}(\varphi(0)\, V(0))\, dx \tag{3.5}$$

where $\varphi(0)(x) \stackrel{\mathrm{def}}{=} \varphi(0,x)$ and $\varphi'(0)(x) \stackrel{\mathrm{def}}{=} \partial\varphi/dt(0,x)$. If, in addition, Ω is an open domain with a Lipschitzian boundary Γ, then

$$dJ_V(0) = \int_\Omega \varphi'(0)\, dx + \int_\Gamma \varphi(0)\, V(0) \cdot n\, dx. \tag{3.6}$$

3.2 Basic Formula for Boundary Integrals

Given ψ in $H^2_{\mathrm{loc}}(\mathbb{R}^N)$, consider for some bounded open Lipschitzian domain Ω in $\mathbb{R}^N$ the shape functional

$$J(\Omega) \stackrel{def}{=} \int_\Gamma \psi\, d\Gamma. \tag{3.7}$$

This integral is invariant with respect to homeomorphisms which map Ω onto itself (and hence Γ onto itself). Given V and $t \geq 0$, consider the expression

$$J(\Omega_t(V)) \stackrel{def}{=} \int_{\Gamma_t(V)} \psi\, d\Gamma_t.$$

Using the change of variable $T_t(V)$

$$J(\Omega_t(V)) \stackrel{def}{=} \int_{\Gamma_t} \psi\, d\Gamma_t = \int_\Gamma \psi \circ T_t\, \omega_t\, d\Gamma, \quad \omega_t = |M(DT_t)n|, \tag{3.8}$$

n is the outward normal on Γ and $M(DT_t)$ is the cofactor matrix of DT_t,

$$M(DT_t) = J_t\, {}^*(DT_t)^{-1} \quad \Rightarrow \omega_t = J_t\, |\, {}^*(DT_t)^{-1} n|. \tag{3.9}$$

From (3.8)–(3.9) the function $t \mapsto \omega_t$ is differentiable in $C^0(\Gamma)$ and that

$$\omega' = \lim_{t \searrow 0} \frac{1}{t}(\omega_t - \omega) = \operatorname{div} V(0) - DV(0)n \cdot n \tag{3.10}$$

(limit in the $C^0(\Gamma)$ norm) is itself linear and continuous with respect to $V(0)$ in the $C^1_{\text{loc}}(\mathbb{R}^N, \mathbb{R}^N)$ topology. Hence

$$dJ(\Omega; V) = \int_\Gamma \nabla\psi \cdot V(0) + \psi\,(\operatorname{div} V(0) - DV(0)n \cdot n)\, d\Gamma. \tag{3.11}$$

From Corollary 2.10 to the Structure Theorem 2.9, $dJ(\Omega; V)$ only depends on the normal component v_n of the velocity field $V(0)$ on Γ

$$v_n \stackrel{def}{=} v \cdot n, \quad v \stackrel{def}{=} V(0)|_\Gamma \tag{3.12}$$

through Hadamard's formula (2.40). In view of this property any other velocity field with the same smoothness and normal component on Γ will yield the same limit. Given $k > 0$ consider the *tubular neighborhood*

$$S_k(\Gamma) \stackrel{def}{=} \{x \in \mathbb{R}^N \,:\, |b(x)| < k\} \tag{3.13}$$

of Γ in $\mathbb{R}^N$ for the oriented distance function $b = b_\Omega$ of Ω. Assuming that Γ is compact and C^2, $\exists h > 0$ such that $b \in C^2(S_{2h}(\Gamma))$. Let $\varphi \in \mathcal{D}(\mathbb{R}^N)$ be such that $\varphi = 1$ in $S_h(\Gamma)$ and $\varphi = 0$ outside of $S_{2h}(\Gamma)$. Consider the velocity field

$$W(t) \stackrel{def}{=} (V(0) \cdot \nabla b)\, \nabla b\, \varphi.$$

Clearly the normal component of $W(0)$ on Γ coincides with v_n. In $S_h(\Gamma)$

$$\nabla\psi \cdot W = \nabla\psi \cdot \nabla b\, V(0) \cdot \nabla b \quad \Rightarrow \nabla\psi \cdot W|_\Gamma = \nabla\psi \cdot n\, V(0) \cdot n = \frac{\partial \psi}{\partial n} v_n$$

$$DW = V(0) \cdot \nabla b\, D^2 b + \nabla b\ {}^*\nabla(V(0) \cdot \nabla b)$$

$$\operatorname{div} W = V(0) \cdot \nabla b\, \Delta b + \nabla b \cdot \nabla(V(0) \cdot \nabla b)$$

$$DW \nabla b \cdot \nabla b = V(0) \cdot \nabla b\, D^2 b \nabla b \cdot \nabla b + \nabla(V(0) \cdot \nabla b) \cdot \nabla b = \nabla(V(0) \cdot \nabla b) \cdot \nabla b$$

$$\operatorname{div} W - DW \nabla b \cdot \nabla b = V(0) \cdot \nabla b\, \Delta b$$

$$\Rightarrow \operatorname{div} W - DW \nabla b \cdot \nabla b|_\Gamma = V(0) \cdot n\, H = H\, v_n$$

since $\nabla b|_\Gamma = n$, $D^2 b \nabla b = 0$, and $H = \Delta b$ is the *additive curvature*, that is the sum of the $N-1$ curvatures of Γ or $N-1$ times the mean curvature $\overline{H}$. Finally

$$dJ(\Omega; V) = \int_\Gamma \left(\frac{\partial \psi}{\partial n} + \psi\, H\right) v_n\, d\Gamma. \tag{3.14}$$

We have proved the following result.

THEOREM 3.3. *Let $\psi \in C^1([0, \tau]; H^2_{\text{loc}}(\mathbb{R}^N))$, Γ be the boundary of a bounded open subset Ω of $\mathbb{R}^N$ of class C^2 Assume that $V \in C^0([0, \tau]; C^1_{\text{loc}}(\mathbb{R}^N, \mathbb{R}^N))$. Consider the functional*

$$J_V(t) \stackrel{\text{def}}{=} \int_{\Gamma_t(V)} \psi(t)\, d\Gamma_t.$$

Then the derivative of $J_V(t)$ with respect to t in $t = 0$ is given by the expression

$$\begin{aligned} dJ_V(0) &= \int_\Gamma \psi'(0) + \left(\frac{\partial \psi}{\partial n} + H\psi\right) V(0) \cdot n\, d\Gamma \\ &= \int_\Gamma \psi'(0) + \nabla\psi \cdot V(0) + \psi\,(\operatorname{div} V(0) - DV(0)n \cdot n)\, d\Gamma, \end{aligned} \tag{3.15}$$

where $\psi'(0)(x) \stackrel{\text{def}}{=} \partial\psi/\partial t(0, x)$.

As in the case of the volume integral, we have two formulae and the identity:

$$\int_\Gamma \left(\frac{\partial\psi}{\partial n} + H\psi\right) V(0)\cdot n\, d\Gamma = \int_\Gamma \nabla\psi\cdot V(0) + \psi\,(\operatorname{div} V(0) - DV(0)n\cdot n)\, d\Gamma. \tag{3.16}$$

3.3 Example: Boundary Integral of the Normal Derivative

Let Γ be of class C^2 and $\phi \in H^2_{\text{loc}}(\mathbb{R}^N)$ be given. Consider the following shape functional

$$J(\Omega) \stackrel{\text{def}}{=} \int_\Gamma \left|\frac{\partial\phi}{\partial n}\right|^2 d\Gamma = \int_\Gamma |\nabla\phi\cdot n|^2\; d\Gamma.$$

By change of variable with $\Omega_t = T_t(V)(\Omega)$ and $\Gamma_t = T_t(V)(\Gamma)$, we get

$$J(\Omega_t) \stackrel{\text{def}}{=} \int_{\Gamma_t} |\nabla\phi\cdot n_t|^2\, d\Gamma_t = \int_\Gamma \left\{\, {}^*(DT_t)^{-1}\nabla(\phi\circ T_t)\cdot(n_t\circ T_t)\right\}^2 \omega_t\, d\Gamma, \tag{3.17}$$

where $n_t\circ T_t$ is the transported normal field n_t from Γ_t onto Γ. The derivative can be obtained by using formula (3.15) of Theorem 3.3 and one of the above two expressions. However the first expression first requires the construction of an extension N_t of the normal n_t in a neighborhood of Γ.

THEOREM 3.4. *Let $k \geq 1$ be an integer. Given a velocity field $V(t)$ verifying condition (V) such that $V \in C([0,\tau]; C^k_{\text{loc}}(\mathbb{R}^N,\mathbb{R}^N))$, then*

$$n_t\circ T_t = \frac{{}^*(DT_t)^{-1}n}{|\,{}^*(DT_t)^{-1}n|} = \frac{M(DT_t)n}{|M(DT_t)n|} \tag{3.18}$$

where n and n_t are the respective outward normals to Ω and Ω_t on Γ and Γ_t and $M(DT_t)$ is the cofactor's matrix of DT_t.

Recalling that $\omega_t = |M(DT_t)n| = J(t)\,|\,{}^*(DT_t)^{-1}n|$, our integral becomes

$$\int_{\Gamma_t} |\nabla\phi\cdot n_t|^2\; d\Gamma_t = \int_\Gamma |[A(t)\nabla(\phi\circ T_t)]\cdot n\,|^2\; \omega_t^{-1}\, d\Gamma. \tag{3.19}$$

Using following expression of the derivative $A'(V)$ in the $C^{k-1}_{\text{loc}}(\mathbb{R}^N,\mathbb{R}^{N^2})$ norm

$$A'(V) \stackrel{\text{def}}{=} \frac{\partial}{\partial t}A(V)(t)|_{t=0} = \operatorname{div} V(0)\, I - DV(0) - {}^*DV(0) \tag{3.20}$$

and expression (3.10) of ω' we get for $dJ(\Omega;V)$

$$= 2\int_\Gamma \frac{\partial\phi}{\partial n}\,[A'(V)\nabla\phi\cdot n + \nabla(\nabla\phi\cdot V(0))\cdot n] - \left|\frac{\partial\phi}{\partial n}\right|^2 \omega'\, d\Gamma$$

$$
\begin{aligned}
&= 2\int_\Gamma \frac{\partial\phi}{\partial n}\,\{[\operatorname{div}V(0)\,I - DV(0) - {}^*DV(0)]\,\nabla\phi\cdot n + \nabla(\nabla\phi\cdot V(0))\cdot n\} \\
&\qquad - \left|\frac{\partial\phi}{\partial n}\right|^2 (\operatorname{div}V(0) - DV(0)n\cdot n)\,d\Gamma \\
&= \int_\Gamma 2\frac{\partial\phi}{\partial n}\left\{\operatorname{div}V(0)\,\frac{\partial\phi}{\partial n} - DV(0)\,\nabla\phi\cdot n + D^2\phi\,V(0)\cdot n\right\} \\
&\qquad - \left|\frac{\partial\phi}{\partial n}\right|^2 (\operatorname{div}V(0) - DV(0)n\cdot n)\,d\Gamma \\
&= \int_\Gamma \left|\frac{\partial\phi}{\partial n}\right|^2 (\operatorname{div}V(0) - DV(0)n\cdot n) \\
&\qquad + 2\frac{\partial\phi}{\partial n}\left\{DV(0)n\cdot n\,\frac{\partial\phi}{\partial n} - DV(0)\,\nabla\phi\cdot n + D^2\phi\,V(0)\cdot n\right\}\,d\Gamma.
\end{aligned}
$$

This formula can be somewhat simplified by using identity (3.16) with

$$
\begin{aligned}
\psi = |\nabla\phi\cdot\nabla b|^2 \quad &\Rightarrow \psi|_\Gamma = \left|\frac{\partial\phi}{\partial n}\right|^2 \\
\nabla\psi = 2\,\nabla\phi\cdot\nabla b\;\nabla(\nabla\phi\cdot\nabla b) &= 2\,\nabla\phi\cdot\nabla b\,\left[D^2\phi\nabla b + D^2 b\nabla\psi\right] \\
&\Rightarrow \nabla\psi|_\Gamma = 2\frac{\partial\phi}{\partial n}\left[D^2\phi n + D^2 b\,\nabla\phi\right] \\
&\Rightarrow \nabla\psi\cdot V(0)|_\Gamma = 2\frac{\partial\phi}{\partial n}\left[D^2\phi n + D^2 b\,\nabla\phi\right]\cdot V(0) \\
\Rightarrow \frac{\partial\psi}{\partial n} = \nabla\psi\cdot\nabla b|_\Gamma &= 2\frac{\partial\phi}{\partial n}\left[D^2\phi n + D^2 b\,\nabla\phi\right]\cdot\nabla b = 2\frac{\partial\phi}{\partial n}\,D^2\phi n\cdot n.
\end{aligned}
$$

We obtain

$$
\begin{aligned}
&\int_\Gamma \left(2\frac{\partial\phi}{\partial n}\,D^2\phi n\cdot n + H\left|\frac{\partial\phi}{\partial n}\right|^2\right) V(0)\cdot n\,d\Gamma \\
&= \int_\Gamma 2\frac{\partial\phi}{\partial n}\left[D^2\phi n + D^2 b\,\nabla\phi\right]\cdot V(0) + \left|\frac{\partial\phi}{\partial n}\right|^2 (\operatorname{div}V(0) - DV(0)n\cdot n)\,d\Gamma
\end{aligned}
\tag{3.21}
$$

and hence $dJ(\Omega;V)$ is equal to

$$
\begin{aligned}
&= \int_\Gamma \left(2\frac{\partial\phi}{\partial n}\,D^2\phi n\cdot n + H\left|\frac{\partial\phi}{\partial n}\right|^2\right) V(0)\cdot n \\
&\qquad + 2\frac{\partial\phi}{\partial n}\left\{\frac{\partial\phi}{\partial n}\,DV(0)n\cdot n - \nabla\phi\cdot\left({}^*DV(0)\,n + D^2 b\,V(0)\right)\right\}\,d\Gamma.
\end{aligned}
\tag{3.22}
$$

This formula can be more readily obtained from the first expression (3.17) and

$$
N_t = \frac{{}^*(DT_t)^{-1}\nabla b}{|\,{}^*(DT_t)^{-1}\nabla b|}\circ T_t^{-1} \tag{3.23}
$$

the extension of the normal n_t. To compute N' decompose N_t as follows

$$
\begin{aligned}
N_t &= f(t)\circ T_t^{-1}, \quad f(t) = \frac{g(t)}{\sqrt{g(t)\cdot g(t)}}, \quad g(t) = {}^*(DT_t)^{-1}\nabla b \\
N' &= f' - Df(0)V(0), \quad g' = -{}^*DV(0)\nabla b, \quad f(0) = \nabla b \\
f' &= \frac{g'\,|g(0)| - g(0)\cdot g'\,g(0)/|g(0)|}{|g(0)|^2} \\
&= g' - g'\cdot\nabla b\,\nabla b = DV(0)\nabla b\cdot\nabla b\,\nabla b - {}^*DV(0)\nabla b.
\end{aligned}
$$

So finally

$$N'|_\Gamma = (DV(0)n\cdot n)\,n - {}^*DV(0)n - D^2b\,V(0) \tag{3.24}$$

and

$$
\begin{aligned}
\frac{\partial}{dt}|\nabla\phi\cdot N_t|^2\Big|_{t=0} &= 2\frac{\partial\phi}{\partial n}\nabla\phi\cdot N' \\
&= 2\frac{\partial\phi}{\partial n}\nabla\phi\cdot\left\{(DV(0)n\cdot n)n - {}^*DV(0)\,n - D^2b\,V(0)\right\}.
\end{aligned}
$$

The first part of the integral (3.22) explicitly depends on the normal component of $V(0)$. Yet we know from the Structure Theorem that for this functional, the shape derivative only depends on the normal component of $V(0)$. To make this explicit, it is necessary to introduce some elements of tangential calculus.

3.4 Elements of Tangential Calculus

In this section some basic elements of the differential calculus on a C^2 submanifold of codimension one are introduced. Let Ω be an open domain of class C^2 in $\mathbb{R}^N$ with compact boundary Γ. Therefore $\exists h > 0$ such that $b = b_\Omega \in C^2(S_{2h}(\Gamma))$, where for $k > 0$ $S_k(\Gamma) = \{x \in \mathbb{R}^N : |b_\Omega(x)| < k\}$. Associate with $F \in C^1(S_{2h}\Gamma))$ and $V \in C^1(S_{2h}(\Gamma))^N$

$$f \overset{def}{=} F|_\Gamma, \quad v \overset{def}{=} V|_\Gamma, \quad v_n \overset{def}{=} v\cdot n, \quad v_\Gamma \overset{def}{=} v - v_n\,n, \tag{3.25}$$

where v_Γ and v_n are the respective tangential part and the normal component of v. In view of the previous definitions the following identities are easy to verify

$$\nabla F|_\Gamma = \nabla_\Gamma f + \frac{\partial F}{\partial n}n \quad DV|_\Gamma = D_\Gamma v + DVn\;{}^*n \tag{3.26}$$

$$\operatorname{div} V|_\Gamma = \operatorname{div}_\Gamma v + DVn\cdot n \quad D_\Gamma v\,n = 0, \quad D^2b\,n = 0. \tag{3.27}$$

Decomposing v into its tangential part of and its normal component,

$$D_\Gamma v = D_\Gamma v_\Gamma + v_n D^2 b + n\;{}^*\nabla_\Gamma v_n \tag{3.28}$$

$$\operatorname{div}_\Gamma v = \operatorname{div}_\Gamma v_\Gamma + \Delta b\,v_n = \operatorname{div}_\Gamma v_\Gamma + H\,v_n \quad \nabla_\Gamma v_n = {}^*D_\Gamma v\,n + D^2b\,v_\Gamma. \tag{3.29}$$

Given $v \in C^1(\Gamma)^N$, we also have the *tangential Stokes' formula* with $H = \Delta b$

$$\int_\Gamma \operatorname{div}_\Gamma v\,d\Gamma = \int_\Gamma H\,v\cdot n\,d\Gamma. \tag{3.30}$$

3.5 Back to the Example of Section 3.3

Coming back to formula (3.22) for $dJ(\Omega; V(0))$ of the boundary integral of the square of the normal derivative in § 3.3, the tangential calculus is now used on the term

$$\int_\Gamma 2\frac{\partial\phi}{\partial n}\left\{\frac{\partial\phi}{\partial n} DV(0)n\cdot n - \nabla\phi\cdot\left({}^*DV(0)\,n + D^2b\,V(0)\right)\right\}d\Gamma$$

$$T \stackrel{def}{=} \frac{\partial\phi}{\partial n} DV(0)n\cdot n - \nabla\phi\cdot\left({}^*DV(0)\,n + D^2b\,V(0)\right).$$

From identity (3.26) with $v = V(0)|_\Gamma$ and identity (3.29)

$$\begin{aligned} {}^*DV(0)n &= {}^*D_\Gamma v\, n + DV(0)n\cdot n\, n \\ \nabla\phi\cdot\,{}^*DV(0)n &= \nabla\phi\cdot\,{}^*D_\Gamma v\, n + \frac{\partial\phi}{\partial n}DV(0)n\cdot n \\ \Rightarrow T = -\nabla\phi\cdot\left[{}^*D_\Gamma v\, n + D^2b\, v_\Gamma\right] &= -\nabla\phi\cdot\nabla_\Gamma v_n = -\nabla_\Gamma\phi\cdot\nabla_\Gamma v_n. \end{aligned}$$

Therefore by using the tangential Stokes' formula (3.30)

$$\begin{aligned} \int_\Gamma 2\frac{\partial\phi}{\partial n}T\,d\Gamma &= -\int_\Gamma 2\frac{\partial\phi}{\partial n}\nabla_\Gamma\phi\cdot\nabla_\Gamma v_n\,d\Gamma \\ &= \int_\Gamma -2\,\mathrm{div}_\Gamma\left\{\frac{\partial\phi}{\partial n}v_n\,\nabla_\Gamma\phi\right\} + 2\,\mathrm{div}_\Gamma\left\{\frac{\partial\phi}{\partial n}\nabla_\Gamma\phi\right\}v_n\,d\Gamma \\ &= \int_\Gamma -2H\frac{\partial\phi}{\partial n}v_n\,\nabla_\Gamma\phi\cdot n + 2\,\mathrm{div}_\Gamma\left\{\frac{\partial\phi}{\partial n}\nabla_\Gamma\phi\right\}v_n\,d\Gamma \\ &= \int_\Gamma 2\,\mathrm{div}_\Gamma\left\{\frac{\partial\phi}{\partial n}\nabla_\Gamma\phi\right\}v_n\,d\Gamma = \int_\Gamma 2\,\mathrm{div}_\Gamma\left\{\frac{\partial\phi}{\partial n}\nabla_\Gamma\phi\right\}V(0)\cdot n\,d\Gamma. \end{aligned}$$

Substituting into expression (3.22), we finally get the explicit formula in term of $V(0)\cdot n$ as predicted by the Structure Theorem

$$dJ(\Omega;V) = \int_\Gamma\left\{2\frac{\partial\phi}{\partial n}D^2\phi n\cdot n + H\left|\frac{\partial\phi}{\partial n}\right|^2 + 2\,\mathrm{div}_\Gamma\left(\frac{\partial\phi}{\partial n}\nabla_\Gamma\phi\right)\right\}V(0)\cdot n\,d\Gamma. \tag{3.31}$$

4 Second Order Derivatives

4.1 Basic Formulas for Domain Integrals

Given $f\in C^2_{\mathrm{loc}}(\mathbb{R}^N)$ and a domain Ω of class C^2, consider the functional

$$J(\Omega_{t,s}(V,W)) \stackrel{def}{=} \int_{\Omega_{t,s}(V,W)} f\,dx \tag{4.1}$$

$$\Omega_{t,s}(V,W) \stackrel{def}{=} T_s(W)(\Omega_t(V)) = T_s(W)(T_t(V)(\Omega)),\ J_{V,W}(t,s) \stackrel{def}{=} J(\Omega_{t,s}(V,W))$$

for some pair of autonomous velocity fields V and W verifying condition (2.1) and additional smoothness conditions as necessary. The objective is to compute

$$d^2J_{V,W} \stackrel{def}{=} \frac{\partial}{\partial s}\left\{\frac{\partial}{\partial t}J_{V,W}(t,s)\Big|_{t=0}\right\}\Big|_{s=0}.$$

From formula (3.6) in Theorem 3.2 of the previous section, we already know that

$$\frac{\partial}{\partial t} J_{V,W}(t,s)\bigg|_{t=0} = dJ(\Omega_s(W);V) = \int_{\Gamma_s(W)} f\,V\cdot n_s\,d\Gamma_s = \int_{\Gamma_s(W)} f\,V\cdot N_s\,d\Gamma_s,$$

where $N_s = N_s(W)$ is the extension (3.23) of the normal n_s. So we can readily use formula (3.15) from Theorem 3.3

$$d^2 J_{V,W} = \int_\Gamma f\,V\cdot N'(W) + \left\{\frac{\partial}{\partial n}(f\,V\cdot\nabla b) + H\,f\,V\cdot n\right\} W\cdot n\,d\Gamma \tag{4.2}$$

where $N'(W)$ is the derivative of the extension $N_s = N_s(W)$ in (3.24). This yields

$$\begin{aligned} d^2 J_{V,W} &= \int_\Gamma f\,V\cdot\{(DWn\cdot n)\,n - {}^*DW - D^2b\,W\} \\ &\quad + \left(\frac{\partial}{\partial n}(f\,V\cdot\nabla b) + H\,f\,V\cdot n\right) W\cdot n\,d\Gamma \\ &= \int_\Gamma f\left\{V\cdot\{(DWn\cdot n)\,n - {}^*DWn - D^2b\,W\} + \frac{\partial}{\partial n}(V\cdot\nabla b)\,W\cdot n\right\} \\ &\quad + \left(\frac{\partial f}{\partial n} + H\,f\right) V\cdot n\,W\cdot n\,d\Gamma. \end{aligned}$$

It remains to untangle the following term in the first part of the integral

$$T \stackrel{\text{def}}{=} V\cdot\{(DWn\cdot n)\,n - {}^*DWn - D^2b\,W\} + \frac{\partial}{\partial n}(V\cdot\nabla b)\,W\cdot n.$$

Using the notation $v = V|_\Gamma$ and $w = W|_\Gamma$

$$\begin{aligned} \nabla(V\cdot\nabla b) &= {}^*DV\nabla b + D^2b\,V \\ \nabla(V\cdot\nabla b)\cdot\nabla b &= {}^*DV\nabla b\cdot\nabla b + D^2b\,V\cdot\nabla b = DV\nabla b\cdot\nabla b \\ \Rightarrow \frac{\partial}{\partial n}(V\cdot\nabla b) &= DVn\cdot n \\ V\cdot{}^*DWn &= V\cdot[{}^*D_\Gamma w + n\,{}^*(DWn)]\,n \\ &= V\cdot\left[\nabla_\Gamma w_n - D^2b\,w\right] + DWn\cdot n\,V\cdot n \\ V\cdot\left[{}^*DWn + D^2b\,W\right] &= \nabla_\Gamma w_n\cdot v_\Gamma + DWn\cdot n\,v_n. \end{aligned}$$

Finally

$$\begin{aligned} T &= DWn\cdot n\,v_n - (v_\Gamma\cdot\nabla_\Gamma w_n + DWn\cdot n\,v_n) + DVn\cdot n\,w_n \\ &= DVn\cdot n\,w_n - v_\Gamma\cdot\nabla_\Gamma w_n \end{aligned}$$

and by using the tangential Stokes' formula (3.30)

$$\begin{aligned} \int_\Gamma f\,T\,d\Gamma &= \int_\Gamma f\,\{DVn\cdot n\,w_n - v_\Gamma\cdot\nabla_\Gamma w_n\}\,d\Gamma \\ &= \int_\Gamma f\,DVn\cdot n\,w_n - \operatorname{div}_\Gamma(f w_m v_\Gamma) + \operatorname{div}_\Gamma(f v_\Gamma)\,w_n\,d\Gamma \\ &= \int_\Gamma \{f\,DVn\cdot n + \operatorname{div}_\Gamma(f v_\Gamma)\}\,w_n - f w_n\,v_\Gamma\cdot n\,d\Gamma \\ &= \int_\Gamma \{f\,(DVn\cdot n + \operatorname{div}_\Gamma v_\Gamma) + \nabla_\Gamma f\cdot v_\Gamma\}\,w_n\,d\Gamma. \end{aligned}$$

Finally we get two equivalent expressions of $d^2 J_{V,W}$ are

$$\begin{aligned}&= \int_\Gamma \left(\frac{\partial f}{\partial n} + H\, f\right) v_n\, w_n + f\, (DVn \cdot n\, w_n - v_\Gamma \cdot \nabla_\Gamma w_n)\, d\Gamma \\ &= \int_\Gamma \left\{ \left(\frac{\partial f}{\partial n} + H\, f\right) v_n + f\, (DVn \cdot n + \operatorname{div}_\Gamma v_\Gamma) + \nabla_\Gamma f \cdot v_\Gamma \right\} w_n\, d\Gamma.\end{aligned} \tag{4.3}$$

Since the above expressions involved the composition $T_s(W) \circ T_t(V)$, it is to be expected that the condition for the symmetry of expression (4.3) will involve the *Lie bracket* $[V, W] = DVW - DWV$. Indeed by using identity (3.29)

$$\begin{aligned}DVW \cdot n &= (D_\Gamma v + DVn\, {}^*n\}\; W \cdot n = w \cdot {}^*D_\Gamma v\, n + DVn \cdot n\, w_n \\ &= w_\Gamma \cdot (\nabla_\Gamma v_n - D^2 b\, v_\Gamma) + DVn \cdot n\, w_n\end{aligned}$$

and substituting in the first expression, we get a symmetrical term plus the first half of the Lie bracket: $d^2 J_{V,W}$ is equal to

$$\begin{aligned}= \int_\Gamma \left(\frac{\partial f}{\partial n} + H\, f\right) v_n\, w_n &+ f\, \left(D^2 b\, v_\Gamma \cdot w_\Gamma - v_\Gamma \cdot \nabla_\Gamma w_n - w_\Gamma \cdot \nabla_\Gamma v_n\right) \\ &+ f\, DVW \cdot n\, d\Gamma.\end{aligned} \tag{4.4}$$

Thus

$$d^2_{V,W} = d^2_{W,V} \quad \Longleftrightarrow \quad \int_\Gamma f\, [V, W] \cdot n\, d\Gamma = 0 \tag{4.5}$$

from which either $f\, [V, W] \cdot n = 0$ on Γ or $\operatorname{div}(f\, [V, W]) = 0$ on Ω can be used as sufficient conditions. We have proved the following result.

THEOREM 4.1. *Let $f \in C^2([0, \tau] \times [0, \tau]; H^2_{\mathrm{loc}}(\mathbb{R}^N))$, Γ be the boundary of a bounded open $\Omega \subset \mathbb{R}^N$ of class C^2 Assume that*

$$V \in C^0([0, \tau]; C^2_{\mathrm{loc}}(\mathbb{R}^N, \mathbb{R}^N)) \text{ and } W \in C^0([0, \tau]; C^1_{\mathrm{loc}}(\mathbb{R}^N, \mathbb{R}^N))$$

Consider the functional

$$J_{V,W}(t, s) \stackrel{\text{def}}{=} \int_{\Omega_{t,s}(V,W)} f(t, s)\, dx. \tag{4.6}$$

Then the partial derivative of $J_{V,W}(t, s)$ with respect to t in $t = 0$ is given by

$$\begin{aligned}\frac{\partial}{\partial t} J_{V,W}(t, s)\bigg|_{t=0} &= \int_{\Omega_s(W)} \frac{\partial f}{\partial t}(0, s) + \operatorname{div}(f(0, s) V(0))\, dx \\ &= \int_{\Omega_s(W)} \frac{\partial f}{\partial t}(0, s)\, dx + \int_{\Gamma_s(W)} f(0, s)\, V(0) \cdot n_s\, d\Gamma_s\end{aligned} \tag{4.7}$$

and the second order mixed derivative of $J_{V,W}(t, s)$ in $(t, s) = (0, 0)$

$$d^2 J_{V,W} \stackrel{\text{def}}{=} \frac{\partial}{\partial s} \left\{ \frac{\partial}{\partial t} J_{V,W}(t, s)\bigg|_{t=0} \right\}\bigg|_{s=0} \tag{4.8}$$

is given by the expression

$$
\begin{aligned}
d^2 J_{V,W} &= \int_\Omega \frac{\partial}{\partial s}\left(\frac{\partial f}{\partial t}\right) + \operatorname{div}\left(\frac{\partial f}{\partial s} V(0) + \frac{\partial f}{\partial t} W(0)\right) \\
&\qquad + \operatorname{div}\left[\operatorname{div}\left(f\, V(0)\right) W(0)\right] dx \\
&= \int_\Omega \frac{\partial}{\partial s}\left(\frac{\partial f}{\partial t}\right) dx + \int_\Gamma \left(\frac{\partial f}{\partial s} V(0) + \frac{\partial f}{\partial t} W(0)\right) \cdot n \\
&\qquad + \operatorname{div}\left(f\, V(0)\right) W(0) \cdot n \, d\Gamma
\end{aligned}
\tag{4.9}
$$

The last term in the second integral can be expressed in terms of $v = V(0)|_\Gamma$ *and* $w = W(0)|_\Gamma$ *as follows*

$$
\begin{aligned}
&\int_\Gamma \operatorname{div}\left(f\, V(0)\right) W(0) \cdot n \, d\Gamma \\
&= \int_\Gamma \left\{ \left(\frac{\partial f}{\partial n} + H f\right) v_n + f\, \left(DVn \cdot n + \operatorname{div}_\Gamma v_\Gamma\right) + \nabla_\Gamma f \cdot v_\Gamma \right\} w_n \, d\Gamma \\
&= \int_\Gamma \left(\frac{\partial f}{\partial n} + H f\right) v_n\, w_n + f\, \left(D^2 b v_\Gamma \cdot w_\Gamma - v_\Gamma \cdot \nabla_\Gamma w_n - w_\Gamma \cdot \nabla_\Gamma v_n\right) \\
&\qquad + f\, DVW \cdot n \, d\Gamma.
\end{aligned}
\tag{4.10}
$$

4.2 Definitions and Properties

We have reduced the computation of the Eulerian semiderivative of $J(\Omega)$ to the computation of the derivative of the function

$$
j(t) \overset{def}{=} J(\Omega_t(V))
\tag{4.11}
$$

for a velocity field $V \in C([0,\tau]; C^k_{\text{loc}}(\mathbb{R}^N; \mathbb{R}^N))$. In $t \geq 0$

$$
j'(t) = dJ(\Omega_t(V); V_t)
\tag{4.12}
$$

since $T_{s+t}(V) = T_s(V_t) \circ T_t(V)$, where $V_t(s) \overset{def}{=} V(t+s)$ and $V_t(0) = V(t)$.

This suggests the following definition.

DEFINITION 4.2. *Let* V *and* W *satisfy condition* (V) *and assume that* $\forall t \in [0,\tau]$, $dJ(\Omega_t(W); V_t)$ *exists for* $\Omega_t(W) = T_t(W)(\Omega)$. *The functional* J *is said to have a* second order Eulerian semiderivative *at* Ω *in the directions* (V, W) *if the following limit exists*

$$
\lim_{t \searrow 0} \frac{dJ(\Omega_t(W); V_t) - dJ(\Omega; V)}{t}.
\tag{4.13}
$$

When it exists, it is denoted $d^2 J(\Omega; V; W)$.

If, for all t, J has a Hadamard semiderivative at $\Omega_t(W)$, recall that

$$
dJ(\Omega_t(W); V_t) = d_H J(\Omega_t(W); V_t(0)) = d_H J(\Omega_t(W); V(t)) = dJ(\Omega_t(W); V(t))
$$

and the above definition reduces to

$$
d^2 J(\Omega; V; W) = \lim_{t \searrow 0} \frac{dJ(\Omega_t(W); V(t)) - dJ(\Omega; V(0))}{t}.
\tag{4.14}
$$

The next theorem is the analogue of Theorem 2.4 and provides the canonical structure of the second order Eulerian semiderivative (cf. (2.18) to (2.20) in § 2.2 for the definitions of $\vec{\mathcal{V}}^{m,\ell}$ and $\mathcal{V}^\ell$). Its proof can be found in [3, 4].

THEOREM 4.3. *Let Ω be a subset of $\mathbb{R}^N$ and let $m \geq 0$ and $\ell \geq 0$ be integers. Assume that*

(i) $\forall V \in \vec{\mathcal{V}}^{m+1,\ell}, \forall W \in \vec{\mathcal{V}}^{m,\ell}$, $\quad d^2J(\Omega; V; W)$ exists,

(ii) $\forall W \in \vec{\mathcal{V}}^{m,\ell}$, $\forall t \in [0,\tau]$, J has a shape gradient of order ℓ and is Hadamard semidifferentiable at $\Omega_t(W)$

(iii) $\forall U \in \mathcal{V}^\ell$, the map $W \mapsto d^2J(\Omega; U; W) : \vec{\mathcal{V}}^{m,\ell} \to \mathbb{R}$ is continuous.

Then for all V in $\vec{\mathcal{V}}^{m+1,\ell}$ and all W in $\vec{\mathcal{V}}^{m,\ell}$

$$d^2J(\Omega; V; W) = d^2J(\Omega; V(0); W(0)) + dJ(\Omega; V'(0)), \tag{4.15}$$

$$V'(0)(x) \stackrel{\text{def}}{=} \lim_{t \searrow 0} \frac{V(t,x) - V(0,x)}{t}. \tag{4.16}$$

This important theorem gives the canonical structure of the second order Eulerian semiderivative: a first term which depends on $V(0)$ and $W(0)$ and a second term which is equal to $dJ(\Omega; V'(0))$. When V is autonomous the second term disappears and the semiderivative coincides with $d^2J(\Omega; V; W(0))$ which can be separately studied for autonomous vector fields in $\mathcal{V}^\ell$.

We conclude this section with the explicit computation of the second order Eulerian semiderivative for a shape functional $J(\Omega)$ with respect to two velocity fields V and W verifying the conditions of Theorem 4.3 and such that shape gradient at any t is of the form

$$dJ(\Omega_t(W); V(t)) = \int_{\Gamma_t(W)} g(t)\, V(t) \cdot n_t \, d\Gamma_t \tag{4.17}$$

for some function $g(t) \in C(\Gamma_t(W))$. Further assume that the family of functions $g(t)$ has an extension $Q \in C^1([0,\tau]; C^k_{\text{loc}}(N(\Gamma); \mathbb{R}^N))$ to an open neighborhood $N(\Gamma)$ of Γ such that $\cup \{\Gamma_t(W) : 0 \leq t \leq \tau\} \subset N(\Gamma)$. Therefore using the extension $N_t(W)$ of the normal n_t on $\Gamma_t(W)$, it amounts to differentiate the expression

$$j(t) = \int_{\Gamma_t(W)} Q(t)\, V(t) \cdot N_t(W)\, d\Gamma_t.$$

Apply the first formula (3.15) of Theorem 3.3 to get

$$\begin{aligned} j'(0) = \int_\Gamma & (Q'_W(0)\, V(0) + Q(0)\, V'(0)) \cdot n + Q(0)\, V(0) \cdot N'(W) \\ & + \left(\frac{\partial}{\partial n} (Q(0)\, V(0) \cdot \nabla b) + H\, Q(0)\, V(0) \cdot n \right) W(0) \cdot n \, d\Gamma \end{aligned}$$

where $Q'_W(0)$ only depends on W. But the last three terms have already been computed in several forms. They constitute expression (4.2) in § 4.1 which yields (4.3) and (4.4) with

$f = Q(0)$, $V = V(0)$ and $W = W(0)$. This yields with the notation $v = V(0)|_\Gamma$ and $w = W(0)|_\Gamma$ the following expression for $d^2J(V;W)$

$$\begin{aligned}
&= \int_\Gamma Q'_W(0)\, v_n + Q(0)\, V'(0)\cdot n + \left(\frac{\partial Q(0)}{\partial n} + H\, Q(0)\right) v_n\, w_n \\
&\quad + Q(0)\, \left(D^2 b\, v_\Gamma \cdot w_\Gamma - v_\Gamma \cdot \nabla_\Gamma w_n - w_\Gamma \cdot \nabla_\Gamma v_n\right) \\
&\quad + Q(0)\, DVW \cdot n\, d\Gamma \qquad\qquad (4.18) \\
&= \int_\Gamma Q'_W(0)\, v_n + Q(0)\, V'(0)\cdot n + \left(\frac{\partial Q(0)}{\partial n} + H\, Q(0)\right) v_n\, w_n \\
&\quad + \{Q(0)\, (DVn\cdot n + \mathrm{div}_\Gamma\, v_\Gamma) + \nabla_\Gamma Q(0)\cdot v_\Gamma\}\, w_n\, d\Gamma.
\end{aligned}$$

REMARK 4.1. *When V is autonomous the term in $V'(0)$ disappears. In that case the first half of the Lie bracket can be eliminated by restarting the computation with $\mathcal{V}(t) = V \circ T_t^{-1}(W)$ in place of V since $\mathcal{V}(0) = V$ and*

$$\begin{aligned}
\mathcal{V}'(0) = -DV\, W \quad &\Rightarrow \int_\Gamma Q(0)\, \mathcal{V}'(0)\cdot n + Q(0)\, DVW\cdot n\, d\Gamma = 0 \\
&\Rightarrow d^2J(V,W) = d^2J(\mathcal{V},W) + dJ(\Omega; DVW).
\end{aligned}$$

REMARK 4.2. *Except for the terms which contain the first half of the Lie bracket DVW and $V'(0)$, the only term which might not be symmetrical in the first expression (4.18) is the one in $Q'_W(0)$. In fact according to the second expression and our theorem $Q'_W(0) = Q'_{W(0)}$ only depends on $W(0)$. Choose now autonomous velocity fields V and W. Furthermore assume that W is of the form $W = w_\Gamma \circ p$. Since $W\cdot n = 0$ on Γ, $dJ(\Omega_t(W); V(0)) = dJ(\Omega; V(0))$ and necessarily $d^2J(\Omega; V; w_\Gamma \circ p) = 0$. Therefore $d^2J(\Omega; V; W)$ only depends on w_n and hence $Q'_W(0) = Q'_{w_n}(0)$ and the integral*

$$\int_\Gamma Q'_{w_n}(0)\, v_n\, d\Gamma$$

only depends on w_n and v_n.

The above expressions give valuable information on the structure of the second order Eulerian derivative. Other expressions can also be obtained. For instance if $j(t)$ is transformed into the volume integral

$$j(t) = dJ(\Omega_t(W); V(t)) = \int_{\Gamma_t(W)} Q(t)\, V(t)\cdot n_t\, d\Gamma_t = \int_{\Omega_t(W)} \mathrm{div}\,(Q(t)\, V(t))\, dx$$

we get from formula (3.5) in Theorem 3.2 the equivalent volume expression

$$\begin{aligned}
d^2J(\Omega; V; W) = \int_\Omega \mathrm{div}\, \{&Q'_{W(0)}(0)\, V(0) + Q(0)\, V'(0) \\
&+ \mathrm{div}\,(Q(0)\, V(0))\; W(0)\}\, dx
\end{aligned}$$

which can obviously be transformed into a boundary expression.

4.3 Decomposition of $d^2J(\Omega; V(0), W(0))$

Observe that the lack of symmetry and the appearance of the first half of the Lie bracket in (4.4) for the computation of the second order Eulerian semiderivative of the domain integral (4.1). The same phenomenon was observed in the final form of the basic formula (4.10) for domain integrals (4.6) in Theorem 4.1, and also in § 4.2 for the derivative of the shape gradient (4.18) when it can be represented in integral form (4.17). In this section perturbations of the identity will be used to show that $d^2J(\Omega; V(0), W(0))$ can be further decomposed into a symmetric term plus the gradient applied to the velocity $DV(0)W(0)$

$$d^2J(\Omega; V(0), W(0)) = < d^2J_\Omega(0)V(0), W(0) > + dJ(\Omega; DV(0)W(0)).$$

Furthermore the symmetrical term can be obtained by the Velocity Method

$$d^2J_\Omega(0; V(0), W(0)) = \frac{d}{dt} J\left(\Omega_t(W(0); V(0) \circ T_t^{-1}(W(0))\right)\bigg|_{t=0}$$

THEOREM 4.4. *Let Θ be a Banach subspace of* $\mathrm{Lip}(\mathbb{R}^N; \mathbb{R}^N)$.

(i) Given f, θ and ξ in B_ε, assume that there exists $\tau > 0$ such that

$$\forall t \in [0, \tau], \quad d^2J_\Omega(f + t\xi; \theta) \text{ exists.}$$

Then

$$d^2J_\Omega(f; \theta; \xi) \text{ exists} \iff d^2J(\Omega_f; \mathcal{V}; W_\xi) \text{ exists} \tag{4.19}$$

for $\Omega_f = [I + f](\Omega)$ and the velocity fields

$$W_\xi(t) \stackrel{\text{def}}{=} \xi \circ [I + f + t\xi]^{-1} \text{ and } \mathcal{V}(t) \stackrel{\text{def}}{=} \theta \circ [I + f + t\xi]^{-1}. \tag{4.20}$$

(ii) If f, θ belong to $\mathcal{V}^{\ell+1}$ and ξ to $\mathcal{V}^\ell$, and $\mathcal{V}$ and W_ξ verify the conditions of Theorem 4.3, then

$$\begin{aligned} d^2J_\Omega(f; \theta; \xi) = & d^2J(\Omega_f; \theta \circ [I + f]^{-1}; \xi \circ [I + f]^{-1}) \\ & - dJ(\Omega_f; D(\theta \circ [I + f]^{-1})\, \xi \circ [I + f]^{-1}) \end{aligned} \tag{4.21}$$

$$d^2J(\Omega_f; \theta; \xi) = d^2J_\Omega(f; \theta \circ [I + f]; \xi \circ [I + f]) + dJ(\Omega_f; D\theta\, \xi). \tag{4.22}$$

(iii) If V, W verify the conditions of Theorem 4.3 and $V \in \vec{\mathcal{V}}^{m+1,\ell+1}$, then

$$d^2J(\Omega; V(0); W(0)) = d^2J_\Omega(0; V(0); W(0)) + dJ(\Omega; DV(0)\, W(0)) \tag{4.23}$$

$$d^2J(\Omega; V; W) = d^2J_\Omega(0; V(0); W(0)) + dJ(\Omega; V'(0) + DV(0)\, W(0)) \tag{4.24}$$

and

$$d^2J_\Omega(0; V(0); W(0)) = \frac{d}{dt} dJ(\Omega_t(W(0)); V(0) \circ T_t^{-1}(W(0))\bigg|_{t=0}. \tag{4.25}$$

Proof. (i) Assume that $d^2 J_\Omega(f;\theta;\xi)$ exists and consider the differential quotient

$$q(t) \stackrel{def}{=} \frac{1}{t}\left[dJ_\Omega(f+t\xi;\theta) - dJ_\Omega(f;\theta)\right] \to d^2 J_\Omega(f;\theta;\xi).$$

From Theorem 2.6 (ii)

$$dJ_\Omega(f+t\xi;\theta) = dJ([I+f+t\xi](\Omega);\theta\circ[I+f+t\xi]^{-1}).$$

Define

$$T_t \stackrel{def}{=} [I+f+t\xi]\circ[I+f]^{-1}, \quad \mathcal{V}(t) \stackrel{def}{=} \theta\circ[I+f+t\xi]^{-1}.$$

From Theorem 2.1 $T_t = T_t(W_\xi)$ for the velocity field

$$W_\xi(t) \stackrel{def}{=} \frac{\partial T_t}{\partial t}\circ T_t^{-1} = \xi\circ[I+f+t\xi]^{-1}.$$

Therefore

$$dJ_\Omega(f+t\xi;\theta) = dJ(T_t(W_\xi)(\Omega_f);\mathcal{V}(t))$$
$$q(t) = \frac{1}{t}\left[dJ(T_t(W_\xi)(\Omega_f);\mathcal{V}(t)) - dJ(\Omega_f;\mathcal{V}(0))\right] \to d^2(\Omega_f;\mathcal{V};W_\xi)$$

since $q(t)$ converges as t goes to 0. The converse is obvious.
(ii) is now a direct consequence of Theorem 4.3 and the fact that

$$\mathcal{V}(t) = \theta\circ[I+f]^{-1}\circ T_t^{-1}(W_\xi) \quad \Rightarrow \mathcal{V}'(0) = -D(\theta\circ[I+f]^{-1})\,\xi\circ[I+f]^{-1}.$$

(iii) From Theorem 4.3 and part (ii) with $f=0$, $\theta = V(0)$ and $\xi = W(0)$. □

If $D^2 J_\Omega(f)$ exists in a neighborhood of $f=0$ and is continuous at $f=0$, then $D^2 J_\Omega(0)$ is symmetrical and this completes the decomposition of the shape gradient into a symmetrical operator and the gradient applied to the first half of the Lie bracket $[V(0), W(0)]$

$$d^2 J(\Omega; V(0); W(0)) = < D^2 J_\Omega(0)\, V(0), W(0) > + < G(\Omega), DV(0)\, W(0) > .$$

But this is not the end of the story. From the computation of the Hessian of the domain integral (4.4) in § 4.1,

$$d^2 J_{V,W} = \int_\Gamma \left(\frac{\partial f}{\partial n} + H\, f\right) v_n\, w_n + f\,\left(D^2 b\, v_\Gamma\cdot w_\Gamma - v_\Gamma\cdot\nabla_\Gamma w_n - w_\Gamma\cdot\nabla_\Gamma v_n\right) + f\, DVW\cdot n\, d\Gamma$$

the result of the computation (4.10) in § 4.1,

$$\int_\Gamma \operatorname{div}(f\, V(0))\, W(0)\cdot n\, d\Gamma$$
$$= \int_\Gamma \left(\frac{\partial f}{\partial n} + H\, f\right) v_n\, w_n + f\,\left(D^2 b v_\Gamma\cdot w_\Gamma - v_\Gamma\cdot\nabla_\Gamma w_n - w_\Gamma\cdot\nabla_\Gamma v_n\right) + f\, DVW\cdot n\, d\Gamma$$

and expression (4.18) of the derivative of (4.17) in § 4.2,

$$d^2J(V;W) = \int_\Gamma Q'_{w_n}(0)\, v_n + Q(0)\, V'(0)\cdot n + \left(\frac{\partial Q(0)}{\partial n} + H\, Q(0)\right) v_n\, w_n$$
$$+ Q(0)\,\left(D^2 b\, v_\Gamma \cdot w_\Gamma - v_\Gamma \cdot \nabla_\Gamma w_n - w_\Gamma \cdot \nabla_\Gamma v_n\right)$$
$$+ Q(0)\, DVW \cdot n\, d\Gamma$$

it is readily seen that the symmetrical term $< D^2 J_\Omega(0)\, V(0), W(0) >$ further decomposes into a symmetrical term which only depends on the normal components v_n and w_n of $V(0)$ and $W(0)$ and another symmetrical term which is framed in the above expressions. This last term is the same in all expressions and only depends on the trace of $G(0)$ (here of f and $Q(0)$) on Γ and the group of terms

$$D^2 b\, v_\Gamma \cdot w_\Gamma - v_\Gamma \cdot \nabla_\Gamma w_n - w_\Gamma \cdot \nabla_\Gamma v_n \tag{4.26}$$

involving v_Γ, w_Γ, and tangential derivatives of v_n and w_n on Γ.

It would be interesting to further investigate this structure. In the context of a domain optimization problem if the shape gradient is zero the term (4.26) and the first half of the Lie bracket will both be multiplied by zero so that they won't contribute to the Hessian which will reduce to the symmetrical part which only depends on v_n and w_n, that is for autonomous velocity fields V and W

$$\int_\Gamma \left(\frac{\partial f}{\partial n} + H\, f\right) v_n\, w_n\, d\Gamma$$
$$\int_\Gamma Q'_{w_n}(0)\, v_n + \left(\frac{\partial Q(0)}{\partial n} + H\, Q(0)\right) v_n\, w_n\, d\Gamma$$

and the term

$$\int_\Gamma Q'_{w_n}(0)\, v_n\, d\Gamma = \int_\Gamma Q'_{v_n}(0)\, w_n\, d\Gamma$$

is symmetrical. For earlier results on the structure of the second order shape derivative the reader is referred to page 434 in [12] and [1].

Bibliography

[1] D. Bucur and J.-P. Zolésio, *Anatomy of the shape Hessian via Lie brackets*, Ann. Mat. Pura Appl. (4) 173 (1997), 127–143.

[2] M.C. Delfour, *Tangential differential calculus and functional analysis on a $C^{1,1}$ submanifold*, in "Differential-geometric methods in the control of partial differential equations", R. Gulliver, W. Littman and R. Triggiani , eds., Contemporary Mathematics, AMS Publications, to appear (Proc. of the conference held in Boulder, Colorado, June 28-July 2, 1999) (34 pages).

[3] M.C. Delfour and J.-P. Zolésio, *Velocity method and Lagrangian Formulation for the computation of the shape Hessian*, SIAM J. Control Optim. 29 (1991), no. 6, 1414–1442.

[4] M.C. Delfour and J.-P. Zolésio, *Structure of shape derivatives for nonsmooth domains*, J. Funct. Anal. **104** (1992), no. 1, 1–33.

[5] M.C. Delfour and J.-P. Zolésio, *Differential equations for linear shells: comparison between intrinsic and classical models*, in "Advances in mathematical sciences: CRM's 25 years" (Montreal, PQ, 1994), 41–124, CRM Proc. Lecture Notes, 11, Amer. Math. Soc., Providence, RI, 1997.

[6] M.C. Delfour and J.-P. Zolésio, *Velocity method and Courant metric topologies in shape analysis of partial differential equations*, in "Control of Nonlinear Distributed Parameter Systems", G. Chen, I. Lasiecka, and J. Zhou, eds, Marcel Dekker, New York, to appear.

[7] A.M. Micheletti, *Metrica per famiglie di domini limitati e proprietà generiche degli autovalori*, Ann. Scuola Norm. Sup. Pisa Ser. III, **26** (1972), 683–694.

[8] F. Murat et J. Simon, *Sur le contrôle par un domaine géométrique*, Rapport 76015, Université Pierre et Marie Curie, Paris, 1976.

[9] J. Sokolowski and J.-P. Zolésio, *Introduction to shape optimization. Shape sensitivity analysis*, Springer Series in Computational Mathematics, 16. Springer-Verlag, Berlin, 1992.

[10] J.-P. Zolésio, *Identification de domaines par déformation*, thèse de doctorat d'état, Université de Nice, France, 1979.

[11] J.-P. Zolésio, *The material derivative (or speed) method for shape optimization*, in "Optimization of distributed parameter structures", Vol. II, (E. J. Haug and J. Céa, eds.), pp. 1089–1151, NATO Adv. Study Inst. Ser. E: Appl. Sci., 50, Sijhofff and Nordhoff, Alphen aan den Rijn, 1981 (Nijhoff, The Hague).

[12] J.-P. Zolésio, *Shape optimization problems and free boundary problems*, Shape optimization and free boundaries (M. C. Delfour, ed.), Kluwer Academic Publishers, Dordrecht, Boston, London, 1992, 397–457.

Michel C. Delfour. Centre de recherches mathématiques, and Département de Mathématiques et de statistique, Université de Montréal, C.P. 6128, succ. Centre-ville, Montréal QC, H3C 3J7, Canada
E-mail: delfour@CRM.UMontreal.CA

Jean-Paul Zolésio. Research Director at CNRS, Centre de Mathématiques Appliquées, Ecole des Mines de Paris, 2004 route des Lucioles, BP. 93, 06902 Sophia Antipolis Cedex, France
E-mail: Jean-Paul.Zolesio@sophia.inria.fr

Slope Stability and Shape Optimization: Numerical Aspects

Jean Deteix

Abstract. We propose to solve a shape optimization problem related to soil mechanics, the optimization of the stability of a slope by the use of a reinforcing material. The method presented will not use the derivative of the cost which will lead to technical problem related to the size of the discrete optimization problem. However the method will be quite general (non-differentiable cost could be considered) and we will obtain a generic code for shape problem in linear elasticity. The result will be a code which is simple and relatively general at the expense of the calculation time and memory size. In a sense this approach and code is similar to the one recently presented in [9].

1 Shape Optimization and Slope Stability

Previously, we gave a formulation and existence results for a shape optimization problem associated with porous media, [6]. Here we propose a numerical scheme to solve this problem. The approach used to formulated the shape problem is very general. It has the advantage of not imposing artificial condition on the optimal shape or a priori knowledge of the shape, so the solution is global. The basic elements of this method are: the use of relaxed characteristic functions; the Cacciopoli sets and its topology.

This approach is applicable to a large family of shape optimization problem, it is in fact a generic way to solve shape optimization problem related to elasticity. We will see how this formulation gives, in a very natural way, numerical schemes using only one or two meshes and since this method can be applied to numerous problem s the difficulties encountered to solve the pit slope problem are representative of any other shape problem solved with this approach.

In this paper we try to develop a method to numerically solve a specific problem but in doing so we will try to consider different possible cost functions. For this reason we will not calculate the gradient of the cost function with respect to the design variables or use the differentiability of the cost (except to perform finite differences calculus). This will generate some important technical problem (mainly limitations regarding the size of the discrete problem and calculations times) but will allow us to consider a code with more general cost function.

2 Review of the Formulation

The *stability* of a slope can be viewed as the *capacity of a slope to resist to rupture.* Suppose we have a fixed pit slope composed of a material 2 and a limited quantity of a material 1 of a different nature. Where should we put material 1 in the pit slope so that the stability is greater (maximize)? We call material 1 a reinforcing material since it increases the stability.

Let D_0 be a two-dimensional model of the part of the pit. Let $\Omega \subset D_0$ be the part occupied by the reinforcing material and $\Omega_* = D_0 \backslash \Omega$. We consider $D = D_0 \cup E$ to minimize the anomalies produced by the boundary conditions.

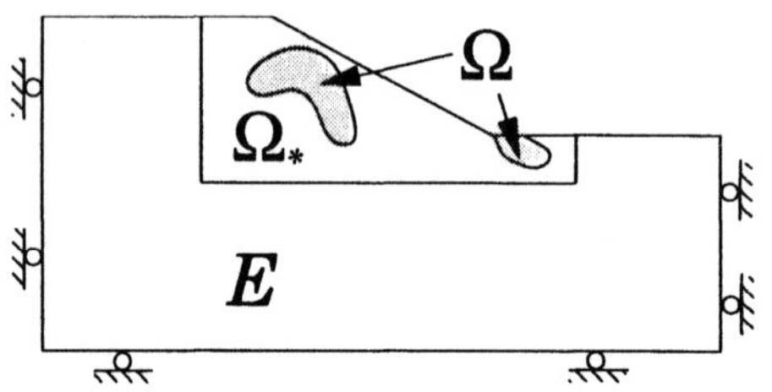

We have two materials, the reinforcing material (index 1) occupying Ω and a material denoted by the index 2 on the rest of D. The characteristic function of Ω is defined as

$$\chi_\Omega(x) = \begin{cases} 0 & \text{if } x \notin \Omega \\ 1 & \text{if } x \in \Omega \end{cases}$$

2.1 Safety Factor

Suppose that all failures are landslides along circular curves.

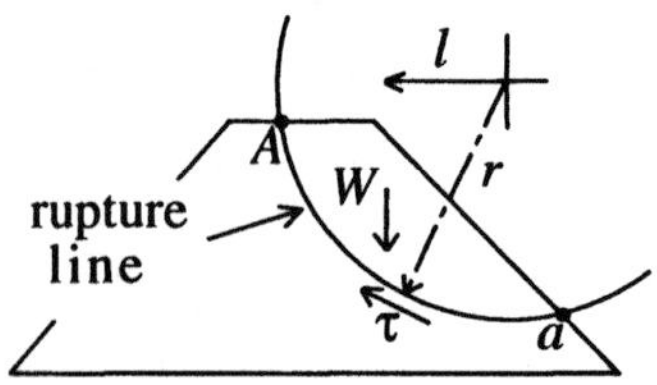

Let $A-a$ be a possible rupture line, let M_M be the moment of motion and M_R the resisting moment. From the Coulomb equation we have

$$M_M = r\int_a^A \tau ds, \qquad M_R = r\int_a^A \tau_r ds = r\int_a^A c + \sigma_n \operatorname{tg}\phi\, ds$$

Where r is the radius of the circle, τ shear stress, σ_n normal stress. The coefficients c and ϕ are only function of the material (size of grain, etc). If c_i ϕ_i $i = 1,\ 2$ are the parameters for the two materials, let $c_\Omega = \chi_\Omega c_1 + (1-\chi_\Omega)c_2$, $\phi_\Omega = \chi_\Omega\phi_1 + (1-\chi_\Omega)\phi_2$ and

$$F = \frac{M_R}{M_M} = \frac{\int_a^A (c_\Omega + \sigma_n \operatorname{tg}\phi_\Omega)\, ds}{\int_a^A \tau\, ds}.$$

$F \leq 1$ indicate a rupture. Let C_{ad} be the set of all circular arcs that are possible rupture line. The slope is *stable* if F is greater then 1 for all the arcs of C_{ad}. We can construct a safety factor by using F and C_{ad}.

$$F_s(\Omega) = \min_{\mathcal{C}\in C_{ad}} F(\mathcal{C}) = \min_{\mathcal{C}\in C_{ad}} \frac{\int_{\mathcal{C}} (c_\Omega + \sigma_n \operatorname{tg}\phi_\Omega)\, ds}{\int_{\mathcal{C}} \tau\, ds}.$$

2.2 Stress

We impose conditions on the base only (on $\Gamma_l \cup \Gamma_r \cup \Gamma_b$). For Ω and $\Omega_* \cup E$ respectively let E^1_{ijkh}, E^2_{ijkh} be the elastic tensors and ρ_1, ρ_2 the densities. Using the characteristic of Ω we have

$$E^\Omega_{ijkh}(x) = \chi_\Omega(x)E^1_{ijkh} + (1-\chi_\Omega(x))E^2_{ijkh} \qquad \rho_\Omega(x) = \rho_1\chi_\Omega(x) + \rho_2(1-\chi_\Omega(x))$$

We pose

$$V = \left\{(v_1, v_2) \in [H(D_+)]^2 \mid v_1 = 0 \text{ on } \Gamma_g \cup \Gamma_d,\ v_2 = 0 \text{ on } \Gamma_b\right\}.$$

The stress tensor σ_{ij} is defined as the solution to

$$\begin{cases} \text{find } u_i(x,y)\ i = 1,\ 2 \text{ in V such that } \forall v \in V \\ \int_D \epsilon_{ij}(v)\sigma_{ij}(u)\, dx = -\int_D v\cdot\rho_\Omega dx, \\ \sigma_{ij}(u) = E^\Omega_{ijkh}\epsilon_{kh}(u) \qquad \epsilon_{ij}(u) = \frac{1}{2}\left(\frac{\partial u_i}{\partial x_j} + \frac{\partial u_j}{\partial x_i}\right) \qquad i,j = 1,\ 2. \end{cases} \tag{2.1}$$

2.3 Volume Constraint

- If the material is reinforcing and we have unlimited access to it then the best solution is to fill the whole pit (D_0) with it!! Since in most case the material is reinforcing we would like to look for non-trivial solution ($\Omega \subsetneq D_0$).

- If the presence of this material diminish the stability we would like to be able to not put any of this material.

To take those considerations into account we will limit the quantity of material at our disposition. This will be formulated as a constraint on the volume of Ω.

$$0 \leq \int_D \chi_\Omega \, dx \leq \alpha < \int_D \chi_{D_0} \, dx$$

2.4 Set of Characteristics

Until now all the hypothesis made were uniquely based on physical considerations. We will now introduce the definition of the set of admissible characteristic. This definition is central to the optimization problem. In particular the existence of a solution will force us to make an hypothesis which is only based on mathematical consideration.

The approach chosen is closely linked to the so-called *power-law* approach,[3]. Contrary to [4], [1], [8] we make no effort to give a physical sense to the composite material. This lead to a simpler method but will necessitate "corrections" for the zone with this "non-physical" material. For a discussion on this and on the constraint for the existence see section 3.5.

Introducing the set $X(D) = \{\chi \in BV(D) \mid \chi(x) \in \{0,1\} \text{ a.e. in D}, \mathcal{P}(\chi) \leq r\}$ we will use the set of relaxed characteristic $X_r(D)$:

$$X(D) \subset X_r(D) = \{\chi \in BV(D) \mid \chi(x) \in [0,1] \text{ a.e. in D}, \mathcal{P}(\chi) \leq r\}.$$

Here r is arbitrary. From a theoretical point of view the perimeter ($\mathcal{P}$) condition is essential since it gives us compactness of the sets. From a numerical point of view, for a fixed mesh, this condition have no effect since it can be arbitrarily large. So it imposes no particular condition on the shape. Of course if we refine the mesh then r will play a role in the definition of the shape (r will play a role in the convergence).

Why Relaxing the Characteristic?

We restrict ourself to strong topology since weak topology would imply complicated theoretical result if any (see [5]). The use of $X_r(D)$ as a major drawback: we could obtain a solution which is not a characteristic. So we have to justify the use of this set. It should be noted that both sets are strongly compact in $L^p(D)$ for any p and any r but $X_r(D)$ is convex.

Theoretically since the two sets can support the same topology using one or the other has no effect from a theoretical point of view. Moreover from $X(D) \subset X_r(D)$, if we have existence of a solution in $X_r(D)$ we will have existence of a solution in $X(D)$ by compactness.

Numerically since $X(D) \subset X_r(D)$ we have an upper bound on the solution in $X(D)$ which is interesting from a practical point of view. By penalizing the cost function we can obtain a solution in $X(D)$. For the numerical part we will discretize the characteristic functions. Using $X_r(D)$ we can solve using gradient method (Sequential Linear Programming, SQP, etc). Clearly discretizing $X(D)$ leads to integer programming. From this we can conclude that there is no major reason not to use $X_r(D)$.

2.5 Modification to the Safety Factor

There is one difficulty with our choice of admissible shape. But this comes from the way we defined the safety factor. For $\chi \in X(D)$ or $X_r(D)$ the integral along a curve in D has no meaning so F_s is not well defined. To solve this we consider

$$C_\delta = \{x \in D \mid d(C,x) < \delta\} \qquad \overline{n} = \frac{x - x_\varrho}{\varrho}$$

where $d(C,x)$ is the distance from C to x, $x_\varrho \notin D$ is the center of the circle. We replace the integrals on C by mean values over C_δ in F and using the characteristic of C_δ we have

$$J(\chi) = \min_{C \in C_{ad}} \frac{\int_D (c_\Omega - \overline{\sigma}_n \operatorname{tg} \phi_\Omega)\chi_{C_\delta}\, dx}{\int_D \overline{\tau}\chi_{C_\delta}\, dx}$$

REMARK 2.1. *For a fixed mesh and certain choice of approximation the perimeter constraint and the modification to F are not active (have no effect). So at the numerical level those considerations made to obtain existence results are transparent.*

2.6 The Problem

We have completed the formulation (mathematical) of the shape optimization problem. What we want to solve now is the following:

$$\max_{\chi \in X_r(D)} J(\chi)$$

subject to

$$0 \leq \int_D \chi\, dx \leq \alpha$$

3 Discretization

We want to produce a discrete version of this problem. To do this we will

- Discretize the elasticity problem.
- Discretize the set $X_r(D)$.

From this we have a discrete cost function J^h and a discrete optimization problem (D.O.P.). Since we will not make use of the gradient of the cost with respect to the design variables (at best we will use a finite difference to approximate the gradient, if it exists) instead of the quality of the solution our main concern will be speed and space size.

3.1 How to Compute the Stress

We will use a finite element method since it is efficient and can give good accuracy. The major problem in that case is that every evaluation of J^h (and there is a lot) implies the calculation of the stress so we need a F.E. code that is *very fast*. One way to solve this problem (this is the way we did it) is to use a standard F.E. code and a low approximation (P^1 approximation) for the displacements (so stress is constant on each elements).

It is clear that the efficiency and speed of the F.E. code is the major element for the use of a higher approximation and/or very fine mesh.

3.2 Discretization of $X_r(D)$.

This is the main point for the numerical scheme. It is clear that:

- The choice of the discretization will give the dimension of the discrete problem.
- The choice of the discretization will influence the type of solution obtained.
- The approximation of the elements of $X_r(D)$ should take into account the fact that we would like something taking only two value (0 and 1).

REMARK 3.1. *Any discretization of $X_r(D)$ gives continuous variables for the discrete case. So the D.O.P. is continuous. For the same choice of approximation but on $X(D)$ we have discontinuous variables and the D.O.P. become an integer programming problem.*

Low Number of Variables

Examples:

- Approx. of $X_r(D)$ by constant in $[0, 1]$ on subset of D.
- Using a spline defined by a small number of points in D (control points).

With these two examples there is no link between the approximation of the stress and $X_r(D)$ so we can increase the quality for each approximation independently.

However those methods will generally give a poor approximation of the solution. To increase the quality one needs to increase the number of points which lead to the next section. For the second method, we have some extra smoothness to the solution which is not always interesting. A good example of this last approach, albeit it is in the case of the boundary variation technique, is given in [7].

High Number of Variables

Examples:

- Discontinuous elements or finite volume (barycentric cell) on the F.E. grid.
- Use of a coarser grid to approximate $X_r(D)$.

This type of approximation gives a D.O.P. with a high number of variables. In the first case we have linked the approximation of $X_r(D)$ to the F.E. mesh so to increase the accuracy of the optimal solution we have to use a finer grid. In the second case we have a second grid to manage.

What we Did

Here we will use constant by barycentric cell on the triangular mesh of the finite element method:

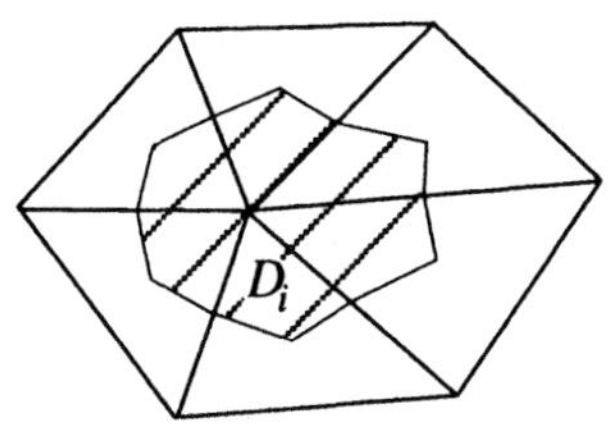

Since we are in 2-d the number of vertices is smaller then the number of triangles so we minimized the number of variables! To limit the problem to size which can be solve in a reasonable amount of time we limit the number of vertices for D_0.

The discrete cost J^h is a function of n (n very high) variables and depends on the F.E. solution. Even if we don't want to include the analytical definition of the gradient (when it exists) we will be interested in this information. So in the case of a differentiable cost function if we want to use the gradient for the optimization we have to keep in mind that *the calculation of the gradient will imply the calculation of n F.E. solutions.* This reinforce the need for a very fast evaluation of the F.E. solution since the finite difference will take a large part of the CPU time.

3.3 Structure of the D.O.P.

Let $X_r^h(D)$ be the discrete version of $X_r(D)$. With the remark on r, $\chi \in X_r^h(D)$ is in fact a vector $(X_1, ..., X_n) \in [0,1]^n$. If $(U_1, ..., U_m)$ denote the displacements then by the definition of F.E.M. we have a matrix A and a vector F depending on X such that

$$A(X)U = F(X) \qquad \text{the finite element problem}$$

If we express the safety factor (or any other function, compliance, weight, etc) as a function a X and U: $G(X,U)$ then the cost function becomes:

$$J^h(X, G(X,U))$$

For the volume constraint we introduce a function H of X and U (it can take into account more than one constraints on X and constraints other then the volume (compliance, weight, Von Mises criteria, etc)). In our case we have:

$$H(X,U) = \sum_{i=1}^{n} Volume(cell_i)X_i = \sum_{i=1}^{n} B_i X_i = (B,X)$$

Then for various shape problem s such as minimization of the weight, maximization of the compliance, minimization of the Von Mises criteria, and maximization of the safety factor of a pit slope the D.O.P. has the following structure:

$$\max_{X \in \mathbf{R}^n} J^h(X, G(X, U))$$

subject to

$$0 \leq X_i \leq 1 \qquad \text{i=1,...,n}$$

$$0 \leq H(X, U) \leq \alpha$$

$$A(X)U = F(X)$$

Since we have chosen a P^1 approximation for V and piecewise constant for $X_r(D)$ if the material parameters c and ϕ have enough regularity then the modification to the safety factor have no effect and we solve the original problem.

It is possible to have large part of the solution which is not in $\{0, 1\}$. There is a large number of techniques to deal with this (in fact the method used in the homogenization method and the power-law method are all applicable). For simplicity we will only look at two of those:

- Take a finer mesh hoping for a concentration of the grey areas to create "black" and "white" cells (see Figure 3.).

- Penalty method to have an integer sub-solution (see Figure 2.).

The easiest way is to penalize the cost by adding

$$-\lambda \int_D \chi(1 - \chi)\, dx, \qquad \lambda > 0,\ \lambda \text{ large}$$

This problem has a solution (the penalty is continuous) satisfying $\chi(1 - \chi) = 0$. The problem here is to choose λ since we can make J^h disappear (ill conditioned cost function). See section 3.5 for the details about the effect of the perimeter and about the zone containing composite material.

3.4 Resolution of the D.O.P.

We have not yet used the differentiability of J (J^h) as a function of χ (X resp.). It is not even necessary to obtain existence of a solution, [5]. The D.O.P. benefit from all the good properties of the original shape optimization problem so that for an u.s.c. J we have an u.s.c. discrete cost and existence of a solution.

In the case where J^h is differentiable with respect to X and U then we can compute its gradient. Using that allows us to take a more effective approach and bypass certain problem s linked to the size of the discrete problem (in fact, in that case the CPU-time is dramatically diminished). However we choose not to use the analytical form of the gradient (this choice penalize us but it will allow us to develop a code more general, at the expense of the calculation time and the size of thc D.O.P.).

It is relatively easy to implement the gradient of the cost and to substitute it to the finite difference. If we consider our D.O.P. as a control problem (X is the control and U the state) then using the adjoint state we can produce the gradient of J^h. The nature of the expression of the gradient depend on the complexity of the cost. We can mention the compliance as one of the simplest case (there is no adjoint state). Adding a water flow in the slope problem (the pressure is solution of a free boundary problem, and the stress will depend on the pressure) will certainly produce one of the hardest case.

It is possible to separate this problem in two families: differentiable and non-differentiable cases. It is even possible to add a third choice which is the integer case (that is to drop the relaxation of χ). The code is then developed in two separate branches. First a treatment for non-differentiable cost and constraint. This means the development of a (large scale) non-differentiable technique for optimization. Second the implementation of a relatively common optimization code with a finite difference scheme to produce the gradient. That way the only information necessary to solve the problem will be the differentiability of the cost and constraints.

Gradient Method

For the case where we have existence of a gradient we chose to apply a gradient method: the sequential quadratic programming. In the case of the safety factor, from the definition of the stress (normal and shear) and the choice made to approximate V we can conclude that J^h is differentiable with respect to X.

Note that in most case (even the simplest: minimal weight, maximal compliance, etc) the problem easily reach the limit of applicability for general SQP code, the solution often come with a non conclusive diagnostic, so we have to play with parameters of the code to obtain solutions that satisfies the optimality criteria.

3.5 Final Remarks

We proposed here a simplification of the power-law method, so it is a simplification of the homogenization method. It inherit most of the flaws of those methods but it can also use most of the tools and remedies developed to overcome those flaws. Any of those methods will give solutions containing composite material ($\chi \in]0,1[$) in some zone (*checkerboard effect*). In our method this effect should ***absolutely be avoid since the material has no physical meaning in those regions.*** But even if this define a "shape" we are interested in more "classical" shape (i.e. $\chi \in \{0,1\}$). To eliminate those zones different techniques can be used but most of them will lead to solution depending on the grid on which they are defined, this is call *mesh-dependency*.

In this last section we will try to answer to three basic questions:

a) What to do with the vanishing perimeter constraint?

b) What to do with the grid scale effect (mesh dependency)?

c) What to do with the sub-scale due to the relaxation (checkerboard)?

a) It is clear that as the grid is refined ($h \to 0$) the perimeter constraint will become active whatever the value of r chosen. In fact as $h \to 0$, since we have to fix r to study the convergence, we will have convergence to a solution respecting the volume constraint *plus* the perimeter constraint.

From a practical point of view we are not interested in the convergence as the grid is refined. What we want is to *control the accuracy of certain physical parameters* (basically u and σ). So what will be done in practice is to choose an expected accuracy for certain physical variables and we will work within this degree of precision. Basically what is done is that we choose a finite number of grid on which the calculation will be done.

Of course in this framework it is possible to choose r big enough so that the perimeter constraint is never active. This constraint is not based on the physics of the problem. It is in fact an artificial constraint imposed for mathematical reasons. The value given to r is absolutely arbitrary and imposing that the solution satisfy this perimeter constraint make no senses, at least physically.

b)-c) The easy answer to b) and c) is to apply the same corrections as for the homogenization or power-law approach. The filtering method, [10], is surely the most efficient since it can be made to control the checkerboard effect and to eliminate the mesh dependency. The penalization of the cost as proposed here however is simpler to implement but lead to grid scale effects, [2].

Since our approach impose important limits on the size of the F.E. grid we will almost always obtain "rough" solution. In that sense a more simple elimination of the checkerboard effect by penalization or by a mesh refinement as proposed in 3.3 can be applied if this solution is used as a pre-processing step (i.e. if we have another step eliminating the grid scale while refining our solution).

The idea is to use the solution obtain this way as a starting point for other methods. Methods based on local variations of the boundary needs an a priori estimate of the optimal shape and cannot create wide variation in the shape (holes,disjoint solution from a connected one, etc). Our solution could be used to generate an initial shape (creating a predictor-corrector shape scheme).

4 Conclusion

Since we did not based our approach on the cost function or some particular property of J we can apply the same method to any cost function which is at least upper semi-continuous (this insure existence of a solution). In that sense we obtained a generic procedure to predict the optimal shape. This procedure gives solutions which are not generally acceptable (relatively to the high quality attainable with more specialized and sophisticated method) but its simplicity and versatility makes it a good candidate as a pre-processing stage of a more refined search for the optimal shape.

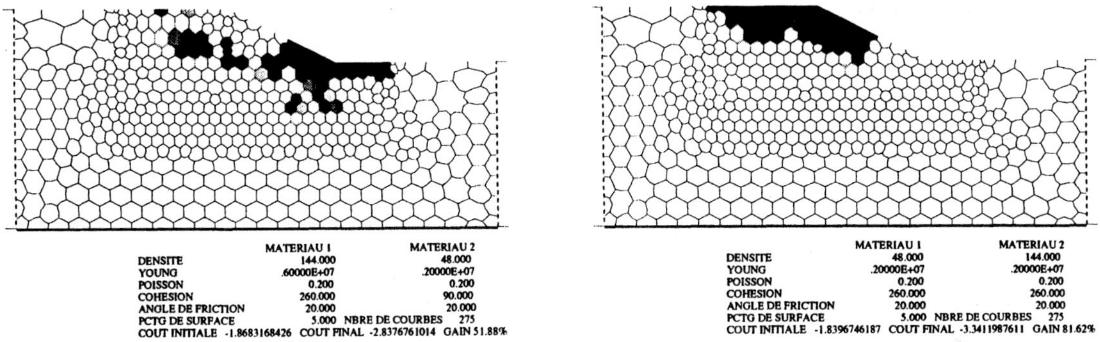

Figure 1: Variation of the densities: on the left $\rho_1 > \rho_2$ and $E_1 > E_2$, on the right no variations for the Young module, the density of the reinforcing material is lower.

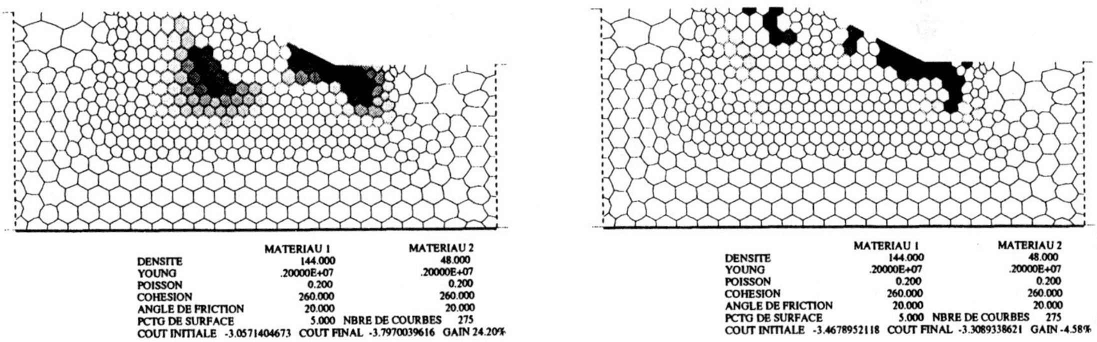

Figure 2: Non characteristic solution: on the left no modification, on the right we penalize the cost for a sub-optimal characteristic (numerically we cannot put $\lambda = \infty$).

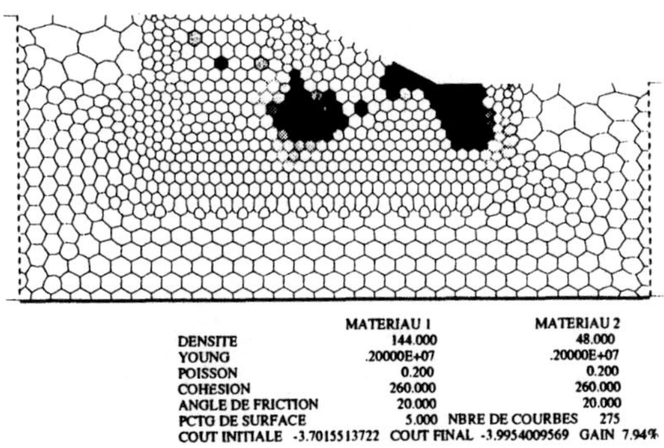

Figure 3: Same problem, to obtain a characteristic we refine the grid (almost double the number of nodes in D_0). Notice how the initial and this solution contains a reinforcing part deep in the slope that is not presented in the penalized one.

Bibliography

[1] G. Allaire, E. Bonnetier, G. Francfort, F. Jouve, *Shape optimization by the homogenization method*, Numerische Mathematatik, 1997, **76**, 27-68.

[2] G. Allaire and R.V. Kohn, *Optimal design for minimum weight and compliance in plane stress using extremal microstrucutre*, Europ. J. Mech. A/Solids, 1993, **12**, 6, 839-878.

[3] M.P. Bendsoe, *Optimal shape design as a material distribution problem*, Structural Optimization, 1989, **1**, 193-202.

[4] M.P. Bendsoe and O. Sigmund, *Material interpolations in topology optimization*, Archive of Applied Mechanics, 1999, **69**, 635-654.

[5] J. Deteix, *Conception optimale d'une digue à deux matériaux*, Thèse de Doctorat,Université de Montréal, 1997.

[6] J. Deteix, *The Dam Problem: Safety Factor and Shape Optimization*, System Modeling and Optimization (Detroit, MI, 1997), 1999, Chapman & Hall/CRC Res. Notes Math., 396, 62-70.

[7] J. Herskovits, G.P. Dias, G. Santos, C.M. Mota Soares, *Shape structural optimization with an interior point mathematical programming algorithm*, Structural Optimization, to appear.

[8] R.V. Kohn and G. Strang, *Optimal design and relaxation of variational problems I-II-III*, Comm. Pure Appl. Math., 1986, **39**, 113-137, 139-182, 353-377.

[9] O. Sigmund, *A 99 line topology optimization code written in Matlab*, Structural Optimization, to appear.

[10] O. Sigmund and J. Peterson, *Numerical instabilities in topology optimization: A survey on procedures dealing with checkerboard, mesh-dependencies and local minima*, Structual Optimization, 1998, **16** 1, page 68-78.

Jean Deteix. CRM, Université de Montréal, CP 6128, Succ. Centre-ville, Montréal (Qc), H3C 3J7, Canada
E-mail: deteix@crm.umontreal.ca

Parallel Solution of Contact Problems

Zdeněk Dostál[1], Francisco A. M. Gomes[1] and Sandra A. Santos[1]

Abstract. An efficient non-overlapping domain decomposition algorithm of the Neumann-Neumann type for solving both coercive and semicoercive frictionless contact problems of elasticity has been recently presented. The method reduces, by the duality theory of convex programming, the discretized problem to a quadratic programming problem with simple bounds and equality constraints on the contact interface. This dual problem is further modified by means of orthogonal projectors to the natural coarse space, and the resulting problem is solved by an augmented Lagrangian type algorithm. The projectors guarantee an optimal rate of convergence for the solution of auxiliary linear problems by the conjugate gradients method. With this approach, it is possible to deal separately with each body or subdomain, so that the algorithm can be implemented in parallel. In this paper, an efficient parallel implementation of this method is presented, together with numerical experiments that indicate the high parallel scalability of the algorithm.

1 Introduction

Duality based domain decomposition methods proved to be practical and efficient tools for parallel solution of large elliptic boundary value problems [15, 16, 23]. Using this approach, a body is partitioned into non-overlapping subdomains, for each subdomain is defined an elliptic problem with Neumann boundary conditions on the subdomain interfaces, and intersubdomain field continuity is enforced via Lagrange multipliers. The Lagrange multipliers are evaluated by solving a relatively well conditioned dual problem of small size that may be efficiently solved by a suitable variant of the conjugate gradient algorithm. The first practical implementations by Farhat and Roux [15, 16] exploited the favorable distribution of the spectrum of the matrix of the smaller problem [22], known also as the dual Schur complement matrix, being an efficient algorithm only with a small number of subdomains. Later, they introduced a "natural coarse problem" whose solution was implemented by auxiliary projectors so that the resulting algorithm became optimal [17, 23]. Recently, the authors have shown how to use the "natural coarse grid"

[1]Research supported by FAPESP 97-12676-4, CNPq, FINEP, FAEP-UNICAMP and by grants CAČR 201/97/0421 and 101/98/0535.

to the solution of a scalar variational inequality [11] and presented an efficient non-overlapping domain decomposition algorithm for solving both coercive and semicoercive frictionless contact problems of elasticity [12].

In this work, we focus on the computational implementation of parallel solution of contact problems. The parallelization is described, analyzed and tested for a model problem.

This paper is organized as follows: in Section 2 we present the discretized problem formulation, from the primal to the modified dual by means of preconditioning with projectors. For completeness, in Section 3 we briefly describe the adopted quadratic programming algorithm, based on the augmented Lagrangian technique and adaptive precision control for solving auxiliary problems. In Section 4, the model problem is described, together with its domain decompositions. In Sections 5 and 6, we present a profile of the algorithm and the parallelization scheme, respectively. Numerical results are shown and discussed in Section 7. Finally, some conclusions and future perspectives are presented in Section 8.

2 Problem Formulation

We consider a domain Ω defined by s homogeneous isotropic elastic bodies in contact, each one occupying, in a reference configuration, a subdomain $\Omega_p \subset \mathbb{R}^d, d = 2, 3$, with sufficiently smooth boundary. Imposing equilibrium conditions, after finite element discretization of $\Omega = \Omega_1 \cup \cdots \cup \Omega_s$, with a suitable numbering of nodes and assuming a secondary decomposition, we obtain the quadratic programming problem:

$$\min \ \frac{1}{2} u^T K u - f^T u \quad \text{s.t.} \quad B_I u \le c \quad \text{and} \quad B_E u = 0, \tag{2.1}$$

where $K \in \mathbb{R}^{n \times n}$ is symmetric positive definite (or semidefinite) block diagonal (i.e. $K = \text{diag}(K_1, \ldots, K_s)$), $B_I \in \mathbb{R}^{m \times n}$ and $B_E \in \mathbb{R}^{\ell \times n}$ are full rank matrices, $f \in \mathbb{R}^n$ and $c \in \mathbb{R}^m$. The matrix B_I and the vector c describe the linearized incremental non-interpenetration conditions, whereas matrix B_E ensures continuity of the displacements across auxiliary interfaces. For more details, see [10, 12]. The vector f describes the nodal forces arising from the volume forces and/or some other imposed tractions. Typically n is large and m, ℓ are much smaller than n. The diagonal blocks K_p that correspond to subdomains Ω_p are positive definite or semidefinite sparse matrices. Moreover, we shall assume that the nodes of the discretization are numbered in such a way that K_p are banded matrices that can be effectively decomposed, possibly after some regularization, by means of the Cholesky factorization.

Even though (2.1) is a standard convex quadratic programming problem, its formulation is not suitable for numerical solution. The reasons are that matrix K is typically ill conditioned, possibly singular and the feasible set is in general so complex that projections into it can hardly be effectively computed. Such difficulties may be essentially reduced by applying the duality theory of convex programming (e.g. [5, 6, 10]). Since the regular case has already been discussed [10] we shall assume that the matrix K has a nontrivial null space that defines the natural coarse grid ([23]).

The Lagrangian associated with problem (2.1) is

$$L(u, \lambda_I, \lambda_E) = \frac{1}{2} u^T K u - f^T u + \lambda_I^T (B_I u - c) + \lambda_E^T B_E u, \tag{2.2}$$

where λ_I and λ_E are the Lagrange multipliers associated with inequalities and equalities, respectively. Introducing notation

$$\lambda = \begin{bmatrix} \lambda_I \\ \lambda_E \end{bmatrix}, \quad B = \begin{bmatrix} B_I \\ B_E \end{bmatrix}, \quad \text{and} \quad \hat{c} = \begin{bmatrix} c \\ 0 \end{bmatrix},$$

we can write the Lagrangian briefly as

$$L(u,\lambda) = \frac{1}{2}u^T K u - f^T u + \lambda^T (Bu - \hat{c}).$$

Problem (2.1) is equivalent to the saddle point problem

$$\text{Find} \quad (\overline{u}, \overline{\lambda}) \quad \text{such that} \quad L(\overline{u}, \overline{\lambda}) = \sup_{\lambda_I \geq 0} \inf_u L(u,\lambda). \tag{2.3}$$

By eliminating u from (2.3) we obtain

$$\min \quad \Theta(\lambda) \quad \text{s.t.} \quad R^T(f - B^T\lambda) = 0 \quad \text{and} \quad \lambda_I \geq 0 \tag{2.4}$$

where

$$\Theta(\lambda) = \frac{1}{2}\lambda^T B K^\dagger B^T \lambda - \lambda^T (BK^\dagger f - \hat{c}), \tag{2.5}$$

R is a matrix whose columns span the null space of K and $K^\dagger$ denotes any matrix that satisfies $KK^\dagger K = K$. The essential fact is that the product of $K^\dagger$ by a vector should be effectively carried out (see e.g. [11, 14]). Once the solution λ of (2.4) is obtained, the vector u that solves (2.3) can be evaluated by an explicit formula (see [6, 10]).

The Hessian of Θ is, under reasonable assumptions, positive definite. Besides, it is closely related to that of the basic FETI method by Farhat and Roux [15, 16], so that its spectrum is relatively favorably distributed for application of the conjugate gradient method [22].

Even though problem (2.4) is much more suitable for computations than (2.1) and was used for efficient solution of contact problems [10], further improvement may be achieved by adapting the results of [17]. Let us denote $F = BK^\dagger B^T$, $\widetilde{d} = BK^\dagger f$, $\widetilde{G} = R^T B^T$, $\widetilde{e} = R^T f$ and let T denote a regular matrix that defines the orthonormalization of the rows of $\widetilde{G}$ so that matrix $G = T\widetilde{G}$ has orthogonal rows. After denoting $e = T\widetilde{e}$, problem (2.4) reads

$$\min \quad \frac{1}{2}\lambda^T F \lambda - \lambda^T \widetilde{d} \quad \text{s.t.} \quad G\lambda = e \quad \text{and} \quad \lambda_I \geq 0. \tag{2.6}$$

Next, the equality constraints may be homogenized by means of an arbitrary $\overline{\lambda}$ that satisfies $G\overline{\lambda} = e$. Denoting $d = \widetilde{d} - F\overline{\lambda}$, the modified problem reads

$$\min \quad \frac{1}{2}\lambda^T F \lambda - \lambda^T d \quad \text{s.t.} \quad G\lambda = 0 \quad \text{and} \quad \lambda_I \geq -\overline{\lambda}_I. \tag{2.7}$$

Further improvement can be obtained based on the decomposition of the augmented Lagrangian for problem(2.7) by the orthogonal projectors $Q = G^T G$ and $P = I - Q$ on the image space of G^T and on the kernel of G, respectively. Indeed, since $P\lambda = \lambda$ for any feasible λ, problem (2.7) is equivalent to

$$\min \quad \frac{1}{2}\lambda^T PFP\lambda - \lambda^T Pd \quad \text{s.t.} \quad G\lambda = 0 \quad \text{and} \quad \lambda_I \geq -\overline{\lambda}_I \tag{2.8}$$

and the Hessian $H = PFP + \rho Q$ of the augmented Lagrangian

$$L(\lambda, \mu, \rho) = \frac{1}{2}\lambda^T(PFP + \rho Q)\lambda - \lambda^T Pd + \mu^T G\lambda \tag{2.9}$$

is decomposed by projectors P and Q whose image spaces are invariant subspaces of H. The analysis of Axelsson [2] and Dostál [8] together with results of the FETI method [17] provide ingredients to show that the rate of convergence for unconstrained minimization of the augmented Lagrangian (2.9) depends on neither the penalization parameter ρ nor the discretization parameter. In fact, provided the aspect ratios of both discretization and decomposition are close to one, the number of conjugate gradient iterations is bounded by the square root of the ratio between subdomain and mesh diameters (see [11, 12]).

3 Algorithm for Quadratic Programming with Equality Constraints and Simple Bounds

Our development of an efficient algorithm for the solution of (2.8) is based on the observation that the solution of such problem may be reduced, by the augmented Lagrangian technique [4, 9], to the solution of a sequence of quadratic programming (QP) problems with simple bounds, and that the latter can be solved much more efficiently than more general QP problems due to the possibility of using projections and results on adaptive precision control in the active set strategy [3, 7, 18, 19, 20]. Here, for completeness, we briefly describe the QP algorithm proposed in [9], conveniently adjusted to problem (2.8).

To simplify our notation, let us denote $F_P = PFP$ so that the augmented Lagrangian for problem (2.8) and its gradient are given by

$$L(\lambda, \mu, \rho) = \frac{1}{2}\lambda^T F_P\lambda - \lambda^T Pd + \mu^T G\lambda + \frac{1}{2}\rho||Q\lambda||^2$$

and

$$g(\lambda, \mu, \rho) = F_P\lambda - Pd + G^T(\mu + \rho G\lambda),$$

respectively. The *projected gradient* $g^P = g^P(\lambda, \mu, \rho)$ of L at λ is then given componentwise by

$$g_i^P = g_i \text{ for } \lambda_i > -\overline{\lambda}_i \text{ or } i \notin I \text{ and } g_i^P = g_i^- \text{ for } \lambda_i = -\overline{\lambda}_i \text{ and } i \in I$$

with $g_i^- = \min(g_i, 0)$, where I is the set of indices of constrained entries of λ.

The algorithm that we describe here may be considered a variant of the one proposed by Conn, Gould and Toint [4] for identification of stationary points of more general problems. However, Algorithm 3.1 is modified to exploit the specific structure of our problem and get improved performance. The most important of such modifications consists in including the adaptive precision control of auxiliary problems in Step 1.

All the parameters that must be defined prior to the application of the algorithm are listed in Step 0, with typical values for our model problem given in section 7.

Algorithm 3.1. (Simple bounded variables and equality constraints)

Step 0. *Initialization of parameters.* Set $0 < \alpha < 1$ for equality precision update, $1 < \beta$ for penalty update, $\rho_0 > 0$ for initial penalty parameter, $\eta_0 > 0$ for initial equality precision, $M > 0$ for balancing ratio, $\varepsilon > 0$ for optimality precision, μ^0 for the Lagrangian multipliers and $k = 0$.

Step 1. Find λ^k so that $||g^P(\lambda^k, \mu^k, \rho_k)|| \leq M||G\lambda^k||$, by solving

$$\min \quad L(\lambda, \mu, \rho) \quad \text{s.t.} \quad \lambda_I \geq -\overline{\lambda}_I.$$

Step 2. If $||g^P(\lambda^k, \mu^k, \rho_k)|| \leq \varepsilon||d||$ and $||G\lambda^k|| \leq \varepsilon||f||$

then λ^k is the solution.

Step 3. If $||G\lambda^k|| \leq \eta_k$

Step 3a. then $\mu^{k+1} = \mu^k + \rho_k G\lambda^k, \quad \rho_{k+1} = \rho_k, \quad \eta_{k+1} = \alpha\eta_k$

Step 3b. else $\rho_{k+1} = \beta\rho_k, \quad \eta_{k+1} = \eta_k$

end if.

Step 4. Increase k and return to Step 1.

The implementation of Step 1 may be carried out by means of any algorithm for quadratic minimization with simple bounds (e.g. [3, 7, 18, 19, 20]). The unique solution $\widetilde{\lambda} = \widetilde{\lambda}(\mu, \rho)$ of this auxiliary problem satisfies the Karush-Kuhn-Tucker conditions $g^P(\widetilde{\lambda}, \mu, \rho) = 0$.

Salient features of this algorithm are that it deals completely separately with each type of constraint and that it accepts inexact solutions of the auxiliary box constrained problems in Step 1. Algorithm 3.1 has been proved to converge for any set of parameters that satisfy the relations prescribed at Step 0 (see [9]). Moreover, the penalty parameter is uniformly bounded and the asymptotic rate of convergence is the same as for the algorithm with exact solution of auxiliary quadratic programming problems (i.e. $M = 0$).

4 A Model Problem and its Domain Decomposition

We consider the model problem that comes from the finite difference discretization of the following continuous problem

$$\begin{aligned} &\text{Minimize} \quad q(u_1, u_2) = \sum_{i=1}^{2} \left(\int_{\Omega_i} |\nabla u_i|^2 d\Omega - \int_{\Omega_i} f u_i d\Omega \right) \\ &\text{subject to} \quad u_1(0, y) \equiv 0 \text{ and } u_1(1, y) \leq u_2(1, y) \text{ for } y \in [0, 1], \end{aligned}$$

where $\Omega_1 = (0,1) \times (0,1)$, $\Omega_2 = (1,2) \times (0,1)$, $f(x,y) = -5$ for $(x,y) \in (0,1) \times [0.75, 1)$, $f(x,y) = 0$ for $(x,y) \in (0,1) \times (0, 0.75)$, $f(x,y) = -1$ for $(x,y) \in (1,2) \times (0, 0.25)$ and $f(x,y) = 0$ for $(x,y) \in (1,2) \times (0.25, 1)$.

The solution $u \equiv (u_1, u_2)$ of the model problem may be interpreted as the displacement of two membranes under the traction f, as shown in Figure 1. The left membrane is fixed on the left and the left edge of the right membrane is not allowed to penetrate below the edge of the left membrane. Moreover, both membranes are stretched by normalized horizontal forces.

This problem is semicoercive due to the lack of Dirichlet data on the boundary of Ω_2, but the solution is unique because the right membrane is pressed down. More details about this model problem, including some other results, may be found in [11, 12].

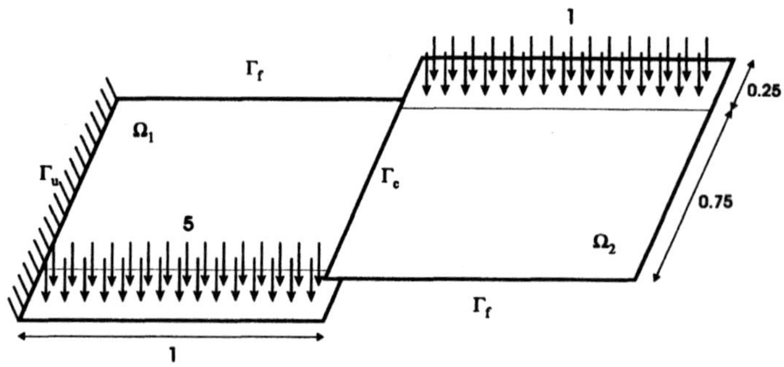

Figure 1: Model problem.

The model problem was discretized by regular grids defined by the stepsize $h = 1/n$ with $n + 1$ nodes in each direction per subdomain Ω_i, $i = 1, 2$. Each subdomain Ω_i was decomposed into $n_x \times n_y$ identical rectangles with dimensions $H_x = 1/n_x$ and $H_y = 1/n_y$. According to the values of n_y, we may have a decomposition into ***strips*** ($n_y = 1$) or into a ***chessboard*** pattern ($n_y > 1$).

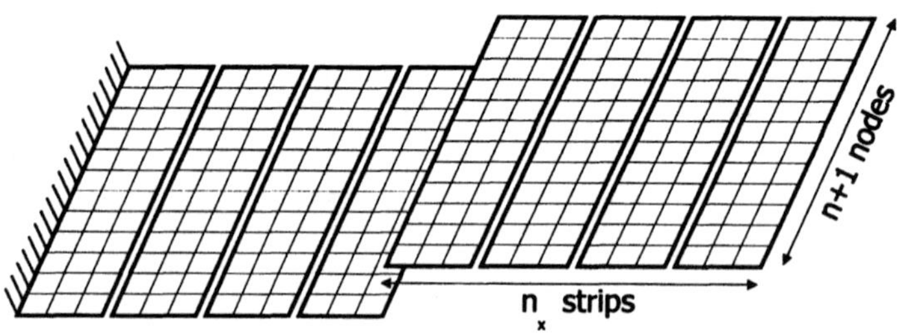

Figure 2: Decomposition into strips.

5 A Profile of the Algorithm

Applying duality theory to (2.1) greatly reduces the dimension of the problem. In fact, it can be shown that the dimension of the dual problem is $O(nn_x)$, while the primal dimension is $O(n^2)$.

Since the size of the dual problem may be still considerable large, a very efficient way to carry out step 1 of Algorithm 3.1 consists in applying a conjugate gradient type method to solve the bound-constrained quadratic problem using a modified lumped preconditioner in the form $C^{-1} = PBKB^TP + (1/\rho)G^TG$ to accelerate the convergence. In this case, the product of $F = BK^\dagger B^T$ by a vector has to be computed at least once per iteration. Observing that F includes $K^\dagger$, the generalized inverse of the primal stiffness matrix, one can conclude that this product might dominate the overall time of Algorithm 3.1.

Fortunately, the product of $K^\dagger$ by a vector can be efficiently performed in parallel, since this matrix is block-diagonal, with $2n_x$ blocks for the decomposition into strips and $2n_xn_y$ blocks if the chessboard decomposition is used. Moreover, for our semicoercive model problem, $K^\dagger$ contains only two different banded blocks, so storing this matrix is not a main concern even if a distributed parallel environment is used.

Besides the product of $K^\dagger$ by a vector, other relevant steps of the computation are

- The Cholesky decomposition of the two distinct diagonal blocks of K, used to compute $K^\dagger$ times a vector.
- The generation of matrix $G = T\widetilde{G}$. In our implementation, this matrix is obtained from the thin QR decomposition of a "condensed" version of $\widetilde{G}^T$ (see [13]).
- The product of $F = BK^\dagger B^T$ by a vector.
- The products of G and G^T by a vector.

To obtain an efficient implementation of Algorithm 3.1, it is important to minimize the time spent on generating matrices $K^\dagger$ and G, since this generation involves matrix decompositions that are difficult to parallelize.

For the model problem, the percentage of the total time that is spent in computing each of the steps described above is presented in Table 1. The chessboard decomposition was defined by setting $n_y = n_x$. Missing results correspond either to problems that are too large for the available memory or problems so small that the results provided by the profiler were not reliable.

Table 1 indicates that, for the decomposition into strips, the generation of $K^\dagger$ is much more expensive than the generation of G, so this last product may be neglected. Besides, the problem can be solved more efficiently in parallel if the number of strips is large, since the dimension of the diagonal blocks of $K^\dagger$ is inversely proportional to the number of strips. Thus, for a reasonably large value of r, the Cholesky decomposition of $K^\dagger$ is almost inexpensive and only the product of $BK^\dagger B^T$ by a vector needs to be performed in parallel.

For the chessboard decomposition, the time spent computing the QR decomposition of $\widetilde{G}^T$ increases exponentially as we increase the number of subdomains, since the number of columns of $\widetilde{G}^T$ is proportional to n_xn_y. In this case, some care must be taken in order to prevent the QR decomposition from dominating the overall time spent by the algorithm because this decomposition cannot be efficiently parallelized. Fortunately, Table 1 suggests that this can be accomplished by keeping ratios n_x/n and n_y/n small. For the model problem, supposing that $n_x = n_y$, this means that the relation $n_x \leq \sqrt{n/2}$ must hold. On the other hand, if n_x and n_y are too small, the decomposition of $K^\dagger$ may reduce the efficiency of the parallelization, so it is important to choose these parameters very carefully.

Table 1.
Percentage of the time spent by each routine.

		decomposition into strips					chessboard decomposition				
n	n_x	*chol* $K^\dagger$	*qr* $\tilde{G}^T$	Fv	Gv, G^Tv	*other*	*chol* $K^\dagger$	*qr* $\tilde{G}^T$	Fv	Gv, G^Tv	*other*
32	2	15.6	0.0	82.9	0.2	1.3	–	–	–	–	–
	4	4.5	0.5	89.5	2.2	3.3	–	–	–	–	–
	8	2.1	1.0	91.2	2.6	3.1	–	–	–	–	–
64	2	27.2	0.1	72.3	0.1	0.3	5.2	0.2	93.0	0.8	0.8
	4	12.8	0.0	86.2	0.5	0.5	0.6	0.4	90.9	4.7	3.4
	8	4.8	0.1	93.1	0.9	1.1	0.1	14.5	63.5	14.7	7.2
	16	1.4	0.0	95.6	1.3	1.7	0.0	85.6	7.3	5.1	2.0
	32	0.3	0.0	94.3	2.1	3.3	0.0	98.9	0.4	0.6	0.1
128	2	30.8	0.0	69.1	0.0	0.1	5.2	0.2	93.0	0.8	0.8
	4	13.7	0.0	86.0	0.1	0.2	0.8	0.1	96.8	1.4	0.9
	8	5.4	0.0	94.0	0.3	0.3	0.1	2.0	89.5	5.0	3.4
	16	1.8	0.0	97.4	0.4	0.4	0.0	64.6	25.6	6.4	3.4
	32	0.5	0.0	97.8	0.7	1.0	0.0	98.2	0.8	0.8	0.2
256	4	16.2	0.0	83.1	0.0	0.7	0.4	0.1	98.4	0.8	0.3
	8	6.8	0.0	93.1	0.0	0.1	0.2	0.4	96.7	1.7	1.0
	16	2.5	0.0	97.1	0.1	0.3	0.0	27.0	65.9	4.4	2.7
	32	0.6	0.0	98.9	0.3	0.2	0.0	95.6	3.0	1.1	0.3
512	4	–	–	–	–	–	2.3	0.0	97.4	0.1	0.2
	8	–	–	–	–	–	0.5	0.0	98.7	0.7	0.1
	16	–	–	–	–	–	0.0	4.3	92.7	1.8	1.2
1024	16	–	–	–	–	–	0.4	0.1	98.4	0.8	0.3

6 The Parallel Scheme

For the model problem, a careful choice of n_x and n_y may ensure that more than 95 percent of the total time of the algorithm will be spent on computing the product of $BK^\dagger B^T$ by a vector.

If the decomposition into strips is used, almost all of the remaining time is spent in computing the Cholesky factors of matrix $K^\dagger$, which means that no other part of the algorithm can be efficiently parallelized. For the chessboard decomposition, though, it is worth considering computing in parallel the product of the entire Hessian $H = PFP + \rho Q$ by a vector, as the products of $Q = G^TG$ or $P = I - Q$ by a vector are also easily performed in parallel.

In this paper, however, we will illustrate the parallel solution of the model problem using only the decomposition into strips. Therefore, we restrict our attention to the computation of product

$$y = BK^\dagger B^T v. \tag{6.1}$$

Product (6.1) can be decomposed into three parts. First, v is "expanded" and stored in a vector z with the same dimension as K. Then $K^\dagger z$ is obtained using the Cholesky factors previously computed. Finally, the resulting vector is compressed to fit in y.

The way this product is computed in parallel depends on the computational model used. In our code, the SPMD (single program, multiple data) model was adopted, which means that the same program is executed by all of the n_{proc} processors. Besides, MPI was chosen as the communication library.

Since vector v is available to all of the processors, each one can pick one part of the vector, expand it, compute the effect of the corresponding diagonal blocks of $K^\dagger$ on it and compress the resulting vector into y. At the end of this procedure, each processor stores a small portion of y, so it is necessary to gather all these parts up and distribute y to all of the processors in order to resume the algorithm. Fortunately, this is the only communication point of the entire algorithm and can be efficiently implemented using routine `MPI_AllGatherV` from MPI. However, some care must be taken when gathering y, since some of the $(2n_x - 1)n$ elements of this vector belong to the interface of two strips and receive contributions from different blocks of $K^\dagger Bv$. To circumvent this problem, a vector with $(2n_x + n_{proc} - 2)n$ elements is used to store all of the n_{proc} parts in which y was divided. After being distributed, this larger vector is compressed by each processor and y is finally generated.

7 Numerical Results

To evaluate the behavior of our parallel algorithm, a FORTRAN code was written. All of the tests were performed on a SGI Origin 2000 shared memory computer, with 4 processors, using MPICH, a portable implementation of MPI developed jointly by the Argonne National Laboratory and the Mississippi State University.

The model problem was solved for a variety of values of n and n_x in order to test experimentally the dependence of the rate of convergence on the discretization parameter. The bound constrained quadratic solver described in [3] was used to compute Step 1. The numerical data used were $\alpha = 0.1$, $\beta = 10$, $\rho_0 = 10^4$, $\eta_0 = 0.1$, $M = 10^4$, $\varepsilon = 10^{-5}$ and $\mu^0 = 0$.

The parallel algorithm attained the same precision as the sequential one. Moreover, both performed the same number of iterations and matrix-vector products. Figure 3 exhibits a typical solution for the strips decomposition.

The performance of the parallel algorithm is shown in Table 2, where S_p and S_a denote, respectively, the "predicted" and the speedup actually obtained in the experiments. For predicted speedup we mean the speedup that could be obtained if the time spent on computing (6.1) using n_{proc} processors was the time spent by one processor divided by n_{proc}, i.e. in the absence of communication costs. We also include the ratio S_a/S_p as a measure of the efficiency of our parallel implementation.

Since a vector with $(2n_x + n_{proc} - 2)n$ components need to be distributed to all of the processors, it should be expected that the efficiency decays as n_{proc} grows. However, for small problems, (6.1) is computed so fast that the time spent on communication becomes more significant, as can be seen in Table 2 for $n = 64$.

The predicted speedup values obtained show that, for each n, the parallel scheme is very efficient for appropriate choices of n_x. The figures for the actual speedup confirm the effectiveness of the algorithm.

Naturally, as the number of processors is increased, other routines than the product (6.1)

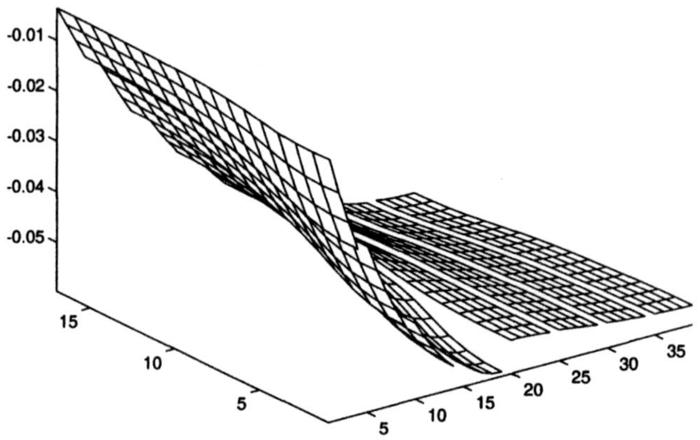

Figure 3: Typical solution.

Table 2.
Parallel performance of the algorithm.

n	n_x	$n_{proc} = 4$			$n_{proc} = 2$		
		S_a	S_p	S_a/S_p	S_a	S_p	S_a/S_p
64	4	2.43	2.83	0.86	1.62	1.76	0.92
	8	2.79	3.31	0.84	1.75	1.87	0.93
	16	2.86	3.53	0.81	1.78	1.92	0.93
	32	2.63	3.42	0.77	1.71	1.89	0.90
128	4	2.82	2.82	1.00	1.75	1.75	1.00
	8	3.28	3.39	0.97	1.88	1.89	1.00
	16	3.36	3.71	0.91	1.87	1.95	0.96
	32	3.30	3.75	0.88	1.87	1.96	0.96
256	4	2.46	2.65	0.93	1.45	1.71	0.85
	8	2.84	3.31	0.86	1.57	1.87	0.84
	16	3.11	3.68	0.85	1.65	1.94	0.85
	32	3.21	3.87	0.83	1.83	1.98	0.93

need also to be implemented in parallel in order to improve efficiency to a better extent.

8 Final Remarks

In this work, we described the computational implementation of a parallel code for solving contact problems. A profile of the algorithm was presented for a model problem with two membranes, using two domain decompositions (strips and chessboard pattern). This profile suggests that an efficient parallel scheme can be obtained. Numerical results that confirm the effectiveness of our implementation were also provided for the decomposition into strips.

Among the possible improvements on the algorithm, it is worth mentioning that, for the strips decomposition, better results could be obtained treating the diagonal blocks $K^{\dagger}$ as general sparse matrices, instead of storing them using a band format as we currently do. With this new approach, we could apply the minimum degree algorithm to permute the columns of $K^{\dagger}$ and reduce the number of nonzero elements in the resulting Cholesky factors.

A 3D contact problem with Signorini type of contact conditions was solved by the sequential version of the algorithm [12]. Future work includes extending the parallel scheme to this more realistic problem motivated by mining engineering, and also to the solution of 2D contact problems with Coulomb friction.

Acknowledgments We would like to acknowledge professor Patricio Letelier, from UNICAMP, who kindly made his parallel computer available for our numerical experiments.

Bibliography

[1] O. Axelsson, A class of iterative methods for finite element equations. *Comp. Meth. in Appl. Mech. and Engng.* 9 (1976) 127-137.

[2] O. Axelsson, *Iterative Solution Methods*, Cambridge Univ. Press, Cambridge, 1995.

[3] R. H. Bielschowsky, A. Friedlander, F. A. M. Gomes, J. M. Martínez, M. Raydan, An adaptive algorithm for bound constrained quadratic minimization, *Investigación Operativa* vol.7, no.1-2 (1997) 67-102.

[4] A. R. Conn, N. I. M.Gould, Ph. L. Toint, A globally convergent augmented Lagrangian algorithm for optimization with general constraints and simple bounds, *SIAM J. Num. Anal.* 28 (1991) 545-572.

[5] Z. Dostál, Duality based domain decomposition with inexact subproblem solver for contact problems. Contact Mechanics II, eds. M.H. Alibiadi, C. Alessandri, Wessex Inst. of Technology, Southampton (1995) 461-468.

[6] Z. Dostál, Duality based domain decomposition with proportioning for the solution of free boundary problems, *J. Comp. Appl. Math.* 63 (1995) 203-208.

[7] Z. Dostál, Box constrained quadratic programming with proportioning and projections, *SIAM J. Opt.* 7 (1997) 871-887.

[8] Z. Dostál, On preconditioning and penalized matrices, *Num. Lin. Alg. Appl.* 6 (1999) 109-114.

[9] Z. Dostál, A. Friedlander, S. A. Santos, Augmented Lagrangians with adaptive precision control for quadratic programming with simple bounds and equality constraints, Technical Report RP 74/96, IMECC-UNICAMP, University of Campinas, October 1996.

[10] Z. Dostál, A. Friedlander, S. A. Santos, Solution of coercive and semicoercive contact problems by FETI domain decomposition, *Contemporary Math.* 218 (1998) 82-93.

[11] Z. Dostál, F. A. M. Gomes, S. A. Santos, Duality-based domain decomposition with natural coarse-space for variational inequalities, Technical Report RP63/98, IMECC - Unicamp, State University of Campinas, October 1998.

[12] Z. Dostál, F. A. M. Gomes, S. A. Santos, Solution of Contact Problems by FETI Domain Decomposition with Natural Coarse Space Projections, Technical Report RP64/98, IMECC - Unicamp, State University of Campinas, October 1998. To appear in *Comp. Meth. in Appl. Math. and Eng.*

[13] Z. Dostál, F. A. M. Gomes, S. A. Santos, On the implementation of an algorithm for the numerical solution of contact problems. In preparation.

[14] C. Farhat, M. Gérardin, On the general solution by a direct method of a large scale singular system of linear equations: application to the analysis of floating structures, *Int. J. Num. Met. Eng.* 41 (1997) 675-696.

[15] C. Farhat, F.-X. Roux, An unconventional domain decomposition method for an efficient parallel solution of large-scale finite element systems, *SIAM J. Sci. Stat. Comput.* 13 (1992) 379-396.

[16] C. Farhat, P. Chen, F.-X. Roux, The dual Schur complement method with well posed local Neumann problems, *SIAM J. Sci. Stat. Comput.* 14 (1993) 752-759.

[17] C. Farhat, J. Mandel, F.-X. Roux, Optimal convergence properties of the FETI domain decomposition method, *Comput. Methods Appl. Mech. Eng.* 115 (1994) 365-385.

[18] A. Friedlander, J. M. Martínez, On the maximization of concave quadratic function with box constraints, *SIAM J. Opt.* 4 (1994) 177-192.

[19] A. Friedlander, J. M. Martínez, M. Raydan, A new method for large scale box constrained quadratic minimization problems, *Optimization Methods and Software* 5 (1995) 57-74.

[20] A. Friedlander, J. M. Martínez, S. A. Santos, A new trust region algorithm for bound constrained minimization, *Applied Mathematics & Optimization* 30 (1994) 235-266.

[21] R. Glowinski, P. Le Tallec, *Augmented Lagrangians and Operator Splitting Methods*, SIAM, Philadelphia 1989.

[22] F.-X. Roux, Spectral analysis of interface operator, Proceedings of the 5th Int. Symp. on Domain Decomposition Methods for Partial Differential Equations, ed. D. E. Keyes et. al., SIAM, Philadelphia, (1992) 73-90

[23] F.-X. Roux, C. Farhat, Parallel Implementation of Direct Solution Strategies for the Coarse Grid Solvers in 2-level FETI Method, *Contemporary Math.* 218 (1998) 158-173.

[24] J. S. Simo, J. A. Laursen, An augmented Lagrangian treatment of contact problems involving friction, *Comp.& Structures* 42 (1992) 97-116.

Zdeněk Dostál. Department of Applied Mathematics, VŠB-Technical University Ostrava, Ostrava, Czech Republic
E-mail: zdenek.dostal@vsb.cz

Francisco A. M. Gomes. Department of Applied Mathematics, IMECC – UNICAMP, University of Campinas, CP 6065, 13081–970 Campinas SP, Brazil
E-mail: chico@ime.unicamp.br

Sandra A. Santos. Department of Applied Mathematics, IMECC – UNICAMP, University of Campinas, CP 6065, 13081–970 Campinas SP, Brazil
E-mail: sandra@ime.unicamp.br

Eulerian Derivative for Non-Cylindrical Functionals

Raja Dziri and Jean-Paul Zolésio

Abstract. We are interested in characterizing Eulerian derivative for non-cylindrical functionals. These functionals are generally associated to evolution problems with time-dependent domains. In the smooth case, a tube (a non-cylindrical domain) can be considered as built by a suitably chosen velocity field. In this paper, we study the structure of the Gâteaux derivative of such a functional with respect to that velocity field. We point-out the special case where the functional depends only on the shape of the tube. Some examples are given to illustrate our main results.

1 Introduction

A non-cylindrical functional is generally related to a non-cylindrical evolution problem. Such problems are defined in tubes (non-cylindrical domains) in the following form

$$Q = \bigcup_{0<t<\tau} (\{t\} \times \Omega_t); \quad \text{at } t = 0,\ \Omega_0 = \Omega$$

In the smooth case, to each tube one can associate a non-autonomous velocity field V such that

$$T_t(V)\Omega = \Omega_t, \quad T_0(V)\Omega = \Omega$$

where $T(V)$ is the flow associated to V. More precisely, the final time $\tau > 0$ and the initial domain Ω being fixed, consider a velocity field $V \in C([0,\tau]; C^k(\mathbb{R}^N, \mathbb{R}^N))$ (the choice of k depends on the regularity of Ω_t, $t \in [0,\tau]$) satisfying

$$V(t).n_{\Omega_t} = v_\nu(t) \quad \text{on } \Sigma \text{ (the lateral boundary of } Q) \tag{1.1}$$

where n_{Ω_t} is the unitary outer normal to Ω_t and $v_\nu(t)$ is in connection with the time-component of the unitary outer normal $\vec{\nu}$ to Q which can be written in a unique way as, cf. [11],

$$\vec{\nu}(t) = \frac{1}{\sqrt{1+v_\nu^2}}(-v_\nu(t), n_{\Omega_t})$$

Then we can define the mapping $t \to T_t(V)$ as the solution of

$$\begin{cases} \frac{d}{dt}T_t & = & V(t) \circ T_t \quad \text{in } (0,\tau) \\ T_0 & = & \text{id}_{\mathbb{R}^N} \end{cases}$$

Condition (1.1) ensures that, at each time $t \in [0, \tau]$,

$$T_t(V)(\Omega) = \Omega_t$$

Conversely, to any sufficiently smooth non-autonomous vector field V one can associate, in the time interval $[0, \tau]$, a tube $Q(V)$ (also denoted Q_V) by setting $\Omega_t = T_t(V)(\Omega)$. Obviously (1.1) is satisfied. For a general abstract setting of tube evolution theory we refer to [1]. Taking the previous into consideration, the purpose of our study is to exhibit the structure of the Eulerian derivative of non-cylindrical functionals in the following form,

$$\mathbf{j}(V) = J(V, Q_V), \quad J \text{ being given}$$

In many examples J depends exclusively on Q_V. So $\mathbf{j}$ is a tube functional. Generally, J is function of Q_V through a boundary value problem and of V through its associated boundary conditions. For a bounded tube Q, there exists a bounded open set D (called hold-all) such that Q and its perturbations remain in $(0, \tau) \times D$. This condition is respected if

$$V(t).n_D = 0 \quad \text{on} \quad (0, \tau) \times \partial D \tag{1.2}$$

since the associated transformations $T_t(V)$ map $\overline{D}$ onto $\overline{D}$. In the computation of the Eulerian derivative we should introduce the *transverse field* $\mathbf{Z}$ solution of

$$\partial_t \mathbf{Z} + [\mathbf{Z}, V] = W \quad \text{in } (0, \tau) \times D \quad \text{and} \quad \mathbf{Z}(0, .) = 0 \quad \text{in } D$$

where V, W are admissible vector fields. The first field is associated to the original tube. The second one represents the direction of the perturbation. For additional properties on the transverse field $\mathbf{Z}$, see [12] where the non-smooth case is considered. In the last section of this paper, we give examples of such derivative in the case of a tube functional and also in the general case.

2 Preliminaries

Let $D \subset \mathbb{R}^N$ $(N \in \mathbb{N}^*)$ be a bounded and smooth open set and

$$V \in C([0, \tau], W^{k,\infty}(D, \mathbb{R}^N)), \; k \geq 1$$

satisfying $V(t).n_D = 0$ on ∂D. Then

PROPOSITION 2.1. *There exists a unique mapping $T \in C^1([0, \tau], W^{k-1,\infty}(D, \mathbb{R}^N))$ solution of*

$$\frac{d}{dt}T(t) = V(t, T(t)) \quad \textit{and} \quad T(0) = Id$$

Proof. The mapping

$$F : [0,\tau] \times W^{k-1,\infty}(D,\mathbb{R}^N) \to W^{k-1,\infty}(D,\mathbb{R}^N)$$

$$(t,X) \to V(t,X)$$

satisfies the Cauchy-Lipschitz conditions. Indeed, $t \to F(t,.)$ is continuous from $[0,\tau]$ to $W^{k-1,\infty}(D,\mathbb{R}^N)$. $X \to F(.,X)$ is Lipschitz-continuous in $W^{k-1,\infty}(D,\mathbb{R}^N)$.

□

The transformation $T(t)$ commonly denoted T_t or $T_t(V)$ (to precise to which vector field it corresponds) is one-to-one and maps $\overline{D}$ into $\overline{D}$. For each $t \in [0,\tau]$ one can consider the transformation T_t^{-1} and notice that it is associated to the vector field V_t defined by $V_t(s) = -V(t-s)$. Also the following result is available

PROPOSITION 2.2. *Mappings T and T^{-1} belong to $C([0,\tau],W^{k,\infty}(D,\mathbb{R}^N))$.*

Proof. It suffices to prove that the Jacobian matrix DT_t belongs to $W^{k-1,\infty}(D,\mathbb{R}^{2N})$ and that the mapping $t \to DT_t$ is C^1 from $[0,\tau]$ to $W^{k-1,\infty}(D,\mathbb{R}^N)$. Notice that DT_t is solution of the linear differential equation

$$\frac{d}{dt}\Xi = DV(t,T_t)\,\Xi \quad \text{and} \quad \Xi(0) = \text{Id}\ .$$

So it can be expressed as follows

$$DT_t = \exp\{\int_0^t DV(s,T_s)\,ds\}.$$

As $s \longrightarrow DV(s)$ (resp. $T_s(V)$) is in

$$C^0([0,\tau];W^{k-1,\infty}(D,\mathbb{R}^N)) \text{ (resp.} C^1([0,\tau];W^{k-1,\infty}(D,\mathbb{R}^N))$$

we obtain the desired conclusion for T. For T^{-1}, we have $T_t^{-1} \circ T_t = Id$. By differentiation we obtain $D(T_t^{-1}) = (DT_t)^{-1} \circ T_t^{-1}$. So

$$D(T_t^{-1}) = \exp\{-\int_0^t DV(s,T_s)\,ds\} \circ T_t^{-1}.$$

□

3 Transverse Field Z

Perturbation of a tube Q_V in a direction W can be obtained by considering transverse fields. At each time $t \in [0,\tau]$ and any s sufficiently small, we consider the moving domain

$$\Omega_t(V+sW) = T_t(V+sW)(\Omega),\ \Omega \subset\subset D \text{ is given.}$$

Let T_s^t be a function which maps $\Omega_t(V)$ onto $\Omega_t(V+sW)$ (and $\overline{D}$ onto $\overline{D}$). A quite natural choice is

$$T_s^t = T_t(V+sW) \circ T_t(V)^{-1}.$$

Under some assumptions, see for instance [2], that transformation can be considered as the flow of the vector field

$$\mathcal{Z}^t(s,.) = \left(\frac{\partial}{\partial s}\mathcal{T}_s^t\right) \circ \mathcal{T}_s^t(\,.\,)^{-1}. \tag{3.1}$$

LEMMA 3.1. *Let I_0 be a neighborhood of zero. The mapping*

$$I_0 \longrightarrow \mathcal{C}([0,\tau]; W^{k,\infty}(D,\mathbb{R}^N))$$

$$s \longrightarrow T(V+sW)$$

is continuously differentiable and $\partial_s(T_t(V+sW))$ satisfies for any $t \in [0,\tau]$,

$$\begin{aligned}\partial_s[T_t(V+sW)] &= \int_0^t D(V+sW)(\mu, T_\mu(V+sW))\partial_s[T_\mu(V+sW)]\,d\mu \\ &+ \int_0^t W(\mu, T_\mu(V+sW))\,d\mu.\end{aligned}$$

Thus $\mathcal{Z}^t(s) = [\partial_s T_t(V+sW)] \circ T_t(V+sW)^{-1}$. From now on we denote $\mathcal{S}^t(s) = \partial_s[T_t(V+sW)]$. Differentiability results in non-cylindrical functionals are expressed using the vector field $\mathbf{Z}(t,x) = \mathcal{Z}^t(0,x)$ which can be characterized by

THEOREM 3.2. *The vector field Z belongs to*

$$\mathcal{C}^1([0,\tau], W^{k-1,\infty}(D,\mathbb{R}^N))$$

and is the unique solution of

$$\partial_t \boldsymbol{Z} + [\boldsymbol{Z}, V] = W \;\; in\; (0,\tau)\times D \tag{3.2}$$

$$\boldsymbol{Z}(0,.) = 0 \;\; in\; D \tag{3.3}$$

where [,] denotes the Lie Brackets.

The proof of this result is based on the next lemma and the fact that $\mathbf{Z}(t) = \mathcal{S}^t(0,.)\circ T_t(V)^{-1}$.

LEMMA 3.3. *The function S, $S(t,.) = \mathcal{S}^t(0,.)$, is the unique vector function, in*

$$\mathcal{C}^1([0,\tau]; W^{k-1,\infty}(D,\mathbb{R}^N))$$

satisfying

$$S(t) = \int_0^t W(\mu, T_\mu(V))d\mu + \int_0^t DV(\mu, T_\mu(V))S(\mu)\,d\mu. \tag{3.4}$$

Proof. Let $\mathcal{F}$ be the mapping defined by

$$\mathcal{F} : [0,\tau]\times W^{k-1,\infty}(D,\mathbb{R}^N) \to W^{k-1,\infty}(D,\mathbb{R}^N)$$

$$(t,\varphi) \to DV(t, T_t(V))\varphi + W(t, T_t(V)).$$

For any $t \in [0,\tau]$ and any $\varphi \in W^{k-1,\infty}(D)$, $\mathcal{F}(t)$ is linear and $\mathcal{F}(.,\varphi)$ is continuous. So the existence and uniqueness are given by the Cauchy-Lipschitz theorem. The solution can be expressed as follows

$$S(t) = \int_0^t W(s, T_s(V)) \exp\{\int_s^t DV(\mu, T_\mu(V))\, d\mu\}\, ds \;\; \forall t \in [0,\tau].$$

□

An important property of $\mathbf{Z}$ is the preservation of volumes if V and W are of free divergence. Indeed

PROPOSITION 3.4. *Assume V and W of free divergence. Then the field $\mathcal{Z}^t(s)$ preserves the volume.*

Proof. Let $f \in L^1(D)$. The transformation $\mathcal{T}_s^t$ maps D onto D. So

$$\int_{\mathcal{T}_s^t(D)} f\, dx = \int_D f \circ \mathcal{T}_s^t \det D\mathcal{T}_s^t\, dx$$

As V and W are of free divergence, it is the same for $V + sW$. Therefore

$$\int_{\mathcal{T}_s^t(D)} f\, dx = \int_{T_t(V+sW)(T_t(V)^{-1}(D))} f\, dx = \int_{T_t(V)^{-1}(D)} f \circ T_t(V+sW)\, dx$$

$$= \int_D f \circ \mathcal{T}_s^t\, dx$$

From these two view-points we deduce (taking in account the continuity of the mapping $x \to \det D\mathcal{T}_s^t(x)$) that

$$\det D\mathcal{T}_s^t(x) = 1 \quad \text{in } D.$$

But

$$\det D\mathcal{T}_s^t(x) = \det D\mathcal{T}_s^t(x) \exp \int_0^s \operatorname{div} \mathcal{Z}^t(r) \circ \mathcal{T}_r^t(x)\, dr.$$

Thus we deduce

$$\operatorname{div} \mathcal{Z}^t(s)(x) = 0 \quad \text{in } D.$$

In particular, for $s = 0$, we obtain $\operatorname{div} \mathbf{Z} = 0$ in D. □

4 Adjoint Problem Associated to Z

As shown in the proof of Theorem 3.2 the solution of (3.2)-(3.3) is obtained via a change of variable and an explicit integral representation. If $H(D)$ is a Banach space of functions defined on D. The same procedure generates the solution of the adjoint problem associated to $\mathbf{Z}$.

THEOREM 4.1. *Let $F \in L^2([0,\tau]; H(D))$, there exists a unique $\Lambda \in C([0,\tau]; L^2(D;\mathbb{R}^N))$ solution of*

$$\partial_t \Lambda + D\Lambda.V + {}^*DV.\Lambda + (\operatorname{div} V)\, \Lambda = F \tag{4.1}$$

$$\Lambda(\tau) = 0. \tag{4.2}$$

Proof. Consider $\theta \in C([0,\tau]; H(D))$ the unique solution of the retrograde problem

$$\begin{aligned} \partial_t \theta + [\,{}^*(DV(t)) \circ T_t + \operatorname{div} V \circ T_t\,]\,\theta &= F \circ T_t \\ \theta(\tau) &= 0. \end{aligned}$$

Multiplying by $\exp \int_0^t [\,{}^*(DV(s)) \circ T_s + (\operatorname{div} V(s)) \circ T_s \mathbf{I}]\, ds$, we get

$$\partial_t \left[\theta(t) \exp\{ \int_0^t [\,{}^*(DV(s)) \circ T_s + (\operatorname{div} V(s)) \circ T_s \mathbf{I}]\, ds\} \right] =$$

$$F \circ T_t \exp\{ \int_0^t [\,{}^*(DV(s)) \circ T_s + (\operatorname{div} V(s)) \circ T_s \mathbf{I}]\, ds\}.$$

By integration we deduce an explicit expression of θ:

$$\theta(t) = -\int_t^\tau F(s) \circ T_s \ \exp\{- \int_s^t [\,{}^*DV(\xi) \circ T_\xi + (\operatorname{div} V(\xi)) \circ T_\xi \mathbf{I}]\, d\xi\}\ ds.$$

Then taking for $\Lambda = \theta \circ T_t^{-1}$, it is easily seen that Λ is the unique solution of (4.1)-(4.2) which , for a suitable right-hand term, represents the adjoint problem associated to $\mathbf{Z}$.

□

4.1 The Case of Right-Hand Term Supported on $\Sigma(V)$

In the sequel we shall be concerned with a right hand term in the following form $F(t) = -\gamma^*_{\Gamma_t}(f(t)n_t)$. We proved in Theorem 4.1 the existence of a unique Λ such that

$$\begin{aligned} &-\partial_t \Lambda - D\Lambda.V - {}^*DV.\Lambda - \operatorname{div} V\ \Lambda = \gamma^*_{\Gamma_t}(f(t)n_t) \\ &\Lambda(\tau) = 0. \end{aligned} \tag{4.3}$$

In the sequel we shall prove that the solution Λ is, in fact, supported by the lateral boundary $\Sigma(V)$ since the right-hand term in this problem is itself supported by $\Sigma(V)$ and there is no diffusion term.

LEMMA 4.2. *Let $f \in L^2(0,\tau; L^2(\Gamma_t))$, there exists a unique solution in $C([0,\tau]; H^1(\Gamma_t))$ of the following problem*

$$\begin{cases} \partial_t \lambda(t) + \nabla_{\Gamma_t} \lambda.V + \lambda \operatorname{div} V = f(t) \ \ \textit{on} \ \ \cup_t (\{t\} \times \Gamma_t) \\ \lambda(\tau) = 0 \ \ \textit{on} \ \ \Gamma_\tau \end{cases}$$

Proof. Note that

$$[\partial_t \lambda(t) + \nabla_{\Gamma_t} \lambda.V]_{|\Gamma_t} \circ T_t = \partial_t(\lambda \circ T_t)_{|\Gamma}.$$

Consider μ the unique solution of

$$\begin{aligned} &\partial_t \mu + (\operatorname{div} V) \circ T_t\ \mu = f(t) \circ T_t \ \text{ on } \ (0,\tau) \times \Gamma \\ &\mu(\tau) = 0 \ \text{ on } \ \Gamma. \end{aligned}$$

which is equivalent to

$$\mu(t) = -\int_t^{\tau} \exp\{\int_t^s (\operatorname{div} V) \circ T_r \, dr\} \, f(s) \circ T_s \, ds.$$

Then λ defined by $\lambda(t) = \mu(t) \circ T_t^{-1}$ is solution of the considered problem. Uniqueness is obvious.

□

THEOREM 4.3. *The solution Λ of (4.3) is supported by $\Sigma(V)$. Precisely*

$$\Lambda(t) = \lambda(t, p(t)) \nabla \chi_{\Omega_t}, \quad \forall t \in (0, \tau) \tag{4.4}$$

where λ is defined in Lemma 4.2 and $p(t)$ is the projection operator on Γ_t, $p(t) : U(\Gamma_t) \to \Gamma_t$, $U(\Gamma_t)$ is a neighborhood of Γ_t.

For further properties of $p(.)$ see [3] or [4].

Proof. Define $\tilde{\lambda}$ by $\tilde{\lambda}(t) = \lambda(t, p(t))$. Set $\mathbf{X} = \tilde{\lambda} \nabla \chi_{\Omega_t}$. We need to identify the distribution $-\partial_t \mathbf{X} - D\mathbf{X}.V - {}^*DV.\mathbf{X} - (\operatorname{div} V)\, \mathbf{X}$. Take $\varphi \in \mathcal{D}((0,\tau) \times D)$. Then

$$
\begin{aligned}
&< -\partial_t \mathbf{X} - D\mathbf{X}.V - {}^*DV.\mathbf{X} - (\operatorname{div} V)\, \mathbf{X} \,,\, \varphi >_{\mathcal{D}',\mathcal{D}} \\
&= \int_0^{\tau} < \tilde{\lambda}\, \nabla \chi_{\Omega_t} \,,\, \partial_t \varphi >_{\mathcal{D}'(D),\mathcal{D}(D)} \, dt \\
&+ \int_0^{\tau} < \tilde{\lambda}\, \nabla \chi_{\Omega_t} \,,\, [D\varphi.V + \operatorname{div} V\, \varphi - DV.\varphi] >_{\mathcal{D}'(D),\mathcal{D}(D)} \\
&\qquad - < \operatorname{div} V\, \mathbf{X} \,,\, \varphi >_{\mathcal{D}'(D),\mathcal{D}(D)} \, dt \\
&= \int_0^{\tau} < \nabla \chi_{\Omega_t} \,,\, \tilde{\lambda}\, \partial_t \varphi > + < \nabla \chi_{\Omega_t} \,,\, \tilde{\lambda}\, [D\varphi.V - DV.\varphi] > dt.
\end{aligned}
$$

The first term $E_1 = \int_0^{\tau} < \tilde{\lambda}\, \nabla \chi_{\Omega_t} \,,\, \partial_t \varphi >_{\mathcal{D}'(D),\mathcal{D}(D)} \, dt$ is treated as follows

$$
\begin{aligned}
E_1 &= \int_0^{\tau} < \nabla \chi_{\Omega_t} \,,\, \tilde{\lambda}\, \partial_t \varphi > dt = -\int_0^{\tau} \int_D \chi_{\Omega_t} \operatorname{div}(\tilde{\lambda}\, \partial_t \varphi) \, dx dt \\
&= -\int_0^{\tau} \int_{\Omega_t} \operatorname{div}(\tilde{\lambda}\, \partial_t \varphi) \, dx dt = -\int_0^{\tau} \int_{\Gamma_t} \lambda\, (\partial_t \varphi).n_t \, d\Gamma_t dt.
\end{aligned}
$$

Using the transformation $T_t(V)$

$$
\begin{aligned}
E_1 &= -\int_0^{\tau} \int_{\Gamma} \lambda \circ T_t \, [(\partial_t \varphi) \circ T_t] \,.\, n_t \circ T_t \, \omega(t) \, d\Gamma dt, \quad \omega(t) = \| \, det(DT_t)\, {}^*DT_t^{-1}.n \, \|_{\mathbb{R}^N} \\
&= -\int_0^{\tau} \int_{\Gamma} < \partial_t(\varphi \circ T_t) - (D\varphi.V) \circ T_t \,,\, n_t \circ T_t > \lambda \circ T_t \, \omega(t) \, d\Gamma dt
\end{aligned}
$$

$$
\begin{aligned}
E_1 &= \int_0^{\tau} \int_{\Gamma} < \varphi \circ T_t \,,\, n_t \circ T_t > \partial_t(\lambda \circ T_t)\, \omega(t) \\
&\qquad + < \varphi \circ T_t \,,\, \partial_t(\omega(t)\, n_t \circ T_t) > \lambda \circ T_t \, d\Gamma dt \\
&\quad + \int_0^{\tau} \int_{\Gamma} < (D\varphi.V) \circ T_t ,\, n_t \circ T_t > \lambda \circ T_t \, \omega(t) \, d\Gamma dt.
\end{aligned}
$$

But $\omega(t)n_t \circ T_t = \gamma(t)\, {}^*DT_t^{-1}n$ so

$$\begin{aligned} E_1 &= \int_0^T \int_\Gamma \partial_t(\lambda \circ T_t) < \varphi \circ T_t\,,\, n_t \circ T_t > \omega(t)\, d\Gamma dt \\ &+ \int_0^T \int_\Gamma < \varphi \circ T_t\,,\, \partial_t(\gamma(t)\, {}^*DT_t^{-1})n) > \lambda \circ T_t\, d\Gamma dt, \\ &+ \int_0^T \int_{\Gamma_t} \lambda < D\varphi.V\,,\, n_t > d\Gamma_t dt, \quad \text{where } \gamma(t) = \det DT_t^{-1} \end{aligned}$$

$$\begin{aligned} E_1 &= \int_0^T \int_\Gamma [\partial_t \lambda + \nabla_{\Gamma_t}\lambda.V] \circ T_t < \omega(t)\, n_t \circ T_t, \varphi \circ T_t > d\Gamma dt \\ &+ \int_0^T \int_\Gamma \lambda \circ T_t < \varphi \circ T_t, \partial_t({}^*DT_t^{-1})n > \gamma(t)\, d\Gamma dt \\ &+ \int_0^T \int_\Gamma \lambda \circ T_t < \varphi \circ T_t, \partial_t(\gamma(t))\, {}^*DT_t^{-1}n > d\Gamma dt \\ &+ \int_0^T \int_{\Gamma_t} \lambda < D\varphi.V\,,\, n_t > d\Gamma_t dt \end{aligned}$$

$$\begin{aligned} E_1 &= \int_0^T \int_{\Gamma_t} [\partial_t \lambda + \nabla_{\Gamma_t}\lambda.V]\, \varphi.n_t\, d\Gamma_t dt + \int_0^T \int_{\Gamma_t} < D\varphi.V, n_t > \lambda\, d\Gamma_t dt \\ &+ \int_0^T \int_\Gamma \lambda \circ T_t < \varphi \circ T_t, n_t \circ T_t > \omega(t)\, (\operatorname{div} V) \circ T_t\, d\Gamma dt \\ &- \int_0^T \int_\Gamma \lambda \circ T_t < \varphi \circ T_t, ,\, {}^*(DV) \circ T_t\, n_t \circ T_t > \omega(t)\, d\Gamma dt \end{aligned}$$

$$\begin{aligned} E_1 &= \int_0^T \int_{\Gamma_t} (\partial_t \lambda + \nabla_{\Gamma_t}\lambda.V)\, \varphi.n_t\, d\Gamma_t dt + \int_0^T \int_{\Gamma_t} \lambda < D\varphi.V, n_t > d\Gamma_t dt \\ &+ \int_0^T \int_{\Gamma_t} \lambda\, \varphi.n_t \operatorname{div} V\, d\Gamma_t dt - \int_0^T \int_{\Gamma_t} \lambda < DV.\varphi, n_t > d\Gamma_t dt. \end{aligned}$$

The second term

$$\begin{aligned} E_2 = \int_0^T < \nabla\chi_{\Omega_t}, \tilde{\lambda}\, D\varphi.V >_{\mathcal{D}'(D),\mathcal{D}(D)} dt - \int_0^T < \nabla\chi_{\Omega_t}, \tilde{\lambda}\, DV.\varphi >_{\mathcal{D}'(D),\mathcal{D}(D)} dt \\ = - \int_0^T \int_{\Omega_t} \operatorname{div}(\tilde{\lambda} D\varphi.V)\, dx dt + \int_0^T \int_{\Omega_t} \operatorname{div}(\tilde{\lambda} DV.\varphi)\, dx dt \\ = - \int_0^T \int_{\Gamma_t} < D\varphi.V, n_t > \lambda\, d\Gamma_t dt + \int_0^T \int_{\Gamma_t} \lambda < DV.\varphi, n_t > d\Gamma_t dt \end{aligned}$$

Finally we obtain

$$\begin{aligned} &< -\partial_t \mathbf{X} - D\mathbf{X}.V - {}^*DV.\mathbf{X} - \operatorname{div} V\, \mathbf{X}, \varphi >_{\mathcal{D}',\mathcal{D}} \\ &= \int_0^T \int_{\Gamma_t} (\partial_t \lambda + \nabla_{\Gamma_t}\lambda.V + \lambda \operatorname{div} V)\, \varphi.n_t\, d\Gamma_t dt = \int_0^T \int_{\Gamma_t} f(t)\, \varphi(t).n_t\, d\Gamma_t dt \end{aligned}$$

which is equivalent, in a distribution sense, to

$$-\partial_t \mathbf{X} - D\mathbf{X}.V - {}^*DV.\mathbf{X} - \operatorname{div} V\, \mathbf{X} = \gamma^*_{\Gamma_t}(f(t)n_t)$$

Moreover it is clear that $\mathbf{X}(\tau) = \lambda(\tau, p(\tau))\nabla\chi_{\Omega_\tau} = 0$. From the uniqueness theorem 4.1, we deduce that $\Lambda(t) = (\lambda(t) \circ p(t))\, \nabla\chi_{\Omega_t}$.

□

4.2 Explicit Expression of $\mathbf{Z}(t).n_{\Omega_t(V)}$

As we will see in applications, the expression of the Eulerian derivative depends on $\mathbf{Z}$. Hence, it is necessary to introduce two adjoint states. One associated to the state and the other to the field $\mathbf{Z}$. In some cases, this unusual situation may be avoided by considering the function $\mathbf{z}$ defined by

$$\mathbf{z}(t) = (\mathbf{Z}(t).n_t) \circ T_t(V) \quad \text{on} \quad (0,\tau) \times \Gamma,\ t \in [0,\tau].$$

PROPOSITION 4.4. *The function z is the unique solution of*

$$\partial_t z(t) - < DV.n_t, n_t > \circ T_t(V)\, z(t) = (W(t).n_t) \circ T_t(V) \quad on \quad (0,\tau) \times \Gamma \tag{4.5}$$

$$z(0) = 0 \quad on \quad \Gamma \tag{4.6}$$

Proof.

$$\partial_t[\mathbf{Z}(t, T_t).\, n_t \circ T_t] = [(\partial_t \mathbf{Z}) \circ T_t + (D\mathbf{Z}.V) \circ T_t]\, n_t \circ T_t + \mathbf{Z}(t, T_t).\partial_t(n_t \circ T_t).$$

But $\partial_t(n_t \circ T_t) =< {}^*DV.n_t, n_t > \circ T_t\, n_t \circ T_t - {}^*DV \circ T_t\, n_t \circ T_t$. Therefore

$$\begin{aligned}
&\partial_t[\mathbf{Z}(t, T_t).\, n_t \circ T_t] = \\
&=< (\partial_t \mathbf{Z} + D\mathbf{Z}.V) \circ T_t, n_t \circ T_t > + < DV.\, n_t, n_t > \circ T_t\, (\mathbf{Z}.n_t) \circ T_t \\
&- < (DV.\mathbf{Z}) \circ T_t,\ n_t \circ T_t > \\
&= (W(t).\, n_t) \circ T_t + < DV.n_t,\ n_t > \circ T_t (\mathbf{Z}(t).\, n_t) \circ T_t.
\end{aligned}$$

Eventually we obtain the desired result.

□

REMARK 4.1. *Let $\alpha(t) = - < DV(t)n_t,\, n_t >$, $\beta(t) = W(t).n_t$. Then*

$$\begin{aligned}
&\partial_t[z \exp(\int_0^t \alpha(s) \circ T_s\, ds)] = [\alpha(t) \circ T_t\, z + \partial_t z] \exp(\int_0^t \alpha(s) \circ T_s\, ds) \\
&= \beta(t) \circ T_t \exp(\int_0^t \alpha(s) \circ T_s\, ds).
\end{aligned}$$

So z can be expressed as follows

$$z(t) = \int_0^t \beta(s) \circ T_s(V) \exp(-\int_s^t \alpha(r) \circ T_r(V)\, dr)\, ds. \tag{4.7}$$

5 Differentiability with Respect to the Field

As mentioned before, we are interested in the structure of the Eulerian derivative of non-cylindrical functionals of the following type

$$\begin{cases} J(Q_V, V) = \int_{Q(V)} F(t,x,u(V)(t,x))\, dxdt \\ Q(V) = \bigcup_{0<t<\tau} (\{t\} \times T_t(V)(\Omega)) \end{cases}$$

where τ is a non-negative scalar and Ω a domain in $\mathbb{R}^N$.
$F : I \times \mathbb{R}^N \times \mathbb{R}^{N'} \to \mathbb{R}$, such that $(t,x) \to F(t,x,y)$ is integrable for a.e. y, u is generally solution of a well-posed non-cylindrical PDE in $Q(V)$. First we give the definition of a *tube functional.*

DEFINITION 5.1. *Let $j(V) = J(Q_V, V)$. If $\forall\, W$ s.t. $W(t).n_{\Omega_t(V)} = 0$ on $\Sigma(V)$,*

$$j(V+W) = j(V),$$

then the functional J depends only on the shape of the tube and is called a tube functional.

The main results stated in this section are available for any admissible field. Nevertheless for simplicity they are proved for free divergence fields. The state u may depend both on the shape of the tube $Q(V)$ and the field V that builds it. The first result we give includes the two situations.

LEMMA 5.2. *Assume that, for any direction W,*

i) $s \to u^s(t,x) = u(V+sW)(t, T_s(\mathcal{Z}^t)(x))$ is differentiable at $s=0$, for a.e. $(t,x) \in I \times \Omega$. It is denoted $\dot{u}(V;W)(t,x)$.

ii) $u'(V;W) = \dot{u}(V;W) - \partial_x u.\mathbf{Z}$ depends linearly on W.

iii) $s \to \int_{Q(V)} F(t, T_s(\mathcal{Z}^t(s))(x), u^s(t,x))\, dxdt$ is differentiable at $s=0$.

Then the functional $j(.)$ is Gâteaux differentiable at V and there exists a time-dependent distribution $G(V) \in L^2(0,\tau;\mathcal{D}'(D,\mathbb{R}^N))$ with $sptG(V)(t) \subset \overline{\Omega_t}(V)$ such that

$$j'(V;W) = \int_0^\tau < G(V)(t), W >_{\mathcal{D}'(D,\mathbb{R}^N),\mathcal{D}(D,\mathbb{R}^N)}\, dt.$$

Proof.

$$\begin{aligned} j(V+sW) &= \int_0^\tau \int_{\Omega_t(V+sW)} F(t,x,u(V+sW)(t,x))\, dxdt \\ &= \int_0^\tau \int_{T_s(\mathcal{Z}^t)(\Omega_t(V))} F(t,x,u(V+sW)(t,x))\, dxdt \\ &= \int_0^\tau \int_{\Omega_t(V)} F(t, T_s(\mathcal{Z}^t)(x), u^s(t,x))\, dxdt \end{aligned}$$

Hypothesis *iii*) ensures the existence of

$$\mathbf{j}'(V;W) = \partial_s \mathbf{j}(V + sW)_{|s=0}.$$

Precisely we have

$$\mathbf{j}'(V;W) = \int_0^T \int_{\Omega_t(V)} \partial_x F(t,x,u(t,x)).\mathbf{Z}(t,x) + \partial_y F(t,x,u(t,x)).\dot{u}(t,x)\,dxdt$$

As $\mathbf{Z}$ is of free divergence we obtain

$$\begin{aligned} \mathbf{j}'(V;W) &= \int_0^T \int_{\Omega_t(V)} \partial_y F(t,x,u(t,x)).[\dot{u}(t,x) - \partial_x u(t,x).\mathbf{Z}(t,x)]\,dxdt \\ &+ \int_0^T \int_{\Omega_t(V)} [\mathrm{div}\, F(t,x,u(t,x))\mathbf{Z}(t)]\,dxdt \end{aligned}$$

which we rewrite, under smoothness assumptions, as follows

$$\mathbf{j}'(V;W) = \int_0^T \int_{\Omega_t(V)} \partial_y F(t,x,u(t,x)).u'(t,x)\,dxdt$$

$$+ \int_0^T \int_{\Gamma_t(V)} F(t,x,u(t,x))\mathbf{Z}(t).n_t\,d\Gamma_t dt.$$

According to assumption ii) and the linear dependence of $\mathbf{Z}$ on W, we obtain the linear dependence of $\mathbf{j}'(V;W)$ on W.

□

Considering Λ the solution of problem (4.1)-(4.2) we can express the boundary integral on $\mathbf{Z}(t).n_t$ explicitly in terms of $W(t).n_t$. It is the object of the following lemma.

LEMMA 5.3. *Let F be a sufficiently smooth function defined on $\Sigma(V)$. Then*

$$\int_0^T \int_{\Gamma_t(V)} F(t)\,\mathbf{Z}(t).n(t)\,d\Gamma_t dt$$

$$= \int_0^T \int_{\Gamma_t(V)} \left\{ \int_t^T F(s) \circ T_s(V) \circ T_t(V)^{-1}\,ds \right\} W(t).n(t)\,d\Gamma_t dt$$

Proof.

$$\int_0^T \int_{\Gamma_t} F(t)\,\mathbf{Z}.n_t\,d\Gamma_t dt = \int_0^T < -\partial_t \Lambda - D\Lambda.V - {}^*DV.\Lambda, \mathbf{Z} > dt$$

$$= \int_0^T < \partial_t \mathbf{Z} + D\mathbf{Z}.V - DV.\mathbf{Z}, \Lambda > dt = \int_0^T \int_D W.\tilde{\lambda}\nabla\chi_{\Omega_t}\,dxdt$$

$$= -\int_0^T \int_{\Omega_t(V)} \mathrm{div}(\tilde{\lambda}W)\,dxdt$$

$$= -\int_0^T \int_{\Gamma_t(V)} \lambda(t)\,W.n_t\,d\Gamma_t dt$$

$$= \int_0^\tau \int_{\Gamma_t(V)} \{ \int_t^\tau F(s) \circ T_s \circ T_t^{-1} \, ds \} \, W(t).n_t \; d\Gamma_t dt$$

$$= \int_0^\tau \int_{\Omega_t(V)} \operatorname{div}\{W(t) \int_t^\tau F(s) \circ T_s \circ T_t^{-1} \, ds\} \; dxdt$$

$$\int_0^\tau \int_{\Gamma_t(V)} \left\{ \int_t^\tau F(s) \circ T_s(V) \circ T_t(V)^{-1} \, ds \right\} W(t).n(t) \, d\Gamma_t dt$$

□

PROPOSITION 5.4. *Under the assumptions of Lemma 5.2 and if u depends only on the trace on the lateral boundary of the field building the tube, then the gradient $G(V)$ is supported on the lateral boundary $\Sigma(V)$.*

Proof. Let W be a field such that $W(t) \in \mathcal{D}(\mathbb{R}^N, \mathbb{R}^N)$ and $\operatorname{spt} W(t) \cap \overline{\Omega_t(V)} = \emptyset$ for any $t \in [0, \tau)$. Thus $T_t(V + sW)\Omega = T_t(V)\Omega \; (= \Omega_t(V))$. Therefore $Q(V + sW) = Q(V)$. On the other-hand, $u(V + sW)(t)_{|\Gamma_t(V)} = u(V)(t)_{|\Gamma_t(V)}$ a.e. in $[0, \tau]$. The well-posedness of the PDE satisfied by u implies its uniqueness. So, in this case, we have $u(V + sW) = u(V)$ a.e. in $Q(V)$. This proves that $\mathbf{j}'(V; W) = 0$. So $\operatorname{spt} G(t, V) \subset \overline{\Omega_t(V)}$ a.e. in $[0, \tau]$. By similar arguments and considering vector fields W such that $W(t) \in \mathcal{D}(\Omega_t(V), \mathbb{R}^N)$ for any $t \in [0, \tau)$, we prove that $\operatorname{spt} G(t, V) \subset \overline{\Omega_t(V)^c}$ for any $t \in [0, \tau)$. Hence, we can conclude that $\operatorname{spt} G(., V) \subset \Sigma(V)$.

□

PROPOSITION 5.5. *Assume the assumptions of Lemma 5.2 and that u is a shape function, i.e. $u(V + W) = u(V)$ for any W s.t. $W(t).n_{\Omega_t(V)} = 0$ in $[0, \tau)$ (so it is denoted $u(Q_V)$). Then*

$$\mathbf{j}'(V; W) = 0 \quad \textit{for any } W \textit{ s.t. } W.n_{\Omega_t(V)} = 0 \;\; \forall t \in [0, \tau).$$

Proof.

$$\mathbf{j}(V + sW) = J(Q(V + sW)) = J\left(\bigcup_{0<t<\tau} [\{t\} \times \mathcal{T}_s^t \Omega_t(V)] \right)$$

where $\mathcal{T}_s^t = T_t(V + sW) \circ T_t(V)^{-1}$. The condition $W(t).n_{\Omega_t(V)} = 0$ for any $t \in [0, \tau)$ implies that $T_t(V + sW)\Omega = T_t(V)\Omega \; (= \Omega_t(V))$. It comes that $Q(V + sW) = Q(V)$. So $u_{Q(V+sW)} = u_{Q(V)}$ and $\mathbf{j}(V + sW) = J(Q(V + sW)) = J(Q(V)) = \mathbf{j}(V)$. Therefore $\mathbf{j}'(V; W) = 0$.

□

COROLLARY 5.6. *Assume Ω of class C^k, $k \geq 1$. Under the assumptions of Lemma 5.2 and that u is a shape function, there exists a time-dependent distribution g, $g(t) \in [\mathcal{D}^{k-1}(\Gamma_t(V))]'$, such that*

$$\mathbf{j}'(V; W) = \int_0^\tau < \gamma^*_{\Gamma_t(V)}(g(V)(t)n_t), W(t) >_{\mathcal{D}^{k-1}(D)', \mathcal{D}^{k-1}(D)} \, dt$$

where $\gamma^*_{\Gamma_t(V)}$ is the adjoint of the trace operator on Γ_t.

REMARK 5.1. *If there exists a vector function R with $R(t) \in L^p(\Gamma_t(V), \mathbb{R}^N)$, $p \geq 1$, such that $G(V)(t) = \gamma^*_{\Gamma_t(V)} R(t)$. Under the assumptions of Lemma 5.2 and if the density g satisfies $g(t) = R(t).n_t \in L^p(\Gamma_t(V))$. Thus we have an integral representation for the derivative*

$$\mathbf{j}'(V; W) = \int_0^\tau \int_{\Gamma_t(V)} g(V)(t) \, W(t).n_t \; d\Gamma_t \, dt.$$

5.1 Tube Functionals with Shape Density

If we consider functionals of the following form

$$\mathrm{j}(V) = J(Q(V)) = \int_0^\tau \rho(\Omega_t(V))\, dt$$

where $\rho(.)$ is a given shape functional.

THEOREM 5.7. *Assume that*

1. *$\rho(\Omega_t(V))$ is shape differentiable at $\Omega_t(V)$ for any $t \in [0,\tau]$*
2. *For any field W, $|\partial_s \rho(\Omega_t(V+sW))| \leq \xi(t)$ where ξ is integrable.*

Therefore the mapping $s \longrightarrow \jmath(V+sW)$ is differentiable at s=0 for any field W and

$$\jmath'(V;W) = \int_0^\tau \int_{\Gamma_t(V)} g(t)\, W(t).n_t\, d\Gamma_t dt.$$

Proof. Referring to Hadamard structure theorem [9], we know that there exists a function $g(t,\Gamma_t(V))$ such that

$$d\rho(\Omega_t(V);\mathbf{Z}(t)) = \int_{\Gamma_t(V)} g(t,\Gamma_t(V))\mathbf{Z}(t).n_t\, d\Gamma_t.$$

So

$$\mathrm{j}'(V;W) = \int_0^\tau \int_{\Gamma_t(V)} g(t,\Gamma_t(V))\mathbf{Z}(t).n_t\, d\Gamma_t dt$$

Using the adjoint state associated to $\mathbf{Z}$ (or to $\mathbf{Z}.n$) we can rewrite the previous expression in terms of $W(t).n_t$.

□

For an example of such a situation see below (subsection 6.2).

6 Applications

Let $\mathbf{H}_0^m(D) = \{v \in H^m(D,\mathbb{R}^3), \operatorname{div} v = 0 \text{ in } D, v.n = 0 \text{ on } \partial D\}$ with $m > (5/2)$ and the set of admissible fields $\mathcal{A}d(D) = H^1(0,\tau;\mathbf{H}_0^m(D))$.

6.1 Non-Cylindrical Heat Equation

For simplicity, assume the domain Ω and the field V sufficiently smooth. The heat equation with Neumann boundary condition in a non-cylindrical domain is

$$\partial_t u - \Delta u = f \quad \text{in} \quad Q \tag{6.1}$$

$$\frac{\partial u}{\partial n_t} = 0 \quad \text{on} \quad \Sigma \tag{6.2}$$

$$u(0) = u_0 \quad \text{in} \quad \Omega \tag{6.3}$$

THEOREM 6.1. *Let $f \in L^2(Q)$, $u_0 \in L^2(\Omega)$. There exists a unique solution of (6.1)-(6.3) in $L^2((0,\tau);H^1(\Omega_t)) \cap L^\infty((0,\tau);L^2(\Omega_t))$.*

Proof. We consider the weak formulation of (6.1)-(6.2) and then with the transformation $T_t(V)$ we deduce the weak formulation in the cylinder $(0,\tau)\times\Omega$.

$$\int_{\Omega_t} \partial_t u\,\varphi\,dx + \int_{\Omega_t} \nabla u \nabla\varphi\,dx = \int_{\Omega_t} f\varphi\,dx \quad \forall\varphi \in H^1(\Omega_t)$$

$$\int_\Omega (\partial_t u)\circ T_t \varphi\circ T_t \omega(t)\,dx + \int_\Omega < A(t)\nabla(u\circ T_t), \nabla\varphi\circ T_t > dx = \int_\Omega f\circ T_t\varphi\circ T_t\omega(t)\,dx$$

$$\begin{aligned}\int_\Omega \omega(t)[\partial_t(u\circ T_t) - < {}^*DT_t^{-1}\nabla(u\circ T_t), V > \circ T_t]\varphi\circ T_t\,dx \\ + \int_\Omega < A(t)\nabla(u\circ T_t), \nabla(\varphi\circ T_t) > dx \\ = \int_\Omega \omega(t) f\circ T_t\,\varphi\circ T_t\,dx\end{aligned}$$

which is equivalent to

$$\int_\Omega \omega(t)[\partial_t(u\circ T_t) - < {}^*DT_t^{-1}\nabla(u\circ T_t), V\circ T_t >]\psi\,dx + \int_\Omega < A(t)\nabla(u\circ T_t), \nabla\psi > dx$$
$$= \int_\Omega \omega(t) f\circ T_t\psi\,dx$$

In a distribution sense, it comes,

$$\partial_t(u\circ T_t) - \omega(t)^{-1}\operatorname{div}[A(t)\nabla(u\circ T_t)] - < DT_t^{-1}V\circ T_t, \nabla(u\circ T_t) > = f\circ T_t \quad \text{in } (0,\tau)\times\Omega$$

and the corresponding boundary condition is

$$< A(t)\nabla(u\circ T_t), n > = 0 \quad \text{on } (0,\tau)\times\Gamma$$

The initial condition remains unchanged.

Thus we are led to consider

$$\partial_t v - \operatorname{div}[\omega(t)^{-1}A(t)\nabla v] + < A(t)\nabla\omega(t)^{-1} - DT_t^{-1}V\circ T_t, \nabla v > = f\circ T_t \tag{6.4}$$
$$< A(t)\nabla v, n > = 0 \quad \text{on } (0,\tau)\times\Gamma \tag{6.5}$$
$$v(0) = u_0 \text{ on } \Omega \tag{6.6}$$

For existence result a priori estimates should be available in

$$L^2(0,\tau;H^1(\Omega)) \cap L^\infty(0,\tau;L^2(\Omega))$$

$$\int_\Omega \partial_t v\, v\, dx + \int_\Omega < \omega(t)^{-1}A(t)\nabla v, \nabla v > + < A(t)\nabla\omega(t)^{-1} - DT_t^{-1}V\circ T_t, \nabla v > v\, dx$$
$$= \int_\Omega f\circ T_t v\, dx$$

$$\frac{1}{2}\int_\Omega \partial_t v^2\, dx + \int_\Omega < \omega(t)^{-1}A(t)\nabla v, \nabla v > - < DT_t^{-1}V\circ T_t - A(t)\nabla\omega(t)^{-1}, \nabla v > v\, dx$$
$$= \int_\Omega f\circ T_t v\, dx$$

$$\frac{1}{2}\frac{d}{dt}\int_\Omega v^2 dx + \int_\Omega < \omega(t)^{-1}A(t)\nabla v, \nabla v > - < DT_t^{-1}V\circ T_t - A(t)\nabla\omega(t)^{-1}, \nabla v > v\, dx$$
$$= \int_\Omega f\circ T_t\, v\, dx.$$

Let $t\in(0,\tau]$. By integration over $(0,t)$, we get

$$\frac{1}{2}\int_\Omega v^2 dx + \int_0^t\int_\Omega < \omega(t)^{-1}A(t)\nabla v, \nabla v > dxdt -$$
$$\int_0^t\int_\Omega < DT_t^{-1}V\circ T_t - A(t)\nabla\omega(t)^{-1}, \nabla v > v\, dxdt$$
$$= \int_0^t\int_\Omega f\circ T_t\, v\, dxdt + \frac{1}{2}\int_\Omega u_0^2 dx.$$

$$|v(t)|^2 + 2\alpha_0\int_0^t |\nabla v|^2\, dt - 2c_1\int_0^t |v||\nabla v|\, dt \le 2\int_0^t |f||v|\, dt + \frac{1}{2}\int_\Omega u_0^2 dx.$$

Using Young inequality with $\varepsilon = \alpha_0/c_1$, it comes

$$|v(t)|^2 + \alpha_0\int_0^t |\nabla v|^2\, dt - \frac{c_1^2}{\alpha_0}\int_0^t |v|^2\, dt \le 2|f|_{L^2(Q)}\int_0^t |v|\, dt + \frac{1}{2}\int_\Omega u_0^2 dx.$$

From this inequality a priori estimates in $L^2((0,\tau), H^1(\Omega)) \cap L^\infty((0,\tau), L^2(\Omega))$, for v, can be derived. Thus we can apply the classical Galerkin method for the existence result. Uniqueness is immediate.

□

PROPOSITION 6.2 (REGULARITY). *Assume that* $f\in L^2(0,\tau;H^2(\Omega_t))\cap H^1(0,\tau;L^2(\Omega_t))$, $u_0\in H^4(\Omega)$ *and V(0)=0. Moreover* u_0 *should satisfy the following compatibility conditions*

$$-\Delta u_0 = f(0,x) \quad in\ \Omega \tag{6.7}$$
$$\frac{\partial u_0}{\partial n} = 0 \quad in\ \Gamma. \tag{6.8}$$
$$\tag{6.9}$$

Then the solution u of (6.1)-(6.3) belongs to $L^2((0,\tau);H^4(\Omega_t))\cap H^2(0,\tau;L^2(\Omega_t))$.

Proof. This result is a particular case of Theorem 6.2, p. 40, given by J.-L. Lions and E. Magenes in [8].

□

Well-posedness and regularity results are ensured, we may now consider the sensitivity analysis of u with respect to perturbations of the field V which constructs the tube. Assume $f \in L^2(0,\tau;H^2(D))$ and $u_0 \in H^4(D)$.

PROPOSITION 6.3. *The mapping $s \to u^s = u(V+sW)\circ T_s(\mathcal{Z}^t(s))$ is differentiable at $s=0$ in $L^2(0,\tau;H^1(\Omega_t))$. Moreover under regularity hypotheses the function $u' = \partial_s u^s_{|s=0} - \nabla u.Z$ is solution of*

$$\partial_t u' - \Delta u' = 0 \quad in\ Q \tag{6.10}$$

$$\frac{\partial u'}{\partial n_t} = (-\partial_t u + f)Z(t).n_t + \mathrm{div}_{\Gamma_t}(Z(t).n_t\nabla_{\Gamma_t} u) \quad on\ \Sigma \tag{6.11}$$

$$u'_0 = 0 \quad in\ \Omega \tag{6.12}$$

Proof. To prove the differentiability result two approaches can be considered. Using the implicit function theorem or else by manipulations of the equation satisfied by the differential quotient, see for example [9] or [6] . The transformation used to come back to $Q(V)$ is $T_s(\mathcal{Z}^t(s))$ since we considered first the equation in $Q(V+sW)$.

□

The purpose of our study is to prove existence and compute the Eulerian derivative of the functional

$$\mathrm{j}(V) = J(Q_V) = \frac{1}{2}\int_{Q_V} |u|^2\, dxd \quad u \text{ is solution of } (6.1)-(6.3).$$

Practically it consists in giving perturbations, in a direction W, to $Q(V)$ and consider

$$\lim_{s\searrow 0}\{J(Q(V+sW)) - J(Q(V))\}/s$$

PROPOSITION 6.4. *The mapping $s \to \jmath(V+sW) = J(Q(V+sW))$ is differentiable at $s=0$ and*

$$\jmath'(V;W) = \int_0^\tau \int_{\Gamma_t(V)} \left\{\int_t^\tau F(s)\circ T_s(V)\circ T_t(V)^{-1}\, ds\right\} W(t).n_t\, d\Gamma_t dt$$

where $F(t) = \frac{1}{2}|u(t)|^2 - \partial_t u + \nabla_{\Gamma_t} u.\nabla_{\Gamma_t} p + f$.

Notice that u, solution of (6.1)-(6.3), depends on V only through $Q(V)$. It is, so, denoted $u_{Q(V)}$.

Proof.

$$\mathrm{j}(V+sW) = \frac{1}{2}\int_{Q(V+sW)} |u_{Q(V+sW)}|^2\, dxdt$$

From Proposition 6.3

$$\frac{d}{ds}\mathrm{j}(V+sW)_{|s=0} = \int_0^\tau \int_{\Omega_t(V)} u\, u'\, dxdt + \frac{1}{2}\int_0^\tau \int_{\Gamma_t(V)} |u|^2 \mathbf{Z}(t).n_t\, d\Gamma_t dt$$

The adjoint state corresponding to that functional is the solution p of the following retrograde equation

$$-\partial_t p - \Delta p = u \quad \text{in} \quad Q \tag{6.13}$$

$$\frac{\partial p}{\partial n_t} - pV(t).n_t = 0 \quad \text{on} \quad \Sigma \tag{6.14}$$

$$p(\tau) = 0 \quad \text{in} \quad \Omega_\tau \tag{6.15}$$

For existence result of (6.14)-(6.15), proceed as in the proof of Proposition 6.1. Therefore

$$\int_0^\tau \int_{\Omega_t(V)} u\, u'\, dxdt = \int_0^\tau \int_{\Omega_t(V)} (-\partial_t p - \Delta p) u'\, dxdt.$$

The calculus leads to

$$\int_0^\tau \int_{\Omega_t(V)} u\, u'\, dxdt = \int_0^\tau \int_{\Gamma_t(V)} \frac{\partial u'}{\partial n_t} p - \frac{\partial p}{\partial n_t} u'\, d\Gamma_t dt + \int_0^\tau \int_{\Gamma_t(V)} pu'V(t).n_t\, d\Gamma_t dt.$$

Then

$$\int_0^\tau \int_{\Omega_t(V)} u\, u'\, dxdt = \int_0^\tau \int_{\Gamma_t(V)} \frac{\partial u'}{\partial n_t} p\, d\Gamma_t dt.$$

By replacing the normal derivative of u' by its expression in terms of u and $\mathbf{Z}$ we obtain

$$\mathbf{j}'(V;W) = \int_0^\tau \int_{\Gamma_t(V)} \left[\frac{1}{2}|u|^2 - \partial_t u + \nabla_{\Gamma_t} u.\nabla_{\Gamma_t} p + f\right] \mathbf{Z}(t).n_t\, d\Gamma_t dt.$$

□

6.2 In Hydrodynamics

Consider, for $V \in H^1(0,\tau;\mathbf{H}_0^m(D))$, the functional

$$\mathrm{j}(V) = \frac{1}{2}\left\{\int_{Q(V)} |\nabla\varphi(V)|^2\, dxdt\right\} \tag{6.16}$$

where $\varphi(V)$, which also may be denoted φ_V, is the solution of

$$\Delta\varphi_V(t) = 0 \quad \text{in } \Omega_t(V) \tag{6.17}$$

$$\frac{\partial\varphi_V}{\partial n_t}(t) = V(t).n_t \quad \text{on } \Gamma_t(V) \tag{6.18}$$

Let $\varphi_s = \varphi(V + sW)$, $s > 0$ in a neighborhood of 0, be the solution of (6.17)-(6.18) when V is replaced by $(V + sW)$. Notice that $\varphi_V(t)$ depends only on $V(t).n_t$. Therefore $\varphi_V = \varphi_{Q(V)}$.

THEOREM 6.5. *The mapping $s \to \varphi_s(t)$ is differentiable at $s = 0$. That derivative, denoted $\varphi'(t)$, is the solution of*

$$\begin{aligned} \Delta\varphi'(t) &= 0 \quad \text{in} \quad \Omega_t(V) \\ \frac{\partial}{\partial n_t}\varphi'(t) &= \mathrm{div}_\Gamma[Z(t).n_t(\nabla_\Gamma\varphi(t) - V_\Gamma)] + Z(t).n_t \quad \text{on} \quad \Gamma_t(V). \end{aligned}$$

Let us now compute the derivative of $\mathbf{j}$ on V in the direction W.

$$\mathbf{j}'(V;W) = \lim_{s\to 0}\frac{\mathbf{j}(V+sW)-\mathbf{j}(V)}{s} = \int_{Q(V)} \nabla\varphi'.\nabla\varphi\, dxdt + \frac{1}{2}\int_0^{\tau}\int_{\Gamma_t(V)} |\nabla\varphi|^2\, \mathbf{Z}.n_t\, d\Gamma dt.$$

But

$$\int_{\Omega_t} \nabla\varphi'.\nabla\varphi\, dx = \int_{\Gamma_t} \frac{\partial\varphi'}{\partial n_t}\varphi\, d\Gamma = \int_{\Gamma_t} \{\mathrm{div}_{\Gamma_t}[\mathbf{Z}.n_t(\nabla_{\Gamma_t}\varphi - V_{\Gamma_t})] + W(t).n_t\}\,\varphi\, d\Gamma_t.$$

$$= -\int_{\Gamma_t} \nabla_{\Gamma_t}\varphi(\nabla_{\Gamma_t}\varphi - V_{\Gamma_t})\,\mathbf{Z}.n_t + W(t).n_t\,\varphi\, d\Gamma_t.$$

Therefore

$$\mathbf{j}'(V;W) = \int_0^{\tau}\int_{\Gamma_t(V)} W(t).n_t\,\varphi\, d\Gamma_t dt - \int_0^{\tau}\int_{\Gamma_t(V)} \nabla_{\Gamma_t}\varphi(\nabla_{\Gamma_t}\varphi - V_{\Gamma_t})\,\mathbf{Z}.n_t\, d\Gamma_t dt$$
$$+\frac{1}{2}\int_0^{\tau}\int_{\Gamma_t(V)} |\nabla\varphi|^2\mathbf{Z}.n_t\, d\Gamma dt.$$

So $F(t) = -\nabla_{\Gamma_t}\varphi(t).(\nabla_{\Gamma_t}\varphi(t) - V_{\Gamma_t}(t)) + \frac{1}{2}|\nabla\varphi(t)|^2$. Following the proof of theorem 5.2, we deduce

$$\mathbf{j}'(V,W) = \int_0^{\tau}\int_{\Gamma_t(V)} \left[\{\int_t^{\tau} F(s)\circ T_s(V)\circ T_t(V)^{-1}\, ds\} + \varphi(t)\right] W(t).n_t\, d\Gamma_t\, dt.$$

Details of the proofs can be found in [5].

6.3 In Viscous Flow

This section is devoted to the study of the differentiability of

$$\mathbf{j}(V) = \frac{1}{2}\left\{\int_{Q(V)} |u(V)|^2\, dxdt\right\} \tag{6.19}$$

where $V \in H^1(0,\tau;\mathbf{H}_0^m(D))$ and $u(V)$ is the solution of

$$\partial_t u - \eta\Delta u + Du.u + \nabla p = f \quad \text{in } Q \tag{6.20}$$
$$\mathrm{div}\, u = 0 \quad \text{in } Q \tag{6.21}$$
$$u = V \quad \text{on } \Sigma, \tag{6.22}$$

where η is the coefficient of kinematic viscosity of the fluid. A domain Ω and a function u_0 being given, we assume that

$$\text{at} \quad t=0, \quad \Omega_0 = \Omega \quad \text{and} \quad u(0) = u_0 \text{ in } \Omega. \tag{6.23}$$

Denote by $U = u - V$ the solution of the associate homogeneous problem.

PROPOSITION 6.6. *The weak derivative* $\dot{U} = \partial_s((D\mathcal{T}_s^t)^{-1}\, U_s \circ \mathcal{T}_s)_{|s=0}$ *exists in*

$$L^2(0,\tau; H^1(\mathrm{div}, \Omega_t(V)))$$

Here we assume Ω of class C^2. Note that since the field V is smooth, $\Omega_t(V)$ has the same regularity than Ω for any $t \in (0,\tau)$. If we assume regularity results for Navier-Stokes equations (see for instance [10]) still valid for the non-cylindrical case i.e. $U(t) \in H^2(\Omega_t(V), \mathbb{R}^3)$ for a.e. $t \in (0, \tau)$. Then

PROPOSITION 6.7. *The field* $U'(t) = \dot{U}(t) - DU(t).\boldsymbol{Z}(t)$, $t \in (0,\tau)$, *satisfies*

$$\begin{aligned}
&\partial_t U' - \eta\Delta U' + DU'.U + DU.U' + DU'.V + DV.U' = -\partial_t W + \eta\Delta W \\
&-DW.U - DU.W - DV.W - DW.V \quad in \quad Q(V) \\
&\operatorname{div} U' = 0 \qquad in \quad Q(V) \\
&U'(t) = - \; < \boldsymbol{Z}(t), n_t > \; DU(t).n_t \quad on \quad \Sigma(V) \\
&U'(0) = 0 \quad in \quad \Omega.
\end{aligned}$$

We set $u' = U' + W$, then

COROLLARY 6.8. *The field* $u'(t) = \dot{U}(t) - DU(t).\boldsymbol{Z}(t) + W$, $t \in (0,\tau)$, *satisfies*

$$\begin{aligned}
&\partial_t u' - \eta\Delta u' + Du'.u + Du.u' = 0 \quad in \quad Q(V) \\
&\operatorname{div} u' = 0 \qquad in \quad Q(V) \\
&u'(t) =< \boldsymbol{Z}(t), n_t > \; (DV(t) - Du(t)).n_t + W(t) \quad on \quad \Sigma(V) \\
&u'(0) = 0 \quad in \quad \Omega.
\end{aligned}$$

The derivative of $\mathbf{j}$ at V in the direction W is

$$\mathbf{j}'(V;W) = \lim_{s\to 0} \frac{J(V+sW) - J(V)}{s}$$

$$\mathbf{j}'(V;W) = \int_{Q(V)} u\, u'\, dxdt + \int_0^\tau \int_{\Gamma_t(V)} |V|^2 \mathbf{Z}.n_t \, d\Gamma_t dt.$$

At this stage we need to introduce the adjoint state associated to U' (we obtain automatically the one associated to u' since $u' = U' + W$). So let $\mathbf{U}$ this adjoint state. It satisfies

$$\begin{aligned}
&-\partial_t \mathbf{U} - \eta\Delta\mathbf{U} - D\mathbf{U}.u + {}^*Du.\mathbf{U} \; = u \quad \text{in } Q(V) \\
&\operatorname{div} \mathbf{U} = 0 \quad \text{in } Q(V) \\
&\mathbf{U} = 0 \quad \text{on } \Sigma(V) \\
&\mathbf{U}(\tau) = 0 \quad \text{in } \Omega_\tau(V).
\end{aligned}$$

Then,

$$\begin{aligned}
\int_{Q(V)} u.u'\, dxdt \;&=\; \int_{Q(V)} (-\partial_t \mathbf{U}, u') - \eta\, \Delta\mathbf{U}.u' - < D\mathbf{U}.u, u' > + <^* Du\mathbf{U}, u' > \, dxdt \\
&= \int_0^\tau \int_{\Gamma_t(V)} < \eta\epsilon(\mathbf{U}).n_t \; - V(t).n_t \; \mathbf{U}\,, u' > \, d\Gamma_t dt
\end{aligned}$$

$\mathbf{j}'(V;W)$ can be expressed as

$$\int_0^T \int_{\Gamma_t(V)} < \eta\, \epsilon(\mathbf{U}).n_t \; - V(t).n_t \; \mathbf{U}, \; \mathbf{Z}(t).n_t \; D(V-u)(t).n_t + W(t) > \; d\Gamma_t dt$$

$$+ \int_0^T \int_{\Gamma_t(V)} |V|^2 \mathbf{Z}.n_t \, d\Gamma_t dt$$

But $\mathbf{U} = 0$ on Σ then the expression is simplified into the following one:

$$\begin{aligned} \mathbf{j}'(V,W) &= \; < \eta\, \epsilon(\mathbf{U}).n_t \,, W(t) > \; d\Gamma_t dt \\ &+ \int_0^T \int_{\Gamma_t(V)} [< \eta\, \epsilon(\mathbf{U}).n_t \,, D(V-u)(t)\,.n_t > + |V|^2] \; \mathbf{Z}.n_t \, d\Gamma_t dt \\ &= \int_0^T \int_{\Gamma_t(V)} \{ \int_t^T F(s) \circ T_s(V) \circ T_t(V)^{-1} \, ds \} \, W(t).n_t \, d\Gamma_t dt \\ &+ \int_0^T \int_{\Gamma_t(V)} < \eta\, \epsilon(\mathbf{U}).n_t \,, W(t) > d\Gamma_t dt. \end{aligned}$$

where $F(s) = < \eta\varepsilon(\mathbf{U}).n_s \,, D(V-u).n_s > + |V(s)|^2$.

Notice here that $\mathbf{j}'(V;W)$ depends on the tangential component of $W(t)$. This comes from the fact that the flow velocity sticks to the boundary. This problem is studied in details in [7].

Bibliography

[1] J.-P. Aubin. *Mutational and morphological analysis : tools for shape evolution and morphogenesis.* Birkhauser, 1998.

[2] M.-C. Delfour and J.-P. Zolésio. Structure of shape derivatives for non smooth domains. *Journal of Functional Analysis*, 104, 1992.

[3] M. C. Delfour and J.-P. Zolésio. Shape analysis via oriented distance functions. *J. Funct. Anal.*, 123(1):1–16, 1994.

[4] F. R. Desaint and J.-P. Zolésio. Manifold derivative in the laplace-beltrami equation. *J. Funct. Anal.*, 151(1):234–269, 1997.

[5] R. Dziri and J.-P. Zolésio. Shape existence in navier-stokes flow with heat convection. *Ann. della Scuola Normale di Pisa*, XXIV:165–192, 1997.

[6] R. Dziri and J.-P. Zolésio. Dynamical shape control in non-cylindrical hydrodynamics. *Inverse Probl.*, 15(1):113–122, 1999.

[7] R. Dziri and J.-P. Zolésio. Dynamical shape control in non-cylindrical navier-stokes equations. *J. Convexe Anal.*, 1999. To appear.

[8] J.-L. Lions and E. Magenes. *Problèmes aux limites non homogènes*, volume 2. Dunod, 1968.

[9] J. Sokolowski and J.-P. Zolésio. *Introduction to Shape Optimization.* SCM 16. Springer-Verlag, 1992.

[10] R. Temam. *Theory and numerical analysis of the Navier-Stokes equations.* North-Holland, 1977.

[11] J.-P. Zolésio. Identification de domaines par déformations. *Ph. D. thesis, Nice*, 1979.

[12] J.-P. Zolésio. Weak form of the shape differential equation. In J. Burns, editor, *Proc. of the AFORS Conf., September 97, Washington, Birkhauser*, 1997.

Raja Dziri. Université de Tunis, Département de Mathématiques, Campus Universitaire 1060, Tunis, Tunisia
E-mail: Raja.Dziri@fst.rnu.tn

Jean-Paul Zolésio. Research Director at CNRS, Centre de Mathématiques Appliquées, Ecole des Mines de Paris, 2004 route des Lucioles, BP. 93, 06902 Sophia Antipolis Cedex, France
E-mail: Jean-Paul.Zolesio@sophia.inria.fr

Simultaneous Exact/Approximate Boundary Controllability of Thermo-Elastic Plates with Variable Transmission Coefficient

Matthias Eller, Irena Lasiecka[1] and Roberto Triggiani[1]

Abstract. We study a controllability problem (exact in the mechanical variables $\{w, w_t\}$ and, simultaneously, approximate in the thermal variable θ) of thermo-elastic plates by means of boundary controls, in both the hinged/Dirichlet, or the clamped/Dirichlet B.C. cases, when the 'thermal expansion' term is variable in space.

1 Thermo-Elastic Systems. Boundary Controllability Problem

Let $\Omega \subset \mathbb{R}^2$ be an open bounded domain with smooth boundary Γ. We shall here consider the following thermo-elastic plate [La.1], [L-L.1] on a finite time interval in the unknown $w(t,x)$ [vertical displacement] and $\theta(t,x)$ [relative temperature about the stress-free state $\theta = 0$]:

$$\begin{cases} w_{tt} - \gamma\Delta w_{tt} + \Delta^2 w + \operatorname{div}(\alpha(x)\nabla\theta) = 0 & \text{in } Q \equiv (0,T] \times \Omega; \quad (1.1a) \\ \theta_t - \Delta\theta - \operatorname{div}(\alpha(x)\nabla w_t) = 0 & \text{in } Q; \quad (1.1b) \\ w(0,\,\cdot\,) = w_0;\ w_t(0,\,\cdot\,) = w_1;\ \theta(0,\,\cdot\,) = \theta_0 & \text{in } \Omega, \quad (1.1c) \end{cases}$$

to be augmented by boundary conditions on $\partial\Omega = \Gamma$. Throughout this paper the constant γ is positive, $\gamma > 0$, so that the model accounts for moments of inertia (rotational forces). The coefficient γ is proportional to the square of the thickness of the plate. The thermal coefficient $\alpha(x)$ represents 'thermal expansion' of the plate material and is assumed to be a function of $x \in \Omega$, and of class $C^2(\bar{\Omega})$. We shall consider two cases of *boundary* controls associated with system (1.1): the case where the boundary controls act in the *hinged mechanical/Dirichlet thermal boundary conditions (B.C.):*

$$w|_\Sigma = u_1; \qquad \Delta w|_\Sigma = u_2; \qquad \theta|_\Sigma = u_3 \quad \text{on } \Sigma = (0,T] \times \Gamma, \tag{1.2}$$

[1]Research partially supported by the National Science Foundation under Grant DMS-9804056 and by the Army Research Office under grant DAAH04-96-1-0059

and the case where the boundary controls act in the *clamped mechanical/Dirichlet thermal B.C.*:

$$w|_\Sigma = u_1; \qquad \left.\frac{\partial w}{\partial \nu}\right|_\Sigma = u_2; \qquad \theta|_\Sigma = u_3 \quad \text{on } \Sigma. \tag{1.3}$$

Well-posedness of the two mixed problems above: (1.1), (1.2), and (1.1), (1.3) is discussed below. [The analysis of the paper works verbatim on R^n, for any $n \geq 2$.

Boundary controllability problem. Qualitatively, the boundary controllability problem studied in this note is as follows. Let $T > 0$ be sufficiently large, depending on the geometry of Ω. Given any initial condition $\{w_0, w_1, \theta_0\}$ and any preassigned target condition $\{w_{0,T}, w_{1,T}, \theta_T\}$ in specified Sobolev spaces, seek boundary controls $\{u_1, u_2, u_3\}$ in specified function spaces (compatible with the regularity of the underlying dynamics) that steer the solution of the corresponding mixed problem (1.1), (1.2), or (1.3) to a state $\{w(T), w_t(T), \theta(T)\}$ at time T, such that: $w(T) = w_{0,T}$, $w_t(T) = w_{1,T}$, while $\theta(T)$ is arbitrarily 'close' to θ_T in the relevant topology. Thus, the above is a problem of *exact controllability* in the mechanical variable and, simultaneously, of *approximate controllability* in the thermal variable. A more precise statement is given in the following theorems.

Main results. Our main results on the above boundary exact/approximate controllability problem follow next.

THEOREM 1.1. (hinged B.C. (1.2)) *Let $\Gamma_1, \Gamma_2, \Gamma_3 \subset \Gamma$ be open subsets of the boundary Γ, with a non-empty intersection of positive measure. (We think of Γ_1 and Γ_3 as being arbitrarily small.) Moreover, regarding Γ_2, we assume that: there exists a point $x_0 \in \mathbb{R}^2$, such that*

$$(x - x_0) \cdot \nu(x) \leq 0 \quad \textit{for } x \in \Gamma \setminus \Gamma_2, \tag{1.4}$$

where here and throughout the paper $\nu(x)$ denotes the unit outward normal at $x \in \Gamma$. Let

$$T_0 \equiv 2\sqrt{\gamma} \max_i \sup_{x \in \Omega} \textit{dist}(x, \Gamma_i), \qquad i = 1, 2, 3. \tag{1.5}$$

Let $\alpha \in C^2(\bar{\Omega})$. Finally, let $\{w_0, w_1, \theta_0\}$ and $\{w_{0,T}, w_{1,T}, \theta_T\}$ be pre-assigned initial and target states, with:

$$\{w_0, w_1\} \textit{ and } \{w_{0,T}, w_{1,T}\} \in [H^2(\Omega) \cap H_0^1(\Omega)] \times H_0^1(\Omega); \qquad \theta_0, \theta_T \in H_0^1(\Omega). \tag{1.6}$$

Then, for any $T > T_0$ and any $\delta > 0$ arbitrarily small, there exist boundary control functions

$$u_1 \equiv \begin{cases} \hat{u}_1 \in C_0^\infty(\Sigma_1) \\ 0 \textit{ on } \Sigma - \Sigma_1 \end{cases}; \quad u_2 \equiv \begin{cases} \hat{u}_2 \in L_2(\Sigma_2) \\ 0 \textit{ on } \Sigma - \Sigma_2 \end{cases}; \quad u_3 \equiv \begin{cases} \hat{u}_3 \in C_0^\infty(\Sigma_3) \\ 0 \textit{ on } \Sigma - \Sigma_3 \end{cases} \tag{1.7}$$

[in particular, with all time derivatives $u_i^{(n)}(0) = u_i^{(n)}(T) = 0$, $i = 1, 3$, and all $n = 0, 1, 2,$ vanishing at $t = 0$, and $t = T$], such that the corresponding solution of the mixed problem (1.1), (1.2) satisfies the following terminal condition at T:

$$w(T) = w_{0,T}; \qquad w_t(T) = w_{1,T}; \qquad \|\theta(T) - \theta_T\|_{H_0^1(\Omega)} \leq \delta. \tag{1.8}$$

THEOREM 1.2. (clamped B.C. (1.3)) *Let $\Gamma_1, \Gamma_3 \subset \Gamma$ be, as in Theorem 1.1, open subsets of the boundary Γ, with non-empty intersection of positive measure, possibly arbitrarily small. Let $T_0 > 0$ be defined by (1.5) with $i = 1, 3$, and let $\alpha \in C^2(\bar{\Omega})$. Finally, let $\{w_0, w_1, \theta_0\}$ and $\{w_{0,T}, w_{1,T}, \theta_T\}$ be preassigned initial and target states, with*

$$\{w_0, w_1\} \text{ and } \{w_{0,T}, w_{1,T}\} \in H_0^1(\Omega) \times \tilde{L}_2(\Omega); \qquad \theta_0, \theta_T \in H_\theta = L_2(\Omega) \tag{1.9}$$

the space $\tilde{L}_2(\Omega)$ being defined in (1.12)-(1.13) below.

Then, for any $T > T_0$ and any $\delta > 0$ arbitrarily small, there exist boundary control functions

$$u_1 \equiv \begin{cases} \hat{u}_1 \in C_0^\infty(\Sigma_1) \\ 0 \text{ on } \Sigma - \Sigma_1 \end{cases}; \quad u_2 \in L_2(\Sigma); \quad u_3 \equiv \begin{cases} \hat{u}_3 \in C_0^\infty(\Sigma_3) \\ 0 \text{ on } \Sigma - \Sigma_3 \end{cases}, \tag{1.10}$$

such that the corresponding solution of the mixed problem (1.1), (1.3) satisfies the terminal condition at T:

$$w(T) = w_{0,T}; \qquad w_t(T) = w_{1,T}; \qquad \|\theta(T) - \theta_T\|_{H_\theta} \leq \delta. \tag{1.11}$$

REMARK 1.1. *The space $\tilde{L}_2(\Omega)$, first noticed in [Las.1], can be defined by duality as follows. Let $\mathcal{A}f = -\Delta f$, $\mathcal{D}(\mathcal{A}) = H^2(\Omega) \cap H_0^1(\Omega)$ and $\mathcal{A}_\gamma = 1 + \gamma\mathcal{A}$ as in (2.4) and (2.1a) below, and let $Af = \Delta^2 f$, $\mathcal{D}(A) = H^4(\Omega) \cap H_0^2(\Omega)$ as in (2.4a) below. Then $\tilde{L}_2(\Omega)$ is defined as*

$$\begin{aligned} \tilde{L}_2(\Omega) &\equiv \text{dual of the space } \mathcal{D}(A^{\frac{1}{2}}) \text{ with respect to the space } \mathcal{D}(\mathcal{A}_\gamma^{\frac{1}{2}}) \\ &\quad \text{as a pivot space, with norm } \|f\|_{\mathcal{D}(\mathcal{A}_\gamma^{\frac{1}{2}})} = \|\mathcal{A}_\gamma^{\frac{1}{2}} f\|_{L_2(\Omega)} && (1.12) \\ &\cong L_2(\Omega)/\mathcal{H} \cong \mathcal{H}^\perp \text{ (isometric identifications)} && (1.13) \\ \mathcal{H} &\equiv \{f \in L_2(\Omega) : (1 - \gamma\Delta)f = 0 \text{ in } H^{-2}(\Omega)\}. && (1.14) \end{aligned}$$

More specifically, we have

$$g \in \tilde{L}_2(\Omega) \Longleftrightarrow A^{-\frac{1}{2}}\mathcal{A}_\gamma g \in L_2(\Omega) \text{ or } g = \left[(A^{-\frac{1}{2}}\mathcal{A}_\gamma)\Big|_{\mathcal{H}^\perp} \right]^{-1} F, \quad F \in L_2(\Omega). \tag{1.15}$$

REMARK 1.2. *We note that, in the above two theorems, the essential, critical control mechanism is provided by the control u_2 in the highest mechanical B.C.: with $u_2 \in L_2(\Sigma_2)$ on a suitable portion of the boundary (see (1.7)) in the hinged case, and $u_2 \in L_2(\Sigma)$ on the full boundary in the clamped case (see (1.10)). The addition of infinitely smooth controllers u_1 and u_3 in the lowest mechanical B.C. and in the thermal B.C., compactly supported on $\Sigma_1 = (0, T] \times \Gamma_1$ and $\Sigma_3 = (0, T] \times \Gamma_3$, with Γ_1, Γ_3* arbitrarily small *open portions of the boundary having non-empty intersection with Γ_2, is only for the purpose of obtaining the property of 'approximate controllability' of the overall thermo-elastic plate. By duality (Hahn-Banach theorem), this latter property is equivalent to the property of* unique continuation *across the boundary of a corresponding over-determined dual or adjoint problem, see Theorem 4.2.2.*

At present, the results of this paper, which improve over the literature [I.1], require that the dual problem be over-determined with all *four boundary conditions on a common, non-empty open portion of the boundary of positive measure, in order to assert that, then, the corresponding solution is identically zero. See the statement of Theorem 4.2.2. This is the reason why we assume three active controls in (1.2), or (1.3), in lieu of just* u_2. *However, any progress in the area of the unique continuation property for thermo-elastic plates will imply corresponding improvements of our results, by allowing us to drop unnecessary controls, such as, possibly,* u_1 *and* u_3. *See Remark 4.2.3 below.*

We explicitly remark that since we think of Γ_1 *and* Γ_3 *as arbitarily small, with* $\Gamma_1 \cap \Gamma_2 \cap \Gamma_3 \neq \emptyset$, *we may always assume without loss of generality that:* $\emptyset \neq \tilde{\Gamma} \equiv \Gamma_1 = \Gamma_3 \subset \Gamma_2$, *where* $\Gamma_2 = \Gamma$ *in the clamped case.*

REMARK 1.3. *Theorem 1.1 covering the hinged B.C. (1.2) was recently announced in [E-L-T.1]. A complete proof of this case is included here.*

REMARK 1.4. *Many, in fact most, steps in the proof of this paper continue to hold true for a much more general thermo-elastic model than (1.1), where now* $(-\Delta)$ *is replaced by a* space *variable coefficient elliptic operator of order two, which is a positive, self-adjoint operator. These steps include: abstract models; the background uniqueness property quoted from [L-R-T.1]; the critical estimates for wave equations used in Section 6 on Eqn. (6.10), and in Section 7, Eqn. (7.12), where the required estimates would follow from [L-T-Y.2], or [L-T-Y.1] plus Appendix E, etc. However, to date, we have not checked in details if the entire proof of the present paper carries over to the fully space variable coefficient case. The Euclidean distance in formula (1.5) will have to be replaced with the distance in a suitable Riemann metric [L-T-Y.1].*

Literature. The problem of controlling thermo-elastic plates has already received attention in the literature [La.1]. Since a thermo-elastic plate consists of a Kirchoff plate coupled with a heat equation, it is natural to view controllability of a thermo-elastic plate as a 'perturbation' of the controllability of an elastic Kirchoff plate. In fact, this strategy works well in the case of *distributed* (internal) control, where the control operator is bounded on the basic state space. Here a classical approach (with roots in finite-dimensional theory) can be used to make the thermo-elastic plate inherit controllability properties from the elastic Kirchoff plate [T-Z.1]. Here, the *distributed* control acts on a layer of the boundary in the mechanical Kirchoff equation and yields exact controllability in the mechanical variables and, simultaneously approximate controllability in the thermal variable in the constant coefficient case. A stronger result—exact controllability for both the mechanical and thermal variables—is obtained in [Av.1], where, moreover, the *distributed* control acts in the thermal equation (only), still in the constant coefficient case. The situation is quite different in the case of *boundary* controls, where then the control operator is highly unbounded. For boundary control a first contribution to the controllability problem of steering an initial state to rest was solved in [H-Z.1], in the one-dimensional case with coupling coefficient $\alpha \equiv$ constant. Then, a direct spectral analysis, based on eigenfunction expansion, yields, via the usual moment problem, the desired result. The present paper aims at the higher (arbitrary) dimensional case, with a thermal coefficient which is variable in space.

To our knowledge, the only papers in the literature in the area of *boundary* controllability of *thermo-elastic plates*—to which we must restrict—are [La.2] and [A-L.1], [Liu.1]. In [La.2], thermo-elastic plates with *constant* coefficients and *boundary* controls in the free B.C. were considered. Here, the main theorem established is an exact controllability result in the mechanical variables $\{w, w_t\}$ *only*, with two mechanical controls in L_2 and with no information about the thermal variable θ, subject, moreover, to the condition that the (constant) thermal parameters be sufficiently *small*. Such smallness of the thermal connection plays an essential role in the arguments of [La.2].

For a thermo-elastic *wave* (rather than plate), paper [Liu.1] originally claimed a result of partial exact controllability with no assumption of smallness on the coupling parameter. However, this claim was later retracted [Liu.1] and corrected with the statement that, in fact, the coupling parameter must be sufficiently small, in the style of [La.2]. The first result on exact/approximate boundary controllability of mechanical/thermal variables for the same thermo-elastic plate with controls in the free B.C., which *does not* require any smallness hypothesis on the model, is [A-L.1]. Here, however, the coefficients, in particular α, are constant. This assumption is also critical to the arguments of [A-L.1], as it allows for the introduction of a certain transformation of variables, to make the problem more tractable. On the other hand, it is known that observability/controllability estimates are sensitive with respect to variable perturbations of the "energy level" terms in the equations. In fact, even for simple *plate* equations or *wave* equations, standard energy methods (multipliers) [Tr.1], [L-T.1], [Li.1], [K.1] used to obtain the right continuous observability estimates are not adequate in the presence of variable coefficients at the energy level (as the one represented by α in (1.1). More sophisticated methods are called for: [B-L-R.1], [L-T.6], [Ta.1], [Ta.2], [L-T-Y.1], [L-T-Y.2], etc. Thus, it is expected that similar difficulties will recur in the study of thermo-elasticity. Thus, the main contribution of this paper is the presence of a *variable* thermal coefficient α without *any smallness* requirement for the present boundary control case where, moreover, the noncritical controls are arbitrarily smooth.

Altogether different is the problem of *exact null-controllability*. In the case of a *one-dimensional* thermoelastic equation with hinged boundary conditions, an exact null controllability result is given in [H-Z.1] by use of the moment problem approach, via a scalar *boundary* control. For a more general thermoelastic model (in any dimension) this time with *distributed* control either in the mechanical (Kirchoff) equation, or else in the thermal equation (only), exact-null controllability is obtained in [L-T.12].

Finally, a very recent exact-null controllability paper (of which we have become aware after completion of the first draft of the present work) is [A-T.1], which deals, however, with a thermoelastic *wave* (rather than *plate*) system, which couples a *wave* equation with a thermal equation, with boundary control in the Dirichlet B.C. of the wave equation. It is likely (though the details may be demanding) that the treatment of [A-T.1] could be adapted to prove exact null controllability of thermoelastic plates with hinged boundary controls. The technical Carleman estimates of [A-T.1] neither imply, nor are implied by, the results of the present paper. In particular, the problem in [A-T.1] does not need backward uniqueness results, which are instead critical for our problem.

2 Thermo-Elastic Well-Posedness and Dual Problems

Well-posedness. Homogeneous problem. Let, at first, $u_1 = u_2 = u_3 = 0$ in both cases of B.C. (1.2) or (1.3).

Hinged B.C. (1.2). We introduce the following operators and spaces (with equivalent norms),

$$\mathcal{A}f = -\Delta f, \ \mathcal{A}_\gamma = (I + \gamma\mathcal{A}), \ \mathcal{D}(\mathcal{A}_\gamma) = \mathcal{D}(\mathcal{A}) = H^2(\Omega) \cap H_0^1(\Omega); \tag{2.1a}$$

$$\mathcal{D}(\mathcal{A}^{\frac{3}{2}}) = \{f \in H^3(\Omega) : \ f|_\Gamma = \Delta f|_\Gamma = 0\}; \tag{2.1b}$$

$$\mathcal{D}(\mathcal{A}_\gamma^{\frac{1}{2}}) = \mathcal{D}(\mathcal{A}^{\frac{1}{2}}) = H_0^1(\Omega), \ (x_1, x_2)_{\mathcal{D}(\mathcal{A}_\gamma^{\frac{1}{2}})} = ((I + \gamma\mathcal{A})x_1, x_2)_H, \quad H = L_2(\Omega); \tag{2.2}$$

$$Y_{\gamma,h} \equiv \mathcal{D}(\mathcal{A}) \times \mathcal{D}(\mathcal{A}_\gamma^{\frac{1}{2}}) \times H \equiv [H^2(\Omega) \cap H_0^1(\Omega)] \times H_0^1(\Omega) \times L_2(\Omega). \tag{2.3}$$

Clamped B.C. (1.3). In this case, we introduce

$$Af = \Delta^2 f, \ \mathcal{D}(A) = \{f \in H^4(\Omega) : \ f|_\Gamma = \left.\frac{\partial f}{\partial \nu}\right|_\Gamma = 0\} = H^4(\Omega) \cap H_0^2(\Omega); \tag{2.4a}$$

$$\mathcal{D}(A^{\frac{1}{2}}) = H_0^2(\Omega); \qquad \mathcal{D}(A^{\frac{1}{4}}) = \mathcal{D}(\mathcal{A}^{\frac{1}{2}}) = H_0^1(\Omega); \tag{2.4b}$$

$$Y_{\gamma,c} \equiv \mathcal{D}(A^{\frac{1}{2}}) \times \mathcal{D}(\mathcal{A}_\gamma^{\frac{1}{2}}) \times H \equiv H_0^2(\Omega) \times H_0^1(\Omega) \times L_2(\Omega). \tag{2.5}$$

LEMMA 2.1. *Let $u_1 = u_2 = u_3 = 0$ in (1.2) or (1.3). Then:*

(i) Problem (1.1), (1.2), or (1.1), (1.3), defines a s.c. semigroup denoted by $e^{\mathbb{A}_\gamma t}$: $y_0 = \{w_0, w_1, \theta_0\} \to e^{\mathbb{A}_\gamma t} y_0 = \{w(t), w_t(t), \theta(t)\}$ in the space Y_γ given by (2.3), or (2.5), respectively.

(i) In the case of hinged *B.C. (1.2), the generator $\mathbb{A}_\gamma$ is given by*

$$\mathbb{A}_\gamma = \mathbb{A}_{\gamma,h} = \begin{bmatrix} 0 & I & 0 \\ -\mathcal{A}_\gamma^{-1}\mathcal{A}^2 & 0 & -\mathcal{A}_\gamma^{-1} div(\alpha\nabla(\,\cdot\,)) \\ 0 & div(\alpha\nabla(\,\cdot\,)) & -\mathcal{A} \end{bmatrix}; \tag{2.6a}$$

$$\mathcal{D}(\mathbb{A}_{\gamma,h}) = \mathcal{D}(\mathcal{A}^{\frac{3}{2}}) \times \mathcal{D}(\mathcal{A}) \times \mathcal{D}(\mathcal{A}); \tag{2.6b}$$

the $Y_{\gamma,h}$-adjoint (see (2.3)) $\mathbb{A}_\gamma^*$ of $\mathbb{A}_\gamma$ is given by

$$\mathbb{A}_\gamma^* = \mathbb{A}_{\gamma,h}^* \begin{bmatrix} 0 & -I & 0 \\ \mathcal{A}_\gamma^{-1}\mathcal{A}^2 & 0 & -\mathcal{A}_\gamma^{-1} div(\alpha\nabla(\,\cdot\,)) \\ 0 & div(\alpha\nabla(\,\cdot\,)) & -\mathcal{A} \end{bmatrix}, \quad \mathcal{D}(\mathbb{A}_{\gamma,h}^*) = \mathcal{D}(\mathbb{A}_{\gamma,h}). \tag{2.7}$$

(ii) In the case of clamped *B.C. (1.3), the generator* $\mathbb{A}_\gamma$ *is given by*

$$\mathbb{A}_\gamma = \mathbb{A}_{\gamma,c} = \begin{bmatrix} 0 & I & 0 \\ -\mathcal{A}_\gamma^{-1}\mathcal{A} & 0 & -\mathcal{A}_\gamma^{-1} div(\alpha\nabla(\cdot)) \\ 0 & div(\alpha\nabla(\cdot)) & -\mathcal{A} \end{bmatrix}; \tag{2.8a}$$

$$\mathcal{D}(\mathbb{A}_{\gamma,c}) = \mathcal{D}(\mathcal{A}^{\frac{3}{4}}) \times \mathcal{D}(\mathcal{A}^{\frac{1}{2}}) \times \mathcal{D}(\mathcal{A}) \tag{2.8b}$$

(by 2.4b); the $Y_{\gamma,c}$*-adjoint (see (2.5) in the* $\mathcal{D}(\mathcal{A}_\gamma^{\frac{1}{2}})$*-topology of (2.2))* $\mathbb{A}_\gamma^*$ *of* $\mathbb{A}_\gamma$ *is given by*

$$\mathbb{A}_\gamma^* = \mathbb{A}_{\gamma,c}^* = \begin{bmatrix} 0 & -I & 0 \\ \mathcal{A}_\gamma^{-1}\mathcal{A} & 0 & \mathcal{A}_\gamma^{-1} div(\alpha\nabla(\cdot)) \\ 0 & -div(\alpha\nabla(\cdot)) & -\mathcal{A} \end{bmatrix}; \ \mathcal{D}(\mathbb{A}_{\gamma,c}^*) = \mathcal{D}(\mathbb{A}_{\gamma,c}). \tag{2.9}$$

In both cases, the dual semigroup dynamics $\bar{y}_0 = \{\phi_0, \phi_1, \eta_0\} \to e^{\mathbb{A}_\gamma^* t}\bar{y}_0 = \{\phi(t), -\phi_t(t), \eta(t)\}$ *is given by the following thermo-elastic (dual) problem:*

$$\begin{cases} \phi_{tt} - \gamma\Delta\phi_{tt} + \Delta^2\phi - div(\alpha(x)\nabla\eta) \equiv 0 & \text{in } Q; & (2.10a) \\ \eta_t - \Delta\eta + div(\alpha(x)\nabla\phi_t) \equiv 0 & \text{in } Q; & (2.10b) \\ \phi(0, \cdot) = \phi_0; \ \phi_t(0, \cdot) = \phi_1; \ \eta(0, \cdot) = \eta_0 & \text{in } \Omega, & (2.10c) \end{cases}$$

with the corresponding homogeneous B.C.:

$$\text{hinged: } \phi = \Delta\phi \equiv 0 \text{ on } \Sigma; \text{ or else clamped: } \phi \equiv \frac{\partial\phi}{\partial\nu} \equiv 0 \text{ on } \Sigma; \text{ and } \eta = 0 \text{ on } \Sigma. \tag{2.10d}$$

Proof. Part (i) is an application of the Lumer-Phillips theorem.

(ii) To compute the Y_γ-adjoint, we use the $\mathcal{D}(\mathcal{A}_\gamma^{\frac{1}{2}})$-inner product in (2.2), as well as the identity

$$(\text{div}(\alpha\nabla\theta_3), y_2)_{L_2(\Omega)} = -\int_\Omega \alpha\nabla y_2 \cdot \nabla\theta_3 d\Omega = (\theta_3, \text{div}(\alpha\nabla y_2))_{L_2(\Omega)}, \ \theta_3, y_2 \in \mathcal{D}(\mathcal{A}), \tag{2.11}$$

so that $\theta_3|_\Gamma = 0$, $y_2|_\Gamma = 0$. Eqn. (2.11) is obtained by using the standard (divergence theorem) identity

$$\int_\Omega \psi \,\text{div}\, h \, d\Omega = \int_\Gamma \psi h \cdot \nu \, d\Gamma - \int_\Omega \nabla\psi \cdot h \, d\Omega, \tag{2.12}$$

first with $\psi = y_2$, $h = \alpha\nabla\theta_3$ to get the middle term in (2.11), and next with $\psi = \theta_3$ and $h = \alpha\nabla y_2$ to get the final right-hand side term in (2.11). □

Well-posedness. Non-homogeneous problems. The following results are relevant [Tr.2, Theorems 1.1 and 4.1]. See also [H-Z.1] in the one-dimensional hinged case.

PROPOSITION 2.2. *With reference to problem (1.1), (1.2), or (1.3), let* $u_1 \equiv u_3 \equiv 0$, $w_0 = w_1 = \theta_0 = 0$. *Then*

***hinged case**, (1.2): the following maps are continuous:*

$$u_2 \in L_2(0,T;L_2(\Gamma)) \Rightarrow \{w, w_t, \theta\} \in C\left([0,T];\mathcal{D}(\mathcal{A}) \times \mathcal{D}(\mathcal{A}^{\frac{1}{2}}) \times \mathcal{D}(\mathcal{A}^{\frac{1}{2}})\right)$$
$$= C\left([0,T];[H^2(\Omega)\cap H_0^1(\Omega)] \times H_0^1(\Omega) \times H_0^1(\Omega)\right); \tag{2.13a}$$

$$w_{tt} \in L_2(0,T;L_2(\Omega)); \quad \theta_t \in L_2(0,T;L_2(\Omega)) \cap C\left([0,T];[\mathcal{D}(\mathcal{A}^{\frac{1}{2}})]'\right). \tag{2.13b}$$

***clamped case**, (1.3): the following maps are continuous (recall (1.13)):*

$$u_2 \in L_2(0,T;L_2(\Gamma)) \Rightarrow \{w, w_t\} \in C\left([0,T];\mathcal{D}(\mathcal{A}^{\frac{1}{2}}) \times \tilde{L}_2(\Omega)\right)$$
$$= C([0,T];H_0^1(\Omega) \times \tilde{L}_2(\Omega)); \tag{2.14a}$$

$$\theta \in C\left([0,T];H_\theta\right). \tag{2.14b}$$

This result, proved in [Tr.1], except for $H^{-\epsilon}(\Omega)$ instead of $H_\theta = L_2(\Omega)$ is noted in Theorem C.4, Appendix C.

The controllability results of Theorems 1.1 and 1.2 are consistent with the regularity results of Proposition 2.2, see (2.2)–(2.4), particularly since $u_1(T) = u_3(T) = 0$ in those theorems.

Abstract model. With reference to problem (1.1), (1.2), or (1.3), let $u_1 \equiv u_3 \equiv 0$, $w_0 = w_1 = \theta_0 = 0$, as in Proposition 2.2. Then the abstract model of the corresponding mixed problem, with $y(t) = [w(t), w_t(t), \theta(t)]$, and $u = [u_1, u_2, u_3]$, is

$$\dot{y} = \mathbb{A}_\gamma y + \mathcal{B}u, \quad y(t) = (\mathcal{L}u)(t) = \int_0^t e^{\mathbb{A}_\gamma (t-\tau)}\mathcal{B}u(\tau)d\tau, \tag{2.15}$$

where the boundary $\to$ interior operator $\mathcal{B}$ is defined by (see Appendix B, Eqns. (B.4) and (B.15), specialized to $u_1 = u_3 = 0$):

$$\mathcal{B} = \mathcal{B}_h = \begin{bmatrix} 0 \\ -\mathcal{A}_\gamma^{-1}\mathcal{A}D \\ 0 \end{bmatrix} \text{ (hinged case)}; \quad \mathcal{B} = \mathcal{B}_c = \begin{bmatrix} 0 \\ \mathcal{A}_\gamma^{-1}\mathcal{A}G_2 \\ 0 \end{bmatrix} \text{ (clamped case)}. \tag{2.16}$$

In (2.16), D is the (Dirichlet) map defined by (3.20) below; while G_2 is the (Green) map defined by (3.32) below.

3 Associated Kirchoff-Equation. Structural Decomposition of the S.C. Semigroup $e^{\mathbb{A}_\gamma t}$, $\gamma > 0$

Associated Kirchoff equation. When $\gamma > 0$, the thermo-elastic plate has a hyperbolic-dominated character, in the sense of the next result. Write $\text{div}(\alpha(x)\nabla\theta) = \alpha(x)\Delta\theta + \nabla\alpha \cdot \nabla\theta$ in the first equation (1.1a), and substitute $\Delta\theta$ from the second Eqn. (1.1b) to obtain

$$w_{tt} - \gamma\Delta w_{tt} + \Delta^2 w - \alpha \ \text{div}(\alpha\nabla w_t) = -\alpha\theta_t - \nabla\alpha \cdot \nabla\theta \quad \text{in } Q. \tag{3.1}$$

This, then, induces one to introduce the purely mechanical Kirchoff equation

$$\begin{cases} v_{tt} - \gamma\Delta v_{tt} + \Delta^2 v - \alpha\text{div}(\alpha\nabla v_t) \equiv 0 \text{ in } Q; & (3.2a) \\ \text{either } v|_\Sigma = u_1,\ \Delta v|_\Sigma = u_2; \text{ or else } v|_\Sigma = u_1,\ \left.\frac{\partial v}{\partial \nu}\right|_\Sigma = u_2. & (3.2b) \end{cases}$$

For use in the analysis below, we introduce the following operator F_α:

$$F_\alpha f \equiv \alpha \text{ div}(\alpha\nabla f), \quad \mathcal{D}(F_\alpha) = H^2(\Omega) \cap H_0^1(\Omega). \tag{3.3}$$

Its adjoint F_α^*, in the sense that $(F_\alpha f, g)_{L_2(\Omega)} = (f, F_\alpha^* g)_{L_2(\Omega)}$, $\forall f \in \mathcal{D}(F_\alpha)$, $g \in \mathcal{D}(F_\alpha^*)$, is given by

$$F_\alpha^* g = \text{div}(\alpha\nabla(\alpha g)), \ \mathcal{D}(F_\alpha^*) = \mathcal{D}(F_\alpha) = H^2(\Omega) \cap H_0^1(\Omega). \tag{3.4}$$

To show (3.4), one starts with (3.3) and applies the divergence theorem (2.12) twice: first with $\psi = g\alpha$ and $h = \alpha\nabla f$; and next, with $\psi = f$ and $h = \alpha\nabla(g\alpha)$; finally, one uses the B.C. $f|_\Gamma = 0$ and $g|_\Gamma = 0$ for membership in $\mathcal{D}(F_\alpha)$ and $\mathcal{D}(F_\alpha^*)$. This way, (3.3) readily yields (3.4).

Homogeneous case: $u_1 = u_2 = 0$ in (3.2b): Hinged B.C. The abstract version of problem (3.2) with $u_1 = u_2 = 0$ in (3.2b) (hinged B.C.) is given, via (2.1a), (3.3), by

$$v_{tt} + \gamma\mathcal{A}v_{tt} + \mathcal{A}^2 v - F_\alpha v_t = 0; \tag{3.5}$$

$$\frac{d}{dt}\begin{bmatrix} v \\ v_t \end{bmatrix} = \mathbb{A}_{1,\gamma}\begin{bmatrix} v \\ v_t \end{bmatrix} \quad \text{on the space } Y_{1,\gamma}, \tag{3.6}$$

where: in the case of (i) *hinged* B.C. (3.2b) (left), with $u_1 = u_2 = 0$, we have

$$\mathbb{A}_{1,\gamma} = \begin{bmatrix} 0 & I \\ -\mathcal{A}_\gamma^{-1}\mathcal{A}^2 & \mathcal{A}_\gamma^{-1}F_\alpha \end{bmatrix}; \quad \mathcal{D}(\mathbb{A}_{1,\gamma}) = \mathcal{D}(\mathcal{A}^{\frac{3}{2}}) \times \mathcal{D}(\mathcal{A}); \tag{3.7}$$

$$Y_{1,\gamma} \equiv \mathcal{D}(\mathcal{A}) \times \mathcal{D}(\mathcal{A}_\gamma^{\frac{1}{2}}) \equiv [H^2(\Omega) \cap H_0^1(\Omega)] \times H_0^1(\Omega), \tag{3.8}$$

with $\mathcal{A}_\gamma = I + \gamma\mathcal{A}$. The $Y_{1,\gamma}$-adjoint $\mathbb{A}_{1,\gamma}^*$ of $\mathbb{A}_{1,\gamma}$ in (3.7) is given by

$$\mathbb{A}_{1,\gamma}^* = \begin{bmatrix} 0 & -I \\ \mathcal{A}_\gamma^{-1}\mathcal{A}^2 & \mathcal{A}_\gamma^{-1}F_\alpha^* \end{bmatrix}, \quad \mathcal{D}(\mathbb{A}_{1,\gamma}^*) = \mathcal{D}(\mathbb{A}_{1,\gamma}). \tag{3.9}$$

Homogeneous case: $u_1 = u_2 = 0$ in (3.2b): Clamped B.C. The abstract version of problem (3.2) with $u_1 = u_2 = 0$ in (3.2b) (clamped B.C.) is given via (2.4a), (3.3) by

$$v_{tt} + \gamma\mathcal{A}v_{tt} + Av - F_\alpha v_t = 0, \tag{3.10a}$$

hence by model (3.6), where now in the case of *clamped* B.C., we have

$$\mathbb{A}_{1,\gamma} = \begin{bmatrix} 0 & I \\ -\mathcal{A}_\gamma^{-1}A & \mathcal{A}_\gamma^{-1}F_\alpha \end{bmatrix}, \; \mathcal{D}(\mathbb{A}_{1,\gamma}) = \mathcal{D}(A^{\frac{3}{4}}) \times \mathcal{D}(A^{\frac{1}{2}}); \tag{3.10b}$$

$$Y_{1,\gamma} = \mathcal{D}(A^{\frac{1}{2}}) \times \mathcal{D}(\mathcal{A}_\gamma^{\frac{1}{2}}) \equiv H_0^2(\Omega) \times H_0^1(\Omega). \tag{3.11}$$

The $Y_{1,\gamma}$-adjoint $\mathbb{A}_{1,\gamma}^*$ of $\mathbb{A}_{1,\gamma}$ in (3.10) [using the $\mathcal{D}(\mathcal{A}_\gamma^{\frac{1}{2}})$-topology in (2.2)] is given by

$$\mathbb{A}_{1,\gamma}^* = \begin{bmatrix} 0 & -I \\ \mathcal{A}_\gamma^{-1}A & \mathcal{A}_\gamma^{-1}F_\alpha^* \end{bmatrix}, \quad \mathcal{D}(\mathbb{A}_{1,\gamma}^*) = \mathcal{D}(\mathbb{A}_{1,\gamma}), \quad [\mathcal{D}(\mathbb{A}_{1,\gamma}^*)]' = \mathcal{D}(A^{\frac{1}{4}}) \times \tilde{L}_2(\Omega), \tag{3.12}$$

recalling (1.13).

Because of the topology on the second component space $\mathcal{D}(\mathcal{A}_\gamma^{\frac{1}{2}})$ given by (2.2), the lower right corner element $\mathcal{A}_\gamma^{-1}F_\alpha$ or $\mathcal{A}_\gamma^{-1}F_\alpha^*$, in (3.7), (3.10b), or in (3.9), (3.12), is a bounded perturbation on $\mathcal{D}(\mathcal{A}_\gamma^{\frac{1}{2}})$ in both cases.

LEMMA 3.1. *(i) The operator $\mathbb{A}_{1,\gamma}$ in (3.7)–(3.9), or (3.10)–(3.12), generates a s.c. group $e^{\mathbb{A}_{1,\gamma}t}$ on $Y_{1,\gamma}$: $\tilde{v}_0 = \{v_0, v_1\} \Rightarrow e^{\mathbb{A}_{1,\gamma}t}\tilde{v}_0 = \{v(t), v_t(t)\}$ where v solves (3.2)–(3.5) or with $u_1 = u_2 = 0$.*

(ii) The adjoint s.c. group $e^{\mathbb{A}_{1,\gamma}^*t}$ describes the following dynamics: $\psi_0 = \{\psi_0 - \psi_1\} \rightarrow e^{\mathbb{A}_{1,\gamma}^*t}\psi_0 = \{\psi(t), -\psi_t(t)\}$, where ψ solves the (dual) Kirchoff problem

$$\begin{cases} \psi_{tt} - \gamma\Delta\psi_{tt} + \Delta^2\psi - div(\alpha\nabla(\alpha\psi_t)) = 0; & (3.13a) \\ either\ \psi = \Delta\psi \equiv 0 \text{ on } \Sigma; \text{ or else } \psi = \frac{\partial\psi}{\partial\nu} \equiv 0 \text{ on } \Sigma, & (3.13b) \end{cases}$$

whose abstract version is:

$$\psi_{tt} + \gamma\mathcal{A}\psi_{tt} + \mathcal{A}^2\psi - F_\alpha^*\psi_t = 0 \quad (\textit{hinged B.C.}); \tag{3.14a}$$

$$\psi_{tt} + \gamma\mathcal{A}\psi_{tt} + A\psi - F_\alpha^*\psi_t = 0 \quad (\textit{clamped B.C.}). \tag{3.14b}$$

We now return to the dual thermo-elastic (ϕ, η)-problem (2.10), and we perform on it the same operations that lead the original thermo-elastic (w, θ)-problem to its Kirchoff-related version (3.1). That is, we likewise write $\text{div}(\alpha\nabla\eta) = \alpha\Delta\eta + \nabla\alpha\cdot\nabla\eta$ in the first equation (2.10a), and substitute $\Delta\eta$ from the second equation (2.10b), to obtain

$$\phi_{tt} - \gamma\Delta\phi_{tt} + \Delta^2\phi - \alpha\ \text{div}(\alpha\nabla\phi_t) = \alpha\eta_t + \nabla\alpha\cdot\nabla\eta. \tag{3.15}$$

By comparison with the dual Kirchoff ψ-problem (3.13a) we see that we need the identity

$$\text{div}(\alpha\nabla(\alpha\phi_t)) = \alpha\ \text{div}(\alpha\nabla\phi_t) + \alpha\nabla\phi_t\cdot\nabla\alpha + \text{div}(\alpha\phi_t\nabla\alpha)$$

to rewrite (3.15) in its definitive form as

$$\phi_{tt} - \gamma\Delta\phi_{tt} + \Delta^2\phi - \text{div}(\alpha\nabla(\alpha\phi_t)) = \alpha\eta_t + \nabla\alpha\cdot\nabla\eta + \alpha\nabla\alpha\cdot\nabla\phi_t + \text{div}(\alpha\phi_t\nabla\alpha), \tag{3.16}$$

with two lower-order terms on the right side of (3.16). We shall return to these equations in Theorem 3.2 below.

Abstract model of non-homogeneous problem (3.2) with hinged boundary controls. The second-order abstract model of problem (3.2a) with non-homogeneous hinged B.C. (3.2b) is given in the style of [L-T.4], [L-T.9], [L-T.10] by (see Appendix A, Eqn. (A.6)),

$$v_{tt} + \gamma \mathcal{A} v_{tt} + \mathcal{A}^2 v - F_\alpha v_t = R(u_1, u_2) \quad \text{on } [\mathcal{D}(\mathcal{A}^2)]'; \tag{3.17}$$

$$R(u_1, u_2) \equiv \mathcal{A}^2 D u_1 - \mathcal{A} D u_2 + \gamma \mathcal{A} D u_{1tt} - F_\alpha D_\alpha u_{1t}, \tag{3.18}$$

where $\mathcal{A}$ and F_α are defined in (2.1a) and (3.3), respectively, while the operators D_α and $D \equiv D_{\alpha=1}$ are defined by [L-T.9],

$$h = D_\alpha g \iff \{F_\alpha h \equiv \alpha \operatorname{div}(\alpha \nabla h) = 0 \text{ in } \Omega;\ h|_\Gamma = g\}; \tag{3.19}$$

$$h = Dg \iff \{\Delta h = 0 \text{ in } \Omega,\ h|_\Gamma = g\},\ D \in \mathcal{L}(H^s(\Gamma);\ H^{s+\frac{1}{2}}(\Omega)),\ s \in \mathbb{R}. \tag{3.20}$$

The first-order model corresponding to (3.17) is then given by

$$\frac{d}{dt}\begin{bmatrix} v \\ v_t \end{bmatrix} = \mathbb{A}_{1,\gamma}\begin{bmatrix} v \\ v_t \end{bmatrix} + \mathcal{B}_{1,h}\begin{bmatrix} u_1 \\ u_2 \end{bmatrix}, \tag{3.21}$$

where $\mathbb{A}_{1,\gamma}$ is defined by (3.7), while $\mathcal{B}_h$ is defined (see (3.18)) by

$$\mathcal{B}_{1,h}\begin{bmatrix} u_1 \\ u_2 \end{bmatrix} = \begin{bmatrix} 0 \\ \mathcal{A}_\gamma^{-1} R(u_1, u_2) \end{bmatrix} = \begin{bmatrix} 0 \\ \mathcal{A}_\gamma^{-1}[\mathcal{A}^2 D u_1 - \mathcal{A} D u_2 + \gamma \mathcal{A} D u_{1tt} - F_\alpha D_\alpha u_{1t} \end{bmatrix}. \tag{3.22}$$

The solution to (3.21) (or (3.17)) is then given, via Lemma 3.1(i), by

$$\begin{bmatrix} v(t) \\ v_t(t) \end{bmatrix} = e^{\mathbb{A}_{1,\gamma} t}\begin{bmatrix} v_0 \\ v_1 \end{bmatrix} + (Lu)(t); \tag{3.23a}$$

$$(Lu)(t) = \left(L\begin{bmatrix} u_1 \\ u_2 \end{bmatrix}\right)(t) = \int_0^t e^{\mathbb{A}_{1,\gamma}(t-\tau)} \mathcal{B}_{1,h}\begin{bmatrix} u_1(\tau) \\ u_2(\tau) \end{bmatrix} d\tau. \tag{3.23b}$$

Trace results. For future reference we note the following

LEMMA 3.2. *With reference to (2.1a), (3.4), (3.19), (3.20), we have*

$$D^* \mathcal{A} f = -\left.\frac{\partial f}{\partial \nu}\right|_\Gamma, \ f \in \mathcal{D}(\mathcal{A});\ D_\alpha^* F_\alpha^* f = \alpha \left.\frac{\partial(\alpha f)}{\partial \nu}\right|_\Gamma = \alpha^2 \left.\frac{\partial f}{\partial \nu}\right|_\Gamma, \ f \in \mathcal{D}(F_\alpha^*). \tag{3.24}$$

Proof. For $f \in \mathcal{D}(F_\alpha^*) = H^2(\Omega) \cap H_0^1(\Omega)$, see (3.4), so that $f|_\Gamma = 0$, and $g \in L_2(\Gamma)$, we compute by using the divergence theorem (2.12) twice: first with $\psi = D_\alpha g$ and $h = \alpha\nabla(\alpha f)$; next, with $\psi = \alpha f$ and $h = \alpha\nabla(D_\alpha g)$,

$$(D_\alpha^* F_\alpha^* f, g)_{L_2(\Gamma)} = (F_\alpha^* f, D_\alpha g)_{L_2(\Omega)} = \int_\Omega (D_\alpha g)\operatorname{div}(\alpha\nabla(\alpha f))d\Omega \tag{3.25}$$

$$= \int_\Gamma (D_\alpha g)\alpha\nabla(\alpha f)\cdot\nu\, d\Gamma - \int_\Omega \alpha\nabla(D_\alpha g)\cdot\nabla(\alpha f)d\Omega \tag{3.26}$$

$$(\text{by } (3.19)) \quad = \int_\Gamma g\alpha\nabla(\alpha f)\cdot\nu\, d\Gamma + \int_\Omega \alpha f \operatorname{div}(\alpha\nabla(D_\alpha g))d\Omega$$

$$- \int_\Gamma \alpha f \alpha\nabla(D_\alpha f)\cdot\nu\, d\Gamma \tag{3.27}$$

$$= \left(\alpha\frac{\partial(\alpha f)}{\partial\nu}, y\right)_{L_2(\Gamma)}, \tag{3.28}$$

recalling (3.19) and (3.4), and then (3.28) proves (3.24) (right). The left-hand side (which is in effect a specialization of the right-hand side when $\alpha \equiv 1$) is well known [L-T.9]. □

Abstract model of non-homogeneous problem (3.2) with clamped boundary controls. The second-order abstract model of problem (3.2a) with non-homogeneous clamped B.C. (3.2b) is given in the style of [L-T.7], [L-T.9] by (Appendix A, Eqn. (A.9))

$$v_{tt} + \gamma\mathcal{A}v_{tt} + Av - F_\alpha v_t = R(u_1, u_2); \tag{3.29}$$

$$R(u_1, u_2) \equiv AG_1 u_1 + AG_2 u_2 + \gamma\mathcal{A}Du_{1tt} - F_\alpha D_\alpha u_{1t}, \tag{3.30}$$

where A, $\mathcal{A}$, F_α, D_α and D are defined in (2.4a), (2.1a), (3.3), (3.19), and (3.20) respectively, while the operators G_i are defined by:

$$h = G_1 g \iff \left\{\Delta^2 h = 0 \text{ in } \Omega;\ h|_\Gamma = g,\ \frac{\partial h}{\partial\nu}|_\Gamma = 0\right\}; \tag{3.31}$$

$$h = G_2 g \iff \left\{\Delta^2 h = 0 \text{ in } \Omega;\ h|_\Gamma = 0,\ \frac{\partial h}{\partial\nu}|_\Gamma = g\right\}. \tag{3.32}$$

The first-order model corresponding to (3.29), (3.30) is then

$$\frac{d}{dt}\begin{bmatrix} v \\ v_t \end{bmatrix} = \mathbb{A}_{1,\gamma}\begin{bmatrix} v \\ v_t \end{bmatrix} + \mathcal{B}_{1,c}\begin{bmatrix} u_1 \\ u_2 \end{bmatrix}; \quad \mathbb{A}_{1,\gamma}^{-1} = \begin{bmatrix} A^{-1}F_\alpha & -A^{-1}\mathcal{A}_\gamma \\ I & 0 \end{bmatrix}, \tag{3.33}$$

where $\mathbb{A}_{1,\gamma}$ is now defined by (3.10), while $\mathcal{B}_{1,c}$ is defined (see (3.30)) by

$$\mathcal{B}_{1,c}\begin{bmatrix} u_1 \\ u_2 \end{bmatrix} = \begin{bmatrix} 0 \\ \mathcal{A}_\gamma^{-1}R(u_1,u_2) \end{bmatrix} = \begin{bmatrix} 0 \\ \mathcal{A}_\gamma^{-1}[AG_1u_1 + AG_2u_2 + \gamma\mathcal{A}Du_{1tt} - F_\alpha D_\alpha u_{1t} \end{bmatrix}. \tag{3.34}$$

The solution of (3.33) (or (3.29)) is then given, via Lemma 3.1(ii), by

$$\begin{bmatrix} v(t) \\ v_t(t) \end{bmatrix} = e^{\mathbb{A}_{1,\gamma}t}\begin{bmatrix} v_0 \\ v_1 \end{bmatrix} + (Lu)(t); \tag{3.35}$$

$$(Lu)(t) = \left(L\begin{bmatrix} u_1 \\ u_2 \end{bmatrix}\right)(t) = \int_0^t e^{\mathbb{A}_{1,\gamma}(t-\tau)}\mathcal{B}_{1,c}\begin{bmatrix} u_1(\tau) \\ u_2(\tau) \end{bmatrix} d\tau. \tag{3.36}$$

Trace results. With reference to the operators G_1 and G_2 in (3.31), the following trace results are known [L-T.7], [L-T.10]:

$$G_1^* A f = \frac{\partial \Delta f}{\partial \nu}, \quad f \in \mathcal{D}(A); \qquad G_2^* A f = -\Delta f|_\Gamma, \quad f \in \mathcal{D}(A). \tag{3.37}$$

We conclude this section by noticing that, via (3.33) and (3.34), we have

$$\mathbb{A}_{1,\gamma}^{-1} \mathcal{B}_{1,c} \begin{bmatrix} u_1 \\ u_2 \end{bmatrix} = \begin{bmatrix} -G_1 u_1 - G_2 u_2 - \gamma A^{-1} \mathcal{A} D u_{1tt} + A^{-1} F_\alpha D_\alpha u_{1t} \\ 0 \end{bmatrix}. \tag{3.38}$$

Structural decomposition of $e^{\mathbb{A}_\gamma t}$. The following result, critical for our present development, holds true.

THEOREM 3.3. *Let Π_m be the projection $Y_\gamma \to Y_{1,\gamma} : [v_1, v_2, v_3] \to [v_1, v_2]$ onto the mechanical space and let $\Pi_m^* : [v_1, v_2] \to [v_1, v_2, 0]$ be its adjoint $Y_{1,\gamma} \to Y_\gamma$.*

(a) Consider the thermo-elastic semigroup $e^{\mathbb{A}_\gamma t}$ on $Y_\gamma = Y_{1,\gamma} \times L_2(\Omega)$ (see (3.8) in the hinged case, and (3.11) in the clamped case, respectively) of Lemma 2.1 associated with the original (w, θ)-problem (1.1) and the corresponding Kirchoff-group $e^{\mathbb{A}_{1,\gamma} t}$ on $Y_{1,\gamma}$ of Lemma 3.1 associated with the v-problem (3.2).

Then, for $u_1 = u_2 = u_3 \equiv 0$, the following structural decomposition holds true for any $t > 0$, and $[w_0, w_1, \theta_0] \in Y_\gamma$:

$$\begin{bmatrix} w(t) \\ w_t(t) \\ \theta(t) \end{bmatrix} = e^{\mathbb{A}_\gamma t} \begin{bmatrix} w_0 \\ w_1 \\ \theta_0 \end{bmatrix} = \begin{bmatrix} e^{\mathbb{A}_{1,\gamma} t} \begin{bmatrix} w_0 \\ w_1 \end{bmatrix} \\ 0 \end{bmatrix} + \mathcal{K}_\gamma(t) \begin{bmatrix} w_0 \\ w_1 \\ \theta_0 \end{bmatrix}; \tag{3.39}$$

$$\begin{bmatrix} w(t) \\ w_t(t) \end{bmatrix} = \Pi_m e^{\mathbb{A}_\gamma t} \begin{bmatrix} w_0 \\ w_1 \\ \theta_0 \end{bmatrix} = e^{\mathbb{A}_{1,\gamma} t} \begin{bmatrix} w_0 \\ w_1 \end{bmatrix} + \Pi_m \mathcal{K}_\gamma(t) \begin{bmatrix} w_0 \\ w_1 \\ \theta_0 \end{bmatrix}, \tag{3.40}$$

where

$$\Pi_m \mathcal{K}_\gamma(t) \begin{bmatrix} w_0 \\ w_1 \\ \theta_0 \end{bmatrix} = \int_0^t e^{\mathbb{A}_{1,\gamma}(t-\tau)} \begin{bmatrix} 0 \\ -A_\gamma^{-1}[\alpha \theta_t(\tau) + \nabla \alpha \cdot \nabla \theta(\tau)] \end{bmatrix} d\tau. \tag{3.41}$$

(b) Consider the dual thermo-elastic semigroup $e^{\mathbb{A}_\gamma^* t}$ on $Y_\gamma = Y_{1,\gamma} \times L_2(\Omega)$ of Lemma 2.1 associated with the dual (ϕ, η)-problem (2.10) and the corresponding Kirchoff-group $e^{\mathbb{A}_{1,\gamma}^* t}$ on $Y_{1,\gamma}$ of Lemma 3.1 associated with the ψ-problem (3.13). Then, the following structural decomposition holds true for any $t > 0$ and $[\phi_0, -\phi_1, \eta_0] \in Y_\gamma$:

$$\begin{bmatrix} \phi(t) \\ -\phi_t(t) \\ \eta(t) \end{bmatrix} = e^{\mathbb{A}_\gamma^* t} \begin{bmatrix} \phi_0 \\ -\phi_1 \\ \eta_0 \end{bmatrix} = \begin{bmatrix} e^{\mathbb{A}_{1,\gamma}^* t} \begin{bmatrix} \phi_0 \\ -\phi_1 \end{bmatrix} \\ 0 \end{bmatrix} + \mathcal{K}_\gamma^*(t) \begin{bmatrix} \phi_0 \\ -\phi_1 \\ \eta_0 \end{bmatrix}; \tag{3.42}$$

$$\begin{bmatrix} \phi(t) \\ -\phi_t(t) \end{bmatrix} = \Pi_m e^{\mathbb{A}_\gamma^* t} \begin{bmatrix} \phi_0 \\ -\phi_1 \\ \eta_0 \end{bmatrix} = e^{\mathbb{A}_{1,\gamma}^* t} \begin{bmatrix} \phi_0 \\ -\phi_1 \end{bmatrix} + \Pi_m \mathcal{K}_\gamma^*(t) \begin{bmatrix} \phi_0 \\ -\phi_1 \\ \eta_0 \end{bmatrix}, \tag{3.43}$$

where

$$\Pi_m \mathcal{K}_\gamma^*(t) \begin{bmatrix} \phi_0 \\ -\phi_1 \\ \eta_0 \end{bmatrix} = \int_0^t e^{\mathbb{A}_{1,\gamma}^*(t-\tau)} \begin{bmatrix} 0 \\ \mathcal{A}_\gamma^{-1}[\alpha\eta_t(\tau) + \nabla\alpha\cdot\nabla\eta(\tau) + \ell.o.t.(\tau)] \end{bmatrix} d\tau; \tag{3.44}$$

$$\ell.o.t.(\tau) = \alpha\nabla\alpha\cdot\nabla\phi_t(\tau) + div(\alpha\phi_t(\tau)\nabla\alpha). \tag{3.45}$$

(c) Let now $\{u_1, u_2, u_3\} \in \mathcal{U}$, where $\mathcal{U}$ is the class of controls in Theorem 1.1 (hinged case) or Theorem 1.2 (clamped case), respectively. Then, with reference to problem (1.1), (1.2), we may write using (3.39):

in the hinged case:

$$\begin{bmatrix} w(T) \\ w_t(T) \end{bmatrix} = \Pi_m \int_0^T e^{\mathbb{A}_\gamma(T-t)} \mathcal{B}_h u(t)dt = \int_0^T e^{\mathbb{A}_{1,\gamma}(T-t)} \Pi_m \mathcal{B}_h u(t)dt + Q_h, \tag{3.46}$$

where $\mathcal{B}_h$ is the control operator defined explicitly in Appendix B, Eqn. (B.4), and where the operator

$$Q_h \equiv \Pi_m \textstyle\int_0^T \mathcal{K}_\gamma(T-t)\mathcal{B}_h u(t)dt \tag{3.47a}$$

$$: compact\ \mathcal{U} \to X_{1,\gamma} \equiv Y_{1,\gamma} = [H^2(\Omega) \times H_0^1(\Omega)] \times H_0^1(\Omega). \tag{3.47b}$$

in the clamped case:

$$\begin{bmatrix} w(T) \\ w_t(T) \end{bmatrix} = \Pi_m \int_0^T e^{\mathbb{A}_\gamma(T-t)} \mathcal{B}_c u(t)dt = \int_0^T e^{\mathbb{A}_{1,\gamma}(T-t)} \Pi_m \mathcal{B}_c u(t)dt + Q_c, \tag{3.48}$$

where $\mathcal{B}_c$ is the control operator defined explicitly in Appendix B, Eqn. (B.15), and where the operator

$$Q_c \equiv \Pi_m \textstyle\int_0^T \mathcal{K}_\gamma(T-t)\mathcal{B}_c u(t)dt \tag{3.49a}$$

$$: compact\ \mathcal{U} \to X_{1,\gamma} \equiv [\mathcal{D}(\mathbb{A}_{1,\gamma}^*)]' \equiv \mathcal{D}(A^{\frac{1}{4}}) \times \tilde{L}_2(\Omega) \equiv H_0^1(\Omega) \times \tilde{L}_2(\Omega). \tag{3.49b}$$

Proof. (a) Eqns. (3.39)–(3.41) are nothing but an abstract rewriting of problems (1.1) and (3.1) [with $u_1 \equiv u_2 \equiv u_3 \equiv 0$]. Similarly, Eqns. (3.42)–(3.44) are nothing but an abstract rewriting of the (boundary) homogeneous dual problems (2.10) and (3.16). [As a matter of fact, the operator $\mathcal{K}_\gamma(t)$ is *compact* on Y_γ for each $t > 0$: in this form, Part (a) above is a special case of a more general structural decomposition result given in [L-T.7, Theorem 1.2.1, Theorem 1.2.2]. See also [L-T.8]. [Actually, [L-T.7] considered explicitly the case of constant α. A variable α

in space produces the additional contribution $\nabla\alpha \cdot \nabla\theta$ in (3.41) [not present in [L-T.7]], which, however, still yields a compact additional contribution to the operator in (3.41)]. Thus, [L-T.7, Theorem 1.2.2] applies to the present case as well.]

However, it is Part (b) of Theorem 3.2 that will be critically used in this paper (see Section 4), based on the compactness of the operators Q_h and Q_c in (3.47) (hinged case) and (3.49) (clamped case), respectively. Compactness of Q_h in the hinged case (3.47) is given in [L-T.7, Proposition 6.2.1, p. 62] with a sketch of the proof. A more detailed expansion of this proof is given below. Compactness of Q_c in the clamped case (3.49) is given below. [The proof in [L-T.7, Proposition 6.2.1] refers to different spaces. We note, preliminarily, that compactness of $\mathcal{K}_\gamma(t)$ does not suffice to claim compactness of Q in either case, as required in (3.47b) and (3.49b), since the control operator $\mathcal{B}$ is (highly) unbounded (*if* $\mathcal{B}$ were *bounded*, as in the *distributed* control case [T-Z.1], compactness of Q would follow by Mazur's Theorem as in [L-T.7]).

Part (b): Compactness of Q_h in (3.47b).

First proof of (3.47b). (After [L-T.7, Proposition 6.2.1, p. 62]) We may set $u_1 \equiv u_3 \equiv 0$ since these controls belong to a C_∞-class. As we concentrate only on $u_2 \in L_2(\Sigma)$, we may set

$$\mathcal{B}_h = \Pi_m^* \mathcal{B}_m, \text{ where } \mathcal{B}_m u = \begin{bmatrix} 0 \\ -\mathcal{A}_\gamma^{-1} A D u_2 \end{bmatrix}, \quad \mathcal{B}_h u = \begin{bmatrix} 0 \\ -\mathcal{A}_\gamma^{-1} A D u_2 \\ 0 \end{bmatrix}, \tag{3.50}$$

see (2.16) (left), or Appendix B, Eqn. (B.4) with $u_1 \equiv u_3 \equiv 0$. In this case, instead of (3.47b), we show equivalently that

$$Q_h^* = \mathcal{B}_m^* \Pi_m K_\gamma^*(\,\cdot\,)\Pi_m^* : \text{ compact } X_{1,\gamma} \equiv Y_{1,\gamma} \to L_2(0,T;L_2(\Gamma)). \tag{3.51}$$

(The notation in (3.51) agrees with that of [L-T.7, Eqn. 6.2.9)], except that in this latter reference we chose—in line with the emphasis of that paper—to start with the dual/adjoint thermo-elastic semigroup, rather than the original one of the present paper.) We now prove (3.51).

Step 1. Let $\bar{y}_0 = [\phi_0, -\phi_1]$, $\Pi_m^* \bar{y}_0 \equiv y_0 = [\phi_0, -\phi_1, 0] \in Y_{\gamma,h}$. Recalling (3.44) in (3.51), we obtain explicitly

$$\begin{aligned}(Q_h^* \bar{y}_0)(T-t) &= \mathcal{B}_m^* \Pi_m K_\gamma^*(t) \Pi_m^* \bar{y}_0 \\ &= \mathcal{B}_m^* \int_0^t e^{\mathbf{A}_{1,\gamma}^*(t-\tau)} \begin{bmatrix} 0 \\ -\mathcal{A}_\gamma^{-1}[\eta_t(\tau; y_0 + \nabla\alpha + \nabla\eta(\tau; y_0)] \end{bmatrix} d\tau + \ell.o.t., \end{aligned} \tag{3.52}$$

where η is the thermal component of the adjoint thermo-elastic problem (2.10).

We next use, critically, the following *trace* regularity for the Kirchoff homogeneous problem (3.13):

$$\mathcal{B}_m^* e^{\mathbf{A}_{1,\gamma}^* t} : \text{ continuous } Y_{1,\gamma} \to L_2(0,T;L_2(\Gamma)). \tag{3.53}$$

By duality [L-T.9], (3.53) is equivalent to the following *interior* regularity of the corresponding Kirchoff non-homogeneous problem (3.2):

$$v_0 = v_1 = 0,\ u_1 \equiv 0,\ u_2 \in L_2(\Sigma) \to \{v, v_t\} = \int_0^t e^{\mathbf{A}_{1,\gamma}(t-\tau)} \mathcal{B}_m u_2(\tau) d\tau \tag{3.54a}$$

$$\in C([0,T]; Y_{1,\gamma} = [H^2(\Omega) \cap H_0^1(\Omega)] \times H_0^1(\Omega)), \tag{3.54b}$$

see (2.3), which is true by [L-T.4]. (Actually, the proof for $\alpha \equiv 1$ in the conservative case of this latter reference extends verbatim to the case of α variable and smooth in the damped case.) Thus, returning to (3.51), (3.52), we estimate by virtue of (3.53) with $U = L_2(\Gamma)$, via Schwarz inequality and a change in the order of integration:

$$\|Q_h^* \bar{y}_0\|^2_{L_2(\Sigma)} = \int_0^T \left\| \int_0^t \mathcal{B}_m^* e^{\mathbf{A}_{1,\gamma}^*(t-\tau)} \begin{bmatrix} 0 \\ -\mathcal{A}_\gamma^{-1}[\eta_t(\tau;y_0) + \nabla\alpha \cdot \nabla\eta(\tau;y_0)] \end{bmatrix} d\tau \right\|_U^2 dt + \ell.o.t. \tag{3.55}$$

$$\leq T \int_0^T \int_0^t \left\| \mathcal{B}_m^* e^{A_{1,\gamma}^*(t-\tau)} \begin{bmatrix} 0 \\ -\mathcal{A}_\gamma^{-1}[\eta_t(\tau;y_0) + \nabla\alpha \cdot \nabla\eta(\tau;y_0)] \end{bmatrix} \right\|_U^2 d\tau\, dt \tag{3.56}$$

$$= T \int_0^T \int_\tau^T \left\| \mathcal{B}_m^* e^{\mathbf{A}_{1,\gamma}^*(t-\tau)} \begin{bmatrix} 0 \\ -\mathcal{A}_\gamma^{-1}[\eta_t(\tau;y_0) + \nabla\alpha \cdot \nabla\eta(\tau;y_0)] \end{bmatrix} \right\|_U^2 dt\, d\tau \tag{3.57}$$

$$(\text{by } (3.53)) \leq C_T \int_0^T \left\| \begin{bmatrix} 0 \\ -\mathcal{A}_\gamma^{-1}[\eta_t(\tau;y_0) + \nabla\alpha \cdot \nabla\eta(\tau;y_0)] \end{bmatrix} \right\|_{Y_{1,\gamma}}^2 d\tau \tag{3.58}$$

$$(\text{by } (2.2)) = C_T \int_0^T \left\| \mathcal{A}_\gamma^{-\frac{1}{2}} [\eta_t(\tau;y_0) + \nabla\alpha \cdot \nabla\eta(\tau;y_0)] \right\|_{L_2(\Omega)}^2 d\tau. \tag{3.59}$$

The key trace estimate (3.53) has been used in going from (3.57) to (3.58). (Eqn. (3.59) coincides with [L-T.7, Eqn. (6.2.10)] for which only a sketch was given, now elaborated in the steps from (3.55) through (3.59).)

Step 2. The map

$$y_0 = [\phi_0, -\phi_1, 0] \in Y_{\gamma,h} \to \mathcal{A}_\gamma^{-\frac{1}{2}}[\eta_t + \nabla\alpha \cdot \nabla\eta] \in L_2(0,T; L_2(\Omega)) \text{ is compact.} \tag{3.60}$$

This step can be proved as in [L-T.7, p. 62] by Aubin's Lemma [A.1], due to the following key regularity properties

$$\left\{ \begin{array}{ll} y_0 = [\phi_0, -\phi_1, 0] \in Y_{\gamma,h} \equiv [H^2(\Omega) \cap H_0^1(\Omega)] \times H_0^1(\Omega) \times L_2(\Omega); & (3.61) \\ \Rightarrow \eta_t \in L_2(0,T; L_2(\Omega)) \text{ and } \eta_{tt} \in L_2(0,T; [\mathcal{D}(\mathcal{A})]'); & (3.62) \\ \eta \in L_2(0,T; H_0^1(\Omega)) \text{ and } |\nabla\eta| \in L_2(0,T; L_2(\Omega)); & (3.63) \\ |\nabla\eta_t| \in L_2(0,T; H^{-1}(\Omega) = [\mathcal{D}(\mathcal{A}^{\frac{1}{2}})]'). & (3.64) \end{array} \right.$$

The above regularity results are known: see [L-T.7, Eqn. (3.3), p. 36] for η in (3.63); [L-T.7, Eqn. (3.41)] for η_t and η_{tt} in (3.62) since the third coordinate of y_0 is zero; then (3.64) follows from (3.62).

The above results are reproved in Lemma 9.2, Eqn. (9.10) as part of a broader program on the analysis of the regularity properties of a thermo-elastic plate. Thus, (3.62) and (3.63) yield

$$\mathcal{A}_\gamma^{-\frac{1}{2}}[\eta_t + \nabla\alpha\cdot\nabla\eta] \in L_2(0,T;\mathcal{D}(\mathcal{A}_\gamma^{\frac{1}{2}})), \text{ injection } \mathcal{D}(\mathcal{A}_\gamma^{\frac{1}{2}}) \to L_2(\Omega) \text{ compact;} \tag{3.65}$$

$$\mathcal{A}_\gamma^{-\frac{1}{2}}[\eta_{tt} + \nabla\alpha\cdot\nabla\eta_t] \in L_2(0,T;[\mathcal{D}(\mathcal{A}_\gamma^{\frac{1}{2}})]' = H^{-1}(\Omega)). \tag{3.66}$$

Then, Aubin's Lemma [A.1, $p = 2$] yields (3.60) from (3.65) and (3.66).

Step 3. Using (3.60) in (3.59) yields (3.51), as desired. Theorem 3.3, Part (b), is proved in the hinged case. □

REMARK 3.1. *The above proof of compactness of Q_h in the hinged case, when applied to the clamped case to show compactness for Q_c, as required in (3.49b), runs into (the usual) technical difficulty because of the different level of topology involved. In short: in the counterpart of Eqn. (3.59), one would obtain the operator $A^{\frac{1}{2}}\mathcal{A}_\gamma^{-1}$ (which is not well defined on $L_2(\Omega)$) instead of $\mathcal{A}_\gamma^{-\frac{1}{2}}$. To avoid this difficulty, we shall estimate the counterpart of (3.55) in one shot, instead of breaking up the estimate into two pieces: in (3.58) first, and in taking care of (3.59) next, as in Step 2. Accordingly, we present a second (more complicated) proof based on this idea in the hinged case as well, for purposes of illustration.*

Proof. (Second one of (3.47b).) Here (unlike the first proof) we shall use explicitly the form of the control operator $\mathcal{B}_m$ in (3.50).

Step 1. First, let, again,

$$\bar{y}_0 = [\phi_0, -\phi_1] \in Y_{1,\gamma} \equiv [H^2(\Omega)\cap H_0^1(\Omega)] \times H_0^1(\Omega), \text{ and } \Pi_m^*\bar{y}_0 = y_0 = [\phi_0, -\phi_1, 0] \in Y_{\gamma,h}. \tag{3.67}$$

Consider the dual thermo-elastic (ϕ,η)-problem (2.10) with initial condition in (2.10c) given by (3.67) and hinged zero B.C. in (2.10d). Call $\{\phi(t;y_0), \eta(t;y_0)\}$ the corresponding solution. This, then, satisfies the associated problem (3.16), whose solution may be written as in (3.43), (3.44). We are interested only on the component (3.44) of this solution. Thus, we let

$$\begin{aligned}\begin{bmatrix} \tilde{\phi}(t;y_0) \\ -\tilde{\phi}_t(t;y_0)\end{bmatrix} &\overset{\text{def}}{\equiv} \Pi_m\mathcal{K}_\gamma^*(t)\begin{bmatrix}\phi_0 \\ -\phi_1 \\ 0\end{bmatrix} \\ &= \int_0^t e^{\mathbf{A}_{1,\gamma}^*(t-\tau)}\begin{bmatrix} 0 \\ \mathcal{A}_\gamma^{-1}[\alpha\eta_t(\tau;y_0) + \nabla\alpha\cdot\nabla\eta(\tau;y_0)] + \ell.o.t.(\tau)\end{bmatrix}, \end{aligned} \tag{3.68}$$

so that $\tilde{\phi}$ solves the problem (compare with (3.16)):

$$\begin{cases} \tilde{\phi}_{tt} - \gamma\Delta\tilde{\phi}_{tt} + \Delta^2\tilde{\phi} - \operatorname{div}(\alpha\nabla(\alpha\tilde{\phi}_t)) = \alpha\eta_t(t;y_0) + \nabla\alpha\cdot\nabla\eta(t;y_0) + \ell.o.t. \text{ in } Q; & (3.69a) \\ \tilde{\phi}(0,\,\cdot\,) = 0,\ \tilde{\phi}_t(0,\,\cdot\,) = 0 \ \text{ in } \Omega; & (3.69b) \\ \tilde{\phi} \equiv 0, \quad \Delta\tilde{\phi} \equiv 0 \ \text{ on } \Sigma; & (3.69c) \\ \ell.o.t. = \alpha\nabla\alpha\cdot\nabla\tilde{\phi}_t + \operatorname{div}(\alpha\tilde{\phi}_t\nabla\alpha). & (3.69d) \end{cases}$$

By the regularity properties in (3.61)–(3.64) and, again, Aubin's Lemma [A.1, $p=2$], we have that the map

$$\begin{cases} y_0 = [\phi_0, -\phi_1, 0] \in Y_{\gamma,h} \\ \qquad \to f \equiv \alpha\eta_t(\,\cdot\,;\bar{y}_0) + \nabla\alpha\cdot\nabla\eta(\,\cdot\,;y_0) \in L_2(0,T;H^{-1}(\Omega)) \text{ is compact.} \end{cases} \tag{3.70}$$

Step 2. Claim. With reference to problem (3.69), we have:

the map

$$y_0 = [\phi_0, -\phi_1, 0] \in Y_{\gamma,h} \equiv [H^2(\Omega)\cap H_0^1(\Omega)] \times H_0^1(\Omega) \times L_2(\Omega)$$

$$\to \tfrac{\partial\tilde{\phi}_t}{\partial\nu} \in L_2(0,T;L_2(\Gamma)) \text{ is compact.} \tag{3.71}$$

Proof of claim. We combine the compactness property in (3.70), with the following regularity property:

the map

$$f \in L_2(0,T;H^{-1}(\Omega)) \to \frac{\partial\tilde{\phi}_t}{\partial\nu} \in L_2(0,T;L_2(\Gamma)) \text{ is continuous,} \tag{3.72}$$

which is a special case of [L-T.11, Theorem 1.3, p. 587] (the proof in [L-T.11] is for $\alpha \equiv 0$, but applies for α nonzero and smooth). This ends the proof of the claim.

Step 3. With reference to Q_h^* in (3.51) we have

$$(Q_h^*\bar{y}_0)(t) = \mathcal{B}_m^*\Pi_m\mathcal{K}_\gamma^*(T-t)y_0 = -\frac{\partial\tilde{\phi}_t}{\partial\nu}(T-t;y_0) \tag{3.73}$$

in the notation of (3.68), where the adjoints $*$ are with respect to $Y_{\gamma,h}$.

Proof of (3.73). By the definition (3.47) or (3.52) of Q_h, we compute this time using explicitly the control operator in (3.50): recalling that $\Pi_m^*\bar{y}_0 = y_0 = [\phi_0, -\phi_1, 0]$:

$$(Q_h u_2, \bar{y}_0)_{Y_{1,\gamma}} = \left(\Pi_m \int_0^T \mathcal{K}_\gamma(T-t)\Pi_m^*\mathcal{B}_m u_2(t)dt, \bar{y}_0\right)_{Y_{1,\gamma}} \tag{3.74}$$

$$= \int_0^T (\mathcal{B}_m u_2(t), \Pi_m K_\gamma^*(T-t)y_0)_{Y_{1,\gamma}} dt \tag{3.75}$$

$$
\begin{aligned}
\text{(by (3.50), (3.68))} \quad &= \int_0^T \left(\begin{bmatrix} 0 \\ -\mathcal{A}_\gamma^{-1}\mathcal{A}Du_2(t) \end{bmatrix}, \begin{bmatrix} \tilde{\phi}(T-t;y_0) \\ -\tilde{\phi}_t(T-t;y_0) \end{bmatrix} \right)_{Y_{1,\gamma}} dt \qquad (3.76) \\
\text{(by (2.2)} \quad &= \int_0^T (\mathcal{A}Du_2(t), \tilde{\phi}_t(T-t;y_0))_{L_2(\Omega)} dt \\
&= \int_0^T (u_2(t), D^*\mathcal{A}\tilde{\phi}_t(T-t;y_0))_{L_2(\Gamma)} dt \qquad (3.77) \\
\text{(by (3.24)} \quad &= \int_0^T \left(u_2(t), -\frac{\partial \tilde{\phi}_t}{\partial \nu}(T-t;y_0) \right)_{L_2(\Gamma)} dt \qquad (3.78) \\
\text{(by (3.74), (3.75))} \quad &= \int_0^T (u_2(t), \mathcal{B}_m^* \Pi_m \mathcal{K}_\gamma^*(T-t)y_0)_{L_2(\Gamma)} dt \\
&= (u_2, Q_h^* \bar{y}_0)_{L_2(0,T;L_2(\Gamma))}. \qquad (3.79)
\end{aligned}
$$

Then, (3.78), (3.79) yield (3.73), as desired. This ends the proof of (3.73).

Step 4. Identity (3.73) in Step 3, combined with the compactness property in (3.71), establish that Q_h^* is compact, as required by (3.51), or that Q_h is compact as required by (3.47b).

Proof of Theorem 3.3, Part (b): Compactness of Q_c in (3.49b). With reference to Remark 3.1, we follow now the approach of the second proof in the hinged case.

Step 1. Let (recall (1.13), (3.12))

$$[x_1, x_2] \in [\mathcal{D}(\mathbb{A}_{1,\gamma}^*)]' \overset{\text{def}}{\equiv} X_{1,\gamma} = \mathcal{D}(A^{\frac{1}{4}}) \times \tilde{L}_2(\Omega) = H_0^1(\Omega) \times \tilde{L}_2(\Omega), \tag{3.80}$$

so that by (3.10b), (3.12), and (2.8b), (2.9), respectively:

$$\bar{y}_0 \equiv [\phi_0, -\phi_1] \equiv \mathbb{A}_{1,\gamma}^{*-1} \mathbb{A}_{1,\gamma}^{-1} [x_1, x_2] \in \mathcal{D}(\mathbb{A}_{1,\gamma}) = \mathcal{D}(\mathbb{A}_{1,\gamma}^*) = \mathcal{D}(A^{\frac{3}{4}}) \times \mathcal{D}(A^{\frac{1}{2}}); \tag{3.81}$$

$$
\begin{aligned}
y_0 \equiv \Pi_m^* \bar{y}_0 \quad &\equiv \quad [\phi_0, -\phi_1, 0] \in \mathcal{D}(\mathbb{A}_{\gamma,c}) \equiv \mathcal{D}(\mathbb{A}_{\gamma,c}^*) = \mathcal{D}(\mathbb{A}_{1,\gamma}^*) \times \mathcal{D}(\mathcal{A}) \\
&= \quad [H^3(\Omega) \cap H_0^2(\Omega)] \times H_0^2(\Omega) \times [H^2(\Omega) \cap H_0^1(\Omega)]. \qquad (3.82)
\end{aligned}
$$

Thus, with reference to the dual thermo-elastic problem (2.10) with initial condition $y_0 = [\phi_0, -\phi_1, 0]$ as in (3.82), we obtain by Lemma 2.1(ii):

$$
\begin{aligned}
[\phi(t), -\phi_t(t), \eta(t)] \quad &\equiv \quad e^{\mathbb{A}_{\gamma,c}^* t} y_0 \in C([0,T]; \mathcal{D}(\mathbb{A}_{\gamma,c}^*)); \qquad (3.83) \\
[\phi_t(t), -\phi_{tt}(t), \eta_t(t)] \quad &\equiv \quad e^{\mathbb{A}_{\gamma,c}^* t} \mathbb{A}_{\gamma,c}^* y_0 \in C([0,T]; Y_{\gamma,c}). \qquad (3.84)
\end{aligned}
$$

We then introduce new variables

$$\hat{\phi}(t) \equiv \phi_t(t), \qquad \hat{\eta}(t) \equiv \eta_t(t), \tag{3.85}$$

so that, by (3.84), (3.85), and (2.5),

$$[\hat{\phi}(t), -\hat{\phi}_t(t), \hat{\eta}(t)] \in C([0,T]; Y_{\gamma,c} = H_0^2(\Omega) \times H_0^1(\Omega) \times L_2(\Omega)) \tag{3.86}$$

solves the dual thermo-elastic problem (2.10). Accordingly, by invoking the usual boost of regularity of $\hat{\eta}$ on top of (3.83) [L-T.11, Eqn. (3.3), p. 36], we can write as $[\hat{\phi}_t, -\hat{\phi}_t, \hat{\eta}_t] \in C([0,T]; [\mathcal{D}(A^*_{\gamma,c})]')$:

$$\hat{\eta} \equiv \eta_t \in C([0,T]; L_2(\Omega)) \cap L_2(0,T; H_0^1(\Omega)), \; \hat{\eta}_t = \eta_{tt} \in C([0,T]; [\mathcal{D}(\mathcal{A})]') \cap L_2(0,T; H^{-1}(\Omega)), \tag{3.87}$$

continuously in $y_0 \in \mathcal{D}(A^*_{\gamma,c})$, hence continuously in $[x_1, x_2] \in X_{1,\gamma} \overset{\text{def}}{\equiv} [\mathcal{D}(A^*_{1,\gamma})]'$.

Step 2. We have

$$\begin{array}{c} [x_1, x_2] \in X_{1,\gamma} \equiv [\mathcal{D}(\mathbb{A}^*_{1,\gamma})]' : \text{ continuous } \rightarrow \hat{\eta} = \eta_t \in C([0,T]; L_2(\Omega)) \cap L_2(0,T; H_0^1(\Omega)) \\ \downarrow \text{ compact} \\ \eta_t \in C([0,T]; H^{-1}(\Omega)) \cap L_2(0,T; L_2(\Omega)), \end{array} \tag{3.88}$$

where the last step follows by Aubin's Lemma [A.1, $p = 2$] or Simon [S.1] on the continuous case, via the regularity in (3.87).

Step 3. The solution $\{\phi(t; y_0), \eta(t; y_0)\}$, given by (3.83), of the thermo-elastic dual problem (2.10) satisfies also the corresponding Kirchoff problem (3.16), whose solution may be written as in (3.43), (3.44). We are interested only in the component (3.44) of this solution. Thus, as in the hinged case [Eqn. (3.68)], we likewise define with $y_0 = [\phi_0, -\phi_1, 0] = \Pi_m^* \bar{y}_0$,

$$\begin{aligned} \begin{bmatrix} \tilde{\phi}(t; y_0) \\ -\tilde{\phi}_t(t; y_0) \end{bmatrix} &\overset{\text{def}}{\equiv} \Pi_m \mathcal{K}^*_\gamma(t) y_0 \\ &= \int_0^t e^{\mathbb{A}^*_{1,\gamma}(t-\tau)} \begin{bmatrix} 0 \\ \mathcal{A}_\gamma^{-1}[\alpha\eta_t(\tau; y_0) + \nabla\alpha \cdot \nabla\eta(\tau; y_0)] + \ell.o.t.(\tau) \end{bmatrix} d\tau, \end{aligned} \tag{3.89}$$

so that $\tilde{\phi}$ solves the problem (compare with (3.16)):

$$\begin{cases} \tilde{\phi}_{tt} - \gamma\Delta\tilde{\phi}_{tt} + \Delta^2\tilde{\phi} - \text{div}(\alpha\nabla(\alpha\tilde{\phi}_t)) = \alpha\eta_t(t; y_0) + \nabla\alpha \cdot \nabla\eta(t; y_0) + \ell.o.t. \text{ in } Q; & \text{(3.90a)} \\ \tilde{\phi}(0, \cdot) = 0, \; \tilde{\phi}_t(0, \cdot) = 0 \text{ in } \Omega; & \text{(3.90b)} \\ \tilde{\phi} \equiv 0, \; \dfrac{\partial\tilde{\phi}}{\partial\nu} \equiv 0 \text{ on } \Sigma, & \text{(3.90c)} \end{cases}$$

(compare with (3.59)) where $\ell.o.t. = \alpha\nabla\alpha \cdot \nabla\tilde{\phi}_t + \text{div}(\alpha\tilde{\phi}_t\nabla\alpha)$. In fact, we shall be interested in the problem which is obtained from (3.90) by differentiating in time; i.e., setting

$$z = \tilde{\phi}_t, \tag{3.91}$$

we shall be interested in the problem:

$$\begin{cases} z_{tt} - \gamma\Delta z_{tt} + \Delta^2 z - \operatorname{div}(\alpha\nabla(\alpha z_t)) = \alpha\eta_{tt}(t;y_0) + \nabla\alpha\cdot\nabla\eta_t(t;y_0) + (\ell.o.t.)'(t); & (3.92a) \\ z(0,\,\cdot\,) = z_0 = 0,\;\; z_t(0,\,\cdot\,) = z_1 \;\text{ in } \Omega; & (3.92b) \\ z \equiv 0,\;\; \dfrac{\partial z}{\partial\nu} \equiv 0 \;\text{ in } \Sigma, & (3.92c) \end{cases}$$

where, by either (3.89) or (3.90), we obtain via (3.91) as well as (3.87),

$$z_1 = z_t(0,\,\cdot\,) = \tilde{\phi}_{tt}(0,\,\cdot\,) = -\mathcal{A}_\gamma^{-1}[\alpha\eta_t(0;y_0) + \nabla\alpha\cdot\nabla\eta(0;y_0)] + (\ell.o.t.)(0) \in \mathcal{D}(\mathcal{A}); \quad (3.93)$$

$$\{z_0, z_1\} = \{z(0,\,\cdot\,), z_t(0,\,\cdot\,)\} \in \mathcal{D}(\mathbb{A}_{1,\gamma}^*). \quad (3.94)$$

Step 4. Claims. (a) We have the following maps with reference to (3.80)-(3.82), (3.92)-(3.94):

$$[x_1, x_2] \in X_{1,\gamma} \equiv [\mathcal{D}(\mathbb{A}_{1,\gamma}^*)]' \quad (3.95)$$

$\Downarrow$ continuous

$$\{z_0, z_1\} \in \mathcal{D}(\mathbb{A}_{1,\gamma}^*), \begin{cases} \eta_t \in C([0,T]; L_2(\Omega)) \cap L_2(0,T; H_0^1(\Omega)) \\ \nabla\eta_t \in L_2(0,T; L_2(\Omega)) \end{cases} \quad (3.96)$$

$\Downarrow$ compact

$$\{z_0, z_1\} \in Y_{1,\gamma}, \begin{cases} \eta_t \in C([0,T]; H^{-1}(\Omega)) \cap L_2(0,T; L_2(\Omega)) \\ \nabla\eta_t \in L_2(0,T; H^{-1}(\Omega)) \end{cases} \quad (3.97)$$

$\Downarrow$ continuous

$$\{z, z_t\} \in C([0,T]; H_0^2(\Omega) \times H_0^1(\Omega)) \quad (3.98)$$

$\Downarrow$ continuous

$$\Delta z|_\Sigma \in L_2(0,T; L_2(\Gamma)). \quad (3.99)$$

(b) Hence, the map

$$[x_1, x_2] \in X_{1,\gamma} \equiv [\mathcal{D}(\mathbb{A}^*_{1,\gamma})]' \to \Delta z|_\Sigma = \Delta\tilde{\phi}_t|_\Sigma \in L_2(0,T;L_2(\Gamma))$$

$$\text{is compact.} \tag{3.100}$$

Proof. The implications (3.95) → (3.96) → (3.97) were given in (3.88). The implication (3.97) → (3.98) → (3.99) follow as in the proof of [L-L.1, Chapter V] with multipliers $h \cdot \nabla z$ applied to problem (3.92). This reference deals only with the initial conditions as in (3.97), to obtain the boundary trace regularity (3.99) by the multiplier $h \cdot \nabla z$. All we need, in order to complete the proof of part (a), is to show that (say, with zero initial conditions $z_0 = z_1 = 0$) each separate forcing term $\alpha\eta_{tt}$ and $\nabla\alpha \cdot \nabla\eta_t$ on the right side of (3.92a) produces a solution $\{z, z_t\}$ with interior regularity as in (3.98).

This is surely the case for $\nabla\alpha \cdot \nabla\eta_t$, since by (3.87), $\nabla\alpha \cdot \nabla\eta_t \in L_2(0,T;L_2(\Omega))$, which is compactly embedded in $L_2(0,T;H^{-1}(\Omega))$ by Aubin's Lemma; while any forcing term $F \in L_1(0,T;H^{-1}(\Omega) \equiv [\mathcal{D}(\mathcal{A}_\gamma^{\frac{1}{2}})]')$ on the right side of (3.92a) always produces a solution $\{z, z_t\} \in C([0,T];Y_{1,\gamma} \equiv H_0^2(\Omega) \times H_0^1(\Omega))$, essentially by semigroup methods [Tr.2, pp. 414–15], since then $\mathcal{A}_\gamma^{-1}F \in L_1(0,T;\mathcal{D}(\mathcal{A}_\gamma^{\frac{1}{2}}))$, $\mathcal{A}_\gamma^{-\frac{1}{2}}F \in L_1(0,T;L_2(\Omega))$ as needed. See also Theorem C.4 in connection with the implication (3.97) → (3.99).

It remains to show that the term $\alpha\eta_{tt}$ on the right-hand side of (3.92a) produces a solution $\{z, z_t\}$ as in (3.98), hence a trace as in (3.99). Upon multiplying problem (3.92) by the multiplier $h \cdot \nabla z$ we obtain the following additional contribution over [L-L.1, Chapter V],

$$\int_0^T (h \cdot \nabla z, \alpha\eta_{tt})_{L_2(\Omega)} = \left[(h \cdot \nabla z, \alpha\eta_t)_{L_2(\Omega)}\right]_0^T - \int_0^T (h \cdot \nabla z_t, \alpha\eta_t)_{L_2(\Omega)} dt, \tag{3.101}$$

and the energy method proof in [L-L.1, Chapter V], then continues to work by using the right side of (3.101), with η_t as in (3.97) to produce (3.98). One may, alternatively, using (3.101), obtain the trace (3.99) directly from (3.97) without explicitly passing through (3.98).

Part (b) is a direct consequence of Part (a) via (3.91).

Step 4. To show compactness of Q_c in (3.49b), it suffices to set $u_1 \equiv u_3 \equiv 0$, since these controls belong to a C_∞-class, and work with $u_2 \in L_2(\Sigma)$. As in (3.50), we set

$$\mathcal{B}_c = \Pi_m^* \mathcal{B}_m, \text{ where now } \mathcal{B}_m u = \begin{bmatrix} 0 \\ \mathcal{A}_\gamma^{-1} A G_2 u_2 \end{bmatrix}, \; \mathcal{B}_c = \begin{bmatrix} 0 \\ \mathcal{A}_\gamma^{-1} A G_2 u_2 \\ 0 \end{bmatrix}, \tag{3.102}$$

see (2.16) (right) or Appendix B, Eqn. (B.15) for $u_1 = u_3 = 0$.

Step 5. With reference to Q_c in (3.49a), let $Q_c^\#$ denote its adjoint with respect to the $X_{1,\gamma} \equiv [\mathcal{D}(\mathbb{A}^*_{1,\gamma})]'$-topology. Then, if $[x_1, x_2] \in X_{1,\gamma} \equiv [\mathcal{D}(\mathbb{A}^*_{1,\gamma})]'$, we have

(a)

$$\left(Q_c^\# \begin{bmatrix} x_1 \\ x_2 \end{bmatrix}\right)(t) = \mathcal{B}_m^* \Pi_m \mathcal{K}_\gamma^*(T-t) y_0 = \Delta\tilde{\phi}_t(T-t; y_0)|_\Gamma, \tag{3.103}$$

where y_0 is defined by (3.81), (3.82) in terms of $[x_1, x_2]$, where $\tilde{\phi}$ is defined by (3.89) or (3.90), and where $*$ denotes the adjoint with respect to the space $Y_{1,\gamma}$.

(b)

$$Q_c^{\#} : X_{1,\gamma} \equiv [\mathcal{D}(\mathbb{A}_{1,\gamma}^*)]' \to L_2(0,T; L_2(\Gamma)) \text{ is compact,} \tag{3.104}$$

or Q_c is compact, as stated in (3.49b).

Proof. By the definition (3.49a) of Q_c, by (3.102), and by (3.89), we compute

$$\left(Q_c u_2, \begin{bmatrix} x_1 \\ x_2 \end{bmatrix}\right)_{X_{1,\gamma}} = \left(\Pi_m \int_0^T \mathcal{K}_\gamma(T-t)\Pi_m^* \mathcal{B}_m u_2(t)dt, \begin{bmatrix} x_1 \\ x_2 \end{bmatrix}\right)_{[\mathcal{D}(\mathbb{A}_{1,\gamma}^*)]'} \tag{3.105}$$

$$= \left(\Pi_m \int_0^T \mathcal{K}_\gamma(T-t)\Pi_m^* \mathcal{B}_m u_2(t)dt, \mathbb{A}_{1,\gamma}^{*-1}\mathbb{A}_{1,\gamma}^{-1} \begin{bmatrix} x_1 \\ x_2 \end{bmatrix}\right)_{Y_{1,\gamma}} \tag{3.106}$$

$$(\text{by } (3.81)) = \int_0^T \left(\mathcal{B}_m u_2(t), \Pi_m \mathcal{K}_\gamma^*(T-t)\Pi_m^* \begin{bmatrix} \phi_0 \\ -\phi_1 \end{bmatrix}\right)_{Y_{1,\gamma}} dt \tag{3.107}$$

$$(\text{by } (3.89), (3.102)) = \int_0^T \left(\begin{bmatrix} 0 \\ A_\gamma^{-1} A G_2 u_2(t) \end{bmatrix}, \begin{bmatrix} \tilde{\phi}(T-t; y_0) \\ -\tilde{\phi}_t(T-t; y_0) \end{bmatrix}\right)_{Y_{1,\gamma}} dt \tag{3.108}$$

$$(\text{by } (2.2)) = \int_0^T (AG_2 u_2(t), -\tilde{\phi}_t(T-t; y_0))_{L_2(\Omega)} dt \tag{3.109}$$

$$= \int_0^T (u_2(t), -G_2^* A \tilde{\phi}_t(T-t; y_0))_{L_2(\Gamma)} dt \tag{3.110}$$

$$(\text{by } (3.37)) = \int_0^T (u_2(t), \Delta\tilde{\phi}_t(T-t; y_0)|_\Gamma)_{L_2(\Gamma)} dt \tag{3.111}$$

$$(\text{by } (3.107)) = \int_0^T (u_2(t), \mathcal{B}_m^* \Pi_m \mathcal{K}_\gamma^*(T-t) y_0)_{L_2(\Gamma)} dt$$

$$= \left(u_2, Q_c^{\#} \begin{bmatrix} x_1 \\ x_2 \end{bmatrix}\right)_{X_{1,\gamma} \equiv [\mathcal{D}(\mathbb{A}_{1,\gamma}^*)]'}. \tag{3.112}$$

Then, (3.111), (3.112) prove (3.103). From here, compactness as in (3.104) follows by (3.100).

The proof of Theorem 3.2 is complete. □

4 Consequence of the Structural Decomposition: A Strategy for the Controllability Problem

4.0 A Two-Step Strategy

Section 6 of [L-T.7] presents a strategy, essentially already used in [L-T.3, p. 119–120], to obtain an exact controllability (surjectivity) result. In the present case of thermo-elastic plates, its applicability is based on the structural decomposition Theorem 3.3, combined with a soft argument as in Appendix D. This is amply elaborated in [L-T.7, Section 6.2].

The target spaces of (the sought-after) controllability—of either the thermo-elastic mixed problems (1.1), (1.2), or (1.3); or else the corresponding Kirchoff mixed problems (3.2a–b)—are different in the two (hinged or clamped) boundary cases under consideration: see the statement of Theorems 1.1 and 1.2. Accordingly, to unify the exposition of the present section, we shall introduce the following spaces. First, we recall the target spaces already defined in the proof of Theorem 3.2 [(3.47b) in the hinged case, and (3.49b) in the clamped case]:

$$X_{1,\gamma} \equiv \begin{cases} Y_{1,\gamma} \equiv \mathcal{D}(\mathcal{A}) \times \mathcal{D}(\mathcal{A}_\gamma^{\frac{1}{2}}) \\ \qquad = [H^2(\Omega) \cap H_0^1(\Omega)] \times H_0^1(\Omega) \text{ (hinged case)} & (4.0.1a) \\ [\mathcal{D}(\mathbb{A}_{1,\gamma}^*)]' \equiv \mathcal{D}(A^{\frac{1}{4}}) \times \tilde{L}_2(\Omega) \equiv H_0^1(\Omega) \times \tilde{L}_2(\Omega) \text{ (clamped case),} & (4.0.1b) \end{cases}$$

which are the target spaces of (the sought-after) exact controllability of the Kirchoff boundary control problems (3.2a–b). For $Y_{1,\gamma}$ in (4.0.1a), we recall (3.8). For the dual space in (4.0.1b), we recall (3.12). Next, we introduce the spaces X_γ,

$$X_\gamma \equiv \begin{cases} X_{1,\gamma} \times \mathcal{D}(\mathcal{A}^{\frac{1}{2}}) \equiv [H^2(\Omega) \cap H_0^1(\Omega)] \times H_0^1(\Omega)] \times H_0^1(\Omega) \subset Y_{\gamma,h} \\ \qquad = Y_{1,\gamma} \times L_2(\Omega) \text{ (hinged case)} & (4.0.2a) \\ X_{1,\gamma} \times H_\theta \equiv H_0^1(\Omega) \times \tilde{L}_2(\Omega) \times H_\theta \text{ (clamped case),} & (4.0.2b) \end{cases}$$

H_θ defined in (1.9), which are the target spaces of (the sought-after) exact controllability in the mechanical variables $\{w, w_t\}$, and, simultaneously, of approximate controllability of the thermal variable, for the whole thermo-elastic boundary control problem (1.1), (1.2), or (1.3), according to Theorems 1.1 and 1.2.

We have [recall (2.5), (2.8b), (2.9), (1.13)]

$$X_\gamma = X_{1,\gamma} \times H_\theta \subset [\mathcal{D}(\mathbb{A}_{\gamma,c}^*)]' \equiv H_0^1(\Omega) \times \tilde{L}_2(\Omega) \times [\mathcal{D}(\mathcal{A})]', \tag{4.0.2c}$$

where the first duality is with respect to $Y_{\gamma,c}$ as a pivot space, while the second duality is with respect to $L_2(\Omega)$ as a pivot space.

We note the following property:

$$\begin{cases} \text{the space } X_\gamma \text{ is invariant under the action of the thermo-elastic semi-} \\ \text{group of Lemma 2.1: } e^{\mathbb{A}_\gamma t} : X_\gamma \to X_\gamma \text{, see Lemma D.2 (hinged case)} \\ \text{and Lemma D.3 (clamped case).} \end{cases} \tag{4.0.3}$$

Having introduced the space X_γ, we then recall the input-solution operator $\mathcal{L}_T$ at the terminal time T [from (3.46) and (3.48), in the two cases, ultimately from Appendix B]:

$$\mathcal{L}_T u = \{w(T), w_t(T), \theta(T)\} = \int_0^T e^{\mathbb{A}_\gamma (T-t)} \mathcal{B} u(t) dt \tag{4.0.4a}$$

$$: \text{continuous } \mathcal{U} \to X_\gamma, \tag{4.0.4b}$$

$\mathcal{U}$ being the space of controls $\{u_1, u_2, u_3\}$ of the class specified in Theorem 1.1 (hinged case) and Theorem 1.2 (clamped case). In particular, the regularity noted in (4.0.4b) follows by Proposition 2.2 on $u_2 \in L_2(\Sigma)$, plus the fact that u_1, u_3 belong to a C_∞-class.

The precise form of the boundary → interior operator $\mathcal{B}$ (which extends the operator $\mathcal{B}$ given by (2.16) to the case where $u_1, u_3 \neq 0$) is given in Appendix B: see (B.4) and (B.15) for both cases. Let Π_m be the projection $X_\gamma \to X_{1,\gamma}$, the spaces in (4.0.2), (4.0.1): $[v_1, v_2, v_3] \to [v_1, v_2]$ onto the mechanical state space and let $\Pi_m^* : [v_1, v_2] \to [v_1, v_2, 0]$ be its adjoint $X_{1,\gamma} \to X_\gamma$. The strategy for controllability, as stated in Theorems 1.1 and 1.2, hinges on the following two steps.

Henceforth, we let $y_0 = [w_0, w_1, \theta_0] \in X_\gamma$ be an arbitrary initial condition in the thermo-elastic controllability space.

Step 1. Show *exact controllability* from an arbitrary initial point $y_0 \in X_\gamma$ at $t = 0$ to an arbitrary target state in $X_{1,\gamma}$ at time $t = T$ of the *thermo-elastic plate problem* (1.1), (1.2), or (1.3), in *the mechanical variables*; in symbols, with reference to (4.0.2), it suffices to show that

$$\Pi_m \mathcal{L}_T : \text{ surjective } \mathcal{U} \text{ onto } X_{1,\gamma}, \tag{4.0.5}$$

where $\mathcal{U}$ is the preassigned space of controls $u = [u_1, u_2, u_3]$, specified in Theorems 1.1 and 1.2.

For then, for any $y_0 \in X_\gamma$, property (4.0.5) then implies

$$\Pi_m e^{\mathbf{A}_\gamma T} y_0 + \Pi_m \mathcal{L}_T \mathcal{U} = X_{1,\gamma}, \tag{4.0.6}$$

since $\Pi_m e^{\mathbf{A}_\gamma T} y_0 \in X_{1,\gamma}$ by the invariance property (4.0.3).

Step 2. Show *approximate controllability* at time $t = T$ in the space X_γ defined in (4.0.2), from any initial condition $y_0 \in X_\gamma$, of the *thermo-elastic plate* (1.1), (1.2), or (1.3): in symbols, it suffices to show that the range of $\mathcal{L}_T$ is dense in X_γ,

$$\overline{\mathcal{L}_T \mathcal{U}} \equiv \overline{\mathcal{R}(\mathcal{L}_T)} = X_\gamma; \qquad \mathcal{R} = \text{range}. \tag{4.0.7}$$

For then, if $y_0 \in X_\gamma$ and if $x_T \in X_\gamma$, given $\epsilon > 0$, (4.0.7) implies that there exists $u \in \mathcal{U}$ such that

$$\|[e^{\mathbf{A}_\gamma T} y_0 + \mathcal{L}_T u] - x_T\|_{X_\gamma} < \epsilon, \tag{4.0.8}$$

since $[x_T - e^{\mathbf{A}_\gamma T} y_0] \in X_\gamma$ by the invariant property (4.0.3).

Once Step 1 and Step 2 are accomplished, a soft argument as in Appendix D, Theorem D.1, which in particular applies when $\mathcal{L}_T$ is continuous as noted in (4.0.4b), then shows the following:

Desired conclusion: Step 1 and Step 2 imply *exact controllability* in the space $X_{1,\gamma}$ at time $t = T$ from any point $y_0 \in X_\gamma$ of the thermo-elastic plate (1.1), (1.2), or (1.3) in the *mechanical variables* and, simultaneously, *approximate controllability* in the third (thermal) component space of X_γ at time $t = T$ *in the thermal variable*; i.e., precisely, the statements of Theorem 1.1 or, respectively, Theorem 1.2.

4.1 Implementation of Step 1: Exact Controllability of the Thermo-Elastic Plate Problem in the Mechanical Variables

As explained in [L-T.7, Section 6], it is at the level of implementing Step 1 that the structural decomposition of the thermo-elastic semigroup as in Theorem 3.2 is critically used. The key is the following simple result, essentially already used in [L-T.3, pp. 119–120], from approximate to exact controllability.

PROPOSITION 4.1.1. [L-T.7, Proposition 6.1.1]. *Let $J = S + Q$, where:*

(i) J is a closed operator $\mathcal{U} \subset \mathcal{D}(J) \to X$ with dense range $\overline{\mathcal{R}(J)} = X$ (approximate controllability); equivalently, with trivial null space of the adjoint $J^* : \mathcal{N}(J^*) = \{0\}$;

(ii) S is a closed, surjective operator: $\mathcal{U} \subset \mathcal{D}(S)$ onto X, where $\mathcal{D}(S) = \mathcal{D}(J)$;

(iii) Q is a compact operator: $\mathcal{U} \to X$.

Then, J is surjective $\mathcal{U} \subset \mathcal{D}(J)$ onto X (exact controllability).

To implement Step 1, i.e., (4.0.5), take $X = X_{1,\gamma}$, and with reference to the decomposition (3.12) of Theorem 3.2, we return to (3.39) and take

$$Ju \equiv \Pi_m \mathcal{L}_T u \equiv \Pi_m \int_0^T e^{\mathbf{A}_\gamma (T-t)} \mathcal{B}u(t)dt; \tag{4.1.1}$$

$$Su \equiv \int_0^T e^{\mathbf{A}_{1,\gamma}(T-t)} \Pi_m \mathcal{B}u(t)dt \stackrel{\text{def}}{\equiv} \mathcal{L}_{m,T}u; \tag{4.1.2}$$

$$Qu \equiv \Pi_m \int_0^T \mathcal{K}_\gamma(T-t)\mathcal{B}u(t)dt. \tag{4.1.3}$$

Assumption (i) of Proposition 4.1.1 then means that $\Pi_m \mathcal{L}_T$ has dense range in $X_{1,\gamma}$: $\overline{\mathcal{R}(\Pi_m \mathcal{L}_T)} = X_{1,\gamma}$: but this is, *a fortiori*, assured by the more demanding condition of Step 2.

Assumption (iii) of Proposition 4.1.1 on compactness of the corresponding operator $Q : \mathcal{U} \to X_{1,\gamma}$ was verified in Theorem 3.2.

Finally, one needs to verify Assumption (ii) of Proposition 4.1.1. More precisely, one needs to establish the following *exact controllability results* of the Kirchoff equation (3.2).

THEOREM 4.1.2. (hinged B.C.) *Let $\emptyset \neq \Gamma_1$, $\Gamma_2 \subset \Gamma$ be open subsets of the boundary Γ, with non-empty intersection of positive measure (we think of Γ_1 as arbitrarily small, so that, without loss of generality, we may always assume that $\emptyset \neq \Gamma_1 \subset \Gamma_2$). Moreover, regarding Γ_2, we assume that: there exists a point $x_0 \in \mathbb{R}^2$, such that*

$$(x - x_0) \cdot \nu(x) \leq 0 \quad \textit{for } x \in \Gamma \setminus \Gamma_2, \tag{4.1.4}$$

with $\nu(x)$ the unit outward normal at $x \in \Gamma$. Let

$$T_{0,h} = 2\sqrt{\gamma} \max_i \sup_{x \in \Omega} \; dist(x, \Gamma_i), \quad i = 1, 2 \tag{4.1.5}$$

(h stands for 'hinged'). Let $\alpha \in C^2(\overline{\Omega})$. *Finally, let* $\{v_0, v_1\}$ *and* $\{v_{0,T}, v_{1,T}\}$ *be pre-assigned initial and target states of the (mechanical) v-problem (3.2), with*

$$\{v_0, v_1\} \text{ and } \{v_{0,T}, v_{1,T}\} \in [H^2(\Omega) \cap H_0^1(\Omega)] \times H_0^1(\Omega). \tag{4.1.6}$$

Then, for any $T > T_{0,h}$, *there exist control functions*

$$u_1 = \begin{cases} \bar{u}_1 \in C_0^\infty(\Sigma_1) \\ 0 \text{ on } \Sigma - \Sigma_1 \end{cases}; \quad u_2 = \begin{cases} \bar{u}_2 \in L_2(\Sigma_2) \\ 0 \text{ on } \Sigma - \Sigma_2 \end{cases}, \tag{4.1.7}$$

such that the corresponding solution to Eqn. (3.2a) with controls $\{u_1, u_2\}$ *in (4.1.7) in the* hinged B.C. *(3.2b) (left), satisfies the terminal condition*

$$v(T) = v_{0,T}, \qquad v_t(T) = v_{1,T}. \tag{4.1.8}$$

The proof of Theorem 4.1.2 is given in two steps as announced in [E-L-T.1, Section 5].

First, in Section 5.1, the property of exact controllability for the v-problem (3.2) claimed in Theorem 4.1.2 is reformulated, by duality, as an equivalent continuous observability inequality (Eqn. (5.1.11)), for the dual homogeneous ψ-problem (3.13). Next, Section 6 establishes validity of the continuous observability inequality (5.1.11) for ψ: see Theorem 6.3.

THEOREM 4.1.3. (clamped B.C.) *Let* $\emptyset \neq \Gamma_1 \subset \Gamma$ *be a non-empty open subset of the boundary* Γ *with positive measure (we think of* Γ_1 *as arbitrarily small). Let*

$$T_{0,c} = 2\sqrt{\gamma} \sup_{x \in \Omega} dist(x, \Gamma_1) \tag{4.1.9}$$

(c stands for 'clamped'). Let $\alpha \in C^2(\bar{\Omega})$. *Finally, let* $\{v_0, v_1\}$ *and* $\{v_{0,T}, v_{1,T}\}$ *be pre-assigned initial and target states of the Kirchoff v-problem (3.2), with*

$$\{v_0, v_1\} \text{ and } \{v_{0,T}, v_{1,T}\} \in H_0^1(\Omega) \times \tilde{L}_2(\Omega), \tag{4.1.10}$$

recalling (1.13). Then, for any $T > T_{0,c}$, *there exist control functions*

$$u_1 = \begin{cases} \hat{u}_1 \in C^\infty(\Sigma_1) \\ 0 \text{ on } \Sigma - \Sigma_1 \end{cases}; \qquad u_2 \in L_2(\Sigma), \tag{4.1.11}$$

such that the corresponding solution to Eqn. (3.2a) with controls $\{u_1, u_2\}$ *in (4.1.11) in the* clamped *B.C. (3.2b) (right), satisfies the terminal condition*

$$v(T) = v_{0,T}, \qquad v_t(T) = v_{1,T}. \tag{4.1.12}$$

The proof of Theorem 4.1.3 is likewise given in two steps. First, in Section 5.2, the property of exact controllability for the v-problem (3.2) claimed in Theorem 4.1.3 is reformulated, by duality, as an equivalent continuous observability inequality (Eqn. (5.2.11)) for the dual homogeneous ψ-problem (3.13). Next, Section 7 establishes the validity of the continuous observability inequality (5.2.11) for ψ: see Theorem 7.5.

REMARK 4.1.1. *We note that Theorems 4.1.2 and 4.1.3 do* not *follow from known results [La.1], [L-T.4], [L-L.1], [Li.1], [K.1], and references therein. Indeed, the two main novelties of Theorems 4.1.2, 4.1.3 over known literature are: (i) the coefficient α is space variable dependent; and, consequently, (ii) the control function u_1 has arbitrarily small support on Γ. These two factors contribute additional technical difficulties and the techniques/methods in the quoted literature are no longer directly applicable.*

Once Theorem 4.1.2 or Theorem 4.1.3 are established—in Sections 6 and 7, respectively—application of the abstract Proposition 4.1.1 specialized to (4.1.1)–(4.1.3) yields the desired *exact controllability of the thermo-elastic plate in the mechanical variables only.*

THEOREM 4.1.4. *Let $T > T_{0,h}$ or $T > T_{0,c}$, see (4.1.5) and (4.1.9), in the hinged or clamped cases, respectively. Under the assumptions and setting of Theorem 4.1.2, or Theorem 4.1.3, respectively, we have that the thermo-elastic plate problem (1.1), (1.2) is exactly controllable on the space $X_{1,\gamma}$ defined in (4.0.1a) of the mechanical variables $\{w, w_t\}$, by means of the boundary controls specified in (1.7) or (1.10), respectively.*

We next provide the continuous observability inequality, for the dual problem (2.10), which corresponds to the property of exact controllability (*a-fortiori* established in Theorem 4.1.4) of the therm-elastic problem (1.1), (1.2), with boundary controls $\{u_1, u_2, u_3\}$ of a clean, larger class than that specified in (1.7), (1.10), respectively. We distinguish the hinged and the clamped cases in two separate statements.

COROLLARY 4.1.5. (hinged B.C.) *Let $T > T_{0,h}$, see (4.1.5). Then, according to Theorem 4.1.4, the thermo-elastic mixed problem (1.1), (1.2) (hinged case) is exactly controllable on $[0,T]$ in the mechanical variables $\{w, w_t\}$, in the state space $X_{1,\gamma} \equiv Y_{1,\gamma} = \Pi_m Y_{\gamma,h} = [H^2(\Omega) \cap H_0^1(\Omega)] \times H_0^1(\Omega)$, see (4.0.1a), (2.3), within the following class $\mathcal{U}$ of controls [compare with (1.7) and with Appendix C, Eqn. (C.2b)]*

$$u_1 \in H_0^2(0,T; L_2(\Gamma_1)); \;\; u_2 \in L_2(0,T; L_2(\Gamma_2)); \;\; u_3 \in L_2(0,T; L_2(\Gamma_3)), \tag{4.1.13}$$

with $\emptyset \neq \tilde{\Gamma} \equiv \Gamma_1 = \Gamma_3 \subset \Gamma_2$, as in the last statement of Remark 1.2. Equivalently, *the following continuous observability inequality holds true for the dual thermo-elastic problem (2.10) (hinged case), with initial condition $\{\phi_0, \phi_1\} \in Y_{1,\gamma}$ and $\eta_0 = 0$ at $t = T$: there exists a constant $C_T > 0$ such that*

$$\left\| \frac{\partial \phi_t}{\partial \nu} \right\|^2_{L_2(0,T;L_2(\Gamma_2))} + \left\| \frac{\partial \eta}{\partial \nu} \right\|^2_{L_2(0,T;L_2(\tilde{\Gamma}))} + \left\| \frac{\partial \Delta \phi_t}{\partial \nu} \right\|^2_{[H_0^2(0,T;L_2(\tilde{\Gamma}))]'}$$

$$\geq C_T \|\{\phi_0, \phi_1\}\|^2_{Y_{1,\gamma}} \qquad Y_{1,\gamma} = [H^2(\Omega) \cap H_0^1(\Omega)] \times H_0^1(\Omega) \tag{4.1.14}$$

(duality with respect to $L_2(0,T; L_2(\tilde{\Gamma}))$).

Proof. **Step 1.** The surjectivity condition (4.0.5) (a restatement of the exact controllability property on $X_{1,\gamma}$ of the present corollary), with $\mathcal{U}$ as in (4.1.13), is *equivalent*, by a standard result [T-L.1, p. 235], to the following inequality: there exists $C_T > 0$ such that

$$\left\| \mathcal{L}_T^* \Pi_m^* \begin{bmatrix} \phi_0 \\ \phi_1 \end{bmatrix} \right\|_{[\mathcal{U}]'} \geq C_T \|\{\phi_0, \phi_1\}\|_{X_{1,\gamma} \equiv Y_{1,\gamma}}, \tag{4.1.15}$$

where, as usual, $\Pi_m^*[\phi_0, \phi_1] = [\phi_0, \phi_1, 0]$, see below (4.0.4), as well as (4.0.1a).

Step 2. By identity (C.5) of Lemma C.1 in Appendix C, then inequality (4.1.15) is equivalent to: there exists a constant $C_T > 0$ such that the solution of the dual thermo-elastic problem (2.10) (hinged case) with initial condition $\{\phi_0, \phi_1\} \in Y_{1,\gamma} \equiv X_{1,\gamma}$, see (4.0.1a), and $\eta_0 = 0$ at time $t = T$, satisfies for $T > T_{0,h}$ and $\Sigma_i = (0, T] \times \Gamma_i$, $i = 2, 3$:

$$\begin{aligned} \left\| \frac{\partial \phi_t}{\partial \nu} \right\|^2_{L_2(\Sigma_2)} &+ \left\| \frac{\partial \Delta \phi_t}{\partial \nu} - \alpha \frac{\partial \eta_t}{\partial \nu} - \gamma \frac{\partial \phi_{ttt}}{\partial \nu} \right\|^2_{[H_0^2(0,T;L_2(\Gamma_1))]'} \\ &+ \left\| \frac{\alpha \partial \phi_t}{\partial \nu} + \frac{\partial \eta}{\partial \nu} \right\|^2_{L_2(\Sigma_3)} \geq C_T \|\{\phi_0, \phi_1\}\|^2_{Y_{1,\gamma}}. \end{aligned} \tag{4.1.16}$$

Step 3. Inequality (4.1.16) is, in turn, equivalent to the claimed inequality (4.1.14). This is so since with $\emptyset \neq \tilde{\Gamma} = \Gamma_1 = \Gamma_3 \subset \Gamma_2$, and thus $\tilde{\Sigma} = \Sigma_1 = \Sigma_3 \subset \Sigma_2$, as assumed, we have

$$\begin{aligned} &\left\| \frac{\partial \Delta \phi_t}{\partial \nu} - \alpha \frac{\partial \eta_t}{\partial \nu} - \gamma \frac{\partial \phi_{ttt}}{\partial \nu} \right\|_{[H_0^2(0,T;L_2(\Gamma_1))]'} + \left\| \frac{\alpha \partial \phi_t}{\partial \nu} + \frac{\partial \eta}{\partial \nu} \right\|^2_{L_2(\Sigma_3)} \\ &\leq C_{\alpha,\gamma} \left\{ \left\| \frac{\partial \Delta \phi_t}{\partial \nu} \right\|_{[H_0^2(0,T;L_2(\Gamma_1))]'} + \left\| \frac{\partial \eta}{\partial \nu} \right\|_{L_2(\tilde{\Sigma})} + \left\| \frac{\partial \phi_t}{\partial \nu} \right\|^2_{L_2(\Sigma_2)} \right\} \end{aligned} \tag{4.1.17}$$

in one direction (4.1.16) $\Rightarrow$ (4.1.14). As to the other direction, we first note that

$$\left\| \frac{\partial \eta}{\partial \nu} \right\|_{L_2(\tilde{\Sigma})} \leq \left\| \alpha \frac{\partial \phi_t}{\partial \nu} + \frac{\partial \eta}{\partial \nu} \right\|_{L_2(\tilde{\Sigma})} + C_\alpha \left\| \frac{\partial \phi_t}{\partial \nu} \right\|_{L_2(\Sigma_2)}; \tag{4.1.18}$$

$$\left\| \frac{\partial \Delta \phi_t}{\partial \nu} \right\|_{[H_0^2(0,T;L_2(\tilde{\Gamma}))]'} \leq \left\| \frac{\partial \Delta \phi_t}{\partial \nu} - \alpha \frac{\partial \eta_t}{\partial \nu} - \gamma \frac{\partial \phi_{ttt}}{\partial \nu} \right\|_{[H_0^2(0,T;L_2(\tilde{\Gamma}))]'} + \left\| \alpha \frac{\partial \eta}{\partial \nu} + \gamma \frac{\partial \phi_t}{\partial \nu} \right\|_{L_2(\tilde{\Sigma})}, \tag{4.1.19}$$

where the last term in (4.1.19) satisfies

$$\begin{aligned} \left\| \alpha \frac{\partial \eta}{\partial \nu} + \gamma \frac{\partial \phi_t}{\partial \nu} \right\|_{L_2(\tilde{\Sigma})} &= \left\| \alpha \left[\frac{\partial \eta}{\partial \nu} - \alpha \frac{\partial \phi_t}{\partial \nu} \right] + (\alpha^2 + \gamma) \frac{\partial \phi_t}{\partial \nu} \right\|_{L_2(\tilde{\Sigma})} \\ &\leq C_{\alpha,\gamma} \left\{ \left\| \frac{\partial \eta}{\partial \nu} - \alpha \frac{\partial \phi_t}{\partial \nu} \right\|_{L_2(\tilde{\Sigma})} + \left\| \frac{\partial \phi_t}{\partial \nu} \right\|_{L_2(\Sigma_2)} \right\}. \end{aligned} \tag{4.1.20}$$

Inserting (4.1.20) into (4.1.19), and using (4.1.18), yields then

$$\left\|\frac{\partial\eta}{\partial\nu}\right\|_{L_2(\tilde\Sigma)} + \left\|\frac{\partial\Delta\phi_t}{\partial\nu}\right\|_{[H_0^2(0,T;L_2(\tilde\Gamma))]'} \leq C_{\alpha,\gamma}\left\{\left\|\frac{\partial\eta}{\partial\nu} - \alpha\frac{\partial\phi_t}{\partial\nu}\right\|_{L_2(\tilde\Sigma)} + \left\|\frac{\partial\phi_t}{\partial\nu}\right\|_{L_2(\Sigma_2)}\right\}$$

$$+ \left\|\frac{\partial\Delta\phi_t}{\partial\nu} - \alpha\frac{\partial\eta_t}{\partial\nu} - \gamma\frac{\partial\phi_{ttt}}{\partial\nu}\right\|_{[H_0^2(0,T;L_2(\tilde\Gamma))]'} \tag{4.1.21}$$

and then (4.1.14) implies (4.1.16), as desired. Corollary 4.1.5 is proved. □

COROLLARY 4.1.6. (clamped B.C.) *Let $T > T_{0,c}$, see (4.1.9). Then, according to Theorem 4.1.4, the thermo-elastic mixed problem (1.1), (1.2) (clamped case) is exactly controllable on $[0,T]$ in the mechanical variables $\{w, w_t\}$ in the state space $X_{1,\gamma} \equiv [\mathcal{D}(\mathbb{A}_{1,\gamma}^*)]' = H_0^1(\Omega) \times \tilde L_2(\Omega)$ (see (4.0.1b)), within the following class $\mathcal{U}$ of controls [compare with (1.10) and with Appendix C, Eqn. (C.15)]*

$$u_1 \in H_0^2(0,T;L_2(\Gamma_1));\; u_2 \in L_2(0,T;L_2(\Gamma));\; u_3 \in L_2(0,T;L_2(\Gamma_3)), \tag{4.1.22}$$

with $\emptyset \neq \tilde\Gamma = \Gamma_1 = \Gamma_3 \subset \Gamma$, as in the last statement of Remark 1.2. Equivalently, *the following continuous observability inequality holds true for the dual thermo-elastic problem (2.10) (clamped case), with initial condition*

$$\{\phi_0, \phi_1\} \in \mathcal{D}(A^{\frac{3}{4}}) \times \mathcal{D}(A^{\frac{1}{2}}) = [H^3(\Omega) \cap H_0^2(\Omega)] \times H_0^2(\Omega) \text{ and } \eta_0 = 0 \text{ and } t = T, \tag{4.1.23}$$

there exists a constant $C_T > 0$ such that, with $\Sigma = (0,T] \times \Gamma$, $\tilde\Sigma = (0,T] \times \tilde\Gamma$, we have:

$$\|\Delta\phi_t\|^2_{L_2(\Sigma)} + \left\|\frac{\partial\eta}{\partial\nu}\right\|^2_{L_2(\tilde\Sigma)} + \left\|\frac{\partial\Delta\phi_t}{\partial\nu}\right\|^2_{[H_0^2(0,T;L(\tilde\Gamma))]'} \geq C_T\|\{\phi_0,\phi_1\}\|^2_{[H^3(\Omega)\cap H_0^2(\Omega)]\times H_0^2(\Omega)}. \tag{4.1.24}$$

Proof. (of Corollary 4.1.6.) As in the proof of Corollary 4.1.5, the surjectivity condition (4.0.5) (a restatement of the exact controllability property on $X_{1,\gamma} \equiv [\mathcal{D}(\mathbb{A}_{1,\gamma}^*]')$, with $\mathcal{U}$ as in (4.1.23) is *equivalent* to the following inequality: there exists $C_T > 0$ such that

$$\left\|\mathcal{L}_T^{\#}\Pi_m^*\begin{bmatrix} z_1 \\ z_2 \end{bmatrix}\right\|_{[\mathcal{U}]'} \geq C_T\|\{z_1,z_2\}\|_{X_{1,\gamma}\equiv[\mathcal{D}(\mathbb{A}_{1,\gamma}^*)]'} \tag{4.1.25}$$

where $\mathcal{L}_T^{\#}$ denotes the $[\mathcal{D}(\mathbb{A}_\gamma^*)]'$-adjoint of $\mathcal{L}_T$.

Let $\Pi_m^*[z_1,z_2] \equiv [z_1,z_2,0] \in [\mathcal{D}(\mathbb{A}_{\gamma,c}^*)]'$ and define

$$\begin{aligned} \bar y_0 &= [\phi_0, -\phi_1, \theta_0] = \mathbb{A}_{\gamma,c}^{*-1}\mathbb{A}_{\gamma,c}^{-1}[z_1,z_2,0] \in \mathcal{D}(\mathbb{A}_{\gamma,c}^*) = \mathcal{D}(\mathbb{A}_{\gamma,c}) \\ &= \mathcal{D}(\mathbb{A}_{1,\gamma}) \times \mathcal{D}(\mathcal{A}) = \mathcal{D}(A^{\frac{3}{4}}) \times \mathcal{D}(A^{\frac{1}{2}}) \times \mathcal{D}(\mathcal{A}). \end{aligned} \tag{4.1.26}$$

Thus,

$$\|\{z_1,z_2\}\|_{X_{1,\gamma}} = \|\{\phi_0,-\phi_1\}\|_{\mathcal{D}(A^{\frac{3}{4}})\times\mathcal{D}(A^{\frac{1}{2}})}. \tag{4.1.27}$$

Let $[\phi, -\phi_t, \eta] = e^{\mathbf{A}^*_{\gamma,c}t}\bar{y}_0$ be the solution of the dual thermo-elastic problem (2.10) (clamped case) with initial condition as in (4.1.26). Using the explicit expression for $(\mathcal{L}^{\#}_T[z_0, z_1, 0])(t)$ given by identity (C.17) of Lemma C.2 in Appendix C, as well as (4.1.27) and (4.1.22), we rewrite the continuous observability inequality (4.1.25) as

$$\|\Delta\phi_t\|^2_{L_2(\Sigma)} + \left\|\frac{\partial\eta}{\partial\nu}\right\|^2_{L_2(\Sigma_3)} + \left\|\frac{\partial\Delta\phi_t}{\partial\nu} + \alpha\frac{\partial\eta_t}{\partial\nu}\right\|^2_{[H^2_0(0,T;L_2(\Gamma_1))]'}$$
$$\geq C_T\|\{\phi_0, \phi_1\}\|^2_{[H^3(\Omega)\cap H^2_0(\Omega)]\times H^2_0(\Omega)}, \tag{4.1.28}$$

and (4.1.28) is, in turn, readily equivalent to Eqn. (4.1.24) for $\tilde{\Gamma} = \Gamma_1 = \Gamma_3 \subset \Gamma$. □

4.2 Implementation of Step 2 by Duality: Unique Continuation of Over-Determined Dual Thermo-Elastic Systems

Approximate controllability from the origin at $t = T$ of the thermo-elastic plate (1.1), (1.2), (1.3). Our task in this section is to address the denseness condition (4.0.7): $\overline{\mathcal{R}(\mathcal{L}_T)} = X_\gamma$ of Steps 2, and recast it in an *equivalent*, more amenable form, by duality.

Dual version: Uniqueness property of an over-determined plate problem. By duality, via the input-solution operator $\mathcal{L}_T$ in (4.0.4), condition (4.0.7) is *equivalent* the following injectivity (observability) condition:

$$\mathcal{N}\{\mathcal{L}^{\#}_T\} = \{0\}; \text{ or } \begin{cases} \{\mathcal{L}^{\#}_T\bar{y}_0\}(t) \equiv 0, \ 0 \leq t \leq T, \ \bar{y}_0 \in X_\gamma \\ \Rightarrow \bar{y}_0 = 0, \end{cases} \tag{4.2.1}$$

where $\mathcal{N}$ denotes the null space and # denotes the X_γ-dual. A precise version of the abstract injectivity condition (4.2.1) in terms of the traces of the solution $\{\phi, \eta\}$ of problem (2.10), with the respective B.C. (2.10d), whether hinged or clamped, is given in Appendix C, Corollary C.3a and Corollary C.3b, respectively. They provide the following result:

PROPOSITION 4.2.1. *In either case, hinged or clamped, let the operator $\mathcal{L}_T$ in (4.0.4) be defined as*

$$\mathcal{L}_T : H^2_0(0, T; L_2(\Gamma)) \times L_2(\Sigma) \times L_2(\Sigma) \to X_\gamma, \tag{4.2.2}$$

with X_γ given by (4.0.2). Then condition (4.2.1) is equivalently *restated as follows:*

(hinged case): here $\emptyset \neq \tilde{\Gamma} \equiv \Gamma_1 = \Gamma_3 \subset \Gamma_2$ (Remark 1.2); $X_\gamma \subset Y_\gamma$.

Let $\{\phi, \eta\} \in C([0,T]; Y_\gamma)$ *be a solution of the following over-determined problem:*

$$\begin{cases} \phi_{tt} - \gamma\Delta\phi_{tt} + \Delta^2\phi + div(\alpha(x)\nabla\eta) \equiv 0 \quad in\ Q; & (4.2.3a) \\ \eta_t - \Delta\eta - div(\alpha(x)\nabla\phi_t) \equiv 0 \quad in\ Q; & (4.2.3b) \\ \phi|_\Sigma \equiv 0,\ \Delta\phi|_\Sigma \equiv 0,\ \eta|_\Sigma \equiv 0\ on\ \Sigma;\ \left.\frac{\partial\Delta\phi_t}{\partial\nu}\right|_{\Sigma_1} \equiv 0\ on\ \Sigma_1; & \\ \left.\frac{\partial\phi_t}{\partial\nu}\right|_{\Sigma_2} \equiv 0\ on\ \Sigma_2;\ \left.\frac{\partial\eta}{\partial\nu}\right|_{\Sigma_3} \equiv 0\ on\ \Sigma_3; & (4.2.3c) \end{cases}$$

(clamped case): here $\emptyset \neq \tilde{\Gamma} = \Gamma_1 = \Gamma_3$ *(Remark 1.2);* $X_\gamma \subset \mathcal{D}(\mathbb{A}_{\gamma,c})$.
Let $\{\phi, \eta\} \in C([0,T]; \mathcal{D}(\mathbb{A}_\gamma))$ *be a solution of the following over-determined problem:*

$$\begin{cases} \phi_{tt} - \gamma\Delta\phi_{tt} + \Delta^2\phi + div(\alpha(x)\nabla\eta) \equiv 0 \quad in\ Q; & (4.2.4a) \\ \eta_t - \Delta\eta - div(\alpha(x)\nabla\phi_t) \equiv 0 \quad in\ Q; & (4.2.4b) \\ \phi|_\Sigma \equiv 0,\ \left.\frac{\partial\phi}{\partial\nu}\right|_\Sigma \equiv 0,\ \eta|_\Sigma \equiv 0\ on\ \Sigma;\ \left.\frac{\partial\Delta\phi_t}{\partial\nu}\right|_{\Sigma_1} \equiv 0\ on\ \Sigma_1; & \\ \Delta\phi_t|_\Sigma \equiv 0\ on\ \Sigma;\ \left.\frac{\partial\eta}{\partial\nu}\right|_{\Sigma_3} \equiv 0\ on\ \Sigma_3. & (4.2.4c) \end{cases}$$

Then, in either case, the initial condition vanishes:

$$\{\phi_0, -\phi_1, \eta_0\} = 0 \ \textit{and hence}\ \phi \equiv 0,\ \eta \equiv 0\ in\ Q. \tag{4.2.5}$$

Proof. Lemma C.1, Lemma C.2, Corollary C.3 in Appendix C *a fortiori* prove this result. □

REMARK 4.2.1. *We recall, e.g., [L-T.10, Proposition 3C.6 of Appendix C to Chapter 3, p. 305 that if* $\phi \in C^2(\bar{\Omega})$ *and* Γ *is of class* C^1, *then*

$$\Delta\phi|_\Gamma = \frac{\partial^2\phi}{\partial\nu^2} + \frac{\partial^2\phi}{\partial\tau^2} + \left(\frac{\partial\phi}{\partial\nu}\right) div\ \nu \quad on\ \Gamma, \tag{4.2.6}$$

where ν *and* τ *are outward unit normal and unit tangential vectors, respectively, the latter oriended counterclockwise. Thus, under clamped B.C., we have* $\Delta\phi|_\Gamma = \frac{\partial^2\phi}{\partial\nu^2}|_\Gamma$, *and one could then interchange* $\frac{\partial^2\phi}{\partial\nu^2}|_\Gamma = 0$ *with* $\Delta\phi|_\Gamma = 0$.

Similarly, if $\phi \in C^3(\bar{\Omega})$ *and* Γ *is of class* C^1, *then [L-T.10, Proposition 3C.11 of Appendix C to Chapter 3, p. 309],*

$$\left.\frac{\partial\Delta\phi}{\partial\nu}\right|_\Gamma = \frac{\partial}{\partial\nu}\left[\frac{\partial^2\phi}{\partial\nu^2} + \frac{\partial^2\phi}{\partial\tau^2}\right] + \frac{\partial}{\partial\nu}\left[\frac{\partial\phi}{\partial\nu}\ div\ \nu\right] \quad on\ \Gamma. \tag{4.2.7}$$

Thus, if $\phi|_\Gamma = \frac{\partial\phi}{\partial\nu}|_\Gamma = \frac{\partial^2\phi}{\partial\nu^2}|_\Gamma = 0$, *then* $\frac{\partial\Delta\phi}{\partial\nu}|_\Gamma = \frac{\partial^3\phi}{\partial\nu^3}|_\Gamma$ *and one could then interchange* $\frac{\partial^3\phi}{\partial\nu^3}|_\Gamma = 0$ *with* $\frac{\partial\Delta\phi}{\partial\nu}|_\Gamma = 0$.

With reference to the over-determined problem (4.2.3) or (4.2.4), the following uniqueness result, Theorem 4.2.2, holds true for T sufficiently large.

REMARK 4.2.2. To put Theorem 4.2.2 below into perspective, we remark that the only uniqueness result for thermo-elastic plates available in the literature is the recent one due to Isakov [I.1]: this requires, however, zero Cauchy data on *all* of $\Sigma = (0,T] \times \Gamma$, and concludes with the statement that, for $T > 0$ sufficiently large, then a solution $\{\phi, \eta\} \in H^3(Q) \times H^1(Q)$ of problem (4.2.5a-b-c) satisfies

$$\phi(T/2, \,\cdot\,) = \phi_t(T/2, \,\cdot\,) = \eta(T/2, \,\cdot\,) = 0. \tag{4.2.8}$$

This is not enough for our present controllability purposes, where, moreover, one needs to conclude that, in fact,

$$\{\phi_0, -\phi_1, \eta_0\} = 0, \text{ and, hence } \phi \equiv 0, \ \eta \equiv 0 \text{ in } Q \text{ (actually in } (0,\infty) \times \Omega). \tag{4.2.9}$$

Fortunately, the passage from (4.2.8) to (4.2.9) holds true. It is *a-fortiori* implied by a recent *backward uniqueness theorem* [L-R-T] which applies to general thermo-elastic plate equations with space variable coefficients (even in the principal parts) to include problem (4.2.3a-b), (4.2.4a-b), and under all canonical B.C.; in particular, hinged and clamped B.C. Paper [L-R-T] was precisely motivated by the result (4.2.8) in [I.1]. □

THEOREM 4.2.2. (Unique continuation for (4.2.3) or (4.2.4)) *Assume that $\{\phi, \eta\} \in H^4(Q) \times H^2(Q)$ is a solution to either the over-determined problem (4.2.3) in the hinged case, or else to the over-determined problem (4.2.4) in the clamped case, with zero Cauchy data in $\tilde{\Sigma} = (0,T] \times \tilde{\Gamma}$, $\emptyset \neq \tilde{\Gamma} \subset \Gamma = \partial\Omega$, $\tilde{\Gamma}$ being open and of positive measure, as in (4.2.3) or (4.2.4). [We could, alternatively, start with a pair $\{\phi, \eta\} \in H^3(Q) \times H^1(Q)$ of lower regularity which either solves (4.2.3a-b), as well as the B.C. (4.2.3c) with the time derivative sign* omitted*; or else, which solves (4.2.4a-b), as well as the B.C. (4.2.4c) with the time derivative sign* omitted.*] Let $T > \tilde{T}$, where $\tilde{T}$ is defined by*

$$\tilde{T} = 2\sqrt{\gamma} \sup_{x \in \Omega} dist(x, \tilde{\Gamma}). \tag{4.2.10}$$

Then:

(i) the vanishing at $t = T/2$: $\phi(T/2, \,\cdot\,) = \phi_t(T/2, \,\cdot\,) = \eta(T/2, \,\cdot\,) = 0$ as in (4.2.8) holds true.

(ii) Moreover, since either problem (4.2.3) with hinged/Dirichlet B.C. for $\{\phi, \eta\}$ on all of Σ (see (4.2.3c), or else problem (4.2.4) with clamped/Dirichlet B.C. for $\{\phi, \eta\}$ on all of Σ (see (4.2.4c)) generates a s.c. semigroup by Lemma 2.1, then the result of [L-R-T] applies, and the unique continuation conclusion (4.2.9) holds true.

The proof of Theorem 4.2.2(i) will be given in Section 11.

COROLLARY 4.2.3. (Unique continuation for (4.2.3) or (4.2.4)) *Let $\{\phi, \eta\} \in C([0,T]; Y_\gamma)$ be a solution of the over-determined problem (4.2.3) in the hinged case, and let $\{\phi, \eta\} \in$*

$C([0,T];\mathcal{D}(\mathbb{A}_{\gamma,c}))$ be a solution of the over-determined problem (4.2.4) in the clamped case, with zero data on $\tilde{\Sigma}$ as in (4.2.3c) or (4.2.4c), where $\tilde{\Gamma} = \Gamma_1 = \Gamma_3 \subset \Gamma_2$ in the hinged case and $\tilde{\Gamma} = \Gamma_1 = \Gamma_3$ in the clamped case. Assume the geometrical condition (1.4) in the hinged case. Then:

(a) we have, in fact, that $\{\phi_t, \eta_t\} \in H^3(Q) \times H^1(Q)$;

(b) consequently, for $T > \tilde{T}$ defined by (4.2.10), we have that $\phi \equiv \eta \equiv 0$ in Q as in (4.2.9).

Proof. (a) The boost of regularity is proved in Section 9. (b) In either case, we differentiate in time the equations [(4.2.3a–b) or (4.2.4a–b)], as well as the B.C. where the time derivative does not yet appear. We then apply Theorem 4.2.2 to the resulting problem. Theorem 4.2.2 then yields $\phi_t \equiv \eta_t \equiv 0$ in Q. Then, the boundary conditions $\phi \equiv 0$ and $\eta \equiv 0$ in $\tilde{\Sigma}$ imply that $\phi \equiv \eta \equiv 0$ in Q, as desired. □

REMARK 4.2.3. *In problems (4.2.3a–c) (hinged case) and (4.2.4a–c) (clamped case), the boundary conditions on the arbitrarily small portions of the boundaries Σ_1 and Σ_3 (where we may take $\Sigma_1 \equiv \Sigma_3$ without loss of generality, by the last statement of Remark 1.2) arise precisely by virtue of the presence of the non-zero controls u_1 and u_3 (see Appendix C). If one could prove the unique continuation result of Theorem 4.2.2 without the B.C. on Σ_1 and Σ_3, then one could dispense of the controls u_1 and u_3, and then one could take $u_1 \equiv u_3 \equiv 0$.*

5 Step 1: Continuous Observability Inequalities for the Homogeneous Kirchoff Problem (3.13)

In this section we return to the homogeneous Kirchoff problem (3.13) and seek, in both cases of hinged and clamped B.C., the corresponding continuous observability inequality. By duality, this inequality is then *equivalent* to the required exact boundary controllability property of the v-problem (3.2), with hinged, respectively clamped boundary controls, as stated by Theorem 4.1.2 (hinged case) or Theorem 4.1.3 (clamped case).

5.1 Hinged Case. The Proof of Theorem 4.1.2 Begins

Identifying the continuous observability inequality for problem (3.13) is the first step in the proof of Theorem 4.1.2. To this end, we return to the non-homogeneous problem (3.2) with boundary controls in the hinged B.C., whose abstract model is given by (3.17), or (3.21)–(3.23). Thus, with reference to the map L in (3.23), we define

$$L_T u \equiv (Lu)(T) = \int_0^T e^{\mathbb{A}_{1,\gamma}(T-t)} \mathcal{B}_{1,h} \begin{bmatrix} u_1(t) \\ u_2(t) \end{bmatrix} dt, \tag{5.1.1}$$

for $u = [u_1, u_2] \in H_0^k(\Sigma_1) \times L_2(\Sigma_2)$, where $\emptyset \neq \Gamma_2 \subset \Gamma_1$ are the subsets of Γ defined in the assumptions of Theorem 4.1.2. We seek to establish exact controllability (from the origin) of the v-problem (3.2) at the time T, on the state space $Y_{1,\gamma} = \mathcal{D}(\mathcal{A}) \times \mathcal{D}(\mathcal{A}_\gamma^{\frac{1}{2}})$ in (3.8), by means of boundary controls $[u_1, u_2] \in H_0^k(\Sigma_1) \times L_2(\Sigma_2)$, any $k \geq 2$, as stated in Theorem 4.1.2. *Equivalently,* we seek to establish surjectivity of the map L_T:

$$L_T : H_0^k(\Sigma_1) \times L_2(\Sigma_2) \to \text{ onto } X_{1,\gamma} \equiv Y_{1,\gamma} \equiv \mathcal{D}(\mathcal{A}) \times \mathcal{D}(\mathcal{A}_\gamma^{\frac{1}{2}}). \tag{5.1.2}$$

It is a standard result [T-L.1, p. 235] that the surjectivity condition (5.1.2) is, in turn, *equivalent* to the condition (traditionally referred to as a Continuous Observability Inequality)

$$\left\| L_T^* \begin{bmatrix} x_1 \\ x_2 \end{bmatrix} \right\|_{H^{-k}(\Sigma_1) \times L_2(\Sigma_2)} \geq c_T \left\| \begin{bmatrix} x_1 \\ x_2 \end{bmatrix} \right\|_{Y_{1,\gamma} = \mathcal{D}(\mathcal{A}) \times \mathcal{D}(\mathcal{A}_\gamma^{\frac{1}{2}})} \tag{5.1.3}$$

where $H^{-k}(\Sigma_1) = [H_0^k(\Sigma_1)]'$, for some constant $c_T > 0$, where the adjoint L_T^* of L_T in (5.1.1) is defined by the following identity involving duality pairings

$$\left(L_T \begin{bmatrix} u_1 \\ u_2 \end{bmatrix}, \begin{bmatrix} x_1 \\ x_2 \end{bmatrix} \right)_{Y_{1,\gamma}} = \left(\begin{bmatrix} u_1 \\ u_2 \end{bmatrix}, L_T^* \begin{bmatrix} x_1 \\ x_2 \end{bmatrix} \right)_{L_2(\Sigma) \times L_2(\Sigma)}, \tag{5.1.4}$$

where $[u_1, u_2] \in H_0^k(\Sigma_1) \times L_2(\Sigma_2)$ and $L_T^*[x_1, x_2] \in H^{-k}(\Sigma_1) \times L_2(\Sigma_2)$.

LEMMA 5.1.1. *Let $x = [\psi_0, \psi_1] \in Y_{1,\gamma}$. With reference to (5.1.1), (5.1.4), we have:*
(i)

$$\left(L_T^* \begin{bmatrix} \psi_0 \\ \psi_1 \end{bmatrix} \right)(t) = \begin{bmatrix} (1)(t) \\ (2)(t) \end{bmatrix}, \ (2)(t) = D^* A \psi_t(T-t;x) = -\frac{\partial \psi_t}{\partial \nu}(T-t;x); \tag{5.1.5}$$

$$\begin{aligned} -(1)(t) &= D^* A^2 \psi_t(T-t;x) + \gamma \frac{d^2}{dt^2} D^* A \psi_t(T-t;x) + \frac{d}{dt} D_\alpha^* F_\alpha^* \psi_t(T-t;x) && (5.1.6) \\ &\equiv \frac{\partial \Delta \psi_t}{\partial \nu}(T-t;x) + \alpha^2(x) \frac{d}{dt} \frac{\partial \psi_t}{\partial \nu}(T-t;x) - \gamma \frac{d^2}{dt^2} \frac{\partial \psi_t}{\partial \nu}(T-t;x), && (5.1.7) \end{aligned}$$

where $\psi(\cdot\,; x)$ solves the adjoint problem (3.13) with $\psi(0;x) = \psi_0$, $\psi_t(0;x) = \psi_1$; that is

$$\begin{bmatrix} \psi(T-t;x) \\ \psi_t(T-t;x) \end{bmatrix} = e^{\mathbf{A}_{1,\gamma}^*(T-t)} \begin{bmatrix} \psi_0 \\ \psi_1 \end{bmatrix} \in C([0,T]; Y_{1,\gamma}), \tag{5.1.8}$$

so that $\psi_t(T-t;x) \in C([0,T]; H_0^1(\Omega) \equiv \mathcal{D}(\mathcal{A}^{\frac{1}{2}}))$.
(ii) For $k \geq 2$, and $\emptyset \neq \Gamma_1 \subset \Gamma_2$:

$$\left\| L_T^* \begin{bmatrix} \psi_0 \\ \psi_1 \end{bmatrix} \right\|^2_{H^{-k}(\Sigma_1) \times L_2(\Sigma_2)} \equiv \|(1)\|^2_{H^{-k}(\Sigma_1)} + \left\| \frac{\partial \psi_t}{\partial \nu}(T-\cdot\,;x) \right\|^2_{L_2(\Sigma_2)} \tag{5.1.9}$$

$$\sim \left\{ \left\| \frac{\partial \Delta \psi_t}{\partial \nu}(T-\cdot\,;x) \right\|^2_{H^{-k}(\Sigma_1)} + \left\| \frac{\partial \psi_t(T-\cdot\,;x)}{\partial \nu} \right\|^2_{L_2(\Sigma_2)} \right\}, \tag{5.1.10}$$

where $\sim$ denotes the norm-equivalence sign.

(iii) The continuous observability inequality (5.1.3) is then equivalently rewritten as: there exists $c_T > 0$ such that

$$\left\{\int_0^T\int_{\Gamma_2}\left(\frac{\partial\psi_t(T-t;x)}{\partial\nu}\right)^2 d\Sigma_2 + \left\|\frac{\partial\Delta\psi_t}{\partial\nu}(T-\cdot;x)\right\|^2_{H^{-k}(\Sigma_1)}\right\}$$

$$\geq c_T\|\{\psi_0,\psi_1\}\|^2_{[H^2(\Omega)\cap H_0^1(\Omega)]\times H_0^1(\Omega)} \tag{5.1.11}$$

Proof. (i) Via (5.1.1), (3.22), (2.2), (5.1.8), and (3.8), we compute with $\{\psi_0,\psi_1\}\in Y_{1,\gamma}$:

$$\left(L_T\begin{bmatrix}u_1\\u_2\end{bmatrix},\begin{bmatrix}\psi_0\\\psi_1\end{bmatrix}\right)_{Y_{1,\gamma}} = \left(\int_0^T e^{\mathbf{A}_{1,\gamma}(T-t)}\mathcal{B}_{1,h}\begin{bmatrix}u_1(t)\\u_2(t)\end{bmatrix}dt,\begin{bmatrix}\psi_0\\\psi_1\end{bmatrix}\right)_{Y_{1,\gamma}} \tag{5.1.12}$$

$$= \int_0^T\left(\mathcal{B}_{1,h}\begin{bmatrix}u_1(t)\\u_2(t)\end{bmatrix},e^{\mathbf{A}^*_{1,\gamma}(T-t)}\begin{bmatrix}\psi_0\\\psi_1\end{bmatrix}\right)_{Y_{1,\gamma}}dt \tag{5.1.13}$$

(by (3.22), (5.1.8))

$$= \int_0^T\left(\begin{bmatrix}0\\\mathcal{A}_\gamma^{-1}[\mathcal{A}^2Du_1(t)-\mathcal{A}Du_2(t)+\gamma\mathcal{A}Du_{1tt}(t)-F_\alpha D_\alpha u_{1t}(t)\end{bmatrix},\right.$$

$$\left.\begin{bmatrix}\psi(T-t;x)\\-\psi_t(T-t;x)\end{bmatrix}\right)_{\mathcal{D}(\mathcal{A})\times\mathcal{D}(\mathcal{A}_\gamma^{\frac12})}dt \tag{5.1.14}$$

(by (3.8), (2.2))

$$= \int_0^T([\mathcal{A}^2Du_1(t)-\mathcal{A}Du_2(t)+\gamma\mathcal{A}Du_{1tt}(t)-F_\alpha D_\alpha u_{1t}(t)],-\psi_t(T-t;x))_{L_2(\Omega)}dt \tag{5.1.15}$$

$$= \int_0^T(u_1(t),-D^*\mathcal{A}^2\psi_t(T-t;x))_{L_2(\Gamma)}dt+\int_0^T(u_2(t),D^*\mathcal{A}\psi_t(T-t;x))_{L_2(\Gamma)}dt$$

$$-\int_0^T(u_{1tt}(t),\gamma D^*\mathcal{A}\psi_t(T-t;x))_{L_2(\Gamma)}dt+\int_0^T(u_{1t}(t),D_\alpha^*F_\alpha^*\psi_t(T-t;x))_{L_2(\Gamma)}dt. \tag{5.1.16}$$

Since, by assumption on the class of controls u_1, we have $u_1^{(n)}(T)=u_1^{(n)}(0)=0$ for all $n=0,1,\dots,k$, integrating by parts in t on the last two integral terms of (5.1.16) yields then, by (5.1.4) and (5.1.16):

$$\left(L_T\begin{bmatrix}u_1\\u_2\end{bmatrix},\begin{bmatrix}\psi_0\\\psi_1\end{bmatrix}\right)_{Y_{1,\gamma}} = \left(\begin{bmatrix}u_1\\u_2\end{bmatrix},L_T^*\begin{bmatrix}\psi_0\\\psi_1\end{bmatrix}\right)_{L_2(\Sigma)\times L_2(\Sigma)} \tag{5.1.17}$$

$$= \int_0^T\left(u_1(t),-\left[D^*\mathcal{A}^2\psi_t(T-t;x)+\gamma\frac{d^2}{dt^2}D^*\mathcal{A}\psi_t(T-t;x)+\frac{d}{dt}D_\alpha^*F_\alpha^*\psi_t(T-t;x)\right]\right)_{L_2(\Gamma)}dt$$

$$+\int_0^T(u_2(t),D^*\mathcal{A}\psi_t(T-t;x))_{L_2(\Gamma)}dt \tag{5.1.18}$$

$$= \left(\begin{bmatrix}u_1\\u_2\end{bmatrix},\begin{bmatrix}(1)\\(2)\end{bmatrix}\right)_{L_2(\Sigma)\times L_2(\Sigma)}, \tag{5.1.19}$$

where (1) and (2) are defined in (5.1.5) (left form) and (5.1.6). Then, invoking the trace results in (3.24) for ψ (which satisfies the B.C. (3.13b)) as well as (2.1), we see that we can rewrite (1) and (2) in the trace forms of (5.1.5) (right form) and (5.1.7). Part (i) is proved.

(ii) With reference to (5.1.5), (5.1.7), we now establish the inequality

$$\|(1)\|_{H^{-k}(\Sigma_1)} + \left\| \frac{\partial \psi_t}{\partial \nu}(T-\cdot;x) \right\|_{L_2(\Sigma_1)} \geq c \left\| \frac{\partial \Delta \psi_t}{\partial \nu}(T-\cdot;x) \right\|_{H^{-k}(\Sigma_1)}, \tag{5.1.20}$$

for some $c > 0$ and $k \geq 2$. Indeed, recalling (5.1.7), we estimate

$$\|(1)\|_{H^{-k}(\Sigma_1)} = \left\| \frac{\partial \Delta \psi_t}{\partial \nu}(T-\cdot;x) + \alpha^2 \frac{d}{dt}\frac{\partial \psi_t}{\partial \nu}(T-\cdot;x) - \gamma \frac{d^2}{dt^2}\frac{\partial \psi_t}{\partial \nu}(T-\cdot;x) \right\|_{H^{-k}(\Sigma_1)} \tag{5.1.21}$$

$$\geq \left\| \frac{\partial \Delta \psi_t}{\partial \nu} \right\|_{H^{-k}(\Sigma_1)} - c_\alpha \left\| \frac{d}{dt}\frac{\partial \psi_t}{\partial \nu} \right\|_{H^{-k}(\Sigma_1)} - \gamma \left\| \frac{d^2}{dt^2}\frac{\partial \psi_t}{\partial \nu} \right\|_{H^{-k}(\Sigma_1)} \tag{5.1.22}$$

$$\geq \left\| \frac{\partial \Delta \psi_t}{\partial \nu} \right\|_{H^{-k}(\Sigma_1)} - c_1 \left\| \frac{\partial \psi_t}{\partial \nu} \right\|_{L_2(\Sigma_1)}, \tag{5.1.23}$$

since, for $k \geq 2$ and $c_1 > 0$:

$$\left\| \frac{d}{dt}\frac{\partial \psi_t}{\partial \nu} \right\|_{H^{-k}(\Sigma_1)}, \left\| \frac{d^2}{dt^2}\frac{\partial \psi_t}{\partial \nu} \right\|_{H^{-k}(\Sigma_1)} \leq c_1 \left\| \frac{\partial \psi_t}{\partial \nu} \right\|_{L_2(\Sigma_1)}. \tag{5.1.24}$$

Then (5.1.23) yields (5.1.20). We now establish the norm-equivalence in (5.1.10). First, returning to (5.1.21) and using (5.1.24), we obtain:

$$\|(1)\|^2_{H^{-k}(\Sigma_1)} \leq c \left\{ \left\| \frac{\partial \Delta \psi_t(T-\cdot;x)}{\partial \nu} \right\|^2_{H^{-k}(\Sigma_1)} + \left\| \frac{\partial \psi_t(T-\cdot;x)}{\partial \nu} \right\|^2_{L_2(\Sigma_1)} \right\}. \tag{5.1.25}$$

Then, since $\Gamma_1 \subset \Gamma_2$ by assumption, then (5.1.25) readily yields

$$\|(1)\|^2_{H^{-k}(\Sigma_1)} + \left\| \frac{\partial \psi_t}{\partial \nu}(T-\cdot;x) \right\|^2_{L_2(\Sigma_2)}$$

$$\leq c \left\{ \left\| \frac{\partial \Delta \psi_t(T-\cdot;x)}{\partial \nu} \right\|^2_{H^{-k}(\Sigma_1)} + \left\| \frac{\partial \psi_t}{\partial \nu}(T-\cdot;x) \right\|^2_{L_2(\Sigma_2)} \right\}, \tag{5.1.26}$$

and one direction in (5.1.10) is established. To obtain the opposite direction, we use inequality (5.1.20):

$$4\|(1)\|^2_{H^{-k}(\Sigma_1)} + 4 \left\| \frac{\partial \psi_t(T-\cdot;x)}{\partial \nu} \right\|^2_{L_2(\Sigma_2)}$$

$$\geq 2\|(1)\|^2_{H^{-k}(\Sigma_1)} + 2 \left\| \frac{\partial \psi_t(T-\cdot;x)}{\partial \nu} \right\|^2_{L_2(\Sigma_2)} + c^2 \left\| \frac{\partial \Delta \psi_t}{\partial \nu}(T-\cdot;x) \right\|^2_{H^{-k}(\Sigma_1)}, \tag{5.1.27}$$

where in the right-hand side of (5.1.24), we ultimately drop the (1)-term. The equivalence (5.1.10), and thus Part (ii) are proved.

Finally, Parts (i) and (ii) yield Part (iii) at once, via (5.1.3) and (3.8) on $Y_{1,\gamma}$. □

In Section 6 we shall establish two inequalities which, in turn, will imply the continuous observability inequality (5.1.11): see Theorem 6.3.

5.2 Clamped Case. The Proof of Theorem 4.1.3 Begins

Identifying the continuous observability inequality for problem (3.13) is the first step in the proof of Theorem 4.1.3. To this end, we return to the non-homogeneous problem (3.2) with boundary controls in the clamped B.C., whose abstract model is given by (3.29), or (3.33)–(3.36). Thus, with reference to the map L in (3.36), (3.34), we defined

$$L_T u = (Lu)(T) = \int_0^T e^{\mathbb{A}_{1,\gamma}(T-t)} \mathcal{B}_{1,c} \begin{bmatrix} u_1(t) \\ u_2(t) \end{bmatrix} dt, \tag{5.2.1}$$

for $u = [u_1, u_2] \in H^k(\Sigma_1) \times L_2(\Sigma)$, where the subset Γ_1 of Γ is defined by the assumption of Theorem 4.1.3. As stated in Theorem 4.1.3, we seek to establish exact controllability (from the origin) of the v-problem (3.2) at the time T, by means of boundary controls $[u_1, u_2] \in H^k(\Sigma_1) \times L_2(\Sigma)$ on the space, see (4.0.1b):

$$X_{1,\gamma} \equiv \mathcal{D}(A^{\frac{1}{4}}) \times \tilde{L}_2(\Omega) \equiv H_0^1(\Omega) \times \tilde{L}_2(\Omega) \equiv [\mathcal{D}(\mathbb{A}_{1,\gamma}^*)]', \tag{5.2.2}$$

larger than the original space $Y_{1,\gamma}$, see (3.10), (3.11):

$$Y_{1,\gamma} \equiv \mathcal{D}(A^{\frac{1}{2}}) \times \mathcal{D}(\mathcal{A}_\gamma^{\frac{1}{2}}) \equiv H_0^2(\Omega) \times H_0^1(\Omega); \quad \mathcal{D}(\mathbb{A}_{1,\gamma}^*) = \mathcal{D}(A^{\frac{3}{4}}) \times \mathcal{D}(A^{\frac{1}{2}}), \tag{5.2.3}$$

where $Y_{1,\gamma}$ is the state space for $u_1 = u_2 = 0$. In (5.2.2), $[\ \ ]'$ denotes duality with respect to $Y_{1,\gamma}$. The validity of (5.2.2) stems from (5.2.3). *Equivalently*, we seek to establish surjectivity of the map L_T,

$$L_T : H^k(\Sigma_1) \times L_2(\Sigma) \to \text{ onto } X_{1,\gamma} \equiv [\mathcal{D}(\mathbb{A}_{1,\gamma}^*)]'. \tag{5.2.4}$$

It is a standard result [T-L.1, p. 235] that the surjectivity condition (5.2.4) is, in turn, *equivalent* to the condition (continuous observability inequality):

$$\left\| L_T^{\#} \begin{bmatrix} x_1 \\ x_2 \end{bmatrix} \right\|_{[H^k(\Sigma_1)]' \times L_2(\Sigma)} \geq c_T \left\| \begin{bmatrix} x_1 \\ x_2 \end{bmatrix} \right\|_{[\mathcal{D}(\mathbb{A}_{1,\gamma}^*)]'}, \tag{5.2.5}$$

for some constant $c_T > 0$, where the $X_{1,\gamma}$-adjoint $L_T^{\#}$ of L_T in (5.2.1) is defined by

$$\left(L_T \begin{bmatrix} u_1 \\ u_2 \end{bmatrix}, \begin{bmatrix} x_1 \\ x_2 \end{bmatrix} \right)_{[\mathcal{D}(\mathbb{A}_{1,\gamma}^*)]'} = \left(\begin{bmatrix} u_1 \\ u_2 \end{bmatrix}, L_T^{\#} \begin{bmatrix} x_1 \\ x_2 \end{bmatrix} \right)_{L_2(\Sigma) \times L_2(\Sigma)}, \tag{5.2.6}$$

where $[u_1, u_2] \in H^k(\Sigma_1) \times L_2(\Sigma_2)$ and $L_T^*[x_1, x_2] \in [H^k(\Sigma_1)]' \times L_2(\Sigma)$, duality with respect to the pivot space $L_2(\Sigma_1)$.

LEMMA 5.2.1. *Let $x = [x_1, x_2] \in X_{1,\gamma}$ in (5.2.2), and let*

$$y_0 \equiv \begin{bmatrix} \psi_0 \\ \psi_1 \end{bmatrix} = \mathbb{A}_{1,\gamma}^{-1} \begin{bmatrix} x_1 \\ x_2 \end{bmatrix} = \begin{bmatrix} A^{-1} F_\alpha x_1 - A^{-1} \mathcal{A}_\gamma x_2 \\ x_1 \end{bmatrix} \in Y_{1,\gamma} \equiv H_0^2(\Omega) \times H_0^1(\Omega), \tag{5.2.7}$$

recalling $\mathbb{A}_{1,\gamma}^{-1}$ from (3.10b), (3.33).

(i) Then, with reference to (5.2.1), (5.2.6), we have

$$\left(L_T^{\#}\begin{bmatrix} x_1 \\ x_2 \end{bmatrix}\right)(t) = \begin{bmatrix} -\left[\frac{\partial \Delta\psi}{\partial\nu}(T-t;y_0)\right]_\Gamma \\ [\Delta\psi(T-t;y_0)]_\Gamma \end{bmatrix}, \tag{5.2.8}$$

where $\psi(\,\cdot\,;y_0)$ solves the adjoint problem (3.13) with $\psi(0;y_0)=\psi_0$, $\psi_t(0;y_0)=\psi_1$; that is

$$\begin{bmatrix} \psi(T-t;y_0) \\ -\psi_t(T-t;y_0) \end{bmatrix} = e^{\mathbb{A}_{1,\gamma}^*(T-t)}y_0 = e^{\mathbb{A}_{1,\gamma}^*(T-t)}\begin{bmatrix} \psi_0 \\ -\psi_1 \end{bmatrix}. \tag{5.2.9}$$

(ii) We have

$$\left\| L_T^{\#}\begin{bmatrix} x_1 \\ x_2 \end{bmatrix}\right\|^2_{[H^k(\Sigma_1)]'\times L_2(\Sigma)} = \left\|\frac{\partial\Delta\psi}{\partial\nu}(T-t;y_0)\right\|^2_{[H^k(\Sigma_1)]'} + \|\Delta\psi(T-t;y_0)\|^2_{L_2(\Sigma)}. \tag{5.2.10}$$

(iii) The continuous observability inequality (5.2.5) for the adjoint ψ-problem (3.13) (clamped B.C.), i.e., (5.2.9), is then rewritten as

$$\left\|\frac{\partial\Delta\psi}{\partial\nu}(T-\,\cdot\,;y_0)\right\|^2_{[H^k(\Sigma_1)]'} + \int_0^T\int_\Gamma |\Delta\psi(T-t;y_0)|_\Gamma|^2\,d\Sigma \geq c_T\|\{\psi_0,\psi_1\}\|^2_{H_0^2(\Omega)\times H_0^1(\Omega)}. \tag{5.2.11}$$

Proof. (i) Let $[x_1,x_2]\in X_{1,\gamma}\equiv[\mathcal{D}(\mathbb{A}_{1,\gamma}^*)]'$, see (5.2.4). Recalling (3.34), (5.2.1), (3.38), (5.2.3), (5.2.7), (5.2.9), we compute

$$\left(L_T\begin{bmatrix} u_1 \\ u_2 \end{bmatrix},\begin{bmatrix} x_1 \\ x_2 \end{bmatrix}\right)_{[\mathcal{D}(\mathbb{A}_{1,\gamma}^*)]'} = \left(\int_0^T e^{\mathbb{A}_{1,\gamma}(T-t)}\mathcal{B}_{1,c}\begin{bmatrix} u_1(t) \\ u_2(t) \end{bmatrix}dt,\begin{bmatrix} x_1 \\ x_2 \end{bmatrix}\right)_{[\mathcal{D}(\mathbb{A}_{1,\gamma}^*)]'} \tag{5.2.12}$$

$$= \int_0^T\left(e^{\mathbb{A}_{1,\gamma}(T-t)}\mathbb{A}_{1,\gamma}^{-1}\mathcal{B}_{1,c}\begin{bmatrix} u_1(t) \\ u_2(t) \end{bmatrix},\mathbb{A}_{1,\gamma}^{-1}\begin{bmatrix} x_1 \\ x_2 \end{bmatrix}\right)_{Y_{1,\gamma}}dt \tag{5.2.13}$$

(by (3.38))

$$= \int_0^T\left(\begin{bmatrix} -G_1u_1(t)-G_2u_2(t)-\gamma A^{-1}\mathcal{A}Du_{1tt}(t)+A^{-1}F_\alpha D_\alpha u_{1t}(t) \\ 0 \end{bmatrix},\right.$$

$$\left. e^{\mathbb{A}_{1,\gamma}^*(T-t)}\mathbb{A}_{1,\gamma}^{-1}\begin{bmatrix} x_1 \\ x_2 \end{bmatrix}\right)_{Y_{1,\gamma}}dt \tag{5.2.14}$$

(by (5.2.9))

$$= \int_0^T \left(\begin{bmatrix} -AG_1u_1(t) - AG_2u_2(t) - \gamma ADu_{1tt}(t) + F_\alpha D_\alpha u_{1t}(t) \\ 0 \end{bmatrix}, \right.$$

$$\left. \begin{bmatrix} \psi(T-t;y_0) \\ -\psi_t(T-t;y_0) \end{bmatrix} \right)_{L_2(\Omega)\times L_2(\Omega)} dt, \quad (5.2.15)$$

recalling (5.2.7), (5.2.9), as well as the first component space $\mathcal{D}(A^{\frac{1}{2}})$ of the space $Y_{1,\gamma}$ in (5.2.3), to obtain (5.2.15) from (5.2.14). Thus, (5.2.15) yields by (5.2.6):

$$\left(L_T \begin{bmatrix} u_1 \\ u_2 \end{bmatrix}, \begin{bmatrix} x_1 \\ x_2 \end{bmatrix} \right)_{[\mathcal{D}(A_{1,\gamma}^*)]'} = \left(\begin{bmatrix} u_1 \\ u_2 \end{bmatrix}, L_T^* \begin{bmatrix} x_1 \\ x_2 \end{bmatrix} \right)_{L_2(\Sigma)\times L_2(\Sigma)} \quad (5.2.16)$$

$$= \int_0^T [-AG_1u_1(t) - AG_2u_2(t) - \gamma ADu_{1tt}(t) + F_\alpha D_\alpha u_{1t}(t)], \psi(T-t;y_0))_{L_2(\Omega)} dt \quad (5.2.17)$$

$$= \int_0^T (u_1(t), -G_1^*A\psi(T-t;y_0))_{L_2(\Gamma)} dt + \int_0^T (u_2(t), -G_2^*A\psi(T-t;y_0))_{L_2(\Gamma)} dt$$

$$+ \int_0^T (u_{1tt}(t), -\gamma D^*A\psi(T-t;y_0))_{L_2(\Gamma)} dt + \int_0^T (u_{1t}(t), D_\alpha^* F_\alpha^* \psi(T-t;y_0))_{L_2(\Gamma)} dt \quad (5.2.18)$$

$$= \int_0^T \left(u_1(t), -\frac{\partial \Delta\psi}{\partial \nu}(T-t;y_0)|_\Gamma \right)_{L_2(\Gamma)} dt + \int_0^T (u_2(t), [\Delta\psi(T-t;y_0)]_\Gamma)_{L_2(\Gamma)} dt. \quad (5.2.19)$$

In going from (5.2.18) to (5.2.19), we have recalled the trace results (3.37) for G_1^*A and G_2^*A for the first two terms in (5.2.18), where ψ satisfies the B.C. (3.13); and, moreover, we have used the following trace results for the third and fourth terms of (5.2.18), due to (3.24),

$$D^*A\psi(T-t;y_0) = \frac{\partial\psi}{\partial\nu}(T-t;y_0)|_\Gamma \equiv 0; \quad D_\alpha^*F_\alpha^*\psi(T-t;y_0) = \alpha^2\frac{\partial\psi}{\partial\nu}(T-t;y_0)|_\Gamma \equiv 0, \quad (5.2.20)$$

since ψ satisfies clamped B.C., see (3.13b). Then, (5.2.16), (5.2.19) establish (5.2.8), as desired. Part (i) is proved. Then, Part (ii), Eqn. (5.2.10), follows at once from Part (i), Eqn. (5.2.8). Then, Part (iii), Eqn. (5.2.11), is a restatement of (5.2.5) via (5.2.10) and (5.2.7) for the initial conditions. □

In Section 7 we shall establish two inequalities which, in turn, will imply the continuous observability inequality (5.2.11): see Theorem 7.5.

6 Step 1. Exact Controllability of the v-Kirchoff Problem (3.2): Hinged Case. Theorem 4.1.2 by Duality

Orientation. In this section we prove the continuous observability inequality (5.1.11) for the (dual) ψ-problem (3.13) (hinged B.C.); thus, by duality between (5.1.2) and (5.1.3), the equivalent exact controllability statement (5.1.2) for the v-problem (3.2) under boundary controls $\{u_1, u_2\} \in H_0^k(\Sigma_1) \times L_2(\Sigma_2)$, any $k \geq 2$. This will be done by first establishing Proposition 6.1 below, which gives a related inequality polluted by lower-order terms, and then Proposition 6.2 below, which absorbs those lower-order terms. The combination of Propositions 6.1 and 6.2 will imply the continuous observability inequality (5.1.11) in Theorem 6.3, via the uniqueness result in Corollary 8.2.

PROPOSITION 6.1. *With reference to the ψ-dynamics in (5.1.8), that is to the homogeneous problem (3.13) (hinged case), with initial conditions*

$$x = \{\psi(0), \psi_t(0)\} = \{\psi_0, \psi_1\} \in Y_{1,\gamma} \equiv \mathcal{D}(\mathcal{A}) \times \mathcal{D}(\mathcal{A}_\gamma^{\frac{1}{2}}) = [H^2(\Omega) \cap H_0^1(\Omega)] \times H_0^1(\Omega) \tag{6.1}$$

(see (3.8)), the following inequality holds true: there exists a constant $c_T > 0$ such that

$$\int_0^T \int_\Gamma \left(\frac{\partial \psi_t}{\partial \nu}(T-t;x)\right)^2 d\Sigma + \|\{\psi, \psi_t\}\|^2_{L_2(0,T;H_0^{\frac{3}{2}+\epsilon}(\Omega)\times L_2(\Omega))}$$

$$\geq \; c_T \|\{\psi_0, \psi_1\}\|^2_{[H^2(\Omega)\cap H_0^1(\Omega)]\times H_0^1(\Omega)}, \; \epsilon > 0. \tag{6.2}$$

The boundary term $\int_\Gamma$ over all of Γ can be replaced by $\int_{\Gamma_2}$ only over Γ_2, if condition (4.1.4) is assumed.

Proof. **Step (i).** As in [L-T.11], we reduce the problem to the wave equation with variable coefficient $\alpha^2(x)$ in the damping term. We then apply known estimates to this wave equation problem. To do this, we first recall the *a-priori* regularity of the ψ-problem (3.13), or (5.1.8), given by Lemma 3.1(ii):

$$\{\psi_0, \psi_1\} \in Y_{1\gamma} \Rightarrow \{\psi, \psi_t\} \in C([0,T]; \mathcal{D}(\mathcal{A}) \times \mathcal{D}(\mathcal{A}^{\frac{1}{2}})). \tag{6.3}$$

We then note the following obvious identities, see (2.1a) for $\mathcal{A}_\gamma$:

$$\begin{cases} \mathcal{A}_\gamma^{-1}\mathcal{A}g = \frac{g}{\gamma} - \frac{\mathcal{A}_\gamma^{-1}g}{\gamma}, \; g \in L_2(\Omega), \; \mathcal{A}_\gamma = (I + \gamma\mathcal{A}); & (6.4a) \\ \mathcal{A}_\gamma^{-1}\mathcal{A}^2 g = \frac{1}{\gamma}\mathcal{A}g - \frac{g}{\gamma^2} + \frac{\mathcal{A}_\gamma^{-1}g}{\gamma^2}, \; g \in \mathcal{D}(\mathcal{A}); & (6.4b) \end{cases}$$

$$\begin{cases} F_\alpha^* g \equiv \operatorname{div}(\alpha\nabla(\alpha g)) = \operatorname{div}\nabla(\alpha^2 g) - \operatorname{div}(\alpha g \nabla\alpha) \\ \qquad = -\mathcal{A}(\alpha^2 g) - \operatorname{div}(\alpha g(\nabla\alpha)), \; g \in \mathcal{D}(\mathcal{A}); & (6.5a) \\ \mathcal{A}_\gamma^{-1}F_\alpha^* g = -\frac{\alpha^2 g}{\gamma} + \frac{\mathcal{A}_\gamma^{-1}(\alpha^2 g)}{\gamma} - \mathcal{A}_\gamma^{-1}\operatorname{div}(\alpha g\nabla\alpha), \quad g \in L_2(\Omega). & (6.5b) \end{cases}$$

Hence, from (3.14) and (6.3), (6.5b),

$$\psi_{tt} = -(\mathcal{A}_\gamma^{-1}\mathcal{A})\mathcal{A}\psi + \mathcal{A}_\gamma^{-1}F_\alpha^*\psi_t \in C([0,T];L_2(\Omega)), \tag{6.6a}$$

and from (6.4a) and (6.4b), (6.5b), we obtain

$$\psi_{tt} = -\frac{1}{\gamma}\mathcal{A}\psi - \frac{\alpha^2}{\gamma}\psi_t + \ell.o.t. \tag{6.6b}$$

where $\ell.o.t.$ stands for lower-order terms. Next, we introduce a new variable z:

$$z(t) \equiv \psi_t(t;x) \in C([0,T];H_0^1(\Omega)), \tag{6.7}$$

$$z(0) = \psi_t(0;x) = \psi_1 \in H_0^1(\Omega);$$

$$z_t(0) = \psi_{tt}(0;x) = -(\mathcal{A}_\gamma^{-1}\mathcal{A})\mathcal{A}\psi_0 + \mathcal{A}_\gamma^{-1}F_\alpha^*\psi_1 \in L_2(\Omega), \tag{6.8}$$

by (6.1) and (6.6). Next, differentiating (6.6) in t, and using (6.7) yields, by (6.4b), (6.5b):

$$z_{tt} = -\mathcal{A}_\gamma^{-1}\mathcal{A}^2 z + \mathcal{A}_\gamma^{-1}F_\alpha^* z_t = -\frac{1}{\gamma}\mathcal{A}z - \frac{\alpha^2}{\gamma}z_t + f, \tag{6.9}$$

so that we obtain the following wave equation problem in z:

$$\begin{cases} z_{tt} = \frac{\Delta z}{\gamma} - \frac{\alpha^2 z_t}{\gamma} + f & \text{in } Q; \quad (6.10a) \\ z|_\Sigma \equiv 0 & \text{in } \Sigma, \quad (6.10b) \end{cases}$$

with variable coefficient $\alpha^2(x)$ in front of the damping term z_t,

$$f \equiv \frac{z}{\gamma^2} - \mathcal{A}_\gamma^{-1}\text{div}(\alpha z_t \nabla\alpha) + \ell.o.t. = \frac{\psi_t}{\gamma^2} - \mathcal{A}_\gamma^{-1}\text{div}(\alpha\psi_{tt}\nabla\alpha) + \ell.o.t. \tag{6.11}$$

Step (ii). We invoke [L-T.6, Theorem 2.1.2(ii), Eqn. (2.1.10b)] (which applies to a general wave equation with variable coefficients in the space variable in the first-order terms (energy level)). We obtain, in the notation of [L-T.6]:

$$(\overline{BT})|_\Sigma + \frac{C_T}{\gamma^2}\int_Q f^2 dQ + TC_{T,\gamma}\|z\|^2_{L_2(Q)} \geq k_{\gamma,T}E_z(0), \tag{6.12}$$

where in the equivalent norm via (6.8):

$$E_z(0) = \|\{z(0), z_t(0)\}\|^2_{\mathcal{D}(\mathcal{A}^{\frac{1}{2}})\times L_2(\Omega)} \tag{6.13}$$

(the grad-norm is equivalent to the $H_0^1(\Omega)$-norm for the z-variable), and where via [L-T.6, (2.1.11) and (2.1.9)] we have by (6.10b) [so that $h\cdot\nabla z = \frac{\partial z}{\partial\nu}h\cdot\nu, |\nabla z|^2 = \left(\frac{\partial z}{\partial\nu}\right)^2, h = (x-x_0)$]:

$$(\overline{BT})|_\Sigma = \int_\Sigma e^{\tau\phi}\frac{\partial z}{\partial\nu}h\cdot\nabla z\,d\Sigma - \frac{1}{2}\int_\Sigma e^{\tau\phi}|\nabla z|^2 h\cdot\nu\,d\Sigma \tag{6.14}$$

$$= \frac{1}{2}\int_\Sigma e^{\tau\phi}\left(\frac{\partial z}{\partial\nu}\right)^2 h\cdot\nu\,d\Sigma \leq C_\phi\int_0^T\int_{\Gamma_2}\left(\frac{\partial z}{\partial\nu}\right)^2 d\Sigma_2 \tag{6.15}$$

(ϕ is the pseudo-convex function in [L-T.6]). In the last step in (6.15), we have invoked assumption (4.1.4) on Γ_2.

Step (iii). From the definition of f in (6.11), we shall obtain, with $\alpha \in C^1(\bar{\Omega})$, that

$$||f||^2_{L_2(\Omega)} \leq C_\alpha \left\{ ||\psi||^2_{H_0^{\frac{3}{2}+2\epsilon}(\Omega)} + ||\psi_t||^2_{L_2(\Omega)} \right\}, \ \epsilon > 0. \tag{6.16}$$

Indeed, by (6.11), it suffices to show that:

$$||\mathcal{A}_\gamma^{-1} \text{div}(\alpha\psi_{tt}\nabla\alpha)||_{L_2(\Omega)} \leq c_\alpha \left\{ ||\psi||_{H_0^{\frac{3}{2}+2\epsilon}(\Omega)} + ||\psi_t||_{H^{-\frac{1}{2}+2\epsilon}(\Omega)} \right\}. \tag{6.17}$$

To this end, we recall that for the operator $\mathcal{A}_\gamma$ in (2.1a) we have [L-T.10] for any $\epsilon > 0$,

$$\mathcal{D}(\mathcal{A}_\gamma^{\frac{3}{4}-\epsilon}) \equiv H_0^{\frac{3}{2}-2\epsilon}(\Omega), \text{ that } [\mathcal{D}(\mathcal{A}_\gamma^{\frac{3}{4}-\epsilon})]' = H^{-\frac{3}{2}+2\epsilon}(\Omega). \tag{6.18}$$

Thus, returning to (6.6b) and using (6.18), we have

$$\begin{aligned} ||\mathcal{A}_\gamma^{-(\frac{3}{4}-\epsilon)} \text{div}(\alpha\psi_{tt}\nabla\alpha)||_{L_2(\Omega)} &= ||\text{div}(\alpha\psi_{tt}\nabla\alpha)||_{[\mathcal{D}(\mathcal{A}_\gamma^{\frac{3}{4}-\epsilon})]'} \leq ||\text{div}(\alpha\psi_{tt}\nabla\alpha)||_{H^{-\frac{3}{2}+2\epsilon}(\Omega)} \\ &\leq c_\alpha ||\psi_{tt}||_{H^{-\frac{1}{2}+2\epsilon}(\Omega)} \qquad (6.19) \\ \text{(by (6.6b))} \qquad &\leq \frac{C_\alpha}{\gamma} \left\{ ||\mathcal{A}\psi||_{H^{-\frac{1}{2}+2\epsilon}(\Omega)} + ||\psi_t||_{H^{-\frac{1}{2}+2\epsilon}(\Omega)} \right\} \\ &\leq \frac{C_\alpha}{\gamma} \left\{ ||\psi||_{H_0^{\frac{3}{2}+2\epsilon}(\Omega)} + ||\psi_t||_{H^{-\frac{1}{2}+2\epsilon}(\Omega)} \right\}. \qquad (6.20) \end{aligned}$$

Then, (6.20) proves *a-fortiori* (6.17), and this, in turn, establishes (6.16).

Step (iv). Using estimate (6.16) in (6.12) yields the desired estimate (6.2), by recalling (6.15), (6.13), (6.8) and $z = \psi_t$ by (6.7). Proposition 6.1 is proved. □

We note that the term $\{\psi, \psi_t\}$ in (6.2) is a lower-order term with respect to the $H^2(\Omega) \times H^1(\Omega)$-energy level in (6.3). We shall now absorb by the desired traces.

PROPOSITION 6.2. *Let $T > T_{0,h}$, see (4.1.5). Let ψ be a solution of (3.13) (hinged), i.e., of the form (5.1.8), thus satisfying estimate (6.2). Then, the following inequality holds true: there exists a constant $C_T > 0$, $C_T = C_{T,\gamma,k}$ ($\gamma > 0$ fixed), such that*

$$C_T ||\{\psi, \psi_t\}||^2_{L_2(0,T;H_0^{\frac{3}{2}+\epsilon}(\Omega) \times L_2(\Omega))} \leq \left\| \frac{\partial \psi_t}{\partial \nu} \right\|^2_{L_2(\Sigma_2)} + \left\| \frac{\partial \Delta \psi_t}{\partial \nu} \right\|^2_{H^{-k}(\Sigma_1)}. \tag{6.21}$$

Proof. By contradiction, suppose that inequality (6.21) is false. Then, there exists a sequence

$$\begin{cases} \{\psi^{(n)}, \psi_t^{(n)}\} \in C([0,T]; \mathcal{D}(\mathcal{A}) \times \mathcal{D}(\mathcal{A}^{\frac{1}{2}})) \\ \text{continuous in the initial data } \{\psi_0^{(n)}, \psi_1^{(n)}\} \in \mathcal{D}(\mathcal{A}) \times \mathcal{D}(\mathcal{A}^{\frac{1}{2}}), \end{cases} \tag{6.22}$$

of solutions to problem (3.13) (hinged B.C.), i.e., of the form (5.1.8), such that

$$\|\{\psi^{(n)}, \psi_t^{(n)}\}\|_{L_2(0,T; H_0^{\frac{3}{2}+\epsilon}(\Omega) \times L_2(\Omega))} \equiv 1; \tag{6.23}$$

$$\left\| \frac{\partial \psi_t}{\partial \nu} \right\|^2_{L_2(\Sigma_2)} + \left\| \frac{\partial \Delta \psi_t}{\partial \nu} \right\|^2_{H^{-k}(\Sigma_1)} \to 0 \quad \text{as } n \to \infty. \tag{6.24}$$

The sequence $\{\psi^{(n)}\}$ satisfies inequality (6.2). Thus, by (6.23), (6.24), the corresponding Initial Conditions are uniformly bounded:

$$\|\{\psi_0^{(n)}, \psi_1^{(n)}\}\|_{\mathcal{D}(\mathcal{A}) \times \mathcal{D}(\mathcal{A}^{\frac{1}{2}})} \leq \text{const}, \quad \forall\, n. \tag{6.25}$$

Then, by (6.22),

$$\|\{\psi^{(n)}, \psi_t^{(n)}\}\|_{C([0,T]; \mathcal{D}(\mathcal{A}) \times \mathcal{D}(\mathcal{A}^{\frac{1}{2}}))} \leq \text{const}, \quad \forall\, n. \tag{6.26}$$

By (6.6) applied to $\psi^{(n)}$, we have via (6.4b), (6.5b):

$$\|\psi_{tt}^{(n)}\|_{L_2(\Omega)} \leq C\{\|\mathcal{A}\psi^{(n)}\|_{L_2(\Omega)} + \|\psi_t^{(n)}\|_{\mathcal{D}(\mathcal{A}^{\frac{1}{2}})}\}, \tag{6.27}$$

so that, by (6.26),

$$\|\psi_{tt}^{(n)}\|_{C([0,T]; L_2(\Omega))} \leq C, \quad \forall\, n. \tag{6.28}$$

Then, *a fortiori* from (6.26) and (6.28), we can apply Aubin's Lemma [A.1], since $\mathcal{A}^{-1}$ is compact on $L_2(\Omega)$, and obtain for a subsequence

$$\{\psi^{(n)}, \psi_t^{(n)}\} \to \text{ some } \{\tilde{\psi}, \tilde{\psi}_t\} \text{ strongly in } L_2(0,T; H_0^{\frac{3}{2}+\epsilon}(\Omega) \times L_2(\Omega)). \tag{6.29}$$

Then (6.23) and (6.29) yield

$$\|\{\tilde{\psi}, \tilde{\psi}_t\}\|_{L_2(0,T; H_0^{\frac{3}{2}+\epsilon}(\Omega) \times L_2(\Omega))} = 1. \tag{6.30}$$

The limit $\tilde{\psi}$ satisfies problem (3.13) in particular the B.C.,

$$\tilde{\psi}|_\Sigma \equiv 0 \text{ and } \Delta\tilde{\psi}|_\Sigma \equiv 0, \text{ hence } \tilde{\psi}_t|_\Sigma \equiv 0 \text{ and } \Delta\tilde{\psi}_t|_\Sigma \equiv 0. \tag{6.31}$$

Moreover, by (6.24),

$$\left.\frac{\partial \tilde{\psi}_t}{\partial \nu}\right|_{\Sigma_2} \equiv 0 \text{ on } \Sigma_2 \quad \text{and} \quad \left.\frac{\partial \Delta \tilde{\psi}_t}{\partial \nu}\right|_{\Sigma_1} \equiv 0 \text{ on } \Sigma_1. \tag{6.32}$$

Moreover, $\tilde{\psi}_t$ itself satisfies the Kirchoff equation (3.13a) (by differentiating in t). Thus, $\tilde{\psi}$ satisfies the Kirchoff equation (3.13a) with over-determined homogeneous B.C. (6.31) and (6.32).

Recalling that $T > T_{0,h}$, we can invoke Corollary 8.2 in Section 8 and conclude that, in fact, $\tilde{\psi}_t \equiv 0$ in Q, and $\tilde{\psi} \equiv$ const in Q. But $\tilde{\psi}|_\Sigma \equiv 0$ by (6.31), and thus $\tilde{\psi} \equiv 0$ in Q. But this contradicts (6.30). Thus, inequality (6.21) holds true. □

THEOREM 6.3. *Let $T > T_{0,h}$, see (4.1.5). With reference to the dynamics in (5.1.8), that is to the homogeneous problem (3.13) (hinged case) with initial conditions as in (6.1), we have that the continuous observability inequality (5.1.11) holds true.*

Proof. We combine (6.2) of Proposition 6.1 with (6.21) of Proposition 6.2 to obtain (5.1.11). □

7 Step 1: Exact Controllability of the v-Kirchoff Problem (3.2): Clamped Case. Theorem 4.1.3 by Duality

Orientation. In this section we prove, in Theorem 7.5 below, the continuous observability inequality (5.2.11) for the (dual) ψ-problem (3.13) (clamped B.C.): thus, by duality between (5.2.4) and (5.2.5), the equivalent exact controllability statement (5.2.4) for the v-problem (3.2) under boundary controls $\{u_1, u_2\} \in H^k(\Sigma_1) \times L_2(\Sigma)$, any $k \geq 2$. This will be done by first establishing, in Proposition 7.3 below, the desired inequality polluted by lower-order terms.

Next, Proposition 7.4 below will absorb those lower-order terms, by virtue of the uniqueness result in Corollary 8.2. The combination of Propositions 7.3 and 7.4 will then imply the continuous observability inequality (5.2.11) in Theorem 7.5. The initial conditions $\{\psi_0, \psi_1\}$ may be prescribed either at $t = T$, as in (5.2.9), or else at $t = 0$.

LEMMA 7.1. *With reference to the ψ-dynamics in (5.2.9), that is, to the homogeneous problem (3.13) (clamped case), with initial conditions at $t = T$:*

$$x = \{\psi(T), \psi_t(T)\} = \{\psi_0, \psi_1\} \in Y_{1,\gamma} \equiv \mathcal{D}(A) \times \mathcal{D}(\mathcal{A}_\gamma^{\frac{1}{2}}) = H_0^2(\Omega) \times H_0^1(\Omega) \tag{7.1}$$

(recall (3.11)), the following inequality holds true: Given $T > T_{0,c}$, see (4.1.9), there exists a constant $c_T > 0$ such that

$$\left\{ \|(\Delta\psi)|_\Gamma\|^2_{L_2(\Sigma)} + \left\| \frac{\partial \Delta\psi}{\partial \nu}\bigg|_\Gamma \right\|^2_{H^{-1}(\Sigma)} + \int_0^T \left[\|\psi(t)\|^2_{H_0^1(\Omega)} + \|\psi_t(t)\|^2_{L_2(\Omega)} \right] dt \right.$$

$$\left. + \|\psi\|^2_{L_\infty(0,T;L_2(\Omega))} + \|\psi_t\|^2_{L_\infty(0,T;H^{-1}(\Omega))} \right\}$$

$$\geq c_T \|\{\psi_0, \psi_1\}\|^2_{H_0^2(\Omega)\times H_0^1(\Omega)}, \tag{7.2}$$

where we recall that $\Delta\psi|_\Gamma \in L_2(\Sigma)$ by [L-L.1, Ch. 5].

Proof. **Step (i).** As in Section 6, we shall reduce the problem to the wave equation with variable coefficient $\alpha^2(x)$ in the damping term. In doing this, we shall need to recall the *a-priori* regularity of the ψ-problem (3.13), or (5.2.9), given by Lemma 3.1(ii), with $Y_{1,\gamma}$ defined by (3.11):

$$y_0 = \{\psi_0, \psi_1\} \in Y_{1,\gamma} \equiv \mathcal{D}(A^{\frac{1}{2}}) \times \mathcal{D}(\mathcal{A}_\gamma^{\frac{1}{2}}) \equiv H_0^2(\Omega) \times H_0^1(\Omega) \tag{7.3}$$

$$\Rightarrow \{\psi(T-t; y_0), -\psi_t(T-t; y_0)\} = e^{\mathbf{A}_{1,\gamma}^*(T-t)} y_0 \in C([0,T]; Y_{1,\gamma} \equiv H_0^2(\Omega) \times H_0^1(\Omega)). \tag{7.4}$$

Hence, differentiating (7.4) in t:

$$\{\psi_t(T-t;y_0), -\psi_{tt}(T-t;y_0)\} = \mathbb{A}^*_{1,\gamma} e^{\mathbb{A}^*_{1,\gamma}(T-t)} y_0 \in C([0,T];[\mathcal{D}(\mathbb{A}_{1,\gamma})]'), \tag{7.5}$$

where, from (7.3) for $Y_{1,\gamma}$ and $\mathcal{D}(\mathbb{A}_{1,\gamma}) = \mathcal{D}(A^{\frac{3}{4}}) \times \mathcal{D}(A^{\frac{1}{2}})$ by (3.10b), we obtain

$$[\mathcal{D}(\mathbb{A}_{1,\gamma})]' = \mathcal{D}(A^{\frac{1}{4}}) \times L_2(\Omega) \equiv H^1_0(\Omega) \times \tilde{L}_2(\Omega)), \tag{7.6}$$

$[\quad]'$ denoting duality with respect to $Y_{1,\gamma}$ as a pivot space. In particular, (7.5) and (7.6) yield the first regularity property in

$$\psi_{tt}(T-t;y_0) \in C([0,T];\tilde{L}_2(\Omega)) \cap L_2(0,T;L_2(\Omega)) \tag{7.7}$$

continuously in $y_0 \in Y_{1,\gamma}$ while the second, critical regularity property in (7.7) was established in [L-T.7, p. 43] (the proof in this reference for $\alpha \equiv 1$—which critically relies on a sharp trace result in [L-L.1, Ch. 5]—readily extends to the present case of smooth variable α). Next, we introduce a new variable z, whose regularity follows from (7.4):

$$z(t) = -\Delta\psi(t) = \mathcal{A}\psi(t) \in C([0,T];L_2(\Omega)). \tag{7.8}$$

Because of the clamped B.C. in (3.13b), we have by (2.4) and (7.8):

$$\|\psi\|^2_{H^2_0(\Omega)} \sim \|A^{\frac{1}{2}}\psi\|^2_{L_2(\Omega)} = (A\psi,\psi)_{L_2(\Omega)} = \int_\Omega |\Delta\psi|^2 d\Omega = \|\mathcal{A}\psi\|^2_{L_2(\Omega)} = \|z\|^2_{L_2(\Omega)}; \tag{7.9}$$

$$\begin{aligned}\|\psi_t\|^2_{H^1_0(\Omega)} \sim \|\mathcal{A}^{\frac{1}{2}}\psi_t\|^2_{L_2(\Omega)} &= (\mathcal{A}\psi_t,\psi_t)_{L_2(\Omega)} = (z_t, \mathcal{A}^{-1}z_t) \\ &= \|\mathcal{A}^{-\frac{1}{2}}z_t\|^2_{L_2(\Omega)} = \|z_t\|^2_{[\mathcal{D}(\mathcal{A}^{\frac{1}{2}})]'} \sim \|z_t\|^2_{H^{-1}(\Omega)},\end{aligned} \tag{7.10}$$

where $\sim$ denotes norm equivalence: in (7.9) we have applied (2.4a-b) and Green's second theorem, while in (7.10) we have recalled (2.1a), (2.2). In particular, from (7.9), (7.10), (7.1), we obtain

$$\|\{\psi_0,\psi_1\}\|^2_{H^2_0(\Omega)\times H^1_0(\Omega)} \sim \|\{z(T), z_t(T)\}\|^2_{L_2(\Omega)\times H^{-1}(\Omega)}. \tag{7.11}$$

Since

$$\text{div}(\alpha\nabla(\alpha\psi_t)) = \alpha^2\Delta\psi_t + \frac{3}{2}\nabla\psi_t\cdot\nabla(\alpha^2) + (\alpha\Delta\alpha + |\nabla\alpha|^2)\psi_t,$$

by (7.8) and (3.13a) for ψ, we see that then the new variable z satisfies the following wave equation

$$\gamma z_{tt} = \Delta z - \alpha^2(x) z_t + f \quad \text{in } (0,T]\times\Omega \equiv Q; \tag{7.12}$$

$$f = -\psi_{tt} + \frac{3}{2}\nabla\psi_t\cdot\nabla(\alpha^2) + (\alpha\Delta\alpha + |\nabla\alpha|^2)\psi_t \in C([0,T];\tilde{L}_2(\Omega)) \cap L_2(0,T;L_2(\Omega)), \tag{7.13}$$

where the regularity of f in (7.13) follows from (7.4) and (7.7).

Step (ii). With reference to Eqn. (7.12), the following ('negative norm') estimate holds for $T > T_{0,c}$, where $T_{0,c}$ is defined by (4.1.9):

$$\int_0^T \left[\|z\|^2_{L_2(\Omega)} + \|z_t\|^2_{H^{-1}(\Omega)} \right] dt$$

$$\leq c_T \left\{ \|z|_\Gamma\|^2_{L_2(\Sigma)} + \left\| \frac{\partial z}{\partial \nu}\Big|_\Gamma \right\|^2_{H^{-1}(\Sigma)} + \int_0^T \left[\|z\|^2_{H^{-1}(\Omega)} + \|\psi_t\|^2_{L_2(\Omega)} + \||\nabla\psi|\|^2_{L_2(\Omega)} \right] dt \right.$$

$$\left. + \|\psi_t\|^2_{L_\infty(0,T;H^{-1}(\Omega))} + \||\nabla\psi|\|^2_{L_\infty(0,T;H^{-1}(\Omega))} \right\}. \tag{7.14}$$

Estimate (7.14) is a proper specialization of Theorem E.1 in Appendix E, where we take $f_2 = 0$, and where now $f_1 = -\psi_t + \frac{3}{2}\nabla\psi \cdot \nabla(\alpha^2) + (\alpha\Delta\alpha + |\nabla\alpha|^2)\psi \in C([0,T]; H^1(\Omega))$.

Using, in the left-hand side of (7.14), the norm equivalence in (7.9) between ψ in $H_0^2(\Omega)$ and z in $L_2(\Omega)$, as well as the norm equivalence in (7.10) between ψ_t (or ψ) in $H_0^1(\Omega)$ and z_t (or z) in $H^{-1}(\Omega)$, we obtain from (7.14) and (7.8),

$$\int_0^T \left[\|\psi(t)\|^2_{H_0^2(\Omega)} + \|\psi_t(t)\|^2_{H_0^1(\Omega)} \right] dt \;\leq\; c_T \left\{ \|\Delta\psi|_\Gamma\|^2_{L_2(\Sigma)} + \left\| \frac{\partial \Delta\psi}{\partial \nu}\Big|_\Gamma \right\|^2_{H^{-1}(\Sigma)} \right.$$

$$+ \int_0^T \left[\|\psi(t)\|^2_{H_0^1(\Omega)} + \|\psi_t(t)\|^2_{L_2(\Omega)} \right] dt$$

$$\left. + \|\psi_t\|^2_{L_\infty(0,T;H^{-1}(\Omega))} + \|\psi\|^2_{L_\infty(0,T;H^{-1}(\Omega))} \right\}. \tag{7.15}$$

Next, recalling the s.c. group $\{\psi(t), \psi_t(t)\} = e^{\mathbf{A}^*_{1,\gamma} t} y_0$, $y_0 = \{\psi_0, \psi_1\}$ on $Y_{1,\gamma}$ in (7.3), we obtain

$$C_T \|\{\psi_0, \psi_1\}\|^2_{H_0^2(\Omega)\times H_0^1(\Omega)} \;\leq\; C_T \int_0^T \|y_0\|^2_{Y_{1,\gamma}} dt \leq \int_0^T \left\| e^{\mathbf{A}^*_{1,\gamma} t} y_0 \right\|^2_{Y_{1,\gamma}} dt$$

$$\sim \int_0^T \left[\|\psi(t)\|^2_{H_0^2(\Omega)} + \|\psi_t(t)\|^2_{H_0^1(\Omega)} \right] dt, \tag{7.16}$$

and hence, by use of (7.16) in (7.15), we arrive at (7.2), as desired. □

LEMMA 7.2. *With reference to inequality (7.2), we have: given $T > T_{0,c}$, see (4.1.9), there exists a constant $c_T > 0$ such that*

$$\left\| \frac{\partial \Delta\psi}{\partial \nu}\Big|_\Gamma \right\|^2_{H^{-1}(\Sigma)} \leq c_T \left\{ \|(\Delta\psi)|_\Gamma\|^2_{L_2(\Sigma)} + \int_0^T \left[\|\psi(t)\|^2_{H_0^1(\Omega)} + \|\psi_t(t)\|^2_{L_2(\Omega)} \right] dt \right\}. \tag{7.17}$$

Proof. **Step (a).** We shall first establish, from (3.13), the following identity in Q:

$$\Delta\psi = D[(\Delta\psi)|_\Gamma] + \mathcal{A}^{-1}\psi_{tt} + \gamma\psi_{tt} - \mathcal{A}^{-1}\text{div}(\alpha\nabla(\alpha\psi_t)), \tag{7.18}$$

where D is the (Dirichlet) map defined in (3.20), and $\mathcal{A}$ is defined in (2.1a). In fact, returning to the clamped ψ-problem (3.13), we rewrite it by virtue of (3.20) for D as

$$\psi_{tt} - \gamma\Delta\psi_{tt} + \Delta(\Delta\psi - D[(\Delta\psi)|_\Gamma]) - \text{div}(\alpha\nabla(\alpha\psi_t)) = 0, \tag{7.19}$$

since $\Delta D[(\Delta\psi)|_\Gamma] \equiv 0$ in Q. Thus, recalling $\mathcal{A}$ in (2.1a), we rewrite (7.19) as

$$\psi_{tt} + \gamma\mathcal{A}\psi_{tt} - \mathcal{A}(\Delta\psi - D[(\Delta\psi)|_\Gamma]) - \text{div}(\alpha\nabla(\alpha\psi_t)) = 0, \tag{7.20}$$

or applying $\mathcal{A}^{-1}$ throughout

$$\Delta\psi - D[(\Delta\psi)|_\Gamma] = \mathcal{A}^{-1}\psi_{tt} + \gamma\psi_{tt} - \mathcal{A}^{-1}\text{div}(\alpha\nabla(\alpha\psi_t)) = 0, \tag{7.21}$$

which is precisely (7.18).

Step (b). Since $\frac{\partial\psi_{tt}}{\partial\nu} \equiv 0$ on Σ by (3.13b) (clamped), then (7.18) yields on Σ:

$$\frac{\partial\Delta\psi}{\partial\nu} = \frac{\partial D[(\Delta\psi)|_\Gamma]}{\partial\nu} + \frac{\partial\mathcal{A}^{-1}\psi_{tt}}{\partial\nu} - \frac{\partial}{\partial\nu}\mathcal{A}^{-1}\text{div}(\alpha\nabla(\alpha\psi_t)), \text{ on } \Sigma. \tag{7.22}$$

Step (c). With reference to (7.22), we have by elliptic regularity, since any term of the form Dg solves an elliptic problem via (3.20)

$$\left\|\frac{\partial}{\partial\nu}D[(\Delta\psi)|_\Gamma]\right\|_{H^{-1}(\Gamma)} \leq c\|D[(\Delta\psi)|_\Gamma\|_{H^{\frac{1}{2}}(\Omega)} \leq c\|(\Delta\psi)|_\Gamma\|_{L_2(\Gamma)}. \tag{7.23}$$

In the first step we have used elliptic regularity twice: for the normal trace of an elliptic solution [Ke.1], as well as for the harmonic extension $D: L_2(\Gamma) \to H^{\frac{1}{2}}(\Omega)$, see (3.20).

Step (d). With reference to (7.22), we have by brutal majorization and trace theory:

$$\left\|\frac{\partial\mathcal{A}^{-1}\psi_{tt}}{\partial\nu}\Big|_\Gamma\right\|^2_{H^{-1}(\Sigma)} \leq c\int_0^T \left\|\frac{\partial\mathcal{A}^{-1}\psi_t}{\partial\nu}\Big|_\Gamma\right\|^2_{H^{-1}(\Gamma)} dt \tag{7.24}$$

$$\leq c\int_0^T \left\|\frac{\partial\mathcal{A}^{-1}}{\partial\nu}\psi_t\right\|^2_{H^{\frac{1}{2}}(\Gamma)} dt \leq c\int_0^T \|\psi_t\|^2_{L_2(\Omega)}\, dt, \tag{7.25}$$

since $\mathcal{A}^{-1}\psi_t \in \mathcal{D}(\mathcal{A}) \subset H^2(\Omega)$.

Step (e). With reference to (7.22), we have similarly

$$\left\| \frac{\partial}{\partial \nu} \mathcal{A}^{-1} \text{div}(\alpha \nabla (\alpha \psi_t)) \Big|_\Gamma \right\|_{H^{-1}(\Sigma)} \le c \int_0^T \left\| \frac{\partial}{\partial \nu} \mathcal{A}^{-1} \text{div}(\alpha \nabla (\alpha \psi)) \Big|_\Gamma \right\|^2_{H^{-1}(\Gamma)} dt \tag{7.26}$$

$$\le c \int_0^T \left\| \mathcal{A}^{-\frac{1}{2}} \text{div}(\alpha \nabla (\alpha \psi)) \right\|^2_{L_2(\Omega)} dt \tag{7.27}$$

$$\le c \int_0^T \|\text{div}(\alpha \nabla (\alpha \psi))\|^2_{H^{-1}(\Omega)} dt \le c \int_0^T \|\psi\|^2_{H_0^1(\Omega)} dt. \tag{7.28}$$

Step (f). Using (7.23), (7.25), (7.28), in (7.22) yields (7.17), as desired. □

PROPOSITION 7.3. *With reference to the ψ-dynamics in (5.2.9), that is, to the homogeneous problem (3.13) (clamped case) with initial conditions $\{\psi_0, \psi_1\} \in Y_{1,\gamma} = H_0^2(\Omega) \times H_0^1(\Omega)$, see (3.11), the following inequality holds true: given $T > T_{0,c}$, see (4.1.9), there exists a constant $c_T > 0$, such that*

$$\left\{ \|(\Delta\psi)|_\Gamma\|^2_{L_2(\Sigma)} + \int_0^T [\|\psi(t)\|^2_{H_0^1(\Omega)} + \|\psi_t(t)\|^2_{L_2(\Omega)}] dt + \|\psi\|^2_{L_\infty(0,T;L_2(\Omega))} + \|\psi_t\|^2_{L_\infty(0,T;H^{-1}(\Omega))} \right\}$$

$$\ge C_T \|\{\psi_0, \psi_1\}\|^2_{H_0^2(\Omega) \times H_0^1(\Omega)} \tag{7.29}$$

Proof. Use (7.17) of Lemma 7.2 into (7.2) of Lemma 7.1. □

PROPOSITION 7.4. *With reference to the ψ-dynamics in (5.2.9), or (3.13) (clamped case), which therefore satisfies estimate (7.29) for $T > T_{0,c}$, see (4.1.9), the following inequality holds true: given such $T > T_{0,c}$, then*

$$\int_0^T \left[\|\psi(t)\|^2_{H_0^1(\Omega)} + \|\psi_t(t)\|^2_{L_2(\Omega)} \right] dt + \|\psi\|^2_{L_\infty(0,T;L_2(\Omega))} + \|\psi_t\|^2_{L_\infty(0,T;H^{-1}(\Omega))}$$

$$\le C_T \left\{ \|(\Delta\psi)|_\Gamma\|^2_{L_2(\Sigma)} + \left\| \frac{\partial \Delta \psi}{\partial \nu} \Big|_\Gamma \right\|^2_{H^{-k}(\Sigma_1)} \right\}, \tag{7.30}$$

where the number $k \ge 1$ is arbitrary, and $\emptyset \ne \Gamma_1$ is an arbitrary, non-empty open subset of $\Gamma = \partial\Omega$, of positive measure.

Proof. We use a compactness/uniqueness argument, as in similar cases. Assume that (7.30) is false. Then, there exists a sequence

$$\begin{cases} \{\psi^{(n)}, \psi_t^{(n)}\} \in C([0,T]; H_0^2(\Omega) \times H_0^1(\Omega)) \\ \text{continuous in the initial data } \{\psi_0^{(n)}, \psi_1^{(n)}\} \in H_0^2(\Omega) \times H_0^1(\Omega), \end{cases} \tag{7.31}$$

of solutions, of the form given by (5.2.9), to problem (3.13) (clamped case), such that

$$\begin{cases} \int_0^T [\|\psi^{(n)}(t)\|^2_{H_0^1(\Omega)} + \|\psi_t^{(n)}(t)\|^2_{L_2(\Omega)}]dt + \|\psi^{(n)}\|^2_{L_\infty(0,T;L_2(\Omega))} \\ \qquad\qquad + \|\psi_t^{(n)}\|^2_{L_\infty(0,T;H^{-1}(\Omega))} \equiv 1 & (7.32) \\ \left\|(\Delta\psi^{(n)})|_\Gamma\right\|^2_{L_2(\Sigma)} + \left\|\left.\frac{\partial\Delta\psi^{(n)}}{\partial\nu}\right|_\Gamma\right\|_{H^{-k}(\Sigma_1)} \to 0 \text{ as } n \to \infty. & (7.33) \end{cases}$$

The sequence $\{\psi^{(n)}\}$ satisfies inequality (7.29). Thus, by (7.32), (7.33), the corresponding Initial Conditions are uniformly bounded:

$$\|\{\psi_0^{(n)}, \psi_1^{(n)}\}\|_{H_0^2(\Omega)\times H_0^1(\Omega)} \leq \text{const}, \qquad \forall\, n. \tag{7.34}$$

Then, by (7.31) [or else recall (7.15), (7.16)],

$$\|\{\psi^{(n)}, \psi_t^{(n)}\}\|^2_{C([0,T];H_0^2(\Omega)\times H_0^1(\Omega))} \leq \text{const}, \qquad \forall\, n. \tag{7.35}$$

By (7.7) [continuous with respect to the initial data], we obtain that

$$\|\psi_{tt}^{(n)}\|_{C([0,T];L_2(\Omega)} \leq \text{const}, \qquad \forall\, n, \tag{7.36}$$

as well. By (7.34), there exists a subsequence (still denoted by the index n), such that

$$\{\psi_0^{(n)}, \psi_1^{(n)}\} \to \text{ some } \{\tilde{\psi}_0, \tilde{\psi}_1\} \text{ in } H_0^2(\Omega) \times H_0^1(\Omega) \text{ weakly.} \tag{7.37}$$

Let, as in (5.2.9),

$$\begin{bmatrix} \tilde{\psi}(T-t) \\ \tilde{\psi}_t(T-t) \end{bmatrix} = e^{\mathbf{A}^*_{1,\gamma}(T-t)} \begin{bmatrix} \tilde{\psi}_0 \\ \tilde{\psi}_1 \end{bmatrix} \in C\left([0,T]; \begin{bmatrix} H_0^2(\Omega) \\ H_0^1(\Omega) \end{bmatrix}\right), \tag{7.38}$$

be the solution of (3.13),

$$\begin{cases} \tilde{\psi}_{tt} - \gamma\Delta\tilde{\psi}_{tt} + \Delta^2\tilde{\psi} - \operatorname{div}(\alpha\nabla(\alpha\tilde{\psi}_t)) \equiv 0 \quad \text{in } Q; & (7.39a) \\ \tilde{\psi}|_\Sigma \equiv 0, \ \left.\frac{\partial\tilde{\psi}}{\partial\nu}\right|_\Sigma \equiv 0 \quad \text{in } \Sigma, & (7.39b) \end{cases}$$

corresponding to the Initial Conditions $\{\tilde{\psi}_0, \tilde{\psi}_1\}$. Then, recalling (7.35), (7.36), and using that, for Ω bounded, the embedding $H^s(\Omega) \hookrightarrow H^{s-\epsilon}(\Omega)$ is compact for all s, and $\epsilon > 0$, we invoke Aubin [A.1] ($p = 2$), Simon [S.1] ($p = \infty$) results and conclude that, for a subsequence we have

$$\{\psi^{(n)}, \psi_t^{(n)}\} \to \{\tilde{\psi}, \tilde{\psi}_t\} \text{ strongly in } C([0,T]; H_0^1(\Omega) \times L_2(\Omega)), \tag{7.40}$$

hence, *a fortiori*, strongly in $L_2(0,T;H_0^1(\Omega) \times L_2(\Omega))$, as well as in $L_\infty(0,T;L_2(\Omega) \times H^{-1}(\Omega))$. As a consequence of all this, using (7.40) in (7.32), yields

$$\int_0^T \left[\|\tilde{\psi}(t)\|^2_{H_0^1(\Omega)} + \|\tilde{\psi}_t(t)\|^2_{L_2(\Omega)}\right] dt + \|\tilde{\psi}\|^2_{L_\infty(0,T;L_2(\Omega))} + \|\tilde{\psi}_t\|^2_{L_\infty(0,T,H^{-1}(\Omega))} = 1. \tag{7.41}$$

Moreover, by (7.33), $\tilde{\psi}$ in (7.39) satisfies two additional B.C.,

$$\Delta\tilde{\psi} \equiv 0 \text{ in } \Sigma = (0,T] \times \Gamma; \quad \frac{\partial \Delta\tilde{\psi}}{\partial \nu} \equiv 0 \text{ on } \Sigma_1 = (0,T] \times \Gamma_1. \tag{7.42}$$

Thus, $\{\tilde{\psi}, \tilde{\psi}_t\} \in C([0,T]; H_0^2(\Omega) \times H_0^1(\Omega))$ satisfies Eqn. (7.39) with three B.C. $\tilde{\psi} \equiv 0$, $\frac{\partial \tilde{\psi}}{\partial \nu} \equiv 0$, $\Delta\tilde{\psi} \equiv 0$ on all of Σ, and a fourth B.C. $\frac{\partial \Delta \tilde{\psi}}{\partial \nu} \equiv 0$ on Σ_1. Since, by assumption, $T > T_{0,c}$, see (4.1.9), we may then appeal to the uniqueness continuation result of Corollary 8.2 in Section 8 (as well as to Remark 4.2.1) and conclude that, in fact, $\tilde{\psi} \equiv 0$ in $Q = (0,T] \times \Omega$. But this contradicts (7.41). Thus, inequality (7.30) holds true. □

As a corollary we obtain the desired continuous observability inequality for $T > T_{0,c}$ in (5.2.11).

THEOREM 7.5. *Let $T > T_{0,c}$, see (4.1.9). The ψ-dynamics in (5.2.9), that is, to the homogeneous problem (3.13) (clamped case), with initial conditions $\{\psi_0, \psi_1\} \in H_0^2(\Omega) \times H_0^1(\Omega)$ [either at $t = 0$, or else at $t = T$] satisfies the following (continuous observability) inequality in (5.2.11): there exists a constant $c_T > 0$, such that*

$$\|(\Delta\psi)|_\Gamma\|^2_{L_2(\Sigma)} + \left\| \left. \frac{\partial \Delta\psi}{\partial \nu} \right|_\Gamma \right\|^2_{H^{-k}(\Sigma_1)} \geq c_T \|\{\psi_0, \psi_1\}\|^2_{H_0^2(\Omega) \times H_0^1(\Omega)}, \tag{7.43}$$

where $\emptyset \neq \Gamma_1$ is a non-empty open subset of $\Gamma = \partial\Omega$, of positive measure.

8 Step 1: Unique Continuation for the Kirchoff Plate Equation with Variable Coefficients. Main Statements

In this section, we state the unique continuation theorem for the Kirchoff equation (3.13a) [related to the adjoint semigroup $e^{\mathbf{A}^*_{1,\gamma} t}$ of Lemma 3.1(ii)], which is invoked in the proof of Proposition 6.2, below (6.32), in the hinged case; and in the proof of Proposition 7.4 below (7.42) in the clamped case. It requires over-determined zero Cauchy data only on a nonempty, open subportion of the boundary.

THEOREM 8.1. *Assume that $\psi \in H^3(Q)$ is a solution of the over-determined Kirchoff problem (same equation as (3.13a)):*

$$\psi_{tt} - \gamma\Delta\psi_{tt} + \Delta^2\psi - div(\alpha\nabla(\alpha\psi_t)) = 0 \ in \ Q = (0,T] \times \Omega; \tag{8.1a}$$

$$\psi \equiv 0, \ \frac{\partial \psi}{\partial \nu} \equiv 0, \ \frac{\partial^2 \psi}{\partial \nu^2} \equiv 0, \ \frac{\partial^3 \psi}{\partial \nu^3} \equiv 0 \ in \ \tilde{\Sigma} = (0,T] \times \tilde{\Gamma}, \tag{8.1b}$$

where $\tilde{\Gamma}$ is an arbitrary nonempty open subset of $\Gamma = \partial\Omega$. Let $T > \tilde{T}$, where $\tilde{T}$ is defined by (4.2.10): that is

$$\tilde{T} = 2\sqrt{\gamma} \sup_{x \in \Omega} dist(x, \tilde{\Gamma}). \tag{8.2}$$

Then, in fact,

$$\psi(T/2,\cdot) = \psi_t(T/2,\cdot) = 0 \tag{8.3}$$

The proof of Theorem 8.1 will be given in the Section 10. The next corollary relaxes the *a-priori* regularity of the solution ψ to Eqn. (8.1a), at the price of requiring the first three B.C. on *all* of $\Sigma = (0,T] \times \Gamma$: that is

$$\psi \equiv 0, \ \frac{\partial \psi}{\partial \nu} \equiv 0, \ \Delta\psi \equiv 0 \quad \text{on } \Sigma = (0,T] \times \Gamma \tag{8.4}$$

[where, under the first two conditions, we have that the third condition $\Delta\psi \equiv 0$ on Σ is equivalent to $\frac{\partial^2 \psi}{\partial \nu^2} \equiv 0$ on Σ, see Remark 4.2.1].

COROLLARY 8.2. *Assume that $\psi \in H^1(Q)$ is a solution of the over-determined Kirchoff problem consisting of Eqn. (8.1a), the B.C. (8.4) on all of Σ, and $\frac{\partial^3 \psi}{\partial \nu^3} \equiv 0$ on $\tilde{\Sigma} = (0,T] \times \tilde{\Gamma}$. See also Remark 4.2.1. Then, for $T > \tilde{T}$, with $\tilde{T}$ defined by (8.2), then in fact, conclusion (8.3) holds true. Moreover, by backward and forward uniqueness, we conclude that $\psi \equiv 0$ in $Q = (0,T) \times \Omega$.*

Proof. (of Corollary 8.2.) Let $\psi \in H^1(Q)$ be the solution given by the assumption. We need to show that, under the additional assumption (8.4) on all of Σ, then $\psi \in H^3(Q)$, and so Theorem 8.1 applies.

By virtue of the assumed B.C. $\psi \equiv 0$ and $\Delta\psi \equiv 0$ on Σ given by (8.4), we can invoke estimate (6.2) of Proposition 6.1 (hinged case) with $\frac{\partial \psi_t}{\partial \nu} \equiv 0$ on Σ, again by the third B.C. in (8.4), and obtain that $\{\psi(0), \psi_t(0)\} \in [H^2(\Omega) \cap H_0^1(\Omega)] \times H_0^1(\Omega)$. Hence, by Lemma 3.1, we have

$$\{\psi, \psi_t\} \in C([0,T]; [H^2(\Omega) \cap H_0^1(\Omega)] \times H_0^1(\Omega)). \tag{8.5}$$

Next, we differentiate in t, Eqn. (8.1a) and the B.C. (8.4) on Σ: We then apply estimate (6.2) this time to the solution ψ_t of (8.1a) with B.C. $\psi_t \equiv \frac{\partial \psi_t}{\partial \nu} = \Delta\psi_t \equiv 0$ on Σ, thus obtaining $\{\psi_t(0), \psi_{tt}(0)\} \in [H^2(\Omega) \times H_0^1(\Omega)] \times H_0^1(\Omega)$ and hence by Lemma 3.1 applied to $\{\psi_t, \psi_{tt}\}$ we obtain

$$\{\psi_t, \psi_{tt}\} \in C([0,T]; [H^2(\Omega) \cap H_0^1(\Omega)] \times H_0^1(\Omega)). \tag{8.6}$$

Thus (8.4) and (8.6) *a fortiori* imply that $\psi \in H^2(Q)$ a boost of regularity by one unit. The process may be repeated once more yielding then $\psi \in H^3(Q)$, as required by Theorem 8.1.

Thus, Theorem 8.1 applies and yields the vanishing of ψ and ψ_t at time $t = \frac{T}{2}$ as in (8.3). Finally, by virtue of the homogeneous hinged B.C. on all of Σ as in (8.4), the corresponding ψ-problem (8.1a), (8.4) is backward and forward well posed, in the sense that it generates a s.c. group, as guaranteed by Lemma 3.1(ii). Then, the conditions in (8.3) taken as initial conditions, yield $\psi \equiv 0$ in $\mathbb{R} \times \Omega$, as desired. □

9 Step 1: Proof of Corollary 4.2.3a. Boosting the A-Priori Regularity of the Over-Determined Homogeneous Problem to $\{w_t, \theta_t\} \in H^3(Q) \times H^1(Q)$

Problem. We return to problem (4.2.3a–c) (hinged case) and problem (4.2.4a–c) (clamped case), except that, for convenience, throughout this section we use the notation $\{w, w_t, \theta\}$ in place of $\{\phi, \phi_t, \eta\}$ as in Section 4.

Thus, let $\{w, w_t, \theta\}$ be a solution of the following (dual) over-determined *homogeneous* problem:

Hinged case (4.2.3a–c): Here $\emptyset \neq \Sigma_1 = \Sigma_3 \subset \Sigma_2$ (Remark 1.2),

$$\begin{cases} w_{tt} - \gamma\Delta w_{tt} + \Delta^2 w + \operatorname{div}(\alpha(x)\nabla\theta) \equiv 0 \text{ in } Q; & (9.1a) \\ \theta_t - \Delta\theta - \operatorname{div}(\alpha(x)\nabla w_t) \equiv 0 \text{ in } Q; & (9.1b) \\ w|_\Sigma \equiv 0;\ \Delta w|_\Sigma \equiv 0;\ \theta|_\Sigma \equiv 0 \text{ on } \Sigma = (0,T] \times \Gamma; & (9.1c) \\ \left.\dfrac{\partial w_t}{\partial \nu}\right|_{\Sigma_2} \equiv 0 \text{ on } \Sigma_2;\ \left.\dfrac{\partial \Delta w_t}{\partial \nu}\right|_{\Sigma_1} \equiv 0 \text{ on } \Sigma_1;\ \left.\dfrac{\partial \theta}{\partial \nu}\right|_{\Sigma_3} \equiv 0 \text{ on } \Sigma_3; & (9.1d) \end{cases}$$

under the geometrical condition (1.4): there exists $x_0 \in \mathbb{R}^2$ such that

$$(x - x_0) \cdot \nu(x) \leq 0 \quad \text{on } \Gamma \setminus \Gamma_2, \tag{9.2}$$

and with initial conditions (see (2.3)):

$$\{w_0, w_1, \theta_0\} \in Y_{\gamma,h} \equiv \mathcal{D}(\mathcal{A}) \times \mathcal{D}(\mathcal{A}_\gamma^{\frac{1}{2}}) \times L_2(\Omega) = [H^2(\Omega) \times H_0^1(\Omega)] \times H_0^1(\Omega) \times L_2(\Omega). \tag{9.3}$$

Thus, by the semigroup generation of Lemma 2.1(i), the following *a-priori* regularity is available at the outset

$$\begin{cases} w \in C([0,T]; \mathcal{D}(\mathcal{A}) \equiv H^2(\Omega) \cap H_0^1(\Omega)),\ w_t \in C([0,T]; \mathcal{D}(\mathcal{A}^{\frac{1}{2}}) = H_0^1(\Omega)) & (9.4a) \\ \theta \in C([0,T]; L_2(\Omega)). & (9.4b) \end{cases}$$

Analogously, let $\{w, w_t, \theta\}$ be a solution of the following (dual) over-determined *homogeneous* problem:

Clamped case (4.2.4a–c): Here $\emptyset \neq \Sigma_1 = \Sigma_3 \subset \Sigma$ (Remark 1.2),

$$\begin{cases} w_{tt} - \gamma\Delta w_{tt} + \Delta^2 w + \operatorname{div}(\alpha(x)\nabla\theta) \equiv 0 \text{ in } Q; & (9.5a) \\ \theta_t - \Delta\theta - \operatorname{div}(\alpha(x)\nabla w_t) \equiv 0 \text{ in } Q; & (9.5b) \\ w|_\Sigma \equiv 0;\ \dfrac{\partial w}{\partial \nu}|_\Sigma \equiv 0;\ \Delta w_t|_\Sigma \equiv 0;\ \theta|_\Sigma \equiv 0 \text{ on } \Sigma = (0,T] \times \Gamma; & (9.5c) \\ \left.\dfrac{\partial \Delta w_t}{\partial \nu}\right|_{\Sigma_1} \equiv 0 \text{ on } \Sigma_1;\ \left.\dfrac{\partial \theta}{\partial \nu}\right|_{\Sigma_3} \equiv 0 \text{ on } \Sigma_3; & (9.5d) \end{cases}$$

and with initial conditions (see (2.5)):

$$\{w_0, w_1, \theta_0\} \in Y_{\gamma,c} \equiv \mathcal{D}(A^{\frac{1}{2}}) \times \mathcal{D}(\mathcal{A}_\gamma^{\frac{1}{2}}) \times L_2(\Omega) \equiv H_0^2(\Omega) \times H_0^1(\Omega) \times L_2(\Omega). \tag{9.6}$$

Thus, by the semigroup generation of Lemma 2.1(ii), the following *a-priori* regularity is available at the outset

$$\begin{cases} w \in C([0,T]; \mathcal{D}(A^{\frac{1}{2}}) = H_0^2(\Omega)), \; w_t \in C([0,T]; \mathcal{D}(\mathcal{A}^{\frac{1}{2}}) = H_0^1(\Omega)) & \text{(9.7a)} \\ \theta \in C([0,T]; L_2(\Omega)). & \text{(9.7b)} \end{cases}$$

Goal. However, in order to apply the uniqueness result of Corollary 4.2.3b, the above regularity in (9.4) and (9.7) is not enough; rather, in both cases, we seek to boost the regularity of a solution $\{w, w_t, \theta\}$ to problem (9.1)–(9.4) or problem (9.5)–(9.7), to the level $H^3(Q)$ for w_t and $H^1(Q)$ for θ_t, as required by Corollary 4.2.3b, in order to apply Theorem 4.2.2. Indeed, it suffices to apply Corollary 4.2.3 on a time interval $[t_0, T_1]$, $t_0 > 0$, to obtain, accordingly, $w(T_1, \cdot) = w_t(T_1, \cdot) = \theta(T_1, \cdot) = 0$ with $T_1 = \frac{T}{2}$. After this, an application of the backward uniqueness result of [L-R-T.1] will yield $w_0 = 0$, $w_1 = 0$, $\theta_0 = 0$, and thus $w \equiv 0$, $w_t \equiv 0$, $\theta \equiv 0$ in $Q = [0,T] \times \Omega$. Thus, henceforth, we take $t_0 > 0$ arbitrarily small, and define $Q_{t_0} = [t_0, T] \times \Omega$, $\Sigma_{t_0} = [t_0, T] \times \Gamma$. We seek to boost the *a-priori* regularity (9.4) or (9.7) to the following level:

$$w_t \in H^3(Q_{t_0}): \text{ i.e., } w_t \in L_2(t_0, T; H^3(\Omega)); \; w_{tttt} \in L_2(t_0, T; L_2(\Omega)); \tag{9.8}$$

$$\theta_t \in H^1(Q_{t_0}): \text{ i.e., } \theta_t \in L_2(t_0, T; H^1(\Omega)); \; \theta_{tt} \in L_2(t_0, T; L_2(\Omega)). \tag{9.9}$$

That this is possible is shown by the following main result of this section.

THEOREM 9.1. *Let $\{w, w_t, \theta\}$ be a solution of the over-determined problem (9.1) or (9.5), with* a-priori *regularity as in (9.4) or (9.7). In the hinged case, we assume the geometrical condition (9.2). Then, in fact, $\{w, w_t, \theta\}$ satisfies the higher regularity (9.8), (9.9). Accordingly, Corollary 4.2.3 applies and yields $w_0 = 0$, $w_1 = 0$, $\theta_0 = 0$, as desired.*

Orientation. The present issue of boosting the *a-priori* regularity for the thermo-elastic problem is more complicated than the corresponding issue of boosting the *a-priori* regularity for the Kirchoff problem, as obtained in Corollary 8.2. While the basic idea is the same in both cases, in Corollary 8.2 the required estimates (6.2) (hinged case) and (7.2) (clamped case) could be applied directly. In the present thermo-elastic case, the boosting of regularity from (9.4) or (9.7) to (9.8), (9.9) will proceed through several steps. First, a preliminary Step 1 to boost initially the regularity of the thermal variable θ, critically of θ_t, by 1 Sobolev unit (see Lemma 9.2). The most demanding steps are Steps 3 and 6 to boost the regularity of the mechanical variable. Both main steps—in Propositions 9.4 and 9.7—will require the *a-priori* continuous observability estimate [L-T.6, Theorem 2.1.2(ii), Eqn. (2.1.10b)] already invoked in Eqn. (6.12) for a suitable wave equation. The procedure requires, in both instances, identifying a suitable wave equation with a inhomogeneous right-hand side term (based on θ) just in $L_2(Q_{t_0})$, and with the wave equation variable just in $L_2(Q_{t_0})$ as *a-priori* regularity. Due to the vanishing of the boundary terms for the over-determined problem, the counterpart version of estimate (6.2) then yields a boost of regularity by 1 Sobolev unit in both position and velocity of the suitable wave equation variable.

Proof. (of Theorem 9.1.) Throughout this proof, we fix $\epsilon > 0$ arbitrarily small.

Step 1

LEMMA 9.2. *With reference either to problem (9.1), (9.2) with* a-priori *regularity (9.4), or else (9.5), (9.6) with* a-priori *regularity (9.7), we have*

$$\theta \in L_2(0,T;H_0^1(\Omega)) \cap C([\epsilon,T];H_0^1(\Omega)); \quad \theta_t \in L_2(\epsilon,T;L_2(\Omega)) \cap C([\epsilon,T];[\mathcal{D}(\mathcal{A}^{\frac{1}{2}})]'),\ \epsilon > 0,$$

$$\theta_t \in L_2(0,T;L_2(\Omega)) \quad (\epsilon = 0,\ \textit{if}\ \theta_0 \in \mathcal{D}(\mathcal{A}^{\frac{1}{2}})); \tag{9.10}$$

$$w_{tt} \in C([0,T];L_2(\Omega))\ \textit{(hinged B.C.)}; \quad w_{tt} \in L_2(0,T;L_2(\Omega))\ \textit{(clamped B.C.)}. \tag{9.11}$$

(See also Lemma D.2 in Appendix D for θ.)

Proof. This result is essentially already known, see [L-T.7, Eqn. (3.3), p. 36 for θ, Eqn. (3.36), p. 41 for w_{tt}, and Remark 3.4, Eqn. (3.41), p. 41 for θ_t with $\theta_0 = 0$], where it was given for an abstract thermo-elastic system which includes, in particular, either hinged or clamped mechanical B.C. and thermal Dirichlet B.C. as in our present paper. For completeness, we indicate the proof in the present case, also because it provides the setting that needs to be pushed further in the subsequent steps. The abstract version of Eqn. (9.1b) (recall (2.1a), in both hinged or clamped cases, is:

$$\theta_t = -\mathcal{A}\theta + \tilde{F}_\alpha w_t, \quad \tilde{F}_\alpha w_t = \operatorname{div}(\alpha \nabla w_t) = -\alpha \mathcal{A} w_t + \nabla\alpha \cdot \nabla w_t \tag{9.12a}$$

$\mathcal{D}(\tilde{F}_\alpha) = H^2(\Omega) \cap H_0^1(\Omega)$ [same operator as in Eqn. (B.3) of Appendix B]. Its solution is

$$\theta(t) - e^{-\mathcal{A}t}\theta_0 = -\int_0^t e^{-\mathcal{A}(t-\tau)} \mathcal{A} w_t(\tau)\alpha\, d\tau + \ell.o.t. \tag{9.12b}$$

$$= -\int_0^t e^{-\mathcal{A}s} \mathcal{A} w_t(t-s)\alpha\, ds + \ell.o.t., \tag{9.12c}$$

where $e^{-\mathcal{A}t}$ is the s.c. analytic (self-adjoint) semigroup on $L_2(\Omega)$ generated by the positive, self-adjoint operator $-\mathcal{A}$ in (2.1a). By standard analytic semigroup results, we obtain from (9.12a),

$$\mathcal{A}^{\frac{1}{2}}\theta(t) = \mathcal{A}^{\frac{1}{2}} e^{-\mathcal{A}t}\theta_0 - \int_0^t \mathcal{A} e^{\mathcal{A}(t-\tau)} \mathcal{A}^{\frac{1}{2}} w_t(\tau)\alpha\, d\tau + \ell.o.t. \in L_2(0,T;L_2(\Omega)), \tag{9.13}$$

by $\theta_0 \in L_2(\Omega)$ and $w_t \in L_2(0,T;\mathcal{D}(\mathcal{A}^{\frac{1}{2}}))$, *a-fortiori* from (9.4a) (hinged case) or (9.7a) (clamped case). Thus (9.10) is proved for θ. As to w_{tt} in (9.11), we return to Eqn. (9.1a) or (9.5a) respectively, which we rewrite abstractly via the subset of B.C. in (9.1c) or in (9.5c) respectively, and (9.12a) for $\tilde{F}_\alpha$ as

$$w_{tt} = (-\mathcal{A}_\gamma^{-1}\mathcal{A})\mathcal{A}w + \mathcal{A}_\gamma^{-1}(\alpha\mathcal{A}\theta) + \ell.o.t. \in C([0,T];L_2(\Omega)) \text{ (hinged)}, \tag{9.14a}$$

$$w_{tt} = -\mathcal{A}_\gamma^{-1}\mathcal{A}w + \mathcal{A}_\gamma^{-1}(\alpha\mathcal{A}\theta) + \ell.o.t. \in L_2(0,T;L_2(\Omega)) \text{ (clamped)}, \tag{9.14b}$$

in the hinged and clamped cases, respectively, where the regularity in (9.14) follows from (9.4a) on w and (9.4b) on θ, hence $\alpha\mathcal{A}\theta \in C([0,T];[\mathcal{D}(\mathcal{A})]')$, while the regularity in (9.14b) was already noted in (7.7) (and proved in [L-T.7, p. 43] using the sharp regularity in [L-L.1, Chapter 5]). After (9.14) has been established, we can return to (9.12b), integrate by parts and obtain

$$\begin{aligned} \theta(t) - e^{-At}\theta_0 &= -w_t(t)\alpha + e^{-At}w_t(0)\alpha + \textstyle\int_0^t e^{-\mathcal{A}(t-\tau)}w_{tt}(\tau)\alpha\, d\tau + \ell.o.t. \\ &\in C([0,T];H_0^1(\Omega) = \mathcal{D}(\mathcal{A}^{\frac{1}{2}})). \end{aligned} \tag{9.14c}$$

As to θ_t, we differentiate (9.12c) in t, to obtain

$$\begin{aligned} \theta_t(t) &= -e^{-\mathcal{A}(t-\epsilon)}\mathcal{A}e^{-\mathcal{A}\epsilon}\theta_0 - \mathcal{A}^{\frac{1}{2}}e^{-\mathcal{A}t}\mathcal{A}^{\frac{1}{2}}w_t(0)\alpha \\ &- \int_0^t \mathcal{A}e^{-\mathcal{A}(t-\tau)}w_{tt}(\tau)\alpha\, d\tau + \ell.o.t. \in L_2(\epsilon,T;L_2(\Omega)) \cap C([\epsilon,T];[\mathcal{D}(\mathcal{A}^{\frac{1}{2}})]'), \end{aligned} \tag{9.15}$$

with $w_{tt} \in L_2(0,T;L_2(\Omega))$ by (9.14b), by invoking (9.3) or (9.6) on $w_t(0)$, and by standard analytic semigroup results. We note explicitly that, if $\theta_0 \in \mathcal{D}(\mathcal{A}^{\frac{1}{2}})$, we can then take $\epsilon = 0$ in $L_2(\epsilon,T;L_2(\Omega))$ in (9.15) (as already obtained in [L-T.7, Eqn. (3.41), p. 41]). Lemma 9.2 is established.

A similar argument—integrating by parts this time on (9.12c) and using again (9.14b) for w_{tt}—yields $\theta \in C([\epsilon,T];H_0^1(\Omega))$, see also Lemma D.2 in Appendix D. □

Step 2

LEMMA 9.3. *With reference to either problem (9.1)–(9.4), or else problem (9.5)–(9.7), and with $\epsilon > 0$ preassigned in Step 1, we have*

$$\begin{cases} \mathcal{A}^{-1}\theta_{tt} \in L_2(\epsilon,T;L_2(\Omega)),\ \mathcal{A}^{-1}w_{tttt} \in L_2(\epsilon,T;L_2(\Omega)); & (9.16) \\ \mathcal{A}^{-1}\theta_{ttt} \in L_2(2\epsilon,T;L_2(\Omega)). & (9.17) \end{cases}$$

Proof. To handle $\mathcal{A}^{-1}\theta_{tt}$ in both cases, we return to (9.12a), differentiate in t once, apply $\mathcal{A}^{-1}$ throughout, and obtain, as desired,

$$\mathcal{A}^{-1}\theta_{tt} = -\theta_t - \mathcal{A}^{-1}(\alpha\mathcal{A}w_{tt}) + \ell.o.t. \in L_2(\epsilon,T;L_2(\Omega)), \tag{9.18}$$

by virtue of (9.10) for θ_t and (9.11) for w_{tt}.

To handle $\mathcal{A}^{-1}w_{tttt}$, we first note that, because of the B.C. in (9.1c), and in (9.5c), respectively, we may write the abstract equation for w_t as follows:

$$(w_t)_{tt} = (-\mathcal{A}_\gamma^{-1}\mathcal{A})\mathcal{A}w_t + \mathcal{A}_\gamma^{-1}(\alpha\mathcal{A}\theta_t) + \ell.o.t., \tag{9.19}$$

for both the hinged case (9.1)–(9.4) [differentiate (9.14a)], as well as for the clamped case (9.5)–(9.7)[since $\Delta w_t|_\Sigma = 0$, see (9.5c), so that, after differentiation in t, A in (9.14b) can be replaced by $\mathcal{A}^2$]. Thus, model (9.19) applies to *both hinged and clamped cases.* Differentiating further (9.19) in t once, and applying $\mathcal{A}^{-1}$ throughout, we obtain

$$\mathcal{A}^{-1}w_{tttt} = (-\mathcal{A}_\gamma^{-1}\mathcal{A})w_{tt} + \mathcal{A}_\gamma^{-1}(\alpha\theta_{tt}) + \ell.o.t. \in L_2(\epsilon, T; L_2(\Omega)), \tag{9.20}$$

where the regularity in (9.20) follows by (9.18) just proved on $\mathcal{A}^{-1}\theta_{tt}$ and (9.11) on w.

We finally handle $\mathcal{A}^{-1}\theta_{ttt}$. Eqn. (9.18) is not good for this purpose; instead, we return to (9.12b) and obtain for $t \geq \epsilon$:

$$\theta(t) - e^{-\mathcal{A}(t-\epsilon)}\theta(\epsilon) = -\int_\epsilon^t e^{-\mathcal{A}(t-\tau)}\mathcal{A}w_t(\tau)\alpha\, d\tau + \ell.o.t. \tag{9.21a}$$

$$= -\int_0^{t-\epsilon} e^{-\mathcal{A}s}\mathcal{A}w_t(t-s)\alpha\, ds + \ell.o.t., \tag{9.21b}$$

with $\theta(\epsilon) \in L_2(\Omega)$ by (9.4b) and (9.7b) respectively. A first, time differentiation of (9.21b) yields for $t \geq 2\epsilon$:

$$\theta_t + e^{-\mathcal{A}(t-2\epsilon)}\mathcal{A}e^{-\mathcal{A}\epsilon}\theta(\epsilon) + e^{-\mathcal{A}(t-2\epsilon)}\mathcal{A}^{\frac{1}{2}}e^{-\mathcal{A}\epsilon}\mathcal{A}^{\frac{1}{2}}w_t(\epsilon)$$

$$= -\int_0^{t-\epsilon} \mathcal{A}^{\frac{1}{2}}e^{-\mathcal{A}s}\mathcal{A}^{\frac{1}{2}}w_{tt}(t-s)\alpha\, ds \tag{9.22a}$$

$$= -\int_\epsilon^t \mathcal{A}^{\frac{1}{2}}e^{-\mathcal{A}(t-\tau)}\mathcal{A}^{\frac{1}{2}}w_{tt}(\tau)\alpha\, d\tau. \tag{9.22b}$$

A subsequent time differentiation of (9.22a) yields for $t \geq 2\epsilon$:

$$\theta_{tt} - e^{-\mathcal{A}(t-2\epsilon)}\mathcal{A}^2e^{-\mathcal{A}\epsilon}\theta(\epsilon) - e^{-\mathcal{A}(t-2\epsilon)}\mathcal{A}^{\frac{3}{2}}e^{-\mathcal{A}\epsilon}\mathcal{A}^{\frac{1}{2}}w_t(\epsilon) + e^{-\mathcal{A}(t-2\epsilon)}\mathcal{A}e^{-\mathcal{A}\epsilon}w_{tt}(\epsilon)$$

$$= -\int_0^{t-\epsilon} \mathcal{A}e^{-\mathcal{A}s}w_{ttt}(t-s)\alpha\, ds \tag{9.23a}$$

$$= -\int_\epsilon^t \mathcal{A}e^{-\mathcal{A}(t-\tau)}w_{ttt}(\tau)\alpha\, d\tau. \tag{9.23b}$$

A third differentiation of (9.23) yields for $t \geq 2\epsilon$:

$$\theta_{ttt}(t) + I_\epsilon(t) = -\int_0^{t-\epsilon} \mathcal{A}e^{-\mathcal{A}s}w_{tttt}(t-s)\alpha\, ds \tag{9.24a}$$

$$= -\int_\epsilon^t \mathcal{A}e^{-\mathcal{A}(t-\tau)}w_{tttt}(\tau)\alpha\, d\tau. \tag{9.24b}$$

$$I_\epsilon(t) = e^{-\mathcal{A}(t-2\epsilon)}\mathcal{A}^3 e^{-\mathcal{A}\epsilon}\theta(\epsilon) + e^{-\mathcal{A}(t-2\epsilon)}\mathcal{A}^{\frac{5}{2}} e^{-\mathcal{A}\epsilon}\mathcal{A}^{\frac{1}{2}} w_t(\epsilon)$$
$$-e^{-\mathcal{A}(t-2\epsilon)}\mathcal{A}^2 e^{-\mathcal{A}\epsilon} w_{tt}(\epsilon) + \mathcal{A}e^{-\mathcal{A}(t-\epsilon)} w_{ttt}(\epsilon). \tag{9.24c}$$

Hence, applying $\mathcal{A}^{-1}$ throughout (9.24b) yields for $t \geq 2\epsilon$:

$$\mathcal{A}^{-1}\theta_{ttt}(t) = -\mathcal{A}^{-1}I_\epsilon(t) - \int_\epsilon^t \mathcal{A}e^{-\mathcal{A}(t-\tau)}\mathcal{A}^{-1}w_{tttt}(\tau)\alpha\, d\tau \in L_2(2\epsilon, T; L_2(\Omega)). \tag{9.25}$$

By the regularity of $\mathcal{A}^{-1}w_{tttt}$ in (9.20) and standard semigroup theory, we find that the integral in (9.25) is in $L_2(\epsilon, T; L_2(\Omega))$. Moreover, by (9.19), we have

$$\mathcal{A}^{-1}w_{ttt} = (-\mathcal{A}_\gamma^{-1}\mathcal{A})w_t + \mathcal{A}_\gamma^{-1}(\alpha\theta_t) + \ell.o.t. \in C([\epsilon, T]; \mathcal{D}(\mathcal{A}^{\frac{1}{2}})), \tag{9.26}$$

see (9.4a), respectively (9.7a) on w_t, and (9.10) or θ_t.

Then, by (9.24c), (9.26), we have $\mathcal{A}^{-1}I_\epsilon(t) \in C([2\epsilon, T]; L_2(\Omega))$. Hence, the regularity in (9.25) follows, as claimed in (9.16). Lemma 9.3 is proved. □

Step 3

PROPOSITION 9.4. *With reference to either problem (9.1)–(9.4), or else problem (9.5)–(9.7), and with $\epsilon > 0$ preassigned in Step 1, we have*

$$w_{tt} \in L_2(2\epsilon, T; \mathcal{D}(\mathcal{A}^{\frac{1}{2}}) = H_0^1(\Omega)), \quad w_{ttt} \in L_2(2\epsilon, T; L_2(\Omega)). \tag{9.27}$$

Proof. Writing $\text{div}(\alpha\nabla\theta) = \alpha\Delta\theta + \nabla\alpha\cdot\nabla\theta$ in the first equation (9.1a), and substituting $\Delta\theta$ from the second Eqn. (9.1b), we obtain, as in (3.1):

$$w_{tt} - \gamma\Delta w_{tt} + \Delta^2 w - \alpha\text{div}(\alpha\nabla w_t) = -\alpha\theta_t - \nabla\alpha\cdot\nabla\theta. \tag{9.28}$$

In the *hinged* case, this Eqn. (9.28), along with the B.C. in(9.1c), can be rewritten abstractly as

$$w_{tt} = -\mathcal{A}_\gamma^{-1}\mathcal{A}^2 w + \mathcal{A}_\gamma^{-1}F_\alpha w_t - \mathcal{A}_\gamma^{-1}(\alpha\theta_t) + \ell.o.t., \tag{9.29}$$

after recalling $\mathcal{A}_\gamma$ from (2.1a) and $F_\alpha w_t = \alpha\ \text{div}(\alpha\nabla w_t) = \alpha^2\Delta w_t + \alpha\nabla\alpha\cdot\nabla w_t$ (see (3.3)).

In the *clamped* case, we obtain similarly

$$w_{tt} = -\mathcal{A}_\gamma^{-1}Aw + \mathcal{A}_\gamma^{-1}F_\alpha w_t - \mathcal{A}_\gamma^{-1}(\alpha\theta_t) + \ell.o.t., \tag{9.30}$$

see (2.4a) for A. Differentiating (9.29) in t once yields

$$w_{ttt} = -\mathcal{A}_\gamma^{-1}\mathcal{A}^2 w + \mathcal{A}_\gamma^{-1}F_\alpha w_{tt} - \mathcal{A}_\gamma^{-1}(\alpha\theta_{tt}) + \ell.o.t., \tag{9.31}$$

Eqn. (9.31) holds true not only in the *hinged* case [Eqns. (9.1)–(9.4)], but also in the *clamped* case [Eqns. (9.5)–(9.7), as $\Delta w_t|_\Sigma \equiv 0$ as well in this case]; that is, after differentiating (9.30) we may replace A by $\mathcal{A}^2$. We differentiate (9.31) in t once more and obtain

$$(w_{tt})_{tt} = -\mathcal{A}_\gamma^{-1}\mathcal{A}^2(w_{tt}) + \mathcal{A}_\gamma^{-1}F_\alpha(w_{tt})_t - \mathcal{A}_\gamma^{-1}(\alpha\theta_{ttt}) + \ell.o.t., \tag{9.32}$$

or setting a new variable

$$q \equiv w_{tt} \in C([0,T]; L_2(\Omega)) \tag{9.33}$$

from (9.11), we obtain

$$q_{tt} = -\mathcal{A}_\gamma^{-1}\mathcal{A}^2 q + \mathcal{A}_\gamma^{-1} F_\alpha q_t - \mathcal{A}_\gamma^{-1}(\alpha\theta_{ttt}) + \ell.o.t., \tag{9.34}$$

or, upon using identity (6.4b) for $\mathcal{A}_\gamma^{-1}\mathcal{A}^2$ as well as

$$F_\alpha q_t = \alpha \text{ div}(\alpha \nabla q_t) = -\mathcal{A}(\alpha^2 q_t) - 2\nabla\alpha \cdot \nabla(\alpha q_t) - \alpha \text{ div}(q_t \nabla \alpha) + \ell.o.t., \tag{9.35}$$

because $q|_\Sigma \equiv 0$, we rewrite (9.34) as the following mixed problem for a wave equation

$$\begin{cases} q_{tt} = \dfrac{\Delta q}{\gamma} - \dfrac{\alpha^2 q_t}{\gamma} + f \text{ in } (2\epsilon, T] \times \Omega = Q_{2\epsilon}; & (9.36a) \\ f \equiv -\mathcal{A}_\gamma^{-1}(\alpha\theta_{ttt}) + \frac{q}{\gamma^2} + \frac{\mathcal{A}_\gamma^{-1}(\alpha^2 q_t)}{\gamma} + \ell.o.t. \in L_2(2\epsilon, T; L_2(\Omega)), & (9.36b) \\ \text{with boundary conditions, either} & \\ q|_\Sigma \equiv 0, \ \left.\dfrac{\partial q}{\partial \nu}\right|_{\Sigma_2} \equiv 0 \text{ (hinged case) } \Sigma_2 = (0,T] \times \Gamma_2 & (9.37) \\ \text{or else} & \\ q|_\Sigma \equiv 0, \ \left.\dfrac{\partial q}{\partial \nu}\right|_{\Sigma} \equiv 0 \text{ (clamped case).} & (9.38) \end{cases}$$

The regularity in (9.36b) for f follows from (9.17) for $\mathcal{A}^{-1}\theta_{ttt}$, (9.33) for q, and *a-fortiori* from (9.19) for $\mathcal{A}^{-1}q_t = \mathcal{A}^{-1}w_{ttt}$. Moreover, in the hinged case, Eqn. (9.37) follows from $w|_\Sigma \equiv 0$ and $\frac{\partial w_t}{\partial \nu}|_{\Sigma_2} \equiv 0$ in (9.1c–d); while in the clamped case, Eqn. (9.38) follows from $w|_\Sigma \equiv 0$ and $\frac{\partial w}{\partial \nu}|_\Sigma \equiv 0$ in (9.5c), both via the variable q in (9.33). Thus, with problem (9.36a–b) and either (9.37), or else (9.38), we are in the same situation as with problem (6.10), except for the over-determined B.C. for q. We likewise invoke [L-T.6, Theorem 2.1.2(ii), Eqn. (2.1.10b) (which actually applies to a general wave equation with space variable coefficients in the first-order terms, thus at the energy level). We then obtain in the notation of [L-T.6],

$$\overline{(BT)}_q|_{\Sigma_{2\epsilon}} + \frac{C_T}{\gamma^2}\int_{Q_{2\epsilon}} f^2 dQ + TC_T\|q\|^2_{L_2(Q_{2\epsilon})} \geq \int_{2\epsilon}^T \|\{q(t), q_t(t)\}\|^2_{H^1(\Omega)\times L_2(\Omega)} dt, \tag{9.39}$$

where, via $q|_\Sigma \equiv 0$ in both cases [so that $h \cdot \nabla q = \frac{\partial q}{\partial \nu} h \cdot \nu$; $|\nabla q|^2 = \left(\frac{\partial q}{\partial \nu}\right)^2$, $h = (x - x_0)$], we have

$$\overline{(BT)}_q|_{\Sigma_{2\epsilon}} = \int_{\Sigma_{2\epsilon}} e^{\tau\phi} \frac{\partial q}{\partial \nu} h \cdot \nabla q \, d\Sigma_{2\epsilon} - \frac{1}{2}\int_{\Sigma_{2\epsilon}} e^{\tau\phi} |\nabla q|^2 h \cdot \nu \, d\Sigma_{2\epsilon} \tag{9.40}$$

$$= \frac{1}{2}\int_{\Sigma_{2\epsilon}} e^{\tau\phi} \left(\frac{\partial q}{\partial \nu}\right)^2 h \cdot \nu \, d\Sigma_{2\epsilon}, \ \Sigma_{2\epsilon} = (2\epsilon, T] \times \Gamma. \tag{9.41}$$

[In (9.40), (9.41) ϕ is the pseudo-convex function in [L-T.6].] From (9.41), we then obtain

$$\overline{(BT)}_q|_{\Sigma_{2\epsilon}} \leq 0 \text{ in the hinged case; } \overline{(BT)}_q|_{\Sigma_{2\epsilon}} = 0 \text{ in the clamped case.} \tag{9.42}$$

Indeed, in the *clamped* case, the B.C. (9.38) used in (9.41) yields (9.42) at once. Instead, in the *hinged case*, assumption (9.2) on $\Gamma \setminus \Gamma_2$, as well as the B.C. (9.37) yield from (9.41)

$$\begin{aligned} 2\overline{(BT)}_q|_{\Sigma_{2\epsilon}} &= \int_{2\epsilon}^{T} \int_{\Gamma_2} e^{\tau\phi} \left(\frac{\partial q}{\partial \nu}\right)^2 h \cdot \nu \, d\Gamma \, d\tau \\ &+ \int_{2\epsilon}^{T} \int_{\Gamma \setminus \Gamma_2} e^{\tau\phi} \left(\frac{\partial q}{\partial \nu}\right)^2 h \cdot \nu \, d\Gamma \, d\tau \leq 0, \end{aligned} \tag{9.43}$$

as desired. Thus, via (9.42) for $\overline{BT}$, (9.36b) for f and (9.33) for q, we obtain

$$q = w_{tt} \in L_2(2\epsilon, T; H_0^1(\Omega) = \mathcal{D}(\mathcal{A}^{\frac{1}{2}})); \ q_t \equiv w_{ttt} \in L_2(2\epsilon, T; L_2(\Omega)), \tag{9.44}$$

and Proposition 9.4 is established. □

Step 4

LEMMA 9.5. *With reference to either problem (9.1)–(9.4), or else problem (9.5)–(9.7), and with $\epsilon > 0$ preassigned in Step 1, we have*

$$\theta_t \in L_2(3\epsilon, T; \mathcal{D}(\mathcal{A}^{\frac{1}{2}}) = H_0^1(\Omega)); \ \theta_{tt} \in L_2(3\epsilon, T; L_2(\Omega)); \tag{9.45}$$

$$w \in L_2(2\epsilon, T; \mathcal{D}(\mathcal{A}^{\frac{3}{2}})) \text{ (hinged case); } w \in L_2(2\epsilon, T; \mathcal{D}(A^{\frac{3}{4}})) \text{ (clamped case);} \tag{9.46}$$

$$\theta \in L_2(2\epsilon, T; \mathcal{D}(\mathcal{A})); \ w_t \in L_2(2\epsilon, T; \mathcal{D}(\mathcal{A}) = H^2(\Omega) \cap H_0^1(\Omega)). \tag{9.47}$$

Proof. Conclusions (9.45) for θ_t and θ_{tt} follow from (9.22b), respectively (9.23b), with ϵ in those integrals changed into 2ϵ, via (9.44) and standard semigroup theory.

As to w_t, we return to Eqn. (9.19) [which holds true both for the hinged and for the clamped case], where θ_t and w_{ttt} have the regularity given by (9.10) and (9.44), respectively. This way, we get $(A_\gamma^{-1}A)\mathcal{A}w_t \in L_2(2\epsilon, T; L_2(\Omega))$ and then (9.47) follows for w_t.

As to w, in the *hinged* case, we return to (9.14a), where w_{tt} and θ have the regularity given by (9.44) and (9.10), respectively. This way, one gets $(A_\gamma^{-1}A)\mathcal{A}w \in L_2(2\epsilon, T; \mathcal{D}(\mathcal{A}^{\frac{1}{2}}))$, and then $w \in L_2(2\epsilon, T; \mathcal{D}(\mathcal{A}^{\frac{3}{2}}))$, as claimed in (9.46). In the *clamped* case, the same argument starting from (9.14b) leads to $(A_\gamma^{-1}A^{\frac{1}{2}})A^{\frac{1}{2}}w \in L_2(2\epsilon, T; \mathcal{D}(\mathcal{A}^{\frac{1}{2}}))$, or equivalently, $(A_\gamma^{-\frac{1}{2}}A^{\frac{1}{4}})A^{\frac{3}{4}}w \in L_2(2\epsilon, T; L_2(\Omega))$. Since $\mathcal{D}(A^{\frac{1}{4}}) = \mathcal{D}(\mathcal{A}^{\frac{1}{2}}) = H_0^1(\Omega)$, it then follows that $A^{\frac{3}{4}}w \in L_2(2\epsilon, L_2(\Omega))$ and (9.45) for w is fully proved. Finally, as to θ, we return to (9.1b) or (9.5b), or their abstract version $\theta_t = -\mathcal{A}\theta - \alpha\mathcal{A}w_t + \ell.o.t.$, with θ_t and w_t having the regularity given by (9.10) and (9.47), respectively. This way we obtain $\mathcal{A}\theta \in L_2(2\epsilon, T; L_2(\Omega))$ and (9.47) for θ follows. □ □

Step 5

LEMMA 9.6. *With reference to either problem (9.1)–(9.4), or else problem (9.5)–(9.7), and with $\epsilon > 0$ preassigned in Step 1, we have*

$$\mathcal{A}^{-1}w_{ttttt} \in L_2(2\epsilon, T; L_2(\Omega)); \quad \mathcal{A}^{-1}\theta_{tttt} \in L_2(3\epsilon, T; L_2(\Omega)), \tag{9.48}$$

involving the fifth and fourth time derivative of w and θ, respectively.

Proof. We return to (9.20) and differentiate in t once to obtain

$$\mathcal{A}^{-1}w_{ttttt} = (-\mathcal{A}_\gamma^{-1}\mathcal{A})w_{ttt} + \mathcal{A}_\gamma^{-1}(\alpha\theta_{ttt}) + \ell.o.t. \in L_2(2\epsilon, T; L_2(\Omega)), \tag{9.49}$$

where the regularity in (9.49) follows from the regularity for w_{ttt} in (9.44) and the regularity for $\mathcal{A}^{-1}\theta_{ttt}$ in (9.17).

Next, we return to (9.26b) with 2ϵ instead of ϵ in the integral term, differentiate in t and obtain for $t \geq 3\epsilon$,

$$\mathcal{A}^{-1}\theta_{tttt}(t) = \mathcal{A}^{-1}J_\epsilon(t) + \int_{2\epsilon}^{t} \mathcal{A}e^{-\mathcal{A}(t-\tau)}\mathcal{A}^{-1}w_{ttttt}(\tau)\alpha\, d\tau \in L_2(3\epsilon, T; L_2(\Omega)); \tag{9.50}$$

$$\begin{aligned}\mathcal{A}^{-1}J_\epsilon(t) &= -e^{-\mathcal{A}(t-3\epsilon)}\mathcal{A}^3 e^{-\mathcal{A}\epsilon}\theta(2\epsilon) - e^{-\mathcal{A}(t-3\epsilon)}\mathcal{A}^{\frac{5}{2}}e^{-\mathcal{A}\epsilon}\mathcal{A}^{\frac{1}{2}}w_t(2\epsilon) \\ &\quad + e^{-\mathcal{A}(t-3\epsilon)}\mathcal{A}^2 e^{-\mathcal{A}\epsilon}w_{tt}(2\epsilon) - \mathcal{A}e^{-\mathcal{A}(t-2\epsilon)}w_{ttt}(2\epsilon) \\ &\in C([3\epsilon, T]; L_2(\Omega)).\end{aligned} \tag{9.51}$$

□

Step 6

PROPOSITION 9.7. *With reference to either problem (9.1)–(9.4), or else problem (9.5)–(9.7), and with $\epsilon > 0$ preassigned in Step 1, we have*

$$w_{ttt} \in L_2(3\epsilon, T; \mathcal{D}(\mathcal{A}^{\frac{1}{2}}) = H_0^1(\Omega)), \; w_{tttt} \in L_2(3\epsilon, T; L_2(\Omega)). \tag{9.52}$$

Proof. We return to problem (9.36)–(9.38), differentiate in t, set a new variable

$$p = q_t \equiv w_{ttt} \in L_2(2\epsilon, T; L_2(\Omega)), \tag{9.53}$$

where the regularity in (9.53) stems from (9.33) and (9.44), and obtain

$$\begin{cases} p_{tt} = \dfrac{\Delta p}{\gamma} - \dfrac{\alpha^2 p_t}{\gamma} + g \text{ in } (3\epsilon, T] \times \Omega \equiv Q_{3\epsilon}; & (9.54a) \\ g = f_t = -\mathcal{A}_\gamma^{-1}(\alpha\theta_{tttt}) + \dfrac{p}{\gamma^2} + \dfrac{\mathcal{A}_\gamma^{-1}(\alpha^2 p_t)}{\gamma} + \ell.o.t. \in L_2(3\epsilon, T; L_2(\Omega)), & (9.54b) \\ \text{with boundary conditions, either} \\ p|_\Sigma \equiv 0, \ \left.\dfrac{\partial p}{\partial \nu}\right|_{\Sigma_2} \equiv 0 \text{ (hinged case)}, \ \Sigma_2 = (0, T] \times \Gamma_2, & (9.55) \\ \text{or else} \\ p|_\Sigma \equiv 0, \ \left.\dfrac{\partial p}{\partial \nu}\right|_{\Sigma} \equiv 0 \text{ (clamped case)}. & (9.56) \end{cases}$$

The regularity in (9.54b) follows from (9.48) on $\mathcal{A}^{-1}\theta_{tttt}$, (9.44) for $p = w_{ttt}$, and (9.20) on $\mathcal{A}^{-1}p_t = \mathcal{A}^{-1}w_{tttt}$. Then, we apply to problem (9.54a–b) and either (9.55) or (9.56) the same inequality (9.39) from [L-T.6], this time in the p-variable, with *a-priori* regularity for p in (9.53) and for g in (9.54b), thus over $[3\epsilon, T]$, rather than $[2\epsilon, T]$, as in (9.39). Exactly the same argument between (9.39) and (9.44) in the proof of Proposition 9.4 now yields

$$p = q_t \equiv w_{ttt} \in L_2(3\epsilon, T; \mathcal{D}(\mathcal{A}^{\frac{1}{2}}) = H_0^1(\Omega)), \quad p_t = q_{tt} = w_{tttt} \in L_2(3\epsilon, T; L_2(\Omega)), \tag{9.57}$$

as claimed in (9.52). □

Step 7

COROLLARY 9.8. *With reference to either problem (9.1)–(9.4), or else problem (9.5)–(9.7), and with $\epsilon > 0$ preassigned in Step 1, we have*

$$w_t \in L_2(3\epsilon, T; \mathcal{D}(\mathcal{A}^{\frac{3}{2}})). \tag{9.58}$$

Proof. We return to Eqn. (9.19) with w_{ttt} and θ_t having the regularity in (9.57) and (9.45), respectively. This way, we obtain $\mathcal{A}_\gamma^{-1}\mathcal{A}\mathcal{A}w_t \in L_2(3\epsilon, T; \mathcal{D}(\mathcal{A}^{\frac{1}{2}}))$, and then (9.58) for w_t follows. □ □

Step 8.

The regularity in (9.58) for w_t and in (9.52) for w_{tttt} prove the desired conclusion (9.8) with $t_0 = 3\epsilon$. Similarly, the regularity in (9.45) for θ_t and for θ_{tt} prove the desired conclusion (9.9) with $t_0 = 3\epsilon$. Theorem 9.1 is fully proved.

REMARK 9.1. *Plainly, the above boot-strap argument can be further continued. Here, however, we limit ourselves to what is needed to apply Corollary 4.2.3.* □

10 Step 1: Proof of Theorem 8.1: Unique Continuation for the Over-Determined Kirchoff Equation

Goal. We begin by rewriting problem (8.1a–b) in the variable w rather than ψ, for notational convenience throughout the present section:

$$\begin{cases} w_{tt} - \gamma\Delta w_{tt} + \Delta^2 w - \operatorname{div}(\alpha\nabla(\alpha w_t)) = 0 \quad \text{in } Q = (0,T] \times \Omega; & (10.1a) \\ w \equiv 0, \ \dfrac{\partial w}{\partial \nu} \equiv 0, \ \dfrac{\partial^2 w}{\partial \nu^2} \equiv 0, \ \dfrac{\partial^3 w}{\partial \nu^3} \equiv 0 \quad \text{in } \tilde{\Sigma} = (0,T] \times \tilde{\Gamma}, & (10.1b) \end{cases}$$

where $\tilde{\Gamma}$ is an arbitrary non-empty open set of $\Gamma = \partial\Omega$ of positive measure. Also, we take $T > \tilde{T}$, with $\tilde{T}$ defined in (8.2).

We recall that, in order to show Theorem 8.1, our *goal is as follows*: we assume that $w \in H^3(Q)$ is a solution of the over-determined problem (10.1a–b), and we want to show that then

$$w\left(\frac{T}{2}\right) = w_t\left(\frac{T}{2}\right) = 0. \tag{10.2}$$

Orientation. Unique continuation for higher order equations and systems poses serious difficulties because of multiple characteristics. The only uniqueness result available for the Kirchoff plate equation with variable non-analytic coefficients was given by V. Isakov in [I.3]. His result, however, requires over-determined Cauchy data on the full lateral boundary Σ, rather than on the portion $\tilde{\Sigma}$ as in (10.1b), with $\tilde{\Gamma}$ open, of positive measure and arbitrarily small. In proving Theorem 8.1, we will use the fact that the coefficient α in (10.1a) is time-independent.

The proof in the present section of Theorem 8.1 is based on Carleman estimates introduced by D. Tataru in [Ta.1]. Since we want to obtain a uniqueness of continuation result across time-like surfaces with respect to the wave operator $\gamma\partial_t^2 - \Delta$, we can not apply the Carleman type estimate directly. We rather use the fact that the principal symbol of the plate operator factors into the Laplacian and the wave operator. Thus, we will start with Carleman estimates for the wave and the Laplace operators.

Preliminaries. First, corresponding to the time variable t and the space variable x, we introduce the Fourier variables ξ_0 and ξ' and set $\xi = (\xi_0, \xi') \in R^3$.

Next, in order to formulate the sought-after Carleman estimates, we introduce the Gaussian regularizer.

$$e^{-\frac{1}{2\tau}D_t^2} u = \frac{1}{(2\pi)^3} \int e^{-\frac{1}{2\tau}\xi_0^2} e^{i(\xi_0 t + \xi_1 x_1 + \xi_2 x_2)} \hat{u}(\xi) d\xi, \tag{10.3}$$

where $\hat{u}$ denotes the Fourier transform of u, and the integral is over $\mathbb{R}^3$.

Finally, we need to introduce some weighted norms in H^s for $s \in \mathbb{R}$. We define

$$|u|_{s,\tau}^2 = \frac{1}{(2\pi)^3} \int (\tau^2 + |\xi|^2)^s |\hat{u}(\xi)|^2 d\xi, \qquad s \in \mathbb{R}, \tag{10.4a}$$

where the integral is over $\mathbb{R}^3$. It is readily seen that, for s positive, an *equivalent* and quite useful, norm, which we shall still denote for convenience with the same symbol $|\cdot|_{s,\tau}$, is given by

$$|u|_{s,\tau}^2 = \sum_{|\beta|\leq s} \tau^{2s-2|\beta|}|D^\beta u|_0^2, \qquad s \in \mathbb{R}^+ \tag{10.4b}$$

for a multi-index β, where $|\ \ |_0$ is the norm of $L_2(\mathbb{R}^3)$. Below, we shall use the following direct consequences of (10.4b): for $\tau > 0$, we have readily

$$\tau|u|_0 \leq |u|_{1,\tau}; \quad \tau|u|_{1,\tau} \leq |u|_{2,\tau} \tag{10.4c}$$

For the notion of strong pseudo-convexity, to be henceforth used, we refer again to D. Tataru's paper [Ta.3, Definition 1.1 and Definition 1.2].

Step 1. Carleman-type estimates for the wave operator and the Laplacian operator.

LEMMA 10.1. *Let φ be a strongly pseudo-convex function with respect to the wave operator $\gamma\partial_t^2 - \Delta$ and with respect to the Laplacian Δ at the point (t_0, x_0) on the level surface $\{\xi_0 = 0\}$.*

Given $\delta > 0$, let $B_\delta(t_0, x_0)$ be the ball in $\mathbb{R}^3$ of radius δ, centered at (t_0, x_0). Then, for all small $\delta > 0$, there exist constants $c > 0$ and $d > 0$ such that, with reference to the norm in (10.4b), we have the following estimates:

$$\begin{aligned} |e^{-\frac{1}{2\tau}D_t^2}e^{\tau\varphi}u|_{2,\tau} \ &\leq \ c\sqrt{\tau}\Big\{|e^{-\frac{1}{2\tau}D_t^2}e^{\tau\varphi}(\gamma\partial_t^2 - \Delta)u|_0 \\ &\quad + e^{-d\tau}|e^{\tau\varphi}(\gamma\partial_t^2 - \Delta)u|_0 + e^{-d\tau}|e^{\tau\varphi}u|_{1,\tau}\Big\}, \end{aligned} \tag{10.5}$$

and

$$|e^{-\frac{1}{2\tau}D_t^2}e^{\tau\varphi}u|_{2,\tau} \leq c\sqrt{\tau}\left\{|e^{-\frac{1}{2\tau}D_t^2}e^{\tau\varphi}\Delta u|_0 + e^{-d\tau}|e^{\tau\varphi}\Delta u|_0 + e^{-d\tau}|e^{\tau\varphi}u|_{1,\tau}\right\} \tag{10.6}$$

provided that τ is large enough, and that u is an $H^1(\mathbb{R}^3)$-function, with compact support in the ball $B_\delta(t_0, x_0)$, such that the right-hand sides are finite [i.e., such that $(\gamma\partial_t^2 - \Delta)u$ and, respectively, Δu are in $L_2(\mathbb{R}^3)$.]

These estimates follow from Theorem 2 in [Ta.4]. We point out that Tataru uses the Gaussian regularizer $e^{-\varepsilon D_t^2/2\tau}$ with $\varepsilon > 0$ small. Since the operators we are using have time-independent coefficients, we can set $\varepsilon = 1$.

Relying on those two estimates, we can now derive a Carleman type estimate for the Kirchoff plate equation.

Step 2. Carleman-type estimate for the (principal part of the) Kirchoff operator.

LEMMA 10.2. *Let φ be a second degree polynomial which is a strongly pseudo-convex function with respect to the wave operator $\gamma\partial_t^2 - \Delta$ and with respect to the Laplacian Δ at the point*

(t_0, x_0) on the level surface $\{\xi_0 = 0\}$. Then, for all small $\delta > 0$, there exist constants $c > 0$ and $d > 0$ such that, in the norm of (10.4b), we have the following estimate

$$|e^{-\frac{1}{2\tau}D_t^2} e^{\tau\varphi} w|_{2,\tau} + |e^{-\frac{1}{2\tau}D_t^2} e^{\tau\varphi} \partial_t^2 w|_{2,\tau} + |e^{-\frac{1}{2\tau}D_t^2} e^{\tau\varphi} \Delta w|_{2,\tau} + |e^{-\frac{1}{2\tau}D_t^2} e^{\tau\varphi} \nabla w|_{2,\tau}$$

$$\leq c\sqrt{\tau}\Big(|e^{-\frac{1}{2\tau}D_t^2} e^{\tau\varphi}(\gamma\partial_t^2 - \Delta)\Delta w|_0 + e^{-d\tau}|e^{\tau\varphi}(\gamma\partial_t^2 - \Delta)\Delta w|_0 + e^{-d\tau}|e^{\tau\varphi}\nabla\Delta w|_0$$

$$+ e^{-d\tau}|e^{\tau\varphi}\Delta w|_{1,\tau} + e^{-d\tau}|e^{\tau\varphi}\partial_t^2 w|_{1,\tau} + e^{-d\tau}|e^{\tau\varphi} w|_{1,\tau} + e^{-d\tau}|e^{\tau\varphi}\nabla w|_{1,\tau}\Big), \tag{10.7}$$

provided that τ is large enough, and w is an $H^3(\mathbb{R}^3)$-function, with compact support in the ball $B_\delta(t_0, x_0)$, such that the right-hand side is finite.

Proof. **Step (i).** With $w \in H^3(\mathbb{R}^3)$ compactly supported in $B_\delta(t_0, x_0)$, as assumed, we first use estimate (10.5) with $u = \Delta w \in H^1(\mathbb{R}^3)$ compactly supported in $B_\delta(t_0, x_0)$, to get:

$$|e^{-\frac{1}{2\tau}D_t^2} e^{\tau\varphi}\Delta w|_{2,\tau} \leq c\sqrt{\tau}\Big\{|e^{-\frac{1}{2\tau}D_t^2} e^{\tau\varphi}\Delta(\gamma\partial_t^2 - \Delta)w|_0$$

$$+ e^{-d\tau}|e^{\tau\varphi}\Delta(\gamma\partial_t^2 - \Delta)w|_0 + e^{-d\tau}|e^{\tau\varphi}\Delta w|_{1,\tau}\Big\}, \tag{10.8}$$

and next use estimate (10.6) with $u = (\gamma\partial_t^2 - \Delta)w \in H^1(\mathbb{R}^3)$ compactly supported in $B_\delta(t_0, x_0)$ to get:

$$|e^{-\frac{1}{2\tau}D_t^2} e^{\tau\varphi}(\gamma\partial_t^2 - \Delta)w|_{2,\tau}$$

$$\leq c\sqrt{\tau}\Big\{|e^{-\frac{1}{2\tau}D_t^2} e^{\tau\varphi}\Delta(\gamma\partial_t^2 - \Delta)w|_0 + e^{-d\tau}|e^{\tau\varphi}\Delta(\gamma\partial_t^2 - \Delta)w|_0$$

$$+ e^{-d\tau}|e^{\tau\varphi}(\gamma\partial_t^2 - \Delta)w|_{1,\tau}\Big\}. \tag{10.9}$$

Next, adding and subtracting, yields with $c = \frac{1}{\gamma}$:

$$|e^{-\frac{1}{2\tau}D_t^2} e^{\tau\varphi}\partial_t^2 w|_{2,\tau} \leq c\Big\{|e^{-\frac{1}{2\tau}D_t^2} e^{\tau\varphi}(\gamma\partial_t^2 - \Delta)w|_{2,\tau} + |e^{-\frac{1}{2\tau}D_t^2} e^{\tau\varphi}\Delta w|_{2,\tau}\Big\}. \tag{10.10}$$

Next, we use estimate (10.9) and (10.8) for the first, respectively, the second term on the right side of (10.10), and obtain

$$|e^{-\frac{1}{2\tau}D_t^2} e^{\tau\varphi}\partial_t^2 w|_{2,\tau} + |e^{-\frac{1}{2\tau}D_t^2} e^{\tau\varphi}\Delta w|_{2,\tau}$$

$$\leq c\sqrt{\tau}\Big\{|e^{-\frac{1}{2\tau}D_t^2} e^{\tau\varphi}(\gamma\partial_t^2 - \Delta)\Delta w|_0 + e^{-d\tau}|e^{\tau\varphi}(\gamma\partial_t^2 - \Delta)\Delta w|_0 + e^{-d\tau}|e^{\tau\varphi}\Delta w|_{1,\tau}$$

$$+ e^{-d\tau}|e^{\tau\varphi}\partial_t^2 w|_{1,\tau}\Big\}. \tag{10.11}$$

Step (ii). We first not that, *a fortiori* from (10.4c), the following estimate holds for $\tau \geq 1$:

$$c\sqrt{\tau}\left|e^{-\frac{1}{2\tau}D_t^2}e^{\tau\varphi}\Delta w\right|_0 \leq C\left|e^{-\frac{1}{2\tau}D_t^2}e^{\tau\varphi}\Delta w\right|_{2,\tau}. \tag{10.12}$$

Next, we use estimate (10.6) with $u = w$ and apply (10.12) and (10.8) to the first term on the right-hand side of (10.6). This leads to the estimate

$$\begin{aligned}|e^{-\frac{1}{2\tau}D_t^2}e^{\tau\varphi}w|_{2,\tau} \leq\ & c\sqrt{\tau}\Big\{|e^{-\frac{1}{2\tau}D_t^2}e^{\tau\varphi}\Delta(\gamma\partial_t^2-\Delta)w|_0 + e^{-d\tau}|e^{\tau\varphi}\Delta(\gamma\partial_t^2-\Delta)w|_0 \\ & + e^{-d\tau}|e^{\tau\varphi}\Delta w|_{1,\tau} + e^{-d\tau}|e^{\tau\varphi}w|_{1,\tau}\Big\},\end{aligned} \tag{10.13}$$

for $\tau \geq 1$. Finally, we use (10.6) with $u = \partial_j w$ $(j = 1, 2)$ and get since $\Delta\partial_j w = \partial_j \Delta w$:

$$|e^{-\frac{1}{2\tau}D_t^2}e^{\tau\varphi}\nabla w|_{2,\tau} \leq c\sqrt{\tau}\left\{|e^{-\frac{1}{2\tau}D_t^2}e^{\tau\varphi}\nabla\Delta w|_0 + e^{-d\tau}|e^{\tau\varphi}\nabla\Delta w|_0 + e^{-d\tau}|e^{\tau\varphi}\nabla w|_{1,\tau}\right\}. \tag{10.14}$$

Step (iii). We wish to estimate the first term on the right-hand side of (10.14) by means of (10.8). However, in order to do so, we need to commute the operators ∂_j and $e^{-D_t^2/2\tau}e^{\tau\varphi}$. Since φ is a second degree polynomial by assumption, we can rely on the following formula (see [Ta.1, Lemma 2.1] or [E.1, Lemma 2.5])

$$e^{-\frac{1}{2\tau}D_t^2}e^{\tau\varphi}\partial_j w = (\partial_j - \tau\partial_j\varphi - \partial_{jt}^2\varphi\partial_t)e^{-\frac{1}{2\tau}D_t^2}e^{\tau\varphi}w \tag{10.15}$$

which, for w replaced by Δw, provides the estimate

$$|e^{-\frac{1}{2\tau}D_t^2}e^{\tau\varphi}\nabla\Delta w|_0 \leq c|e^{-\frac{1}{2\tau}D_t^2}e^{\tau\varphi}\Delta w|_{1,\tau}, \tag{10.16}$$

after recalling the definition of the norm $|\cdot|_{1,\tau}$ in (10.4b). Thus, (10.16) yields for $\tau \geq 1$, by virtue of (10.4c) and (10.8):

$$\begin{aligned}c\sqrt{\tau}\left|e^{-\frac{1}{2\tau}D_t^2}e^{\tau\varphi}\nabla\Delta w\right|_0 \leq\ & c\sqrt{\tau}\left|e^{-\frac{1}{2\tau}D_t^2}e^{\tau\varphi}\Delta w\right|_{1,\tau} \leq c\left|e^{-\frac{1}{2\tau}D_t^2}e^{\tau\varphi}\Delta w\right|_{2,\tau} \\ \leq\ & c\sqrt{\tau}\Big\{\left|e^{-\frac{1}{2\tau}D_t^2}e^{\tau\varphi}\Delta(\gamma\partial_t^2-\Delta)w\right|_0 \\ & + e^{-d\tau}\left|e^{\tau\varphi}\Delta(\gamma\partial_t^2-\Delta)w\right|_0 + e^{-d\tau}\left|e^{\tau\varphi}\Delta w\right|_{1,\tau}\Big\}.\end{aligned} \tag{10.17}$$

Using now (10.17) for the first term on the right side of (10.14), we obtain

$$\begin{aligned}|e^{-\frac{1}{2\tau}D_t^2}e^{\tau\varphi}\nabla w|_{2,\tau} \leq\ & c\sqrt{\tau}\Big(|e^{-\frac{1}{2\tau}D_t^2}e^{\tau\varphi}\Delta(\gamma\partial_t^2-\Delta)w|_0 + e^{-d\tau}|e^{\tau\varphi}\nabla\Delta w|_0 \\ & + e^{-d\tau}|e^{\tau\varphi}\nabla w|_{1,\tau} + e^{-d\tau}|e^{\tau\varphi}\Delta(\gamma\partial_t^2-\Delta)w|_0 + e^{-d\tau}|e^{\tau\varphi}\Delta w|_{1,\tau}\Big).\end{aligned} \tag{10.18}$$

Adding now estimate (10.18) with (10.11) and (10.13) yields estimate (10.7), as desired, and finishes the proof. □

Step 3. Local unique continuation of the Kirchoff Eqn. (10.1a).
Estimate (10.7) can be used to obtain a local unique continuation result.

THEOREM 10.3. *Let S be a C^2-surface which is time-like with respect to the wave operator $\gamma\partial_t^2 - \Delta$ at $(t_0, x_0) \in Q = (0,T] \times \Omega$. Assume that $w \in H^3(Q)$ is a solution to the Kirchoff equation (10.1a) which vanishes on one side of S. Then $w \equiv 0$ in a neighborhood of (t_0, x_0).*

Proof. Without loss of generality we assume that $(t_0, x_0) = 0$ and that $B_{2\delta}(0) \subset Q$.

Let $\phi \in C^2(\overline{Q})$ be a function such that the surface S is described by the level surface: $\{\phi = 0\}$. Since S is assumed to be time-like with respect to wave operator $\gamma\partial_t^2 - \Delta$ at 0 the surface is non-characteristic with respect to the wave operator and with respect to the Laplacian at 0. Consequently, the surface is strongly pseudo-convex with respect to the operators $\gamma\partial_t^2 - \Delta$ and Δ at 0 on the level surface $\{\xi_0 = 0\}$ (see [Ta.3, p. 882 and [H.2, Section 28.4]). Moreover, for sufficiently large $\lambda > 0$ the function

$$\varphi(x) = e^{\lambda\phi} - 1$$

is strongly pseudo-convex with respect to the operators $\gamma\partial_t^2 - \Delta$ and Δ at 0 on $\{\xi_0 = 0\}$. Moreover, S is likewise given as the level surface: $\varphi \equiv 0$. Note that ϕ and φ have the same level surfaces. We denote the two sides of S by S^+ and S^-, respectively. We assume that

$$w = 0 \text{ in } S^+ = \{\varphi > 0\}. \tag{10.19}$$

□

REMARK 10.1. *Throughout the proof of uniqueness of the Kirchoff problem (10.1a-b) in the present Section 10, $\lambda > 0$ is a large but* fixed *parameter, chosen once and for all to have the function $\varphi(x) = e^{\lambda\phi} - 1$ strongly pseudo-convex, as stated above. Accordingly, all the constants in the subsequent estimates of the present Section 10 do depend on λ. But we do not note such a dependence, as λ is fixed. By contrast, our setting in the more demanding uniqueness proof of a corresponding thermoelastic problem in Section 11 will include a* variable *parameter λ, to be eventually chosen "large enough." See Remark 11.1.*

Step (i). (Perturbation of φ and localization of w.) The Carleman estimate (10.7) will be applied after perturbing φ and localizing w. Consider the following second degree polynomial in (t, x):

$$\psi(t,x) = \sum_{|\alpha|\le 2} \partial^\alpha \varphi(0)\frac{(t,x)^\alpha}{\alpha!} - 3\epsilon(t^2 + |x|^2). \tag{10.20}$$

The following two claims provide insight on the geometry of the level surfaces of $\psi(t, x)$.

Claim 1. The level surface $\psi(t,x) = 0$ is contained in $\{\varphi > 0\}$ for $0 \ne (t,x) \in B_\delta(0)$, for $\delta > 0$ small enough. [See Figure 1.]

Proof of Claim 1. The condition $\psi(t,x) \equiv 0$ implies

$$\sum_{|\alpha|\le 2} \partial^\alpha \varphi(0)\, \frac{(t,x)^\alpha}{\alpha!} = 3\epsilon(t^2 + |x|^2)$$

by (10.20). Thus, the LHS of this identity is equal to $\varphi(t,x) + \mathcal{O}(|(t,x)|^3)$, while the RHS of this identity is positive for $(t,x) \neq 0$ and of second order. We conclude that $\varphi(t,x) > 0$ for $\delta > 0$ small enough, as claimed.

Claim 2. The level surface $\psi(t,x) = -\beta$, $\beta > 0$, is contained in $\{\varphi < 0\}$ for $t^2 + |x|^2$ small.

Proof of Claim 2. The condition $\psi(t,x) \equiv -\beta$ implies

$$-\beta + 3\epsilon(t^2 + |x|^2) = \sum_{|\alpha|\le 2} \partial^\alpha \varphi(0)\, \frac{(t,x)^\alpha}{\alpha!} = \varphi(t,x) + \mathcal{O}(|(t,x)|^3)$$

by (10.20), hence

$$\varphi(t,x) = -\beta + 3\epsilon(t^2 + |x|^2) + \mathcal{O}(|(t,x)|^3) < 0,$$

for $t^2 + |x|^2$ small, and the claim is established.

Next, choose $\epsilon > 0$ such that $\psi(x)$ is still strongly pseudo-convex with respect to $\gamma\partial_t^2 - \Delta$ and Δ at 0 and that

$$\psi(t,x) \le \varphi(t,x) - 2\epsilon(t^2 + |x|^2), \quad \forall\ (t,x) \in B_{2\delta}(0). \tag{10.21}$$

This can always be achieved by choosing $\delta > 0$ sufficiently small, if necessary. Moreover, we use the fact that the strong pseudo-convexity condition is stable under small perturbation.

Next, let χ be a smooth cutoff function such that

$$\chi(s) = \begin{cases} 1 & \text{if} \quad s \ge -\epsilon\delta^2, & (10.22\text{a}) \\ 0 & \text{if} \quad s \le -2\epsilon\delta^2, & (10.22\text{b}) \end{cases}$$

and define

$$\tilde{w}(t,x) = \begin{cases} \chi(\psi(t,x))w(t,x) & \text{if } (t,x) \in B_{2\delta}(0) & (10.23\text{a}) \\ 0 & \text{otherwise} & (10.23\text{b}) \end{cases} .$$

Observe that, because of the definition of $\tilde{w}$ in (10.23) and ψ in (10.20), we have

$$\begin{aligned} & supp\ \tilde{w} && \subset\ supp\ w \cap supp\ \chi(\psi) \cap B_{2\delta}(0) && (10.24\text{a}) \\ & \text{(by (10.19), (10.22b))} && \subset\ \{\varphi \le 0\} \cap \{\psi \ge -2\epsilon\delta^2\} \cap B_{2\delta}(0) && (10.24\text{b}) \\ & \text{(by (10.21))} && \subset\ \{\varphi \le 0\} \cap \{\varphi - 2\epsilon(t^2 + |x|^2) > -2\epsilon\delta^2\} && (10.24\text{c}) \\ & && \subset\ B_\delta(0), && (10.24\text{d}) \end{aligned}$$

where in the last step we have used that: $-\frac{\varphi}{2\epsilon} + (t^2 + |x|^2) < \delta^2$ implies $t^2 + |x|^2 < \delta^2$, since $\varphi \leq 0$.

Step (ii). Using the fact that w is a solution to the Kirchoff equation (10.1a), we will derive an equation for $\tilde{w}$. Notice preliminarily that

$$(-\gamma\partial_t^2 + \Delta)\Delta\tilde{w} = \chi(\psi)(-\gamma\partial_t^2 + \Delta)\Delta w + F_1 = \chi(\psi)(-\partial_t^2 w + \text{div}(\alpha\nabla(\alpha\partial_t w))) + F_1, \quad (10.25)$$

where F_1 is given by

$$F_1 = [(-\gamma\partial_t^2 + \Delta)\Delta\chi(\psi)]w \in L_2, \; w \in H^3, \quad (10.26)$$

$[\ ,\ ]$ denoting a commutator, as usual; in the present case $[\ ,\ ]$ is a third-order operator $(4+0-1)$ in all variables. Hence $F_1 \in L_2$ for $w \in H^3$, as assumed. Moreover, it follows from (10.26a) that $F_1 \equiv 0$ whenever $\chi(\psi) \equiv 1$ and whenever $\chi(\psi) \equiv 0$, and whenever $w \equiv 0$. Hence, recalling definition (10.22), we have that, *a-fortiori*

$$\textit{supp}\, F_1 \subset \{\psi \leq -\epsilon\delta^2\} \cap B_{2\delta}(0)\ . \quad (10.27)$$

Next, we expand via (10.23):

$$\chi(\psi)\partial_t^2 w = \partial_t^2\tilde{w} + [\chi(\psi), \partial_t^2]w = \partial_t^2\tilde{w} - F_2; \quad (10.28)$$

$$\chi(\psi)\text{div}(\alpha\nabla(\alpha\partial_t w)) = (|\nabla\alpha|^2 + \alpha\Delta\alpha)\partial_t\tilde{w} + 3\alpha\nabla\alpha \cdot \nabla\partial_t\tilde{w} + \alpha^2\Delta\partial_t\tilde{w} + F_3; \quad (10.29)$$

$$F_2 = -[\chi(\psi), \partial_t^2]w; \quad F_3 = [\chi(\psi), \text{div}(\alpha\nabla(\alpha\partial_t\ \cdot\ ))]w. \quad (10.30)$$

Proceeding for F_2 and F_3 in the same way as we did for F_1 in obtaining (10.27), we can write likewise

$$\begin{aligned} F &\equiv F_1 + F_2 + F_3;\ \text{supp}\, F \subset \{\psi \leq -\epsilon\delta^2\} \cap B_{2\delta}(0);\ F \in L_2 \text{ for } w \in H^3 \\ F &= \text{third-order operator.} \end{aligned} \quad (10.31)$$

Finally, using (10.28) and (10.29) in the right side of (10.25), we obtain via (10.31) the sought-after equation satisfied by $\tilde{w}$:

$$(-\gamma\partial_t^2 + \Delta)\Delta\tilde{w} = -\partial_t^2\tilde{w} + (|\nabla\alpha|^2 + \alpha\Delta\alpha)\partial_t\tilde{w} + 3\alpha\nabla\alpha \cdot \nabla\partial_t\tilde{w} + \alpha^2\Delta\partial_t\tilde{w} + F \quad (10.32)$$

Step (iii). (Carleman estimate (10.7) applied to $\tilde{w}$, ψ in place of w, φ: first form) We now apply the Carleman estimate (10.7) of Lemma 10.2 with respect to ψ and $\tilde{w}$, instead of φ, w. We have already noted in correspondence of (10.20), that ψ is a second-degree polynomial which is a strongly pseudo-convex function with respect to the operators $\gamma\partial_t^2 - \Delta$ and Δ at $0 = (t_0, x_0)$, as required by Lemma 10.2. As a preliminary step, we estimate from (10.32) by the triangle inequality

$$
\begin{aligned}
&|e^{-\frac{1}{2\tau}D_t^2}e^{\tau\psi}(-\gamma\partial_t^2+\Delta)\Delta\tilde{w}|_0 \\
&\quad\le |e^{-\frac{1}{2\tau}D_t^2}e^{\tau\psi}\partial_t^2\tilde{w}|_0+|e^{-\frac{1}{2\tau}D_t^2}e^{\tau\psi}(|\nabla\alpha|^2+\alpha\Delta\alpha)\partial_t\tilde{w}|_0 \\
&\quad\quad+|e^{-\frac{1}{2\tau}D_t^2}e^{\tau\psi}3\alpha\nabla\alpha\cdot\nabla\partial_t\tilde{w}|_0+|e^{-\frac{1}{2\tau}D_t^2}e^{\tau\psi}\alpha^2\partial_t\Delta\tilde{w}|_0+|e^{-\frac{1}{2\tau}D_t^2}e^{\tau\psi}F|_0 \\
&\quad\le C\Big(|e^{-\frac{1}{2\tau}D_t^2}e^{\tau\psi}\partial_t^2\tilde{w}|_0+|e^{-\frac{1}{2\tau}D_t^2}e^{\tau\psi}\partial_t\tilde{w}|_0 \\
&\quad\quad+|e^{-\frac{1}{2\tau}D_t^2}e^{\tau\psi}\nabla\partial_t\tilde{w}|_0+|e^{-\frac{1}{2\tau}D_t^2}e^{\tau\psi}\partial_t\Delta\tilde{w}|_0+|e^{-\frac{1}{2\tau}D_t^2}e^{\tau\psi}F|_0\Big). \qquad (10.33)
\end{aligned}
$$

In the last step we made also use of the fact that α is independent of t.

We now use, as announced, Carleman estimate (10.7) with respect to ψ and $\tilde{w}$, instead of φ and w, and in doing so we invoke estimate (10.33) for the first term on the right side of (10.7). We thus obtain

$$
\begin{aligned}
&|e^{-\frac{1}{2\tau}D_t^2}e^{\tau\psi}\tilde{w}|_{2,\tau}+|e^{-\frac{1}{2\tau}D_t^2}e^{\tau\psi}\partial_t^2\tilde{w}|_{2,\tau}+|e^{-\frac{1}{2\tau}D_t^2}e^{\tau\psi}\Delta\tilde{w}|_{2,\tau}+|e^{-\frac{1}{2\tau}D_t^2}e^{\tau\psi}\nabla\tilde{w}|_{2,\tau} \\
&\le c\sqrt{\tau}\Big\{|e^{-\frac{1}{2\tau}D_t^2}e^{\tau\psi}\partial_t^2\tilde{w}|_0+|e^{-\frac{1}{2\tau}D_t^2}e^{\tau\psi}\partial_t\tilde{w}|_0+|e^{-\frac{1}{2\tau}D_t^2}e^{\tau\psi}\nabla\partial_t\tilde{w}|_0 \\
&\quad+|e^{-\frac{1}{2\tau}D_t^2}e^{\tau\psi}\partial_t\Delta\tilde{w}|_0+|e^{-\frac{1}{2\tau}D_t^2}e^{\tau\psi}F|_0+e^{-d\tau}|e^{\tau\psi}(\gamma\partial_t^2-\Delta)\Delta\tilde{w}|_0+e^{-d\tau}|e^{\tau\psi}\Delta\tilde{w}|_{1,\tau} \\
&\quad+e^{-d\tau}|e^{\tau\psi}\partial_t^2\tilde{w}|_{1,\tau}+e^{-d\tau}|e^{\tau\psi}\tilde{w}|_{1,\tau}+e^{-d\tau}|e^{\tau\psi}\nabla\tilde{w}|_{1,\tau}+e^{-d\tau}|e^{\tau\psi}\nabla\Delta\tilde{w}|_0\Big\}. \qquad (10.34)
\end{aligned}
$$

Step (iv). (Carleman estimate for $\tilde{w}$ and ψ: second form) In the second, third, and fourth terms in the right-hand side of (10.34), we now commute the pseudo-differential weight and the differentiation with respect to t, using formula (10.15). Hence, for such second term, we have with ∂_j replaced by ∂_t:

$$
\begin{aligned}
e^{-\frac{1}{2\tau}D_t^2}e^{\tau\psi}\partial_t\tilde{w} &= (\partial_t-\tau\partial_t\psi-\partial_t^2\psi\partial_t)e^{-\frac{1}{2\tau}D_t^2}e^{\tau\psi}\tilde{w} \\
&= (1-\partial_t^2\psi)\partial_t e^{-\frac{1}{2\tau}D_t^2}e^{\tau\psi}\tilde{w}-\tau\partial_t\psi e^{-\frac{1}{2\tau}D_t^2}e^{\tau\psi}\tilde{w}\,. \qquad (10.35)
\end{aligned}
$$

This leads to the estimate

$$|e^{-\frac{1}{2\tau}D_t^2}e^{\tau\psi}\partial_t\tilde{w}|_0\le c|e^{-\frac{1}{2\tau}D_t^2}e^{\tau\psi}\tilde{w}|_{1,\tau}, \qquad (10.36)$$

after making use of definition (10.4b) for $\tau\ge1$; or if (10.4c) (left side).

Similarly, for the third term on the right side of (10.34), we use Eqn. (10.35) with $\tilde{w}$ replaced by $\partial_j\tilde{w}$ and obtain

$$e^{-\frac{1}{2\tau}D_t^2}e^{\tau\psi}\partial_t\partial_j\tilde{w}=(1-\partial_t^2\psi)\,\partial_t e^{-\frac{1}{2\tau}D_t^2}e^{\tau\psi}\partial_j\tilde{w}-\tau\partial_t\psi e^{-\frac{1}{2\tau}D_t^2}e^{\tau\psi}\partial_j\tilde{w}. \qquad (10.37)$$

This leads to the estimate (as ∂_j and ∂_t commute)

$$|e^{-\frac{1}{2\tau}D_t^2}e^{\tau\psi}\nabla\partial_t\tilde{w}|_0 \leq c|e^{-\frac{1}{2\tau}D_t^2}e^{\tau\psi}\nabla\tilde{w}|_{1,\tau}, \tag{10.38}$$

counterpart of (10.36). In a similar fashion, replacing $\tilde{w}$ with $\Delta\tilde{w}$ in (10.36), yields

$$|e^{-\frac{1}{2\tau}D_t^2}e^{\tau\psi}\partial_t\Delta\tilde{w}|_0 \leq c|e^{-\frac{1}{2\tau}D_t^2}e^{\tau\psi}\Delta\tilde{w}|_{1,\tau}\,. \tag{10.39}$$

Thus, using estimates (10.36), (10.38), and (10.39) for the second, third, and fourth term in the Carleman inequality (10.34), we may rewrite it as

$$\begin{aligned}
&|e^{-\frac{1}{2\tau}D_t^2}e^{\tau\psi}\tilde{w}|_{2,\tau} + |e^{-\frac{1}{2\tau}D_t^2}e^{\tau\psi}\partial_t^2\tilde{w}|_{2,\tau} + |e^{-\frac{1}{2\tau}D_t^2}e^{\tau\psi}\Delta\tilde{w}|_{2,\tau} + |e^{-\frac{1}{2\tau}D_t^2}e^{\tau\psi}\nabla\tilde{w}|_{2,\tau} \\
&\leq c\sqrt{\tau}\Big\{|e^{-\frac{1}{2\tau}D_t^2}e^{\tau\psi}\partial_t^2\tilde{w}|_0 + |e^{-\frac{1}{2\tau}D_t^2}e^{\tau\psi}\tilde{w}|_{1,\tau} + |e^{-\frac{1}{2\tau}D_t^2}e^{\tau\psi}\nabla\tilde{w}|_{1,\tau} \\
&\quad + |e^{-\frac{1}{2\tau}D_t^2}e^{\tau\psi}\Delta\tilde{w}|_{1,\tau} + |e^{-\frac{1}{2\tau}D_t^2}e^{\tau\psi}F|_0 + e^{-d\tau}|e^{\tau\psi}(\gamma\partial_t^2 - \Delta)\Delta\tilde{w}|_0 + e^{-d\tau}|e^{\tau\psi}\Delta\tilde{w}|_{1,\tau} \\
&\quad + e^{-d\tau}|e^{\tau\psi}\partial_t^2\tilde{w}|_{1,\tau} + e^{-d\tau}|e^{\tau\psi}\tilde{w}|_{1,\tau} + e^{-d\tau}|e^{\tau\psi}\nabla\tilde{w}|_{1,\tau} + e^{-d\tau}|e^{\tau\psi}\nabla\Delta\tilde{w}|_0\Big\},
\end{aligned} \tag{10.40}$$

which is the second version of our Carleman estimate for $\tilde{w}$ and ψ.

Step (v). Carleman estimate for $\tilde{w}$ and ψ: third form. We shall now show that by choosing $\tau > 4c^2$, we can cancel the first four terms on the right-hand side of (10.40) against the last three terms of the left-hand side of (10.40) and obtain

$$\begin{aligned}
|e^{-\frac{1}{2\tau}D_t^2}e^{\tau\psi}\tilde{w}|_{2,\tau} \leq\; & c\sqrt{\tau}\Big\{|e^{-\frac{1}{2\tau}D_t^2}e^{\tau\psi}F|_0 + e^{-d\tau}|e^{\tau\psi}(\gamma\partial_t^2 - \Delta)\Delta\tilde{w}|_0 \\
& + e^{-d\tau}|e^{\tau\psi}\Delta\tilde{w}|_{1,\tau} + e^{-d\tau}|e^{\tau\psi}\partial_t^2\tilde{w}|_{1,\tau} + e^{-d\tau}|e^{\tau\psi}\tilde{w}|_{1,\tau} \\
& + e^{-d\tau}|e^{\tau\psi}\nabla\tilde{w}|_{1,\tau} + e^{-d\tau}|e^{\tau\psi}\nabla\Delta\tilde{w}|_0\Big\},
\end{aligned} \tag{10.41}$$

which is the third version of our Carleman estimate for $\tilde{w}$ and ψ [Eqn. (10.41) is precisely Eqn. (10.40) after we have omitted the last three terms of its left-hand side, as well as the first four terms of its right-hand side.] More precisely, we shall compare the following four terms on the left-hand side (LHS) of (10.40) with the following four terms on the right-hand side (RHS) of (10.40):

LHS	RHS	
$\left\|e^{-\frac{1}{2\tau}D_t^2}e^{\tau\psi}\tilde{w}\right\|_{2,\tau}$	$c\sqrt{\tau}\left\|e^{-\frac{1}{2\tau}D_t^2}e^{\tau\psi}\tilde{w}\right\|_{1,\tau}$	(10.42a)
$\left\|e^{-\frac{1}{2\tau}D_t^2}e^{\tau\psi}\Delta\tilde{w}\right\|_{2,\tau}$	$c\sqrt{\tau}\left\|e^{-\frac{1}{2\tau}D_t^2}e^{\tau\psi}\Delta\tilde{w}\right\|_{1,\tau}$	(10.42b)
$\left\|e^{-\frac{1}{2\tau}D_t^2}e^{\tau\psi}\nabla\tilde{w}\right\|_{2,\tau}$	$c\sqrt{\tau}\left\|e^{-\frac{1}{2\tau}D_t^2}e^{\tau\psi}\nabla\tilde{w}\right\|_{1,\tau}$	(10.42c)
$\left\|e^{-\frac{1}{2\tau}D_t^2}e^{\tau\psi}\partial_t^2\tilde{w}\right\|_{2,\tau}$	$c\sqrt{\tau}\left\|e^{-\frac{1}{2\tau}D_t^2}e^{\tau\psi}\partial_t^2\tilde{w}\right\|_0$	(10.42d)

To this end, and thus in order to obtain (10.41) from (10.40), we shall use the following inequality

$$\frac{1}{\sqrt{2}}\sqrt{a-2b} \leq \sqrt{a}-\sqrt{b}, \text{ equivalent to } (\sqrt{a}-2\sqrt{b})^2 \geq 0, \tag{10.43}$$

for $a \geq 2b$. Next, by means of (10.43), we shall prove the following estimates:

$$|u|_{2,\tau} - c\sqrt{\tau}|u|_{1,\tau} \geq \frac{\tau}{2}|u|_{1,\tau} \geq 0 \text{ for } \sqrt{\tau} > 2c, \tag{10.44}$$

$$|u|_{2,\tau} - c\sqrt{\tau}|u|_0 \geq \frac{\tau(2\tau^2-1)^{\frac{1}{2}}}{2}|u|_0 \geq 0 \text{ for } \sqrt{\tau} > 2c,\ \tau > \frac{1}{\sqrt{2}}, \tag{10.45}$$

$$|v|_{2,\tau} - c\sqrt{\tau}|v|_{1,\tau} \geq \frac{1}{2}|v|_{2,\tau} \text{ for } \sqrt{\tau} > 2c. \tag{10.46}$$

Then, (10.40) yields (10.41) by using:

$$\begin{cases} (10.44) \text{ for } u = e^{-\frac{1}{2\tau}D_t^2}e^{\tau\psi}\Delta\tilde{w} \text{ and } u = e^{-\frac{1}{2\tau}D_t^2}e^{\tau\psi}\nabla\tilde{w}; \\ (10.45) \text{ for } u = e^{-\frac{1}{2\tau}D_t^2}e^{\tau\psi}\partial_t^2\tilde{w}; \\ \text{and } (10.46) \text{ for } v = e^{-\frac{1}{2\tau}D_t^2}e^{\tau\psi}\tilde{w}. \end{cases} \tag{10.47}$$

It remains to show (10.44)–(10.46), by means of (10.42).

To establish (10.44), we take $\sqrt{\tau} > 2c$ as assumed, and estimate by use of (10.42) with $\sqrt{a} = |u|_{2,\tau}$ and $\sqrt{b} = \frac{\tau}{2}|u|_{1,\tau}$, to obtain

$$|u|_{2,\tau} - c\sqrt{\tau}|u|_{1,\tau} \geq |u|_{2,\tau} - \frac{\tau}{2}|u|_{1,\tau} \tag{10.48}$$

$$\text{(by (10.42))} \qquad \geq \frac{1}{\sqrt{2}}\left(|u|_{2,\tau}^2 - 2\frac{\tau^2}{4}|u|_{1,\tau}^2\right)^{\frac{1}{2}} \tag{10.49}$$

$$\text{(by (10.4c))} \qquad \geq \frac{1}{\sqrt{2}}\left(\tau^2|u|_{1,\tau}^2 - \frac{\tau^2}{2}|u|_{1,\tau}^2\right)^{\frac{1}{2}} = \frac{\tau}{2}|u|_{1,\tau}, \tag{10.50}$$

and (10.44) is proved. To establish (10.46), we likewise estimate for $\sqrt{\tau} > 2c$ via (10.49):

$$|v|_{2,\tau} - c\sqrt{\tau}|v|_{1,\tau} \geq \frac{1}{\sqrt{2}}\left(|v|_{2,\tau}^2 - \frac{\tau^2}{2}|v|_{1,\tau}^2\right)^{\frac{1}{2}} \tag{10.51}$$

$$\text{(by (10.4c))} \qquad \geq \frac{1}{\sqrt{2}}\left(|v|_{2,\tau}^2 - \frac{1}{2}|v|_{2,\tau}^2\right)^{\frac{1}{2}} = \frac{1}{2}|v|_{2,\tau}, \tag{10.52}$$

and (10.46) is proved. The proof of (10.45) is similar by using both estimates in (10.4c). Thus, (10.41) is fully proved.

Step (vi). Analysis of the RHS of (10.41). Final Carleman estimate for $\tilde{w}$ and ψ. Conclusion.

We next analyze the right-hand side of Eqn. (10.41). As to its first term, we use the facts that

$$F \in L_2 \text{ and } supp\, F \subset \{\psi \leq -\epsilon\delta^2\} \cap B_{2\delta}(0) \text{ from (10.31),}$$

which lead to the desired estimate

$$|e^{-\frac{1}{2\tau}D_t^2} e^{\tau\psi} F|_0 = \frac{1}{(2\pi)^3} |e^{-\frac{1}{2\tau}\xi_0^2} \widehat{e^{\tau\psi} F}|_0 \leq |e^{\tau\psi} F|_0 \leq ce^{-\epsilon\delta^2\tau}, \tag{10.53}$$

where we have set $c = |F|_0$ (depending on w) and where we have used Parseval's identity twice. We next estimate the remaining six terms on the right-hand side of (10.41). Here, we use that $\tilde{w} \in H^3$ (see (10.23)), since by assumption $w \in H^3$. Hence we can estimate the last six terms in (10.41) and we get

$$e^{-d\tau}|e^{\tau\psi}(\gamma\partial_t^2 - \Delta)\Delta\tilde{w}|_0 + e^{-d\tau}|e^{\tau\psi}\Delta\tilde{w}|_{1,\tau} + e^{-d\tau}|e^{\tau\psi}\partial_t^2\tilde{w}|_{1,\tau} + e^{-d\tau}|e^{\tau\psi}\tilde{w}|_{1,\tau}$$
$$+e^{-d\tau}|e^{\tau\psi}\nabla\tilde{w}|_{1,\tau} + e^{-d\tau}|e^{\tau\psi}\nabla\Delta\tilde{w}|_0 \leq c\tau e^{-d\tau}. \tag{10.54}$$

In obtaining (10.54), we have used several facts: Eqn. (10.32) whose right-hand side show that $(-\gamma\partial_t^2 + \Delta)\Delta\tilde{w} \in L_2$ for $w \in H^3$, since F is a third-order operator, see (10.31); the definition (10.4b) of the $|\cdot|_{1,\tau}$-norm, which accounts for pulling out a factor τ in the right-hand side of (10.54); and, finally, the property that $\tilde{w} \equiv 0$ in $\{\psi > 0\}$ [since $w \equiv 0$ for $\{\varphi > 0\}$ by assumption] so that the factor $e^{\tau\psi}$ is active only for $\psi \leq 0$ and hence does not produce exponential growth. Moreover, of course, the constant "c" in (10.54) depends (again) on w [as the constant "c" in (10.53)]. Using the last two inequalities on the right-hand side of (10.41), we obtain the estimate

$$\tau^2 \left|e^{-\frac{1}{2\tau}D_t^2} e^{\tau\psi}\tilde{w}\right|_0 \leq \left|e^{-\frac{1}{2\tau}D_t^2} e^{\tau\psi}\tilde{w}\right|_{2,\tau} \leq c\sqrt{\tau}\left\{e^{-\epsilon\delta^2\tau} + \tau e^{-d\tau}\right\}, \tag{10.55}$$

where on the left-hand side the inequality follows from appplying the estimates in (10.4c). Dividing estimate (10.41) by $\tau^{3/2}$ we finally obtain

$$\tau^{\frac{1}{2}} \left|e^{-\frac{1}{2\tau}D_t^2} e^{\tau\psi}\tilde{w}\right|_0 \leq c(e^{-\epsilon\delta^2\tau} + e^{-d\tau}), \tag{10.56}$$

which in turn yields

$$\left|e^{-\frac{1}{2\tau}D_t^2} e^{\tau(\psi+\beta)}\tilde{w}\right|_0 \leq \frac{2c}{\sqrt{\tau}}. \tag{10.57}$$

Here $\beta = \min\{\epsilon\delta^2, d\}$. Finally, an application of Proposition 4.1 in [Ta.3] shows that $\tilde{w} \equiv 0$ in $\psi > -\beta$, which implies that $w \equiv 0$ in a neighborhood of $0 = (t_0, x_0)$. Theorem 10.3 is proved. □

Step 4. Global unique continuation of Kirchoff problem (10.1a–b).

Our next step is to extend our local unique continuation result, Theorem 10.3, to a global result. Since our local result, Theorem 10.3, guarantees unique continuation across any surface that is time-like with respect to the wave operator $\gamma\partial_t^2 - \Delta$, we can use the procedure which is used for obtaining John's global Holmgren theorem (see, e.g., [R.1, p. 41]).

We introduce a continuous family of surfaces S_λ which are time-like with respect to $\gamma\partial_t^2 - \Delta$ for $\lambda \in [0,1]$. We assume that

- the set $\mathcal{O} \subset\subset \mathbb{R}^2$ is open and the mapping $\sigma : ([0,1] \times \overline{\mathcal{O}}) \longrightarrow \overline{Q}$ is continuous.
- For each $\lambda \in [0,1]$ the image of the mapping $\sigma(\lambda, \cdot) : \mathcal{O} \longrightarrow \overline{Q}$ is a C^1 hypersurface S_λ which is time-like with respect to $\gamma\partial_t^2 - \Delta$.
- $S_0 = \tilde{\Sigma}$.
- $\sigma([0,1] \times \partial\mathcal{O}) \subset S_0$, which expresses the fact that the edge $\sigma(\lambda, \partial\mathcal{O})$ of S_λ lies in S_0.

COROLLARY 10.4. *Let $w \in H^3(Q)$ be a solution to the Kirchoff equation (10.1a) such that*

$$\frac{\partial^k}{\partial\nu^k} w = 0 \qquad \text{on } S_0 = \tilde{\Sigma}, \qquad k = 0,1,2,3, \tag{10.58}$$

as assumed in (10.1b). Then, $w \equiv 0$ on $\sigma([0,1] \times \overline{\mathcal{O}})$.

Proof. We extend the domain Q to a larger domain Q^* across S_0. Then we extend w to a new function w^* to all of Q^* as follows. Let

$$w^* = \begin{cases} w & \text{in } Q, \\ 0 & \text{in } Q^* \setminus Q. \end{cases} \tag{10.59}$$

Since w has zero Cauchy data on S_0 by (10.58), we have that $w^* \in H^3(Q^*)$ is a solution to the plate equation (10.1a) in Q^*.

Applying Theorem 10.3 at each $(t,x) \in S_0$, we obtain that $w^* \equiv 0$ in a neighborhood ω_0 of S_0 in Q^*. Since S_λ is a continuous family of surfaces, there exists a $\varepsilon(0) > 0$ such that $S_{\varepsilon(0)} \subset \omega_0$.

This step can be generalized. Assume $\lambda \in [0,1]$ and $w^* \equiv 0$ on one side of S_λ. According to Theorem 10.3 there exists a neighborhood ω_λ of S_λ in Q^* such that $w^* \equiv 0$ in ω_λ. Furthermore, there exists a $\varepsilon(\lambda) > 0$ such that $S_{\lambda+\varepsilon(\lambda)} \subset \omega_\lambda$.

The proof is concluded by a compactness argument. The intervals $(\lambda, \varepsilon(\lambda))$ cover the interval $[0,1]$. Consequently, there exists a finite subcover $(\lambda_j, \varepsilon(\lambda_j))$ for $j = 0,..,N$ and without loss of generality we assume that $\lambda_0 = 0$ and $\lambda_j < \lambda_{j+1}$ for $j = 0,..,N$. Then, after a finite number of steps, we obtain that $w^* \equiv 0$ in $\sigma([0,1] \times \overline{\mathcal{O}})$. However, $w = w^*$ in Q and that means the same result holds true for w. □

From this result one can derive a statement on unique continuation with respect to a special geometry.

LEMMA 10.5. *Let A be a triangle in in the tx_1 plane with vertices at $(0,0)$, $(T,0)$ and $(T/2, T/(2\sqrt{\gamma}))$. Furthermore, let $D = A \times \{|x_2| < \varepsilon\}$ be a triangular slice with thickness 2ε. If $w \in H^3(D)$ solves the plate equation (10.1a) and has zero Cauchy data on $\partial D \cap \{x_1 = 0\}$ then $w \equiv 0$ in D.*

Proof. The proof is essentially the same as the proof of Lemma 3.4.6 in [I.2]. We like to point out that this lemma is formulated for H^2 solutions to the wave equation. However, the proof relies only on the unique continuation property across non-characteristic surfaces which we established in Corollary 10.4. □

We can now complete the proof of Theorem 8.1. Choose $x \in \Omega$ arbitrarily. Then, according to the definition of the distance $\text{dist}(x, \tilde{\Gamma})$, there exists a finite collection of intervals $I_1, I_2, ..., I_k \subset \overline{\Omega}$ where $I_j = [y_{j-1}, y_j]$ such that $y_0 \in \tilde{\Gamma}$, $y_j \in \overline{\Omega}$, $y_k = x$ and

$$\sum_{j=1}^{n} |y_{j-1} - y_j| = dist(x, \tilde{\Gamma}) \,. \tag{10.60}$$

We extend Ω across $\tilde{\Gamma}$ to a larger domain Ω^*. Then we set $Q^* = (0,T) \times \Omega^*$ and we extend w to all of Q^* by introducing w^* as in (10.60). See Figure 2.

Consider now the interval $I_1 = [y_0, y_1]$. We choose cartesian coordinates $\{x_1, x_2\}$ such that $y_0 = (\delta_1, 0)$ and $y_1 = (|y_1 - y_0| + \delta_1, 0)$. Here $\delta_1 > 0$ is a small number which will be specified later. For $y_0 \in \tilde{\Gamma}$ there exists a $\varepsilon_1 > 0$ such that the parallelepiped $(0,T) \times (0, |y_1 - y_0| + \delta_1) \times (-\varepsilon_1, \varepsilon_1) \subset Q^*$ and w^* has zero Cauchy data on $(0,T) \times \{0\} \times (-\varepsilon_1, \varepsilon_1)$. The number $\delta_1 > 0$ depends on ε_1 and $\tilde{\Gamma}$. It has to be large enough to guarantee zero Cauchy data on the lateral boundary $(0,T) \times \{0\} \times (-\varepsilon_1, \varepsilon_1)$. See Figure 3 and Figure 4.

Now we apply Lemma 10.5. Here we consider the triangle with the vertices $(0,0)$, $(T,0)$ and $(T/2, T/(2\sqrt{\gamma}))$ in the tx_1 plane. Note that this triangle does not have to be a subset of Q^*. This guarantees that $w^* \equiv 0$ in a neighborhood of

$$\{(y_1, t) \ : \ \sqrt{\gamma}(|y_1 - y_0| + \delta_1) < t < T - \sqrt{\gamma}(|y_1 - y_0| + \delta_1)\} \,. \tag{10.61}$$

This set is never empty since $T > 2\sqrt{\gamma}(|y_1 - y_0| + \delta_1)$ which is a consequence of definition (8.2) for $\tilde{T}$, where $T > \tilde{T}$. See Figure 5.

This step will be repeated in each interval I_j. However, the triangles will become smaller and smaller.

When we consider the interval $I_2 = [y_1, y_2]$, we choose the coordinates such that $y_1 = (\delta_2, 0)$ and $y_2 = (|y_2 - y_1| + \delta_2, 0)$. Again $\delta_2 > 0$ is a small number which will guarantee that the parallelepiped

$$(\sqrt{\gamma}(|y_1 - y_0| + \delta_1), T - \sqrt{\gamma}(|y_1 - y_0| + \delta_1)) \times (0, |y_2 - y_1| + \delta_1) \times (-\varepsilon_2, \varepsilon_2) \subset Q^* \tag{10.62}$$

has zero Cauchy data on

$$(\sqrt{\gamma}(|y_1 - y_0| + \delta_1), T - \sqrt{\gamma}(|y_1 - y_0| + \delta_1)) \times \{0\} \times (-\varepsilon_2, \varepsilon_2) \,. \tag{10.63}$$

For that we have to guarantee that this set is a subset of the triangular slice constructed in the previous step. Using now a triangle with the vertices at

$$(\sqrt{\gamma}(|y_1 - y_0| + \delta_1), 0) \quad (T - \sqrt{\gamma}(|y_1 - y_0| + \delta_1), 0) \quad \left(\frac{T}{2}, \frac{T}{2\sqrt{\gamma}} - (|y_1 - y_0| + \delta_1)\right) \tag{10.64}$$

we obtain that $w^* \equiv 0$ in a neighborhood of

$$\{(y_2,t) : \sqrt{\gamma}(|y_1 - y_0| + |y_2 - y_1| + \delta_1 + \delta_2) < t < T - \sqrt{\gamma}(|y_1 - y_0| + |y_2 - y_1| + \delta_1 + \delta_2)\} . \tag{10.65}$$

Repeating this step k times we obtain $w^* \equiv 0$ in a neighborhood of

$$\{(x,t) \ : \ \sqrt{\gamma}(dist(x,\tilde{\Gamma}) + \delta) < t < T - \sqrt{\gamma}(dist(x,\tilde{\Gamma}) + \delta), \tag{10.66}$$

see Fig. 6, where we used (10.60), and we set $\delta = \sum \delta_j$. Next we choose $\delta > 0$ so small that for some $\epsilon > 0$ we have

$$T - \epsilon = 2\sqrt{\gamma}(dist(x,\tilde{\Gamma}) + \delta) . \tag{10.67}$$

This can be always done due to condition (1.1.1) which defines $\tilde{T}$, where $T > \tilde{T}$. Consequently the set (10.66) can be represented as

$$\left\{(x,t) \ : \ \frac{T-\epsilon}{2} < t < T - \frac{T-\epsilon}{2}\right\} \tag{10.68}$$

which in turn implies

$$w^* \equiv 0 \quad \text{in} \quad \left(\frac{T-\epsilon}{2}, \frac{T+\epsilon}{2}\right) \times \Omega, \tag{10.69}$$

which establishes the desired goal (10.2).

REMARK 10.2. *After writing this section, the authors have become aware of a recent preprint by W. Littman [Lit.1], which yields global unique continuation results for linear PDE's starting from a local version of it. It is possible that Littman's result may be invoked to replace our Step 4.*

11 Step 2: Proof of Theorem 4.2.2(i): Unique Continuation of the Over-Determined Thermoelastic System

Goal. We begin by writing the over-determined thermoelastic problem of interests

$$\begin{cases} w_{tt} - \gamma \Delta w_{tt} + \Delta^2 w + \operatorname{div}(\alpha(x)\nabla\theta) \equiv 0 \ \text{ in } Q = (0,T] \times \Omega; & \text{(11.1a)} \\ \theta_t - \Delta\theta - \operatorname{div}(\alpha(x)\nabla w_t) \equiv 0 \ \text{ in } Q; & \text{(11.1b)} \\ w \equiv 0; \ \frac{\partial w}{\partial \nu} \equiv 0, \ \frac{\partial^2 w}{\partial \nu^2} \equiv 0, \ \frac{\partial^3 w}{\partial \nu^3} \equiv 0 \ \text{ on } \tilde{\Sigma} = (0,T] \times \Gamma; & \text{(11.1c)} \\ \theta = 0, \ \frac{\partial \theta}{\partial \nu} = 0 \ \text{ on } \tilde{\Sigma}, & \text{(11.1d)} \end{cases}$$

where $\tilde{\Gamma}$ is an arbitrarily small, non-empty open set of $\Gamma = \partial\Omega$ of positive measure. Also, we take $T > \tilde{T}$, where $\tilde{T}$ is defined by (4.2.10). We recall that, in order to show (*a-fortiori*) Theorem 4.2.2, our goal is as follows: we assume that $w \in H^3(Q)$, $\theta \in H^1(Q)$ is a solution of the over-determined problem (11.1a–d), and we want to show that then

$$w\left(\frac{T}{2}\right) = w_t\left(\frac{T}{2}\right) = \theta\left(\frac{T}{2}\right) = 0. \tag{11.2}$$

Orientation. The proof of the uniqueness property for the above thermoelastic system is much harder than the proof of the uniqueness property for the corresponding Kirchoff system in Section 10. This is so since the thermoelastic system is strongly coupled. In particular, the coupling term $\operatorname{div}(\alpha(x)\nabla\theta)$ is not a lower-order term; rather, it is of the same order as $\theta_t - \Delta\theta$. The proof will be based again on Carleman estimates, with critical differences, however, pointed out below. The only uniqueness result available so far is by V. Isakov [I.1]. The crucial fact in his uniqueness argument is based on Carleman estimates carrying a second large parameter. However, this result requires zero Cauchy data on all of Σ. By contrast, in the present proof, we wish to prescribe zero Cauchy data only on part of the boundary $\tilde{\Gamma}$, which is an open set of positive measure, arbitrarily small, as in (11.1c–d). In order to state the relevant estimates, we need to introduce some new functions and some new norms. Given an analytic function ψ, we set

$$\varphi(t,x) = e^{\lambda\psi(t,x)} - 1 \quad \text{and} \quad \kappa(t,x) = \tau\lambda e^{\lambda\psi(t,x)} . \tag{11.3}$$

Here λ and τ denote two (large) positive parameters. Note that the level surfaces of ψ are the same as the level surfaces of φ; in particular, $\varphi \equiv 0$ whenever $\psi \equiv 0$.

On the Sobolev space H^m we introduce two weighted norms

$$|u|^2_{m,\kappa} = \sum_{|\alpha|\le m} \int |\kappa(t,x)^{2m-2|\alpha|}|D^\alpha u|^2; \quad |u|^2_{m,\kappa*} = \sum_{|\alpha|\le m} \int |\kappa(t,x)^{2m-2|\alpha|-1}|D^\alpha u|^2, \tag{11.4a}$$

where m is a non-negative integer. The integrations are in t and x over $\mathbb{R}^3$. We observe, and shall use below in (11.28), that by choosing τ and λ sufficiently large so that $\kappa(t,x) = \tau\lambda e^{\lambda\psi(t,x)} \geq 1$, we can always obtain that

$$|u|_{1,\kappa} \leq |u|_{2,\kappa^*}, \quad \kappa \geq 1. \tag{11.4b}$$

We shall also introduce the anisotropic norm

$$|u|^2_{N,s,\tau} = \frac{1}{(2\pi)^3} \int (\tau^2 + \xi_0^2)^N (\tau^2 + |\xi|^2)^s |\hat{u}(\xi)|^2 d\xi \tag{11.5}$$

[compare with (10.4a)].

Differences with respect to Section 10. We have already noted in Remark 10.1 that the proof of uniqueness in Section 10 for the Kirchoff system uses a parameter $\lambda > 0$ large and fixed once and for all. For the uniqueness proof for the thermoelastic system of the present section, we need to use more precise Carleman estimates. More precisely, we need a type of estimate where λ is a second variable parameter, in addition to τ, to be eventually selected sufficiently large. These estimates were developed in [E.2]. Only the presence of the additional parameter λ will allow us to finally absorb "lower-order terms" in a step similar to that yielding from (10.40) to (10.41) in Section 10.

Step 1. Carleman estimates for the Laplacian, the wave operator, and the heat operator.

The following lemma states the Carleman estimates for the Laplacian, the wave operator and the heat operator. It is a generalization of Lemma 10.1 and involves now two parameters λ and τ.

LEMMA 11.1. *Let ψ be an analytic function which is non-characteristic with respect to the operators $\gamma\partial_t^2 - \Delta$ and Δ at $(t_0, x_0) \in Q = (0,T] \times \Omega$. Let φ be defined by (11.3). For small $\delta > 0$, there exist constants $c > 0$ and $\lambda_0 > 0$, as well as two positive functions $d(\lambda) > 0$ and $\tau_0(\lambda)$ for $\lambda \geq \lambda_0$, such that the following estimates hold true, in the norms defined by (11.3)–(11.5) and (10.4), provided $\lambda \geq \lambda_0$ and $\tau \geq \tau_0(\lambda)$:*

$$\sqrt{\lambda}|e^{-\frac{1}{2\tau}D_t^2}e^{\tau\varphi}u|_{2,\kappa*} \leq c\left\{|e^{-\frac{1}{2\tau}D_t^2}e^{\tau\varphi}\Delta u|_0 + e^{-d\tau}|e^{\tau\varphi}\Delta u|_0 + e^{-d\tau}|e^{\tau\varphi}u|_{1,\tau}\right\} \tag{11.6}$$

$$\begin{aligned}\sqrt{\lambda}|e^{-\frac{1}{2\tau}D_t^2}e^{\tau\varphi}u|_{2,\kappa*} \leq\ & c\Big\{|e^{-\frac{1}{2\tau}D_t^2}e^{\tau\varphi}(\gamma\partial_t^2 - \Delta)u|_0 \\ & + e^{-d\tau}|e^{\tau\varphi}(\gamma\partial_t^2 - \Delta)u|_0 + e^{-d\tau}|e^{\tau\varphi}u|_{1,\tau}\Big\}\end{aligned} \tag{11.7}$$

$$\begin{aligned}\sqrt{\lambda}|e^{-\frac{1}{2\tau}D_t^2}e^{\tau\varphi}u|_{2,\kappa} \leq\ & c\Big\{|\sqrt{\kappa}e^{-\frac{1}{2\tau}D_t^2}e^{\tau\varphi}(\partial_t - \Delta)u|_0 \\ & + e^{-d\tau}|e^{\tau\varphi}(\partial_t - \Delta)u|_0 + e^{-d\tau}|e^{\tau\varphi}u|_{1,\tau}\Big\},\end{aligned} \tag{11.8}$$

for $u \in H^1(\mathbb{R}^3)$ with compact support in the ball $B_\delta(t_0, x_0)$ such that the right-hand sides are finite.

Comment on the proof: This lemma is a consequence of Theorem 1.1, Corollary 1.2, and Remark 1.1 in [E.2].

An important ingredient of the proof of these Carleman estimates (11.6)–(11.8) is the commutation of the operator $P(t, x, D + i\tau\nabla\varphi(t, x))$ with the Gaussian regularizer. This can be done with the following estimate, see Corollary 2.8 in [E.2].

Assume that $P(x, D)$ is an operator of order m with time-independent coefficients and assume that ψ is analytic in a neighborhood of $B_\delta(0)$. Given $\delta > 0$ and having found λ_0, we further have that for every non-negative integer N, there exist constants $C = C(\lambda, \delta)$ and $\tau_0(N, \lambda, \delta)$ such that for all $\tau \geq \tau_0(N, \lambda, \delta)$ and provided $w \in C_0^\infty(B_\delta(0))$, the following estimate holds true:

$$\begin{aligned}&|P(t, x, D + i\tau\nabla\varphi(t, x))\mu e^{-\frac{1}{2\tau}D_t^2}w - e^{-\frac{1}{2\tau}D_t^2}P(t, x, D + i\tau\nabla\varphi(t, x))w|_0 \\ &\leq \frac{C(\lambda,\delta)}{\tau}|\mu e^{-\frac{1}{2\tau}D_t^2}w|_{m,\tau} + 2e^{-\frac{\delta^2\tau}{4}}|w|_{-N,m,\tau}.\end{aligned} \tag{11.9a}$$

Here $\mu(t)$ is a smooth cutoff function such that $\mu \equiv 1$ in $(-2\delta, 2\delta)$ and $\mu \equiv 0$ on the complement of $(-3\delta, 3\delta)$.

Estimate (11.9a) will be used in conjunction with the following commutation identity

$$e^{\tau\varphi}P(x, D)u = P(x, D + i\tau\nabla\varphi(x))(e^{\tau\varphi}u). \tag{11.9b}$$

Step 2. Carleman estimate for the Kirchoff operator.

As in Section 10 we can now derive a Carleman estimate for the Kirchoff plate operator, see Lemma 10.2.

LEMMA 11.2. *Let ψ be an analytic function which is non-characteristic with respect to the operators $\gamma\partial_t^2 - \Delta$ and Δ at $(t_0, x_0) \in Q$. For $\delta > 0$ small, there exist constants $c > 0$ and*

$\lambda_0 > 0$, as well as two positive functions $d(\lambda) > 0$ and $\tau_0(\lambda)$, for $\lambda \geq \lambda_0$, such that the following estimate holds true provided $\lambda \geq \lambda_0$ and $\tau \geq \tau_0(\lambda)$:

$$\sqrt{\lambda}\left(|e^{-\frac{1}{2\tau}D_t^2}e^{\tau\varphi}w|_{2,\kappa*} + |e^{-\frac{1}{2\tau}D_t^2}e^{\tau\varphi}\partial_t^2 w|_{2,\kappa*} + |e^{-\frac{1}{2\tau}D_t^2}e^{\tau\varphi}\Delta w|_{2;\kappa*} + |e^{-\frac{1}{2\tau}D_t^2}e^{\tau\varphi}\nabla w|_{2,\kappa*}\right)$$

$$\leq C\Big\{|e^{-\frac{1}{2\tau}D_t^2}e^{\tau\varphi}(\gamma\partial_t^2 - \Delta)\Delta w|_0 + e^{-d\tau}|e^{\tau\varphi}(\gamma\partial_t^2 - \Delta)\Delta w|_0 + e^{-d\tau}|e^{\tau\varphi}\nabla\Delta w|_0$$

$$+ e^{-d\tau}|e^{\tau\varphi}\Delta w|_{1,\tau} + e^{-d\tau}|e^{\tau\varphi}\partial_t^2 w|_{1,\tau} + e^{-d\tau}|e^{\tau\varphi}w|_{1,\tau} + e^{-d\tau}|e^{\tau\varphi}\nabla w|_{1,\tau}\Big\}, \quad (11.10)$$

for $w \in H^3(\mathbb{R}^3)$ with compact support in $B_\delta(t_0, x_0)$ such that the right side is finite.

Proof. **Step 1.** The proof of this lemma is similar to that of Lemma 10.2. However, instead of the exact commutation formula (10.15), one has to use the approximate commutation formula (11.9). Clearly, the analogue of formula (10.8) is

$$\sqrt{\lambda}|e^{-\frac{1}{2\tau}D_t^2}e^{\tau\varphi}\Delta w|_{2,\kappa*} \leq c\Big\{|e^{-\frac{1}{2\tau}D_t^2}e^{\tau\varphi}\Delta(\gamma\partial_t^2 - \Delta)w|_0$$

$$+ e^{-d\tau}|e^{\tau\varphi}\Delta(\gamma\partial_t^2 - \Delta)w|_0 + e^{-d\tau}|e^{\tau\varphi}\Delta w|_{1,\tau}\Big\}. \quad (11.11)$$

Moreover, the counterpart of (10.14) is

$$\sqrt{\lambda}|e^{-\frac{1}{2\tau}D_t^2}e^{\tau\varphi}\nabla w|_{2,\kappa*} \leq c\Big\{|e^{-\frac{1}{2\tau}D_t^2}e^{\tau\varphi}\nabla\Delta w|_0$$

$$+ e^{-d\tau}|e^{\tau\varphi}\nabla\Delta w|_0 + e^{-d\tau}|e^{\tau\varphi}\nabla w|_{1,\tau}\Big\}. \quad (11.12)$$

As in the proof of Lemma 10.2, we need to estimate the first term in the right-hand side of the last estimate (11.12) using estimate (11.11). In order to do so, we need to commute the operators ∂_j and $e^{-\frac{1}{2\tau}D_t^2}e^{\tau\varphi}$. Using $\partial_j(e^{\tau\varphi}\Delta w) = \tau\partial_j\varphi e^{\tau\varphi}\Delta w + e^{\tau\varphi}\partial_j(\Delta w)$ in the first step below, and invoking Eqn. (11.9) with $m = 1$ and $(e^{\tau\varphi}\Delta w)$ in place of w in the second step below, we estimate the first term on the RHS of (11.12) as follows:

$$|e^{-\frac{1}{2\tau}D_t^2}e^{\tau\varphi}\nabla\Delta w|_0 = |e^{-\frac{1}{2\tau}D_t^2}(\nabla - \tau\nabla\varphi(x))e^{\tau\varphi}\Delta w|_0 \quad (11.13)$$

$$\text{(by (11.9))} \quad \leq |(\nabla - \tau\nabla\varphi(x))\mu e^{-\frac{1}{2\tau}D_t^2}e^{\tau\varphi}\Delta w|_0$$

$$+ \frac{C(\lambda,\delta)}{\tau}|\mu e^{-\frac{1}{2\tau}D_t^2}e^{\tau\varphi}\Delta w|_{1,\tau} + 2e^{-\frac{\delta^2\tau}{4}}|e^{\tau\varphi}\Delta w|_{-N,1,\tau} \quad (11.14)$$

$$\leq c|e^{-\frac{1}{2\tau}D_t^2}e^{\tau\varphi}\Delta w|_{1,\kappa} + \frac{C(\lambda,\delta)}{\tau}|e^{-\frac{1}{2\tau}D_t^2}e^{\tau\varphi}\Delta w|_{1,\tau} \quad (11.15)$$

$$+2e^{-\frac{\delta^2\tau}{4}}|e^{\tau\varphi}\Delta w|_{-N,1,\tau}.$$

In going from (11.14) to (11.15), we have invoked the norm in (11.4), since $\tau\nabla\varphi = \tau\lambda e^{\lambda\psi}\nabla\psi$ from the definition of φ. Using now estimate (11.15) with $N = 0$ for the first term in the RHS of (11.12), we obtain

$$\begin{aligned}\sqrt{\lambda}|e^{-\frac{1}{2\tau}D_t^2}e^{\tau\varphi}\nabla w|_{2,\kappa*} &\leq c\Big\{|e^{-\frac{1}{2\tau}D_t^2}e^{\tau\varphi}\Delta w|_{1,\kappa} + e^{-d\tau}|e^{\tau\varphi}\nabla\Delta w|_0 \\ &\quad + e^{-d\tau}|e^{\tau\varphi}\nabla w|_{1,\tau} + e^{-d\tau}|e^{\tau\varphi}\Delta w|_{1,\tau}\Big\} \\ &\quad + \frac{C(\lambda,\delta)}{\tau}|e^{-\frac{1}{2\tau}D_t^2}e^{\tau\varphi}\Delta w|_{1,\tau}, \end{aligned} \tag{11.16}$$

with possibly a new function $d(\lambda)$. Summing up (11.11) and (11.16) yields

$$\begin{aligned}&\sqrt{\lambda}\left\{\left|e^{-\frac{1}{2\tau}D_t^2}e^{\tau\varphi}\Delta w\right|_{2,\kappa^*} + \left|e^{-\frac{1}{2\tau}D_t^2}e^{\tau\varphi}\nabla w\right|_{2,\kappa^*}\right\} \\ &\leq C\Big\{\left|e^{-\frac{1}{2\tau}D_t^2}e^{\tau\varphi}\Delta(\gamma\partial_t^2 - \Delta)w\right|_0 + \left|e^{-\frac{1}{2\tau}D_t^2}e^{\tau\varphi}\Delta w\right|_{1,\kappa} + e^{-d\tau}\left|e^{\tau\varphi}\Delta(\gamma\partial_t^2 - \Delta)w\right|_0 \\ &\quad + e^{-d\tau}\left|e^{\tau\varphi}\Delta w\right|_{1,\tau} + e^{-d\tau}\left|e^{\tau\varphi}\nabla\Delta w\right|_0 + e^{-d\tau}\left|e^{\tau\varphi}\nabla w\right|_{1,\tau}\Big\} \\ &\quad + \frac{C(\lambda,\delta)}{\tau}\left|e^{-\frac{1}{2\tau}D_t^2}e^{\tau\varphi}\Delta w\right|_{1,\tau}. \end{aligned} \tag{11.17}$$

Step (ii). We next shall show that:
Claim.

$$\sqrt{\lambda}|u|_{2,\kappa^*} - \frac{C(\lambda,\delta)}{\tau}|u|_{s,\tau} \geq \frac{\sqrt{\lambda}}{2}|u|_{2,\kappa^*}, \quad s = 1,2, \tag{11.18}$$

provided τ is sufficiently large as in (11.25) below. Thus, the claim, used with

$$u = |e^{-\frac{1}{2\tau}D_t^2}e^{\tau\varphi}\Delta w|$$

permits one to absorb the last term on the RHS of (11.17) by $\frac{1}{2}$ of the first term on the LHS of (11.17), so that (11.17) yields

$$\begin{aligned}&\sqrt{\lambda}\left\{\left|e^{-\frac{1}{2\tau}D_t^2}e^{\tau\varphi}\Delta w\right|_{2,\kappa^*} + \left|e^{-\frac{1}{2\tau}D_t^2}e^{\tau\varphi}\nabla w\right|_{2,\kappa^*}\right\} \\ &\leq 2C\Big\{\left|e^{-\frac{1}{2\tau}D_t^2}e^{\tau\varphi}\Delta(\gamma\partial_t^2 - \Delta)w\right|_0 + \left|e^{-\frac{1}{2\tau}D_t^2}e^{\tau\varphi}\Delta w\right|_{1,\kappa} \\ &\quad + e^{-d\tau}\left|e^{\tau\varphi}\Delta(\gamma\partial_t^2 - \Delta)w\right|_0 + e^{-d\tau}\left|e^{\tau\varphi}\Delta w\right|_{1,\tau} \\ &\quad + e^{-d\tau}\left|e^{\tau\varphi}\nabla\Delta w\right|_0 + e^{-d\tau}\left|e^{\tau\varphi}\nabla w\right|_{1,\tau}\Big\}. \end{aligned} \tag{11.19}$$

Proof of Claim (11.18). Recalling from (10.43) the inequality $\sqrt{a}-\sqrt{b}\geq\frac{1}{\sqrt{2}}\sqrt{a-2b}$ for $a\geq 2b$, we estimate, recalling also definigion (11.4) for $m=2$ (right) and (10.4b) for $s=1,2$:

$$\sqrt{\lambda}\,|u|_{2,\kappa^*}-\frac{C(\lambda,\delta)}{\tau}|u|_{s,\tau}$$

$$\geq \frac{1}{\sqrt{2}}\left\{\lambda|u|^2_{2,\kappa^*}-2\frac{C^2(\lambda,\delta)}{\tau^2}|u|^2_{s,\tau}\right\}^{\frac{1}{2}} \tag{11.20}$$

$$= \frac{1}{\sqrt{2}}\left\{\lambda\sum_{|\alpha|\leq 2}\int\kappa(t,x)^{3-2|\alpha|}|D^\alpha u|^2-2\frac{C^2(\lambda,\delta)}{\tau^2}\sum_{|\alpha|\leq s}\tau^{2-2|\alpha|}\int|D^\alpha u|^2\right\}^{\frac{1}{2}} \tag{11.21}$$

$$\geq \frac{1}{\sqrt{2}}\left\{\sum_{|\alpha|\leq 2}\int\left[\lambda\,\kappa(t,x)^{3-2|\alpha|}-2\frac{C^2(\lambda,\delta)}{\tau^2}\tau^{2-2|\alpha|}\right]|D^\alpha u|^2\right\}^{\frac{1}{2}}. \tag{11.22}$$

Regarding the kernel in (11.22), we next estimate by recalling that $\kappa(t,x)=\tau\lambda e^{\lambda\psi(t,x)}$ from (11.3):

$$\lambda\kappa(t,x)^{3-2|\alpha|}-2\frac{C(\lambda,\delta)^2}{\tau^2}\tau^{2-2|\alpha|} \tag{11.23}$$

$$= \lambda\kappa(t,x)^{3-2|\alpha|}\left(1-\frac{2C(\lambda,\delta)^2}{\tau^3\lambda^{4-2|\alpha|}e^{(3-2|\alpha|)\lambda\psi(t,x)}}\right)\geq\frac{\lambda\kappa(t,x)^{3-2|\alpha|}}{2}, \tag{11.24}$$

provided that

$$\tau\geq\tau_0(\lambda,\delta)=\left\{\frac{4C(\lambda,\delta)^2}{\lambda^{4-2|\alpha|}e^{(3-2|\alpha|)\lambda\min\psi(t,x)}}\right\}^{\frac{1}{3}}, \tag{11.25}$$

where the min is taken over $B_\delta(0)$. Using (11.24) in (11.22) and recalling once more definition (11.4) for $m=2$ (right) then yields

$$\sqrt{\lambda}\,|u|_{2,\kappa^*}-\frac{C(\lambda,\delta)}{\tau}|u|_{1,\tau}\geq\frac{\sqrt{\lambda}}{2}\left\{\sum_{|\alpha|\leq 2}\int\kappa(t,x)^{3-2|\alpha|}|D^\alpha u|^2\right\}^{\frac{1}{2}} \tag{11.26}$$

$$= \frac{\sqrt{\lambda}}{2}|u|_{2,\kappa^*}, \tag{11.27}$$

and Claim (11.18) is proved.

Step (iii). Continuing with the proof of Lemma 11.2, we estimate the second term on the

RHS of (11.19) as follows, by invoking (11.4b) and (11.11) for $\lambda \geq \lambda_0$,

$$\left| e^{-\frac{1}{2\tau}D_t^2} e^{\tau\varphi} \Delta w \right|_{1,\kappa} \leq \left| e^{-\frac{1}{2\tau}D_t^2} e^{\tau\varphi} \Delta w \right|_{2,\kappa^*} \tag{11.28}$$

$$\text{(by (11.11))} \quad \leq \frac{C}{\sqrt{\lambda_0}} \left\{ \left| e^{-\frac{1}{2\tau}D_t^2} e^{\tau\varphi} \Delta(\gamma\partial_t^2 - \Delta) w \right|_0 + e^{-d\tau} \left| e^{\tau\varphi} \Delta(\gamma\partial_t^2 - \Delta) w \right|_0 + e^{-d\tau} \left| e^{\tau\varphi} \Delta w \right|_{1,\tau} \right\}. \tag{11.29}$$

Hence, using (11.29) on the RHS of (11.19) yields

$$\sqrt{\lambda} \left\{ \left| e^{-\frac{1}{2\tau}D_t^2} e^{\tau\varphi} \Delta w \right|_{2,\kappa^*} + \left| e^{-\frac{1}{2\tau}D_t^2} e^{\tau\varphi} \nabla w \right|_{2,\kappa^*} \right\}$$

$$\leq C \left\{ \left| e^{-\frac{1}{2\tau}D_t^2} e^{\tau\varphi} \Delta(\gamma\partial_t^2 - \Delta) w \right|_0 + e^{-d\tau} \left| e^{\tau\varphi} \Delta(\gamma\partial_t^2 - \Delta) w \right|_0 + e^{-d\tau} \left| e^{\tau\varphi} \Delta w \right|_{1,\tau} + e^{-d\tau} \left| e^{\tau\varphi} \nabla \Delta w \right|_0 + e^{-d\tau} \left| e^{\tau\varphi} \nabla w \right|_{1,\tau} \right\}. \tag{11.30}$$

Eqn. (11.30) gives the desired estimate for the third and fourth terms on the LHS of (11.10). As for the first and second term on the LHS of (11.10), we use Lemma 11.1. More precisely, for the second term on the LHS of (11.10), we proceed as in (10.10), by adding and subtracting, and obtain via Eqn. (11.6) with $u = (\gamma\partial_t^2 - \Delta)w$ and (11.7) with $u = \Delta w$ on the first and second term, respectively, of (11.31) below:

$$\sqrt{\lambda} \left| e^{-\frac{1}{2\tau}D_t^2} e^{\tau\varphi} \partial_t^2 w \right|_{2,\kappa^*} \leq C\sqrt{\lambda} \left\{ \left| e^{-\frac{1}{2\tau}D_t^2} e^{\tau\varphi} (\gamma\partial_t^2 - \Delta) w \right|_{2,\kappa^*} + \left| e^{-\frac{1}{2\tau}D_t^2} e^{\tau\varphi} \Delta w \right|_{2,\kappa^*} \right\} \tag{11.31}$$

$$\text{(by (11.6) and (11.7))} \quad \leq C \left\{ \left| e^{-\frac{1}{2\tau}D_t^2} e^{\tau\varphi} \Delta(\gamma\partial_t^2 - \Delta) w \right|_0 + e^{-d\tau} \left| e^{\tau\varphi} \Delta(\gamma\partial_t^2 - \Delta) w \right|_0 + e^{-d\tau} \left| e^{\tau\varphi} (\gamma\partial_t^2 - \Delta) w \right|_{1,\tau} + e^{-d\tau} \left| e^{\tau\varphi} \Delta w \right|_{1,\tau} \right\}. \tag{11.32}$$

Finally, summing up estimates (11.30), (11.32) and (11.6) for $u = w$ yields, as desired, the final estimate (11.10) since $|u|_0 \leq |u|_{1,\tau}$, Lemma 11.1 is proved. □

Step 3. Local unique continuation of the thermoelastic system (11.1a-b).

The above Carleman estimates for the Kirchhoff plate and for the heat operator are the essential ingredients for the proof of the local uniqueness theorem of the thermoelastic system.

THEOREM 11.3. *Let S be a C^2-surface which is time-like with respect to the wave operator $\gamma\partial_t^2 - \Delta$ at $(t_0, x_0) \in Q$. Assume that $(w, \theta) \in H^3(Q) \times H^1(Q)$ is a solution to the thermoelastic system (11.1a–b) which vanishes on one side of S. Then $w \equiv 0$ and $\theta \equiv 0$ in a neighborhood of (t_0, x_0).*

Proof. The proof is very similar to the proof of Theorem 10.3. However, there are some technical differences which we shall point out.

Step (i). Once more, we assume that (t_0, x_0) is the origin 0. Since S is a C^2-surface, there exists a $\phi \in C^2(\overline{Q})$ such that S is the level surface $\{\phi = 0\}$ and $(w, \theta) \equiv 0$ in $\{\phi > 0\}$ by assumption. This time the function ψ will be a perturbation of ϕ, not of φ as in Section 10, Eqn. (10.20). For $\epsilon > 0$ small, we set

$$\psi(t,x) = \sum_{|\alpha| \le 2} \frac{\partial^\alpha \phi(0)}{\alpha!}(t,x)^\alpha - 3\epsilon(t^2 + |x|^2). \tag{11.33}$$

REMARK 11.1. *Notice here a big difference of procedure with respect to the local uniqueness proof of Theorem 10.3 in Section 10.*

Here we take at first the level surface ϕ, perturb it to get ψ by (11.33), and finally take the exponential $\varphi(t,x) = e^{\lambda\psi(t,x)} - 1$ and $\kappa(t,x) = \tau\lambda e^{\lambda\psi(t,x)}$ in (11.3). In the proof of Theorem 10.3 in Section 10, these steps were reversed: we started with the level surface ϕ, next took the exponential $\varphi(x) = e^{\lambda\phi(x)} - 1$ (for λ large and fixed), and finally we perturbed $\varphi(x)$ to get $\psi(t,x)$ in (10.20). By choosing the present approach, we insure that the underlying geometry is independent of λ.

Step (ii). As in the proof of Theorem 10.3 (see Eqn. (10.32)) we introduce

$$\begin{cases} \tilde{w}(t,x) = \chi(\psi(t,x)w(t,x) \text{ and } \tilde{\theta}(t,x) = \chi(\psi(t,x)\theta(t,x) & \text{if } (t,x) \in B_{2\delta}(0); \\ \tilde{w} \equiv \tilde{\theta} \equiv 0 & \text{otherwise,} \end{cases} \tag{11.34}$$

where χ is defined by (10.22), and obtain

$$(-\gamma\partial_t^2 + \Delta)\Delta\tilde{w} = -\partial_t^2\tilde{w} - \operatorname{div}(\alpha(x)\nabla\tilde{\theta}) + F \tag{11.35}$$

$$\partial_t\tilde{\theta} - \Delta\tilde{\theta} = \operatorname{div}(\alpha(x)\nabla\partial_t\tilde{w}) + G \tag{11.36}$$

where the counterpart of (10.31) is now:

$$F, G \in L_2, \quad \operatorname{supp} F, \quad \operatorname{supp} G \subset \{\psi \le -\epsilon\delta^2\} \cap B_{2\delta}(0). \tag{11.37}$$

Step (iii). Carleman estimate for the thermoelastic system: first version.

Next, we return to the two estimates (11.10) and (11.8), applied this time with respect to $\tilde{w}$ and $\tilde{\theta}$ rather than w and u: on their right-hand sides, however, we then invoke identities (11.35) and (11.36), respectively, for the terms $\left[e^{-\frac{1}{2\tau}D_t^2}e^{\tau\varphi}(\gamma\Delta_t^2-\Delta)\Delta\tilde{w}\right]$ and $\left[e^{-\frac{1}{2\tau}D_t^2}e^{\tau\varphi}(\partial_t-\Delta)\tilde{\theta}\right]$. We thus obtain (11.10) and (11.35)

$$\begin{aligned}
&\sqrt{\lambda}\Bigg\{\left|e^{-\frac{1}{2\tau}D_t^2}e^{\tau\varphi}\tilde{w}\right|_{2,\kappa^*}+\left|e^{-\frac{1}{2\tau}D_t^2}e^{\tau\varphi}\partial_t^2\tilde{w}\right|_{2,\kappa^*}\\
&\qquad+\left|e^{-\frac{1}{2\tau}D_t^2}e^{\tau\varphi}\Delta\tilde{w}\right|_{2;\kappa^*}+\left|e^{-\frac{1}{2\tau}D_t^2}e^{\tau\varphi}\nabla\tilde{w}\right|_{2;\kappa^*}\Bigg\}\\
&\qquad\leq C\Bigg\{\left|e^{-\frac{1}{2\tau}D_t^2}e^{\tau\varphi}\partial_t^2\tilde{w}\right|_0+\left|e^{-\frac{1}{2\tau}D_t^2}e^{\tau\varphi}\operatorname{div}(\alpha(x)\nabla\tilde{\theta})\right|_0+\left|e^{-\frac{1}{2\tau}D_t^2}e^{\tau\varphi}F\right|_0\\
&\qquad+e^{-d\tau}\left|e^{\tau\varphi}(\gamma\partial_t^2-\Delta)\Delta\tilde{w}\right|_0+e^{-d\tau}\left|e^{\tau\varphi}\nabla\Delta\tilde{w}\right|_0\\
&\qquad+e^{-d\tau}\left|e^{\tau\varphi}\Delta\tilde{w}\right|_{1,\tau}+e^{-d\tau}\left|e^{\tau\varphi}\partial_t^2\tilde{w}\right|_{1,\tau}\\
&\qquad+e^{-d\tau}\left|e^{\tau\varphi}\tilde{w}\right|_{1,\tau}+e^{-d\tau}\left|e^{\tau\varphi}\nabla\tilde{w}\right|_{1,\tau},
\end{aligned}\tag{11.38}$$

and by (11.8) and (11.36),

$$\begin{aligned}
\sqrt{\lambda}\left|e^{-\frac{1}{2\tau}D_t^2}e^{\tau\varphi}\tilde{\theta}\right|_{2,\kappa}\leq\ & C\Bigg\{\left|\sqrt{\kappa}e^{-\frac{1}{2\tau}D_t^2}e^{\tau\varphi}\operatorname{div}(\alpha\nabla(\partial_t\tilde{w}))\right|_0+\left|\sqrt{\kappa}e^{-\frac{1}{2\tau}D_t^2}e^{\tau\varphi}G\right|_0\\
&+e^{-d\tau}|e^{\tau\varphi}(\partial_t-\Delta)\tilde{\theta}|_0+e^{-d\tau}|e^{\tau\varphi}\tilde{\theta}|_{1,\tau}\Bigg\}.
\end{aligned}\tag{11.39}$$

Summing up estimates (11.38) and (11.39), we then obtain the sought-after Carleman estimate for the thermoelastic system (first version):

$$\begin{aligned}
&\sqrt{\lambda}\Bigg\{\left|e^{-\frac{1}{2\tau}D_t^2}e^{\tau\varphi}\tilde{w}\right|_{2,\kappa*}+\left|e^{-\frac{1}{2\tau}D_t^2}e^{\tau\varphi}\partial_t^2\tilde{w}\right|_{2,\kappa*}\\
&\qquad+\left|e^{-\frac{1}{2\tau}D_t^2}e^{\tau\varphi}\Delta\tilde{w}\right|_{2,\kappa*}+\left|e^{-\frac{1}{2\tau}D_t^2}e^{\tau\varphi}\nabla\tilde{w}\right|_{2,\kappa*}+\left|e^{-\frac{1}{2\tau}D_t^2}e^{\tau\varphi}\tilde{\theta}\right|_{2,\kappa}\Bigg\}\\
&\qquad\leq c\Bigg\{\left|e^{-\frac{1}{2\tau}D_t^2}e^{\tau\varphi}\partial_t^2\tilde{w}\right|_0+|e^{-\frac{1}{2\tau}D_t^2}e^{\tau\varphi}\operatorname{div}(\alpha\nabla\tilde{\theta})|_0+|\sqrt{\kappa}e^{-\frac{1}{2\tau}D_t^2}e^{\tau\varphi}\operatorname{div}(\alpha\nabla\partial_t\tilde{w})|_0\\
&\qquad+|e^{-\frac{1}{2\tau}D_t^2}e^{\tau\varphi}F|_0+|\sqrt{\kappa}e^{-\frac{1}{2\tau}D_t^2}e^{\tau\varphi}G|_0+e^{-d\tau}|e^{\tau\varphi}(\gamma\partial_t^2-\Delta)\Delta\tilde{w}|_0\\
&\qquad+e^{-d\tau}|e^{\tau\varphi}\nabla\Delta\tilde{w}|_0+e^{-d\tau}|e^{\tau\varphi}\Delta\tilde{w}|_{1,\tau}+e^{-d\tau}|e^{\tau\varphi}\partial_t^2\tilde{w}|_{1,\tau}+e^{-d\tau}|e^{\tau\varphi}\tilde{w}|_{1,\tau}\\
&\qquad+e^{-d\tau}|e^{\tau\varphi}\nabla\tilde{w}|_{1,\tau}+e^{-d\tau}|e^{\tau\varphi}(\partial_t-\Delta)\tilde{\theta}|_0+e^{-d\tau}|e^{\tau\varphi}\tilde{\theta}|_{1,\tau}\Bigg\}.
\end{aligned}\tag{11.40}$$

Step (iv). Carleman estimate for the thermoelastic system: second version.

Using the fact that α is time-independent and estimate (11.9) we bound the second and third term on the right-hand side of (11.40). To begin with, we rewrite the second term on the RHS of (11.40) as $\operatorname{div}(\alpha\nabla\tilde{\theta}) = \alpha\Delta\tilde{\theta} + \nabla\alpha\cdot\nabla\tilde{\theta}$ so that

$$|e^{-\frac{1}{2\tau}D_t^2}e^{\tau\varphi}\operatorname{div}(\alpha\nabla\tilde{\theta})|_0 \leq c\left\{|e^{-\frac{1}{2\tau}D_t^2}e^{\tau\varphi}\nabla\tilde{\theta}|_0 + |e^{-\frac{1}{2\tau}D_t^2}e^{\tau\varphi}\Delta\tilde{\theta}|_0\right\}. \tag{11.41}$$

Next, we write

$$e^{-\frac{1}{2\tau}D_t^2}e^{\tau\varphi}\Delta\tilde{\theta} = e^{-\frac{1}{2\tau}D_t^2}(\nabla - \tau\nabla\varphi)^2(e^{\tau\varphi}\tilde{\theta}) \tag{11.42}$$

[where $(\nabla-\tau\nabla\varphi)^2$ means $(\nabla-\tau\nabla\varphi)(\nabla-\tau\nabla\varphi)$], and both formulas in (11.42a–b) are special cases (the second, an obvious one) of identity (11.9b) with $m = 2$ and $m = 1$, respectively. Starting now with (11.41) and using estimates (11.42a–b) on its right, we obtain

$$\begin{aligned}
&|e^{-\frac{1}{2\tau}D_t^2}e^{\tau\varphi}\operatorname{div}(\alpha\nabla\tilde{\theta})|_0 \\
&\leq c\left\{|e^{-\frac{1}{2\tau}D_t^2}(\nabla - \tau\nabla\varphi)^2(e^{\tau\varphi}\tilde{\theta})|_0 + |e^{-\frac{1}{2\tau}D_t^2}(\nabla - \tau\nabla\varphi)(e^{\tau\varphi}\tilde{\theta})|_0\right\} && (11.43)\\
\text{(by (11.9a))} &\leq \Big\{|(\nabla - \tau\nabla\varphi)^2 e^{-\frac{1}{2\tau}D_t^2}e^{\tau\varphi}\tilde{\theta}|_0 + |(\nabla - \tau\nabla\varphi)e^{-\frac{1}{2\tau}D_t^2}e^{\tau\varphi}\tilde{\theta}|_0 \\
&\quad + \frac{C(\lambda,\delta)}{\tau}|e^{-\frac{1}{2\tau}D_t^2}e^{\tau\varphi}\tilde{\theta}|_{2,\tau} + 4e^{-\frac{\delta^2\tau}{4}}|e^{\tau\varphi}\tilde{\theta}|_{-N,2,\tau}\Big\} && (11.44)\\
&\leq c\left\{|e^{-\frac{1}{2\tau}D_t^2}e^{\tau\varphi}\tilde{\theta}|_{2,\kappa} + e^{-\frac{\delta^2\tau}{4}}|e^{\tau\varphi}\tilde{\theta}|_{-N,2,\tau}\right\} \\
&\quad + \frac{C(\lambda,\delta)}{\tau}|e^{-\frac{1}{2\tau}D_t^2}e^{\tau\varphi}\tilde{\theta}|_{2,\tau}, && (11.45)
\end{aligned}$$

where in obtaining (11.44) we have also used, after a double application of (11.9a) with $m = 2$ and $m = 1$, the obvious majorizations: $|\cdot|_{1,\tau} \leq |\cdot|_{2,\tau}$ and $|\cdot|_{-N,1,\tau} \leq |\cdot|_{-N,2,\tau}$.

In going from (11.44) to (11.45) we have used, in the first term on the RHS of (11.44), that $\tau\nabla\varphi = \kappa\nabla\psi$ from (11.3) and, moreover, that $|\nabla\psi| \leq C$ in $B_\delta(0)$.

As to the third term on the RHS of (11.40), we again use estimate (11.9) twice to obtain

$$\begin{aligned}
|e^{-\frac{1}{2\tau}D_t^2}e^{\tau\varphi}\operatorname{div}(\alpha\nabla\partial_t\tilde{w}|_0) &\leq c\left\{|e^{-\frac{1}{2\tau}D_t^2}e^{\tau\varphi}\nabla\partial_t\tilde{w}|_0 + |e^{-\frac{1}{2\tau}D_t^2}e^{\tau\varphi}\Delta\partial_t\tilde{w}|_0\right\} && (11.46)\\
\text{(by (11.9))} &\leq c\Big\{|e^{-\frac{1}{2\tau}D_t^2}e^{\tau\varphi}\nabla\tilde{w}|_{1,\kappa} + |e^{-\frac{1}{2\tau}D_t^2}e^{\tau\varphi}\Delta\tilde{w}|_{1,\kappa} \\
&\quad + e^{-\frac{\delta^2\tau}{4}}|e^{\tau\varphi}\nabla\tilde{w}|_{-N,1,\tau} + e^{-\frac{\delta^2\tau}{4}}|e^{\tau\varphi}\Delta\tilde{w}|_{-N,1,\tau}\Big\} \\
&\quad + \frac{C(\lambda,\delta)}{\tau}\left\{|e^{-\frac{1}{2\tau}D_t^2}e^{\tau\varphi}\nabla\tilde{w}|_{1,\tau} + |e^{-\frac{1}{2\tau}D_t^2}e^{\tau\varphi}\Delta\tilde{w}|_{1,\tau}\right\} && (11.47)
\end{aligned}$$

Using now estimates (11.45) and (11.47) for the second and third terms on the RHS of (11.40) we obtain a second version of the Carleman estimate

$$\begin{aligned}
&\sqrt{\lambda}\Big\{|e^{-\frac{1}{2\tau}D_t^2}e^{\tau\varphi}\tilde{w}|_{2;\kappa*} + |e^{-\frac{1}{2\tau}D_t^2}e^{\tau\varphi}\partial_t^2\tilde{w}|_{2;\kappa*} \\
&\quad + |e^{-\frac{1}{2\tau}D_t^2}e^{\tau\varphi}\Delta\tilde{w}|_{2;\kappa*} + |e^{-\frac{1}{2\tau}D_t^2}e^{\tau\varphi}\nabla\tilde{w}|_{2;\kappa*} + |e^{-\frac{1}{2\tau}D_t^2}e^{\tau\varphi}\tilde{\theta}|_{2;\kappa}\Big\} \\
\le\ & c\Big\{|e^{-\frac{1}{2\tau}D_t^2}e^{\tau\varphi}\partial_t^2\tilde{w}|_0 + |e^{-\frac{1}{2\tau}D_t^2}e^{\tau\varphi}\tilde{\theta}|_{2,\kappa} + |e^{-\frac{1}{2\tau}D_t^2}e^{\tau\varphi}\nabla\tilde{w}|_{1,\kappa} \\
&\quad + |e^{-\frac{1}{2\tau}D_t^2}e^{\tau\varphi}\Delta\tilde{w}|_{1,\kappa} + |e^{-\frac{1}{2\tau}D_t^2}e^{\tau\varphi}F|_0 + |\sqrt{\kappa}e^{-\frac{1}{2\tau}D_t^2}e^{\tau\varphi}G|_0 \\
&\quad + e^{-d\tau}|e^{\tau\varphi}(\gamma\partial_t^2-\Delta)\Delta\tilde{w}|_0 + e^{-d\tau}|e^{\tau\varphi}\nabla\Delta\tilde{w}|_0 + e^{-d\tau}|e^{\tau\varphi}\Delta\tilde{w}|_{1,\tau} + e^{-d\tau}|e^{\tau\varphi}\partial_t^2\tilde{w}|_{1,\tau} \\
&\quad + e^{-d\tau}|e^{\tau\varphi}\tilde{w}|_{1,\tau} + e^{-d\tau}|e^{\tau\varphi}\nabla\tilde{w}|_{1,\tau} + e^{-d\tau}|e^{\tau\varphi}(\partial_t-\Delta)\tilde{\theta}|_0 + e^{-d\tau}|e^{\tau\varphi}\tilde{\theta}|_{1,\tau} \\
&\quad + e^{-\frac{\delta^2\tau}{4}}|e^{\tau\varphi}\tilde{\theta}|_{-N,2,\tau} + e^{-\frac{\delta^2\tau}{4}}|e^{\tau\varphi}\nabla\tilde{w}|_{-N,1,\tau} + e^{-\frac{\delta^2\tau}{4}}|e^{\tau\varphi}\Delta\tilde{w}|_{-N,1,\tau}\Big\} \\
&+ \frac{C(\lambda,\delta)}{\tau}\Big\{|e^{-\frac{1}{2\tau}D_t^2}e^{\tau\varphi}\tilde{\theta}|_{2,\tau} + |e^{-\frac{1}{2\tau}D_t^2}e^{\tau\varphi}\nabla\tilde{w}|_{1,\tau} + |e^{-\frac{1}{2\tau}D_t^2}e^{\tau\varphi}\Delta\tilde{w}|_{1,\tau}\Big\}. \qquad (11.48)
\end{aligned}$$

Step (v). Analysis of RHS of (11.48). Carleman estimate for the thermoelastic system: final version.

We make the following observations on estimate (11.48).

(a) The terms in the last line of (11.48) can be cancelled by one-half of the left-hand side provided τ is large enough. This can be done as in the proof of Lemma 11.2. See Claim (11.18) for $u = e^{-\frac{1}{2\tau}D_t^2}e^{\tau\varphi}\Delta\tilde{w}$, $u = e^{-\frac{1}{2\tau}D_t^2}e^{\tau\varphi}\nabla\tilde{w}$ and $s = 1$, as well as $u = e^{-\frac{1}{2\tau}D_t^2}e^{\tau\varphi}\theta$ and $s = 2$.

(b) Next, we choose $\sqrt{\lambda} \ge 2c$, and then freeze such value of λ afterwards until the end of the proof. Then, the first four terms on the right-hand side of (11.48) can be put into the left-hand side of (11.48). This is the only point were we need the flexibility of the additional parameter λ, however, it is very critical here.

(c) Next, the three exponentially decaying terms with $e^{-d\tau}$ in front in the line next to the very last line of the RHS of (11.48) can be combined with the other exponentially decaying terms with possibly a smaller d. This is easily done for the two terms containing $\tilde{w}$. For the term containing $\tilde{\theta}$, we use the estimate

$$|e^{\tau\varphi}\tilde{\theta}|_{-N,2,\tau} \le c(\lambda)\left(\sqrt{\tau}|e^{\tau\varphi}(\partial_t-\Delta)\tilde{\theta}|_0 + |e^{\tau\varphi}\tilde{\theta}|_{-N+2,1,\tau}\right) \qquad (11.49)$$

for $N = 2$. This is a consequence of Corollary 2.6 in [E.2].

(d) With $\sqrt{\lambda} \ge 2c$ fixed, once and for all, then $\kappa(t,x)$ in (11.3) becomes only a function of τ, not of λ. Accordingly, the definitions (11.4) yield readily: $|u|^2_{2,\kappa} \ge |u|^2_{2,\kappa*} \ge \tau^3|u|^2_0$, and hence we shall use

$$\begin{cases} \tau^{\frac{3}{2}}|u|_0 & \leq & |u|_{2,\kappa*} \quad \text{for } u = e^{-\frac{1}{2\tau}D_t^2}e^{\tau\varphi}\tilde{w}, \\ \tau^{\frac{3}{2}}|v|_0 & \leq & |v|_{2,\kappa} \quad \text{for } v = e^{-\frac{1}{2\tau}D_t^2}e^{\tau\varphi}\tilde{\theta}, \end{cases} \tag{11.50}$$

on the LHS of (11.48), while we shall drop the other terms on the LHS of (11.48).

Thus, carrying out the program (a) through (d) described above, we see that estimate (11.48) then yields the following final version of the Carleman estimate, once in addition we multiply (11.48) across by $\sqrt{\tau}$:

$$\tau^2\left\{|e^{-\frac{1}{2\tau}D_t^2}e^{\tau\varphi}\tilde{w}|_0 + |e^{-\frac{1}{2\tau}D_t^2}e^{\tau\varphi}\tilde{\theta}|_0\right\}$$

$$\leq c\sqrt{\tau}\Big\{e^{-\frac{1}{2\tau}D_t^2}e^{\tau\varphi}F|_0 + \sqrt{\tau}|e^{-\frac{1}{2\tau}D_t^2}e^{\tau\varphi}G|_0 + e^{-d\tau}|e^{\tau\varphi}(\gamma\partial_t^2 - \Delta)\Delta\tilde{w}|_0$$

$$+ e^{-d\tau}|e^{\tau\varphi}\nabla\Delta\tilde{w}|_0 + e^{-d\tau}|e^{\tau\varphi}\Delta\tilde{w}|_{1,\tau} + e^{-d\tau}|e^{\tau\varphi}\partial_t^2\tilde{w}|_{1,\tau}$$

$$+ e^{-d\tau}|e^{\tau\varphi}\tilde{w}|_{1,\tau} + e^{-d\tau}|e^{\tau\varphi}\nabla\tilde{w}|_{1,\tau} + e^{-d\tau}|e^{\tau\varphi}(\partial_t - \Delta)\tilde{\theta}|_0 + e^{-d\tau}|e^{\tau\varphi}\tilde{\theta}|_{1,\tau}\Big\} \tag{11.51}$$

The factor $\sqrt{\tau}$ in front of the G-term comes from $\sqrt{\kappa}$. We repeat that in (11.51) λ is frozen (once chosen as $\sqrt{\lambda} \geq 2c$).

Step (vi). Conclusion.

We can now finish the proof in a manner similar to the conclusion of the proof of Theorem 10.3. We have

$$|e^{-\frac{1}{2\tau}D_t^2}e^{\tau\varphi}F|_0 \leq |e^{\tau\varphi}F|_0 \leq ce^{\sigma\tau}; \quad |e^{-\frac{1}{2\tau}D_t^2}e^{\tau\varphi}G|_0 \leq |e^{\tau\varphi}G|_0 \leq ce^{\sigma\tau}, \tag{11.52}$$

recalling, in the first majorization, both (10.53), and the property F, $G \in L_2$ by (11.37); and recalling, in the second majorization, $\varphi(t,x) = e^{\lambda\psi(t,x)} - 1$ from (11.3), and again (11.37) for supp F and supp G, where now $\sigma = e^{-\lambda\epsilon\delta^2} - 1 < 0$. In (11.52), the constant c is given by $c = |F|_0$ and $c = |G|_0$, respectively, and hence depends on w [as in (10.53)]. We finally concentrate on the remaining terms on the RHS of (11.51). For $w \in H^3$ and $\theta \in H^1$ as assumed, we have $\tilde{w} \in H^3$ and $\tilde{\theta} \in H^1$ by (11.34). Then all norms in the RHS of (11.51) are well-defined, including the term $(\gamma\partial_t^2 - \Delta)\Delta\tilde{w} \in L_2$ precisely as in Section 2 below (10.54), by recalling Eqn. (10.32). Moreover, as there, we have that $\tilde{w} = 0$ and $\tilde{\theta} \equiv 0$ in $\{\varphi > 0\}$ so that the factor $e^{\tau\varphi}$ is active only for $\varphi \leq 0$ and hence does not produce exponential growth. [The level surfaces of ψ coincide with the level surface of φ, see (11.3).] Then, recalling that $|u|_{1,\tau} \geq \tau|u|_0$ from (10.4c), we see then that all the terms on the RHS of (11.51), other than those containing F and G, are bounded by $\sqrt{\tau}c\tau e^{-d\tau}$, where the constant c again depends on w. In conclusion, the Carleman estimate (11.51) becomes, via also (11.52),

$$\tau^2(|e^{-\frac{1}{2\tau}D_t^2}e^{\tau\varphi}\tilde{w}|_0 + |e^{-\frac{1}{2\tau}D_t^2}e^{\tau\varphi}\tilde{\theta}|_0) \leq c(\tau e^{\sigma\tau} + \tau^{\frac{3}{2}}e^{-d\tau}), \ \sigma < 0, \tag{11.53}$$

which then implies

$$|e^{-\frac{1}{2\tau}D_t^2}e^{\tau(\varphi+\beta)}\tilde{w}|_0 + |e^{-\frac{1}{2\tau}D_t^2}e^{\tau(\varphi+\beta)}\tilde{\theta}|_0 \leq \frac{2c}{\sqrt{\tau}}, \tag{11.54}$$

with $\beta = \min\{-\sigma, d\}$. With the help of Proposition 4.1 in [Ta.3] we then conclude from (11.54) that

$$\tilde{w} \equiv 0 \text{ and } \tilde{\theta} \equiv 0 \text{ in } \varphi > -\beta. \tag{11.55}$$

This implies

$$0 \notin \text{supp } w \text{ and } 0 \notin \text{supp } \theta \quad \text{or } w \equiv 0,\ \theta \equiv 0 \text{ in a neighborhood of } 0 = (t_0, x_0). \tag{11.56}$$

The proof of Theorem 11.3 is complete. □

From here we can finish the proof of Theorem 4.2.2(i) as in Section 10.

Appendix A: Abstract Models of the Non-Homogeneous v-Kirchoff Problem (3.2)

In this appendix, we briefly justify the abstract model (3.17), respectively (3.29) for the v-Kirchoff problem (3.2a) with non-homogeneous B.C. (3.2b) either in the hinged B.C., or else in the clamped B.C.

Hinged B.C. Abstract model (3.17), when $v|_\Sigma = u_1$, $\Delta v|_\Sigma = u_2$. Let (analogously to (3.31), (3.32))

$$h = P_1 g \Longleftrightarrow \{\Delta^2 h = 0 \text{ in } \Omega;\ h|_\Gamma = g,\ \Delta h|_\Gamma = 0\}; \tag{A.1}$$

$$h = P_2 g \Longleftrightarrow \{\Delta^2 h = 0 \text{ in } \Omega;\ h|_\Gamma = 0,\ \Delta h|_\Gamma = g\}. \tag{A.2}$$

It is readily seen [L-T.4], [L-T.9] that, then

$$P_1 = D \text{ and } P_2 = -\mathcal{A}^{-1}D, \tag{A.3}$$

where D and $\mathcal{A}$ are defined in (3.20) and (2.1a), respectively. Next, recall D_α and F_α from (3.19) and (3.3). Then, using the definitions of all these operators, we may rewrite Eqn. (3.2) first as

$$v_{tt} - \gamma\Delta(v - Du_1)_{tt} + \Delta^2(v - P_1u_1 - P_2u_2) - \alpha \text{ div}(\alpha\nabla(v - D_\alpha u_1)_t) \equiv 0 \text{ in } Q, \tag{A.4}$$

and next, abstractly recalling (2.1a), as

$$v_{tt} + \gamma\mathcal{A}(v - Du_1)_{tt} + \mathcal{A}^2(v - P_1u_1 - P_2u_2) - F_\alpha(v - D_\alpha u_1)_t = 0 \text{ in } L_2(\Omega). \tag{A.5}$$

Finally, we recall (A.3) and extend the operators $\mathcal{A}$ and F_α into $L_2(\Omega) \to [\mathcal{D}(\mathcal{A})]'$ and $L_2(\Omega) \to [\mathcal{D}(F_\alpha^*)]'$ respectively, by isomorphism techniques, where $[\ \ ]'$ denotes duality with respect to $L_2(\Omega)$ as a pivot space. This way, (A.5) is rewritten as

$$v_{tt} + \gamma \mathcal{A} v_{tt} + \mathcal{A}^2 v - F_\alpha v_t = \mathcal{A}^2 D u_1 - \mathcal{A} D u_2 + \gamma \mathcal{A} D u_{1tt} - F_\alpha D_\alpha u_{1t} \text{ on } [\mathcal{D}(\mathcal{A}^2)]', \quad \text{(A.6)}$$

which is precisely (3.17), as desired.

Clamped B.C. Abstract model (3.29), when $v|_\Sigma = u_1$ and $\frac{\partial v}{\partial \nu}|_\Sigma = u_2$. In this case, we recall the operators D_α, D, G_1, G_2, F_α defined by (3.19), (3.20), (3.31), (3.32), (3.3), respectively. Then, using their definitions, we may rewrite Eqn. (3.2) first as

$$v_{tt} - \gamma\Delta(v - Du_1)_{tt} + \Delta^2(v - G_1u_1 - G_2u_2) - \alpha \text{ div}(\alpha\nabla(v - D_\alpha u_1)_t) \equiv 0 \text{ in } Q, \quad \text{(A.7)}$$

and next, abstractly, recalling also (2.4a), as

$$v_{tt} + \gamma\mathcal{A}(v - Du_1)_{tt} + A(v - G_1u_1 - G_2u_2) - F_\alpha(v - D_\alpha u_1)_t = 0. \quad \text{(A.8)}$$

Then, upon extending A and F_α on all of $L_2(\Omega)$, as in the previous case, we obtain

$$v_{tt} + \gamma\mathcal{A}v_{tt} + Av - F_\alpha v = AG_1u_1 + AG_2u_2 + \gamma\mathcal{A}Du_{1tt} - F_\alpha D_\alpha u_{1t} = 0 \text{ on } [\mathcal{D}(A)]', \quad \text{(A.9)}$$

which is precisely (3.29), as desired.

Appendix B: Abstract Models of the Non-Homogeneous Thermo-Elastic Problem (1.1), (1.2)

Here we derive the abstract model, in the style of [L-T.7], [L-T.9], [L-T.10] of the thermo-elastic non-homogeneous problem (1.1), under either hinged or else clamped B.C. as in (1.2).

Hinged B.C. Consider problem (1.1) under the hinged B.C.: $w|_\Sigma = u_1$; $\Delta w|_\Sigma = u_2$; $\theta|_\Sigma = u_3$, where u_1 is assumed in $H^2(0,T;L_2(\Gamma))$, $u_2, u_3 \in L_2(\Sigma)$. Then, the abstract model for problem (1.1) may be written as

$$\frac{d}{dt}\begin{bmatrix} w \\ w_t \\ \theta \end{bmatrix} = \mathbb{A}_\gamma \begin{bmatrix} w \\ w_t \\ \theta \end{bmatrix} + \mathcal{B}_h \begin{bmatrix} u_1 \\ u_2 \\ u_3 \end{bmatrix}, \quad \text{(B.1)}$$

where

$$\mathbb{A}_\gamma = \begin{bmatrix} 0 & I & 0 \\ -\mathcal{A}_\gamma^{-1}\mathcal{A}^2 & 0 & -\mathcal{A}_\gamma^{-1}\tilde{F}_\alpha \\ 0 & \tilde{F}_\alpha & -\mathcal{A} \end{bmatrix} \quad \text{(as in (2.6));} \quad \text{(B.2)}$$

$$\tilde{F}_\alpha f \equiv \operatorname{div}(\alpha \nabla f); \; \tilde{F}_\alpha^* = \tilde{F}_\alpha; \; \mathcal{D}(\tilde{F}_\alpha) = H^2(\Omega) \cap H_0^1(\Omega) = \mathcal{D}(F_\alpha), \text{ see (3.3).} \tag{B.3}$$

The self-adjointness of $\tilde{F}_\alpha$ follows from (2.11) and (B.3).

$$\mathcal{B}_h \begin{bmatrix} u_1 \\ u_2 \\ u_3 \end{bmatrix} = \begin{bmatrix} 0 \\ \mathcal{A}_\gamma^{-1}[\mathcal{A}^2 D u_1 - \mathcal{A} D u_2 + \tilde{F}_\alpha \tilde{D}_\alpha u_3 + \gamma \mathcal{A} D u_{1tt}] \\ \mathcal{A} D u_3 - \tilde{F}_\alpha \tilde{D}_\alpha u_{1t} \end{bmatrix}, \tag{B.4}$$

with $\mathcal{A}$, $\mathcal{A}_\gamma$, D defined in (2.1), (3.20), $\tilde{F}_\alpha$ defined by (B.3), and $\tilde{D}_\alpha$ defined accordingly as [compare with (3.19), (3.3), for D_α and F_α]:

$$h = \tilde{D}_\alpha g \Longleftrightarrow \{\tilde{F}_\alpha h = \operatorname{div}(\alpha \nabla h) = 0 \text{ in } \Omega; \; h|_\Gamma = g\}. \tag{B.5}$$

As in Lemma 3.3, Eqn. (3.24), one proves that

$$\tilde{D}_\alpha^* \tilde{F}_\alpha^* f = \tilde{D}_\alpha^* \tilde{F}_\alpha f = \frac{\partial(\alpha f)}{\partial \nu} = \alpha \frac{\partial f}{\partial \nu}, \quad \text{for } f \in \mathcal{D}(F_\alpha); \text{ hence } f|_\Gamma = 0. \tag{B.6}$$

To establish (B.1)–(B.4), we return to problem (1.1) with B.C. (1.2) (hinged) and rewrite it, by (3.20) for D and (A.1), (A.2) for P_1 and P_2, first as

$$\begin{cases} w_{tt} - \gamma \Delta(w - Du_1)_{tt} + \Delta^2(w - P_1 u_1 - P_2 u_2) + \operatorname{div}(\alpha \nabla(\theta - \tilde{D}_\alpha u_3)) = 0 & \text{(B.7)} \\ \theta_t - \Delta(\theta - Du_3) - \operatorname{div}(\alpha \nabla(w - \tilde{D}_\alpha u_1)_t) = 0, & \text{(B.8)} \end{cases}$$

and next, abstractly, recalling (2.1) and (B.3),

$$\begin{cases} w_{tt} + \gamma \mathcal{A}(w - Du_1)_{tt} + \mathcal{A}^2(w - P_1 u_1 - P_2 u_2) + \tilde{F}_\alpha(\theta - \tilde{D}_\alpha u_3) = 0 & \text{(B.9)} \\ \theta_t - \mathcal{A}(\theta - Du_3) - \tilde{F}_\alpha(w - \tilde{D}_\alpha u_1)_t = 0. & \text{(B.10)} \end{cases}$$

Recalling identities (A.3) and performing the usual extension by isomorphism of the operators $\mathcal{A}$, $\mathcal{A}^2$, $\tilde{F}_\alpha$ yields

$$\begin{cases} w_{tt} + \gamma \mathcal{A} w_{tt} + \mathcal{A}^2 w + \tilde{F}_\alpha \theta = \mathcal{A}^2 D u_1 + \gamma \mathcal{A} D u_{1tt} - \mathcal{A} D u_2 + \tilde{F}_\alpha \tilde{D}_\alpha u_3 & \text{(B.11)} \\ \theta_t + \mathcal{A}\theta - \tilde{F}_\alpha w_t = \mathcal{A} D u_3 - \tilde{F}_\alpha \tilde{D}_\alpha u_{1t}. & \text{(B.12)} \end{cases}$$

Then (B.11), (B.12) yields readily to (B.1), (B.2), (B.4).

Clamped B.C. We now consider problem (1.1) under the clamped B.C.: $w|_\Sigma = u_1$; $\left.\frac{\partial w}{\partial \nu}\right|_\Sigma = u_2$; $\theta|_\Sigma = u_3$, where $u_1 \in H^2(0,T; L_2(\Gamma))$, $u_2, u_3 \in L_2(\Sigma)$. Then, the abstract model for problem (1.1) may be written as

$$\frac{d}{dt} \begin{bmatrix} w \\ w_t \\ \theta \end{bmatrix} = \mathbb{A}_\gamma \begin{bmatrix} w \\ w_t \\ \theta \end{bmatrix} + \mathcal{B}_c \begin{bmatrix} u_1 \\ u_2 \\ u_3 \end{bmatrix}, \tag{B.13}$$

where this time

$$\mathbb{A}_\gamma = \begin{bmatrix} 0 & I & 0 \\ -\mathcal{A}_\gamma^{-1}A & 0 & -\mathcal{A}_\gamma^{-1}\tilde{F}_\alpha \\ 0 & \tilde{F}_\alpha & -\mathcal{A} \end{bmatrix} \quad \text{(as in (2.8a));} \tag{B.14}$$

A and $\tilde{F}_\alpha$ as in (2.4a), (B.3), and

$$\mathcal{B}_c \begin{bmatrix} u_1 \\ u_2 \\ u_3 \end{bmatrix} = \begin{bmatrix} 0 \\ \mathcal{A}_\gamma^{-1}[AG_1u_1 + AG_2u_2 + \gamma\mathcal{A}Du_{1tt} + \tilde{F}_\alpha\tilde{D}_\alpha u_3] \\ \mathcal{A}Du_3 - \tilde{F}_\alpha\tilde{D}_\alpha u_{1t} \end{bmatrix}, \tag{B.15}$$

where, now, G_1 and G_2 defined by (3.31), (3.32).

To establish (B.13)–(B.15), we proceed as in the hinged case: we first obtain

$$\begin{cases} w_{tt} - \gamma\Delta(w - Du_1)_{tt} + \Delta^2(w - G_1u_1 - G_2u_2) + \operatorname{div}(\alpha\nabla(\theta - \tilde{D}_\alpha u_3)) = 0 & \text{(B.16)} \\ \theta_t - \Delta(\theta - Du_3) - \operatorname{div}(\alpha\nabla(w - \tilde{D}_\alpha u_1)_t) = 0, & \text{(B.17)} \end{cases}$$

and then the abstract version,

$$\begin{cases} w_{tt} + \gamma\mathcal{A}(w - Du_1)_{tt} + A(w - G_1u_1 - G_2u_2) + \tilde{F}_\alpha(\theta - \tilde{D}_\alpha u_3) = 0 & \text{(B.18)} \\ \theta_t + \mathcal{A}(\theta - Du_3) - \tilde{F}_\alpha(w - \tilde{D}_\alpha u_1)_t = 0, & \text{(B.19)} \end{cases}$$

hence, after the usual extension

$$\begin{cases} w_{tt} + \gamma\mathcal{A} + Aw + \tilde{F}_\alpha\theta = AG_1u_1 + AG_2u_2 + \gamma\mathcal{A}Du_{1tt} + \tilde{F}_\alpha\tilde{D}_\alpha u_3 & \text{(B.20)} \\ \theta_t + \mathcal{A}\theta - \tilde{F}_\alpha w_t = \mathcal{A}Du_3 - \tilde{F}_\alpha\tilde{D}_\alpha u_{1t}, & \text{(B.21)} \end{cases}$$

from which (B.13)–(B.15) follows.

Appendix C: A PDE Formulation of the Observability Condition (4.2.1): $\mathcal{B}^* e^{\mathbb{A}_\gamma^*(T-t)}\, \bar{y}_0 \equiv 0 \Rightarrow \bar{y}_0 = 0$. Proof of (2.14): Regularity of the Thermo-elastic System with Clamped Control in $L_2(\Sigma)$

In this section we provide a PDE version, in terms of the traces of the $\{\phi, \eta\}$-problem (2.10), of the injectivity (observability) condition (4.2.1), in both hinged and clamped cases.

Hinged B.C.

LEMMA C.1. *Let $\bar{y}_0 = \{\phi_0, -\phi_1, \eta_0\} \in Y_{\gamma,h}$ (given by (2.3)), and let $\{\phi, \eta\}$ be the solution of the adjoint problem (2.10) with homogeneous hinged B.C. (2.10d), so that*

$$e^{\mathbb{A}_{\gamma,h}^* t}\bar{y}_0 = \{\phi(t;\bar{y}_0), -\phi_t(t;\bar{y}_0), \eta(t;\bar{y}_0)\}, \tag{C.1}$$

with generator $\mathbb{A}^*_{\gamma,h}$ defined by (2.7) as guaranteed by Lemma 2.1. Consider the operator $\mathcal{L}_T$ defined by (4.0.4), i.e., recalling (B.4) by

$$\mathcal{L}_T u = \mathcal{L}_T \begin{bmatrix} u_1 \\ u_2 \\ u_3 \end{bmatrix} \equiv \int_0^T e^{\mathbb{A}_{\gamma,h}(T-t)} \mathcal{B}_h \begin{bmatrix} u_1(t) \\ u_2(t) \\ u_3(t) \end{bmatrix} dt \tag{C.2a}$$

$$: H_0^2(0,T;L_2(\Gamma)) \times L_2(\Sigma) \times L_2(\Sigma) \to Y_\gamma = \mathcal{D}(\mathcal{A}) \times \mathcal{D}(\mathcal{A}_\gamma^{\frac{1}{2}}) \times L_2(\Omega). \tag{C.2b}$$

Thus, in particular, $u_{1tt} \in L_2(\Sigma)$; $u_1, u_{1t} \in C([0,T];L_2(\Gamma))$ and $u_1(T) = u_1(0) = u_{1t}(T) = u_{1t}(0) = 0$. Then,

(i) its adjoint $\mathcal{L}_T^*$, in the sense that

$$\left(\mathcal{L}_T \begin{bmatrix} u_1 \\ u_2 \\ u_3 \end{bmatrix}, \begin{bmatrix} \phi_0 \\ -\phi_1 \\ \eta_0 \end{bmatrix} \right)_{Y_\gamma} = \left(\begin{bmatrix} u_1 \\ u_2 \\ u_3 \end{bmatrix}, \mathcal{L}_T^* \begin{bmatrix} \phi_0 \\ -\phi_1 \\ \eta_0 \end{bmatrix} \right)_{[L_2(\Sigma)]^3} \tag{C.3}$$

is given by

$$\left(\mathcal{L}_T^* \begin{bmatrix} \phi_0 \\ \phi_1 \\ \eta_0 \end{bmatrix} \right)(t) = \mathcal{B}_h^* e^{\mathbb{A}^*_{\gamma,h}(T-t)} \begin{bmatrix} \phi_0 \\ -\phi_1 \\ \eta_0 \end{bmatrix} \tag{C.4}$$

$$= \begin{bmatrix} -\frac{\partial \Delta \phi_t}{\partial \nu}(T-t;\bar{y}_0) + -\alpha \frac{\partial \eta_t}{\partial \nu}(T-t;\bar{y}_0) + -\gamma \frac{\partial \phi_{ttt}}{\partial \nu}(T-t;\bar{y}_0) - \\ -\frac{\partial \phi_t}{\partial \nu}(T-t;\bar{y}_0) \\ -\alpha \frac{\partial \phi_t}{\partial \nu}(T-t;\bar{y}_0) - -\frac{\partial \eta}{\partial \nu}(T-t;\bar{y}_0) \end{bmatrix}. \tag{C.5}$$

[Regarding the second entry in (C.5)—which is due to u_2—refer also to (3.71), (3.72).]

(ii) Thus, the injectivity (observability) condition (4.2.1), specialized to the present case:

$$\begin{cases} \{\mathcal{L}_T^* \bar{y}_0\}(t) \equiv \mathcal{B}_h^* e^{\mathbb{A}^*_{\gamma,h}(T-t)} \bar{y}_0 \equiv 0, \quad 0 \le t \le T, \ \bar{y}_0 \in Y_{\gamma,h} \\ \Rightarrow \bar{y}_0 = 0, \end{cases} \tag{C.6}$$

is equivalently *rewritten, for $\{\phi, \eta\}$ solution of problem (2.10), as*

$$\begin{cases} \frac{\partial \Delta \phi_t}{\partial \nu}(T-t;\bar{y}_0) \equiv 0; \ \frac{\partial \phi_t}{\partial \nu}(T-t;\bar{y}_0) \equiv 0; \ \frac{\partial \eta}{\partial \nu}(T-t;\bar{y}_0) \equiv 0 \ \text{on } \Sigma = (0,T] \times \Gamma \\ \Rightarrow \bar{y}_0 = \{\phi_0, -\phi_1, \eta_0\} = 0, \end{cases} \tag{C.7}$$

where, of course, by definition, the solution $\{\phi, \eta\}$ of (2.10) satisfies the hinged/Dirichlet B.C. (2.10d):

$$\phi(T-t;\bar{y}_0)|_\Sigma \equiv 0; \quad \Delta\phi(T-t;\bar{y}_0)|_\Sigma \equiv 0; \quad \eta(T-t;\bar{y}_0)|_\Sigma \equiv 0, \ \text{on } \Sigma. \tag{C.8}$$

In (C.4), (C.6), $\mathcal{B}_h^*$ is in the sense of (C.9)–(C.12) below.

Proof. (i) Recalling $\mathcal{B}_h$ in (B.4) and (2.2), we compute with $\{u_1, u_2, u_3\}$ as in (C.2b):

$$\left(\mathcal{B}_h \begin{bmatrix} u_1 \\ u_2 \\ u_3 \end{bmatrix}, e^{\mathbb{A}^*_{\gamma,h}(T-\,\cdot\,)}\bar{y}_0\right)_{Y_\gamma} = \left(\mathcal{B}_h \begin{bmatrix} u_1 \\ u_2 \\ u_3 \end{bmatrix}, \begin{bmatrix} \phi(T-\,\cdot\,;\bar{y}_0) \\ -\phi_t(T-\,\cdot\,;\bar{y}_0) \\ \eta(T-\,\cdot\,;\bar{y}_0) \end{bmatrix}\right)_{Y_\gamma} \tag{C.9}$$

$$\begin{aligned}
\text{(by (B.4))} \quad &= (\mathcal{A}^2 Du_1 - \mathcal{A}Du_2 + \tilde{F}_\alpha \tilde{D}_\alpha u_3 + \gamma \mathcal{A} D u_{1tt}, -\phi_t(T-\,\cdot\,;\bar{y}_0))_{L_2(\Omega)} \\
&\quad + (\mathcal{A}Du_3 - \tilde{F}_\alpha \tilde{D}_\alpha u_{1t}, \eta(T-\,\cdot\,;\bar{y}_0))_{L_2(\Omega)} \qquad \text{(C.10)} \\
&= (u_1, -D^*\mathcal{A}^2\phi_t(T-\,\cdot\,;\bar{y}_0)_{L_2(\Sigma)} + (u_2, D^*\mathcal{A}\phi_t(T-\,\cdot\,;\bar{y}_0))_{L_2(\Sigma)} \\
&\quad - (u_3, \tilde{D}^*_\alpha \tilde{F}^*_\alpha \phi_t(T-\,\cdot\,;\bar{y}_0)_{L_2(\Sigma)} - (u_{1tt}, \gamma D^*\mathcal{A}\phi_t(T-\,\cdot\,;\bar{y}_0))_{L_2(\Sigma)} \\
&\quad + (u_3, D^*\mathcal{A}\eta(T-\,\cdot\,;\bar{y}_0))_{L_2(\Sigma)} - (u_{1t}, \tilde{D}^*_\alpha \tilde{F}^*_\alpha \eta(T-\,\cdot\,;\bar{y}_0))_{L_2(\Sigma)}. \qquad \text{(C.11)}
\end{aligned}$$

Recalling (3.24) for $D^*\mathcal{A}$, $\mathcal{A} = -\Delta$ and (B.6) for $\tilde{D}^*_\alpha \tilde{F}^*_\alpha$ on ϕ and η which both satisfy homogeneous Dirichlet B.C. as in (2.10d), we then obtain from (C.11),

$$\begin{aligned}
\left(\begin{bmatrix} u_1 \\ u_2 \\ u_3 \end{bmatrix}, \mathcal{B}^*_h e^{\mathbb{A}^*_{\gamma,h}(T-\,\cdot\,)}\bar{y}_0\right)_{[L_2(\Sigma)]^3} &= \left(u_1, -\frac{\partial \Delta \phi_t}{\partial \nu}(T-\,\cdot\,;\bar{y}_0)\right)_{L_2(\Sigma)} \\
&- \left(u_2, \frac{\partial \phi_t}{\partial \nu}(T-\,\cdot\,;\bar{y}_0)\right)_{L_2(\Sigma)} - \left(u_3, \alpha\frac{\partial \phi_t}{\partial \nu}(T-\,\cdot\,;\bar{y}_0)\right)_{L_2(\Sigma)} \\
&+ \left(u_1, \gamma\frac{\partial \phi_{ttt}}{\partial \nu}(T-\,\cdot\,;\bar{y}_0)\right)_{L_2(\Sigma)} + \left(u_3, -\frac{\partial \eta}{\partial \nu}(T-\,\cdot\,;\bar{y}_0)\right)_{L_2(\Sigma)} \\
&+ \left(u_1, \alpha\frac{\partial \eta_t}{\partial \nu}(T-\,\cdot\,;\bar{y}_0)\right)_{L_2(\Sigma)}.
\end{aligned} \tag{C.12}$$

In the terms in (C.11) containing u_{1tt} and u_{1t}, we have integrated by parts in t, using the assumed vanishing conditions $u_1(T) = u_1(0) = u_{1t}(T) = u_{1t}(0) = 0$, as desired.

(iii) Part (ii) follows at once from (C.5). □

Clamped B.C

LEMMA C.2. *Let $Z = [z_1, z_2, z_3] \in [\mathcal{D}(\mathbb{A}^*_{\gamma,c})]'$, and let*

$$\bar{y}_0 = \{\phi_0, -\phi_1, \eta_0\} = \mathbb{A}^{*-1}_{\gamma,c}\mathbb{A}^{-1}_{\gamma,c}[z_1, z_2, z_3] \in \mathcal{D}(\mathbb{A}_{\gamma,c}) \subset Y_{\gamma,c}, \tag{C.13a}$$

see (2.8b), be the initial condition of the solution $\{\phi, \eta\}$ of the adjoint problem (2.10) with homogeneous clamped B.C. (2.10d), so that

$$e^{\mathbb{A}^*_{\gamma,c}t}\bar{y}_0 = \{\phi(t;\bar{y}_0), -\phi_t(t;\bar{y}_0); \eta(t;\bar{y}_0)\}, \tag{C.13b}$$

with generator $\mathbb{A}^*_{\gamma,c}$ defined by (2.9). Consider the operator $\mathcal{L}_T$ defined by (4.0.4), i.e., by

$$\mathcal{L}_T u = \mathcal{L}_T \begin{bmatrix} u_1 \\ u_2 \\ u_3 \end{bmatrix} = \int_0^T e^{\mathbb{A}_{\gamma,c}(T-t)}\mathcal{B}_c \begin{bmatrix} u_1(t) \\ u_2(t) \\ u_3(t) \end{bmatrix} dt \tag{C.14}$$

$$: H^2_0(0,T; L_2(\Gamma)) \times L_2(\Sigma) \times L_2(\Sigma) \to X_\gamma \subset [\mathcal{D}(\mathbb{A}^*_{\gamma,c})]', \tag{C.15}$$

with the space X_γ defined in (4.0.2b) and the dual in (C.15) defined in (4.0.2c). Thus, in particular, $u_{1tt} \in L_2(\Sigma)$; $u_1, u_{1t} \in C([0,T]; L_2(\Gamma))$ and $u_1(T) = u_1(0) = u_{1t}(T) = u_{1t}(0) = 0$. Then,

(i) its $[\mathcal{D}(\mathbb{A}^*_{\gamma,c})]'$-adjoint $\mathcal{L}_T^\#$, in the sense that

$$(\mathcal{L}_T u, z)_{[\mathcal{D}(\mathbb{A}^*_{\gamma,c})]'} = (u, \mathcal{L}_T^\# z)_{\mathcal{U}}, \ u \in \mathcal{U}, \ z \in [\mathcal{D}(\mathbb{A}^*_{\gamma,c})]'$$

is given by

$$\left(\mathcal{L}_T^\# \begin{bmatrix} z_1 \\ z_2 \\ z_3 \end{bmatrix} \right)(t) = \mathcal{B}_c^* e^{\mathbb{A}^*_{\gamma,c}(T-t)} \bar{y}_0, \ \bar{y}_0 \text{ as in (C.13a)} \tag{C.16}$$

$$= \begin{bmatrix} -\dfrac{\partial \Delta \phi_t}{\partial \nu}(T-t;\bar{y}_0) + \alpha \dfrac{\partial \eta_t}{\partial \nu}(T-t;\bar{y}_0) \\ \Delta\phi_t(T-t;\bar{y}_0) \\ -\dfrac{\partial \eta}{\partial \nu}(T-t;\bar{y}_0) \end{bmatrix}. \tag{C.17}$$

(ii) Thus, the injectivity (observability) condition (4.2.1), specialized to the present case:

$$\begin{cases} \{\mathcal{L}_T^\# \bar{y}_0\}(t) \equiv \mathcal{B}_c^* e^{\mathbb{A}^*_{\gamma,c}(T-t)} \bar{y}_0 \equiv 0, \quad 0 \le t \le T, \ \bar{y}_0 \in \mathcal{D}(\mathbb{A}_{\gamma,c}) \\ \Rightarrow \bar{y}_0 = 0, \end{cases} \tag{C.18}$$

is equivalently rewritten, for $\{\phi, \eta\}$ solution in (C.13) of problem (2.10), as

$$\begin{cases} \dfrac{\partial \Delta \phi_t}{\partial \nu}(T-t;\bar{y}_0) \equiv 0; \ \Delta\phi_t(T-t;\bar{y}_0) \equiv 0, \ \dfrac{\partial \eta}{\partial \nu}(T-t;\bar{y}_0) \equiv 0, \quad \text{on } \Sigma = (0,T] \times \Gamma \\ \Rightarrow \bar{y}_0 = \{\phi_0, -\phi_1, \eta_0\} = 0, \end{cases} \tag{C.19}$$

where, of course, by definition, the solution $\{\phi, \eta\}$ of (2.10) in (C.13) satisfies the clamped/-Dirichlet B.C. (2.10d):

$$\phi(T-t;\bar{y}_0)|_\Sigma \equiv 0, \quad \frac{\partial \phi}{\partial \nu}(T-t;\bar{y}_0)|_\Sigma \equiv 0, \quad \eta(T-t;\bar{y}_0)|_\Sigma \equiv 0 \text{ on } \Sigma. \tag{C.20}$$

In (C.16), (C.18), $\mathcal{B}_c^*$ is in the sense of (C.24)–(C.29) below.

Proof. (i) Recalling $\mathcal{B}_c$ in (B.15) and (2.2), we compute with $\{u_1, u_2, u_3\}$ as in (C.15), via (C.14),

$$(\mathcal{L}_T u, z)_{[\mathcal{D}(\mathbb{A}^*_{\gamma,c})]'} = \left(\int_0^T e^{\mathbb{A}_{\gamma,c}(T-t)} \mathcal{B}_c u(t) dt, z \right)_{[\mathcal{D}(\mathbb{A}^*_{\gamma,c})]'} \tag{C.21}$$

$$= \left(\int_0^T e^{\mathbb{A}_{\gamma,c}(T-t)} \mathcal{B}_c u(t) dt, \mathbb{A}^{*-1}_{\gamma,c} \mathbb{A}^{-1}_{\gamma,c} z \right)_{Y_{\gamma,c}} \tag{C.22}$$

$$\text{(by (C.13a))} = \int_0^T \left(\mathcal{B}_c u(t), e^{\mathbb{A}^*_{\gamma,c}(T-t)} \bar{y}_0 \right)_{Y_{\gamma,c}} dt. \tag{C.23}$$

Next, by (B.15) on $\mathcal{B}_c$ and (2.2), we compute via (C.13b),

$$\left(\mathcal{B}_c \begin{bmatrix} u_1 \\ u_2 \\ u_3 \end{bmatrix}, e^{\mathbf{A}^*_{\gamma,c}(T-\,\cdot\,)}\bar{y}_0\right)_{Y_{\gamma,c}} = \left(\mathcal{B}_c \begin{bmatrix} u_1 \\ u_2 \\ u_3 \end{bmatrix}, \begin{bmatrix} \phi(T-\,\cdot\,;\bar{y}_0) \\ -\phi_t(T-\,\cdot\,;\bar{y}_0) \\ \eta(T-\,\cdot\,;\bar{y}_0) \end{bmatrix}\right)_{Y_{\gamma,c}} \tag{C.24}$$

$$= (AG_1u_1 + AG_2u_2 + \tilde{F}_\alpha\tilde{D}_\alpha u_3 + \gamma\mathcal{A}Du_{1tt}, -\phi_t(T-\,\cdot\,;\bar{y}_0))_{L_2(\Omega)}$$

$$+ (\mathcal{A}Du_3 - \tilde{F}_\alpha\tilde{D}_\alpha u_{1t}, \eta(T-\,\cdot\,;\bar{y}_0))_{L_2(\Omega)} \tag{C.25}$$

$$= -(u_1, G_1^*A\phi_t(T-\,\cdot\,;\bar{y}_0))_{L_2(\Sigma)} - (u_2, G_2^*A\phi_t(T-\,\cdot\,;\bar{y}_0))_{L_2(\Sigma)}$$

$$- (u_3, \tilde{D}_\alpha^*\tilde{F}_\alpha^*\phi_t(T-\,\cdot\,;\bar{y}_0))_{L_2(\Sigma)} - (u_{1tt}, \gamma D^*\mathcal{A}\phi_t(T-\,\cdot\,;\bar{y}_0))_{L_2(\Sigma)}$$

$$+ (u_3, D^*\mathcal{A}\eta(T-\,\cdot\,;\bar{y}_0))_{L_2(\Sigma)} - (u_{1t}, \tilde{D}_\alpha^*\tilde{F}_\alpha^*\eta(T-\,\cdot\,;\bar{y}_0))_{L_2(\Sigma)}. \tag{C.26}$$

By (C.21)-(C.26), we obtain

$$(\mathcal{L}_Tu, z)_{[\mathcal{D}(\mathbf{A}^*_{\gamma,c})]'} = (u, \mathcal{L}_T^{\#}z)_{[L_2(\Sigma)]^3}$$

$$= \int_0^T \left(u(t), \mathcal{B}_c^* e^{\mathbf{A}^*_{\gamma,c}(T-t)}\bar{y}_0\right)_{[L_2(\Gamma)]^3} dt \tag{C.27}$$

$$= \left(\begin{bmatrix} u_1 \\ u_2 \\ u_3 \end{bmatrix}, \mathcal{B}_c^* e^{\mathbf{A}^*_{\gamma,c}(T-\,\cdot\,)}\bar{y}_0\right)_{[L_2(\Sigma)]^3}$$

$$\text{(by (C.23))} = \left(u_1, -\frac{\partial\Delta\phi_t}{\partial\nu}(T-\,\cdot\,;\bar{y}_0)\right)_{L_2(\Sigma)} + (u_2, \Delta\phi_t(T-\,\cdot\,;\bar{y}_0))_{L_2(\Sigma)}$$

$$+ \left(u_3, \alpha\frac{\partial\phi_t}{\partial\nu}(T-\,\cdot\,;\bar{y}_0)\right)_{L_2(\Sigma)} + \left(u_{1tt}, -\gamma\frac{\partial\phi_t}{\partial\nu}(T-\,\cdot\,;\bar{y}_0)\right)_{L_2(\Sigma)}$$

$$+ \left(u_3, -\frac{\partial\eta(T-\,\cdot\,;\bar{y}_0)}{\partial\nu}\right)_{L_2(\Sigma)} - \left(u_{1t}, \alpha\frac{\partial\eta}{\partial\nu}(T-\,\cdot\,;\bar{y}_0)\right)_{L_2(\Sigma)} \tag{C.28}$$

$$= \left(u_1, -\frac{\partial\Delta\phi_t}{\partial\nu}(T-\,\cdot\,;\bar{y}_0)\right)_{L_2(\Sigma)} + (u_2, \Delta\phi_t(T-\,\cdot\,;\bar{y}_0))_{L_2(\Sigma)}$$

$$+ \left(u_3, -\frac{\partial\eta}{\partial\nu}(T-\,\cdot\,;\bar{y}_0)\right)_{L_2(\Sigma)} + \left(u_1, \alpha\frac{\partial\eta_t}{\partial\nu}(T-\,\cdot\,;\bar{y}_0)\right)_{L_2(\Sigma)} \tag{C.29}$$

To go from (C.26) to (C.28) we have recalled the traces relations in (3.37) for G_1^*A and G_2^*A, and in (B.6) for $\tilde{D}_\alpha^*\tilde{F}_\alpha^*$ on ϕ and η which satisfy the B.C. (2.10d). These yield two cancellations as in (C.28). Finally, (C.29) yields (C.17), as desired. □

COROLLARY C.3. *(a) In the* hinged *case, take, as in (1.7) and the last statement in Remark 1.1:*

$$u_1 \equiv 0 \text{ on } \Sigma - \Sigma_1; \; u_2 \equiv 0 \text{ on } \Sigma - \Sigma_2; \; u_3 \equiv 0 \text{ on } \Sigma - \Sigma_3; \; \emptyset \neq \tilde{\Gamma} \equiv \Gamma_1 = \Gamma_3 \subset \Gamma_2, \tag{C.30}$$

and otherwise let $\{u_1, u_2, u_3\} \in H_0^2(0,T; L_2(\Gamma_1)) \times L_2(\Sigma_2) \times L_2(\Sigma_3)$.

Then, condition (C.7) specializes, with $\bar{y}_0 \in Y_{\gamma,h}$, *to*

$$\begin{cases} \dfrac{\partial \Delta \phi_t}{\partial \nu}(T-t;\bar{y}_0)\Big|_{\Sigma_1} \equiv 0 \text{ on } \Sigma_1; \ \dfrac{\partial \Delta \phi_t}{\partial \nu}(T-t;\bar{y}_0)\Big|_{\Sigma_2} \equiv 0 \text{ on } \Sigma_2; \\ \dfrac{\partial \eta}{\partial \nu}(T-t;\bar{y}_0)\Big|_{\Sigma_3} \equiv 0 \text{ on } \Sigma_3 = \Sigma_1, \ \bar{y}_0 \in Y_{\gamma,h} \\ \Rightarrow \bar{y}_0 = \{\phi_0, -\phi_1, \eta_0\} = 0, \end{cases} \tag{C.31}$$

where the pair $\{\phi, \eta\} \in C([0,T]; Y_{\gamma,h})$, *see (2.3) and Lemma 2.1, satisfies the hinged B.C. on* all *of* Σ, *as in (2.10d). This way, the equivalent injectivity (observability) condition (4.2.1) becomes (4.2.3).*

(b) In the clamped *case, take* u_1 *and* u_3 *and* $\Gamma_1 = \Gamma_3$ *as in part (a) while now* $\Gamma_2 = \Gamma$. *Then, condition (C.19) specializes, with* $\bar{y}_0 \in \mathcal{D}(\mathbb{A}_{\gamma,c})$, *to*

$$\begin{cases} \dfrac{\partial \Delta \phi_t}{\partial \nu}(T-t;\bar{y}_0)\Big|_{\Sigma_1} \equiv 0 \text{ on } \Sigma_1; \ \Delta\phi_t|_{\Sigma} \equiv 0 \text{ on } \Sigma; \\ \frac{\partial \eta}{\partial \nu}\Big|_{\Sigma_3} \equiv 0 \text{ on } \Sigma_3 = \Sigma_1 \\ \Rightarrow \bar{y}_0 = \{\phi_0, -\phi_1, \eta_0\} = 0, \end{cases} \tag{C.32}$$

where now the pair $\{\phi, \eta\} \in C([0,T]; \mathcal{D}(\mathbb{A}_{\gamma,c}))$, *see (2.8b) and Lemma 2.1, satisfies the clamped B.C. on* all *of* Σ, *as in (2.10d). This way, the equivalent injectivity (observability) condition (4.2.1) becomes (4.2.4).*

REMARK C.1. (Clamped B.C.) *The following remark further enlightens Lemma C.2 and, moreover, it provides information to be used in the subsequent regularity result, Theorem C.4 below. We return to (C.14) with* $z \in [\mathcal{D}(\mathbb{A}_{\gamma,c}^*)]'$ *(duality with respect to* $Y_{\gamma,c}$ *in (2.5), as a pivot space), which we now specialize to* $u_1 \equiv 0$, $u_3 \equiv 0$; $u_2 \in L_2(0,T; L_2(\Gamma)) \equiv L_2(\Sigma)$. *We obtain that:*

$$\mathcal{L}_T \begin{bmatrix} 0 \\ u_2 \\ 0 \end{bmatrix} : L_2(\Sigma) \to [\mathcal{D}(\mathbb{A}_{\gamma,c}^*)]' \equiv H_0^1(\Omega) \times \tilde{L}_2(\Omega) \times [\mathcal{D}(\mathcal{A})]' \tag{C.33}$$

[see (4.0.2c)] if and only if

$$\mathcal{L}_T^{\#} : [\mathcal{D}(\mathbb{A}_{\gamma,c}^*)]' \to L_2(\Sigma), \tag{C.34}$$

where now (C.16) is accordingly specialized to

$$(\mathcal{L}_T^{\#} z)(t) = \begin{bmatrix} 0 \\ \Delta\phi_t(T-t;\bar{y}_0) \\ 0 \end{bmatrix}, \ \bar{y}_0 = \{\phi_0, -\phi_1, \eta_0\} = \mathbb{A}_{\gamma,c}^{*-1} \mathbb{A}_{\gamma,c}^{-1} z \in \mathcal{D}(\mathbb{A}_{\gamma,c}), \tag{C.35}$$

see (C.13a). In (C.35) we have that $\{\phi, \eta\}$ is a solution of the adjoint thermoelastic problem (2.10) with homogeneous clamped B.C. We can equivalently rewrite (C.35) as

$$(\mathcal{L}_T^{\#} z)(t) = \begin{bmatrix} 0 \\ \Delta\tilde{\phi}(T-t;\tilde{y}_0) \\ 0 \end{bmatrix}, \quad \tilde{y}_0 = \{\tilde{\phi}_0, -\tilde{\phi}_1, \tilde{\eta}_0\} = \mathbb{A}_{\gamma,c}^{-1} z \in Y_{\gamma,c} \equiv H_0^2(\Omega) \times H_0^1(\Omega) \times L_2(\Omega) \tag{C.36}$$

where $\{\tilde{\phi}(t) \equiv \phi_t(t;\tilde{y}_0), \tilde{\eta}(t) \equiv \eta_t(t;\tilde{y}_0)\}$ solves again the adjoint thermoelastic problem (2.10) with homogeneous clamped B.C.:

$$\begin{cases} \tilde{\phi}_{tt} - \gamma\Delta\tilde{\phi}_{tt} + \Delta^2\tilde{\phi} - div(\alpha(x)\nabla\tilde{\eta}) \equiv 0 & \text{in } Q; \quad (C.37a) \\ \tilde{\eta}_t - \Delta\tilde{\eta} + div(\alpha(x)\nabla\tilde{\phi}_t) \equiv 0 & \text{in } Q; \quad (C.37b) \\ \tilde{\phi}(0,\,\cdot\,) = \tilde{\phi}_0;\ \tilde{\phi}_t(0,\,\cdot\,) = \tilde{\phi}_1;\ \tilde{\eta}(0,\,\cdot\,) = \tilde{\eta}_0 & \text{in } \Omega; \quad (C.37c) \\ \tilde{\phi}|_\Sigma \equiv 0,\ \left.\frac{\partial\tilde{\phi}}{\partial\nu}\right|_\Sigma \equiv 0 & \text{in } \Sigma, \quad (C.37d) \end{cases}$$

but smoother initial conditions as in (C.36). Indeed, via Lemma 2.1(ii):

$$\begin{aligned} [\tilde{\phi}(t), -\tilde{\phi}_t(t), \tilde{\eta}(t)] &= \frac{d}{dt}[\phi(t;\tilde{y}_0), -\phi_t(t;\tilde{y}_0), \eta(t;\tilde{y}_0)] \\ &= \frac{d}{dt} e^{\mathbb{A}_{\gamma,c}^* t}\tilde{y}_0 = e^{\mathbb{A}_{\gamma,c}^* t}\mathbb{A}_{\gamma,c}^*\tilde{y}_0 = e^{\mathbb{A}_{\gamma,c}^* t}\mathbb{A}_{\gamma,c}^{-1} z, \end{aligned} \tag{C.38}$$

and (C.36), (C.37) are established. Of course, the form (C.36) could have emerged directly in the proof of Lemma C.2, by slightly modifying the computations: instead of (C.22)–(C.24), we could have written as follows:

$$\left(\mathcal{L}_T \begin{bmatrix} 0 \\ u_2 \\ 0 \end{bmatrix}, z\right)_{[\mathcal{D}(\mathbb{A}_{\gamma,c}^*)]'} = \left(\int_0^T e^{\mathbb{A}_{\gamma,c}(T-t)}\mathbb{A}_{\gamma,c}^{-1}\mathcal{B}_c u(t)dt, \mathbb{A}_{\gamma,c}^{-1} z\right)_{Y_{\gamma,c}} \tag{C.39}$$

$$= \int_0^T \left(\mathcal{B}_c \begin{bmatrix} 0 \\ u_2(t) \\ 0 \end{bmatrix}, \mathbb{A}_{\gamma,c}^{*-1} e^{\mathbb{A}_{\gamma,c}^*(T-t)}\mathbb{A}_{\gamma,c}^{-1} z\right)_{Y_{\gamma,c}} dt \tag{C.40}$$

$$(\text{by (C.38)}) \quad = \int_0^T \left(\begin{bmatrix} 0 \\ \mathcal{A}_\gamma^{-1} A G_2 u_2(t) \\ 0 \end{bmatrix}, \mathbb{A}_{\gamma,c}^{*-1} \begin{bmatrix} \tilde{\phi}(T-t;\tilde{y}_0) \\ -\tilde{\phi}_t(T-t;\tilde{y}_0) \\ \tilde{\eta}(T-t;\tilde{y}_0) \end{bmatrix}\right)_{Y_{\gamma,c}} dt \tag{C.41}$$

$$(\text{by (2.2)}) \quad = \int_0^T (AG_2 u_2(t), -\tilde{\phi}(T-t;\tilde{y}_0))_{L_2(\Omega)} dt \tag{C.42}$$

$$(\text{by (3.37)}) \quad = \int_0^T (u_2(t), \Delta\tilde{\phi}(T-t;\tilde{y}_0))_{L_2(\Gamma)} dt, \tag{C.43}$$

and (C.36) follows again from (C.43), where $\tilde{y}_0$ is defined by (C.36). In the next result, we shall see that the equivalent relationships (C.33), (C.34), $\Delta\phi_t(T - \cdot\,; \tilde{y}_0) \in L_2(\Sigma)$, and $\Delta\tilde{\phi}(T - \cdot\,; \tilde{y}_0) \in L_2(\Sigma)$ are all true.

THEOREM C.4. (Clamped B.C.) *With reference to the mixed thermoelastic problem (1.1a–c) with data*

$$w_0 = 0,\ w_1 = 0,\ \theta_0 = 0;\ u_1 \equiv 0,\ u_3 \equiv 0,\ u_2 \in L_2(0,T; L_2(\Gamma)), \tag{C.44}$$

the following regularity holds true, continuously in u_2:

$$\{w, w_t, \theta\} \in C([0,T]; H_0^1(\Omega) \times \tilde{L}_2(\Omega) \times H_\theta). \tag{C.45}$$

Proof. We shall prove here the result with $H^{-\epsilon}(\Omega)$ instead of $H_\theta = L_2(\Omega)$, after [Tr.2, Theorem 4.1]. For $\epsilon = 0$ see [L-T.13]. The regularity (C.45) for the thermoelastic problem (1.1a–c) will be deduced from the (sharp) regularity of the corresponding Kirchoff problem. We shall first prove that (C.44) implies

$$\{w, w_t, \theta\} \in C([0,T]; [\mathcal{D}(\mathbb{A}^*_{\gamma,c})]' \equiv H_0^1(\Omega) \times \tilde{L}_2(\Omega) \times [\mathcal{D}(\mathcal{A})]'), \tag{C.46}$$

continuously in u_2, where the first duality is with respect to $Y_{\gamma,c}$ as a pivot space, while the second duality is with respect to $L_2(\Omega)$ as a pivot space. Remark C.1 above will be invoked to establish (C.46).

Proof of (C.46). Step 1. (Duality) We consider the dual $\{\tilde{\phi}, \tilde{\eta}\}$-thermoelastic problem in (C.37) with initial conditions $\{\tilde{\phi}_0, -\tilde{\phi}_1, \tilde{\eta}_0\} \in Y_{\gamma,c} \equiv H_0^2(\Omega) \times H_0^1(\Omega) \times L_2(\Omega)$, as in (C.36), and hence *a-priori* regularity (from Lemma 2.1(ii) for (C.47a) below, and Lemma 9.2 for (C.47b) below):

$$\begin{cases} \{\tilde{\phi}, \tilde{\phi}_t, \tilde{\eta}\} \in C([0,T]; H_0^2(\Omega) \times H_0^1(\Omega) \times L_2(\Omega)); & \text{(C.47a)} \\ \tilde{\eta} \in L_2(0,T; H_0^1(\Omega)); \tilde{\eta}_t \in L_2(0,T; H^{-1}(\Omega)). & \text{(C.47b)} \end{cases}$$

In the style of (3.1), we rewrite problem $\{\tilde{\phi}, \tilde{\eta}\}$ in (C.37) using $\operatorname{div}(\alpha\nabla\tilde{\eta}) = \alpha\Delta\tilde{\eta} + \nabla\alpha \cdot \nabla\tilde{\eta}$ as the following Kirchoff problem:

$$\begin{cases} \tilde{\phi}_{tt} - \gamma\Delta\tilde{\phi}_{tt} + \Delta^2\tilde{\phi} + \alpha \operatorname{div}(\alpha\nabla\tilde{\phi}_t) \equiv F \text{ in } Q; & \text{(C.48a)} \\ F \equiv \alpha\tilde{\eta}_t + \nabla\alpha \cdot \nabla\tilde{\eta} \in L_2(0,T; H^{-1}(\Omega)); & \text{(C.48b)} \\ \tilde{\phi}|_\Sigma \equiv 0,\ \left.\frac{\partial\tilde{\phi}}{\partial\nu}\right|_\Sigma \equiv 0. & \text{(C.48c)} \end{cases}$$

The regularity in (C.48b) follows from (C.47). For problem (C.48), the same (energy method) proof in [L-L.1, Chapter 5, where $\alpha \equiv 0$ and $F \equiv 0$] by use of the multiplier $h \cdot \nabla\tilde{\phi}$ yields the following (sharp) trace regularity:

$$\{\tilde{\phi}_0, \tilde{\phi}_1\} \in H_0^2(\Omega) \times H_0^1(\Omega) \Rightarrow \Delta\tilde{\phi}|_\Gamma \in L_2(0,T, L_2(\Gamma)), \tag{C.49}$$

where of course $\tilde{\eta}_0 \in L_2(\Omega)$ as well. [This is the same as the implication (3.97) $\rightarrow$ (3.99).]

Step 2. The regularity (C.49) establishes, via (C.36) that $\mathcal{L}_T^{\#} : [\mathcal{D}(\mathbb{A}^*_{\gamma,c})]' \rightarrow L_2(\Sigma)$ as in (C.34) and hence establishes (C.33). Eqn. (C.46) is the explicit version of (C.46), via [L-T.9]. Thus, (C.46) is established.

Step 3. We now improve upon the regularity of θ in (C.46). We shall here only show that

$$\theta \in C([0,T]; H^{-\epsilon}(\Omega)),\ \epsilon > 0 \text{ arbitrary,} \tag{C.50}$$

while for $\epsilon = 0$ see [L-T.13]. Eqn. (1.1b) for θ is rewritten as

$$\theta_t - \Delta\theta - \alpha\Delta w_t - \nabla\alpha \cdot \nabla w_t = 0;\ \theta_t = -\mathcal{A}\theta - \alpha\mathcal{A}w_t + \ell.o.t.$$

This can be seen by using the regularity of w_t and the properties of the space $\tilde{L}_2(\Omega)$ in

$$\theta(t) = \int_0^t e^{-\mathcal{A}(t-\tau)}\mathcal{A}w_t(\tau)\alpha\, d\tau + \ell.o.t. \tag{C.51}$$

Details are given in [Tr.2, Theorem 4.1]. □

Appendix D: Simultaneous Exact Controllability on Y_1 and Approximate Controllability on Y_2, Where $Y = Y_1 \times Y_2$

In this appendix we provide the abstract result invoked in Section 4.

Premise. Let $Y = Y_1 \times Y_2$ be a Hilbert space with two components. In the thermo-elastic case of the present paper, Y_1 = space of mechanical components $\{w, w_t\}$, while Y_2 = space of thermal component. Consider the dynamics

$$\begin{cases} y(t) = e^{At}\xi + (\mathcal{L}u)(t), \quad \xi \in Y; & \text{(D.1a)} \\ (\mathcal{L}u)(t) = \int_0^t e^{A(t-\tau)}Bu(\tau)d\tau, & \text{(D.1b)} \end{cases}$$

where e^{At} is a s.c. semigroup on Y. We shall consider the dynamics at time $t = T$:

$$\begin{cases} y(T, \xi; u) = e^{AT}\xi + \mathcal{L}_T u; & \text{(D.2a)} \\ \mathcal{L}_T u = \int_0^T e^{A(T-t)}Bu(\tau)d\tau : L_2(0,T;U) \supset \mathcal{D}(\mathcal{L}_T) \rightarrow Y. & \text{(D.2b)} \end{cases}$$

$\mathcal{L}_T$ need *not* be continuous, but it is closed. In the thermo-elastic case of the present paper, $\mathcal{L}_T$ is actually *continuous* (bounded) with the choice of Y specified in (D.23) (hinged case) and (D.27) (clamped case): see (4.0.4b). However, in the abstract result given below, we shall only assume a much weaker hypothesis, (H.1) below. To state it, we shall let $\Pi_i[y_1, y_2] = y_i$ the projection Y onto Y_i. Thus, $\Pi_i^* Y_i \rightarrow Y$ is: $\Pi_1^* y_1 = [y_1, 0]$, $\Pi_2^* y_2 = [0, y_2]$.

Simultaneous exact controllability on Y_1 and approximate controllability on Y_2 . Let $\xi \in Y$, $0 < T < \infty$, and assume hypotheses (H.1) through (H.3) below.

(H.1) The operator $\Pi_2\Lambda$, $\Lambda = \mathcal{L}_T\mathcal{L}_T^*\Pi_1(\Pi_1\mathcal{L}_T\mathcal{L}_T^*\Pi_1^*)^{-1}$, see (D.10b) below, is bounded $Y \to Y_2$.

(H.2) The dynamics (D.1) is approximately controllable in Y at time T from the initial point ξ within the class of $L_2(0,T;U)$-controls: i.e., the set of attainability at time T in Y is dense in Y:

$$\text{the set } \{e^{AT}\xi + \mathcal{L}_T u, \text{ as } u \text{ runs over } \mathcal{D}(\mathcal{L}_T)\} \text{ is dense in } Y. \tag{D.3a}$$

Thus, given any target $z = [z_1, z_2] \in Y$, and given $\epsilon > 0$ there exists a $u_1 \in \mathcal{D}(\mathcal{L}_T)$ such that

$$\|e^{AT}\xi + \mathcal{L}_T u_1 - z\|_Y < \epsilon. \tag{D.3b}$$

(H.3) The dynamics (D.1) is exactly controllable on the first component space Y_1; i.e., given any $z_1 \in Y_1$, there exists a $u_2 \in \mathcal{D}(\mathcal{L}_T)$ such that

$$\Pi_1[e^{AT}\xi + \mathcal{L}_T u] = z_1. \tag{D.4a}$$

Equivalently, the set of attainability projected onto Y_1 is all of Y_1:

$$\{\Pi_1[e^{AT}\xi + \mathcal{L}_T u], \text{ as } u \text{ runs over } \mathcal{D}(\mathcal{L}_T)\} = Y_1. \tag{D.4b}$$

THEOREM D.1. *Assume (H.1) through (H.3). Then, in fact, system (D.1) is simultaneously exactly controllable on Y_1 and approximately controllable on Y_2, at time T from the initial point ξ, within the class of $L_2(0,T;U)$-controls. This means that: given $\epsilon > 0$ and any target point $z = [z_1, z_2] \in Y$, there exists a control $u \in \mathcal{D}(\mathcal{L}_T)$ such that the corresponding solution $y(T,\xi;u)$ satisfies simultaneously the following two properties:*

$$\Pi_1 y(T,\xi;u) = z_1 \quad \textit{and} \quad \|\Pi_2 y(T,\xi;u) - z_2\|_{Y_2} < \epsilon. \tag{D.5}$$

Proof. **Step 1.** Given $z = [z_1, z_2] \in Y$ and $\epsilon > 0$, by virtue of (H.2), we can pick $u_1 \in \mathcal{D}(\mathcal{L}_T)$ such that the corresponding response

$$y^1(T) \equiv y(T,\xi;u_1) = e^{AT}\xi + \mathcal{L}_T u_1 \tag{D.6}$$

satisfies (D.3b), i.e.,

$$\|y^1(T) - z\|_Y < \frac{\epsilon}{2}. \tag{D.7}$$

Step 2. Let $x = [x_1, x_2] \in Y$ be, as yet, a point to be identified (in (D.17) below regarding the critical first component x_1). By virtue of (H.3) we can hit x_1; i.e., there exists a control $u_2 \in \mathcal{D}(\mathcal{L}_T)$ such that

$$\Pi_1[e^{AT}\xi + \mathcal{L}_T u_2] = x_1 \qquad \text{so that } \Pi_1\mathcal{L}_T u_2 = x_1 - \Pi_1 e^{AT}\xi. \tag{D.8}$$

Indeed, we can take u_2 to be the minimal norm control, so that according to [L-T.1, Appendix B] we have

$$u_2 = \mathcal{L}_T^* \Pi_1 (\Pi_1 \mathcal{L}_T \mathcal{L}_T^* \Pi_1^*)^{-1} \Pi_1 [x - e^{AT}\xi]; \tag{D.9}$$

$$\mathcal{L}_T u_2 = \Lambda \Pi_1 [x - e^{AT}\xi] = \Lambda [x_1 - \Pi_1 e^{AT}\xi]; \tag{D.10a}$$

$$\Lambda = \mathcal{L}_T \mathcal{L}_T^* \Pi_1 (\Pi_1 \mathcal{L}_T \mathcal{L}_T^* \Pi_1^*)^{-1}. \tag{D.10b}$$

The operator $\Pi_1 \Lambda$ is the identity on Y_1. Thus, the operator Λ is bounded, provided that $\Pi_2 \Lambda$ is bounded, as assumed. Since the operator Λ in (D.10) is bounded by (H.1), then, as $\Pi_1 x = x_1$, we have

$$\|\mathcal{L}_T u_2\|_Y \leq \|\Lambda\| \, \|x_1 - \Pi_1 e^{AT}\xi\|_{Y_1}. \tag{D.11}$$

One requirement that we impose on the sought-after point x is that,

$$\|x_1 - \Pi_1 e^{AT}\xi\|_{Y_1} < \frac{\epsilon}{2\|\Lambda\|} \quad \text{so that by (D.8)} \ \|\mathcal{L}_T u_2\|_Y < \frac{\epsilon}{2}. \tag{D.12}$$

Step 3. Using the control $u^* = u_1 + u_2$, the corresponding response is, by (D.2b), (D.6), and (D.8):

$$y^*(T) \equiv y(T, \xi; u^*) = e^{AT}\xi + \mathcal{L}_T(u_1 + u_2) \tag{D.13}$$

$$\text{(by (D.8))} \quad = y^1(T) + \mathcal{L}_T u_2 = y^1(T) + \begin{bmatrix} x_1 - \Pi_1 e^{AT}\xi \\ \Pi_2 \mathcal{L}_T u_2 \end{bmatrix}. \tag{D.14}$$

Step 4. We next demand that $y^*(T)$ due to u^* achieves exact controllability on Y_1, i.e., that $\Pi_1 y^*(T) = z_1$, i.e., specifically from (D.14) and (D.6):

$$\Pi_1 y^*(T) = \Pi_1 y^1(T) + x_1 - \Pi_1 e^{AT}\xi \tag{D.15}$$

$$\text{(by (D.6))} \quad = [\Pi_1 e^{AT}\xi + \Pi_1 \mathcal{L}_T u_1] + x_1 - \Pi_1 e^{AT}\xi = z_1, \tag{D.16}$$

which then identifies x_1 as

$$x_1 = z_1 - \Pi_1 \mathcal{L}_T u_1. \tag{D.17}$$

Thus, the requirement (D.12) imposed on x_1 is now explicitly rewritten via (D.17) as

$$\|\Pi_1 e^{AT}\xi - x_1\|_{Y_1} = \|\Pi_1 [e^{AT}\xi + \mathcal{L}_T u_1] - z_1\|$$

$$\text{(by (D.6))} \quad = \|\Pi_1 y^1(T) - z_1\| < \frac{\epsilon}{2}, \tag{D.18}$$

which is true by the approximate controllability condition in (D.7). Thus, the requirement (D.12) has been achieved.

Step 5. Finally, we seek to achieve approximate controllability of $y^*(T)$ due to u^* on Y_2; i.e., by (D.14) and adding and subtracting

$$\Pi_2 y^*(T) = \Pi_2 y^1(T) + \Pi_2 \mathcal{L}_T u_2 = \Pi_2 y^1(T) - z_2 + z_2 + \Pi_2 \mathcal{L}_T u_2 \tag{D.19}$$

$$= z_2 + [\Pi_2 y^1(T) - z_2] + \Pi_2 \mathcal{L}_T u_2. \tag{D.20}$$

By (D.7) and (D.12), we have, respectively,

$$\|\Pi_2 y^1(T) - z_2\|_{Y_2} < \frac{\epsilon}{2} \text{ and } \|\Pi_2 \mathcal{L}_T u_2\|_{Y_2} < \frac{\epsilon}{2}. \tag{D.21}$$

Thus (D.20), (D.21) yield the sought-after approximate controllability requirement of the second component of $y^*(T)$:

$$\|\Pi_2 y^*(T) - z_2\|_{Y_2} < \epsilon. \tag{D.22}$$

We conclude that: Given $z = [z_1, z_2] \in Y$, with x_1 selected as in (D.17) and x_2 irrelevant, the control $u^* = u_1 + u_2$ yields a solution $y^*(T)$ at $t = T$, which hits z_1: $\Pi_1 y^*(T) = z_1$, see (D.16), and gets arbitrarily close to z_2, see (D.22), as desired. Theorem D.1 is proved. □

Specialization: hinged case. Here we take (see (4.0.1a), (4.0.2a))

$$Y \equiv X_\gamma; \quad Y_1 = Y_{1,\gamma} \equiv X_{1,\gamma} = \mathcal{D}(\mathcal{A}) \times \mathcal{D}(\mathcal{A}_\gamma^{\frac{1}{2}}); \quad Y_2 = \mathcal{D}(\mathcal{A}^{\frac{1}{2}}). \tag{D.23}$$

The following result and proof are essentially contained in [L-T.6, Remark 3.4]. This result was needed in (4.0.3).

LEMMA D.2. *The thermo-elastic (hinged) semigroup $e^{\mathbf{A}_{\gamma,h}t}$ of Lemma 2.1 leaves the space X_γ above invariant: $e^{\mathbf{A}_{\gamma,h}t} X_\gamma \subset X_\gamma$. More specifically, if $[w_0, w_1, \theta_0] \in X_\gamma \subset Y_\gamma$, then*

$$\theta \in C([0,T]; \mathcal{D}(\mathcal{A}^{\frac{1}{2}}) = H_0^1(\Omega)). \tag{D.24}$$

Proof. We return to Eqn. (9.12): after integration by parts, we may rewrite it as (compare with (9.15))

$$\theta(t) = e^{-\mathcal{A}t}\theta_0 - \left[e^{-\mathcal{A}(t-\tau)} w_t(\tau)\right]_{\tau=0}^{\tau=t} + \int_0^t e^{-\mathcal{A}(t-\tau)} w_{tt}(\tau) d\tau; \tag{D.25}$$

$$\theta(t) = e^{-\mathcal{A}t}\theta_0 - w_t(t) + e^{-\mathcal{A}t} w_t(0) + \int_0^t e^{-\mathcal{A}(t-\tau)} w_{tt}(\tau) d\tau. \tag{D.26}$$

Recalling Lemma 2.1 and above all the sharp regularity $w_{tt} \in L_2(0,T; L_2(\Omega))$ from (9.11) [i.e., from [L-T.7, Eqn. (3.36)], as well as $\theta_0 \in \mathcal{D}(\mathcal{A}^{\frac{1}{2}})$, we see by standard semigroup theory that each term in (D.26) is in $C([0,T]; \mathcal{D}(\mathcal{A}^{\frac{1}{2}}))$, and (D.24) follows. □ □

In conclusion: Theorem 1.1 is obtained by applying the general strategy explained in Section 4—Steps 1 and 2—which rests also on the application of Theorem D.1 with the choice of space given by (D.23). As noted in (4.0.4b)—essentially as a consequence of Proposition 2.2 on u_2—the operator $\mathcal{L}_T$ is continuous $\mathcal{U} \to X_\gamma$, and so Theorem D.1 applies.

Specialization: clamped case. Here we take (see (4.0.1b), (4.0.2b))

$$Y \equiv X_\gamma; \; Y_1 \equiv X_{1,\gamma} = [\mathcal{D}(\mathbb{A}^*_{1,\gamma})]' = H^1_0(\Omega) \times \tilde{L}_2(\Omega); \; Y_2 \equiv H_\theta. \tag{D.27}$$

The following result was needed in (4.0.3).

LEMMA D.3. *The thermoelastic (clamped) semigroup $e^{\mathbb{A}_{\gamma,c}t}$ of Lemma 2.1 leaves the space X_γ in (D.27) invariant: $e^{\mathbb{A}_{\gamma,c}t}X_\gamma \subset X_\gamma$. More specifically, if $\{w_0, w_1, \theta_0\} \in X_\gamma$, then*

$$\{w, w_t, \theta\} \in C([0,T]; X_\gamma \equiv H^1_0(\Omega) \times \tilde{L}_2(\Omega) \times H_\theta). \tag{D.28}$$

Proof. Let $y_0 = \{w_0, w_1, \theta_0\} \in X_\gamma$. But $X_\gamma \subset [\mathcal{D}(\mathbb{A}^*_{\gamma,c})]' = H^1_0(\Omega) \times \tilde{L}_2(\Omega) \times [\mathcal{D}(\mathcal{A})]'$. Then $e^{\mathbb{A}_{\gamma,c}t}y_0 = \{w(t), w_t(t), \theta(t)\}| \in C([0,T]; [\mathcal{D}(\mathbb{A}^*_{\gamma,c})]')$. But then the argument of Theorem C.4, Step 3, from (C.51) to (C.54) showing (C.50), establishes that $\theta \in C([0,T]; H_\theta)$, and Lemma D.3 is proved. □

In conclusion: The same conclusion as in the hinged case above applies, *mutatis mutandis*: i.e., with (D.23) replaced by (D.27).

Appendix E: An $L_2(\Omega) \times H^{-1}(\Omega)$-A-Priori Estimate for Wave Equations

In this appendix E, we wish to present a lower-level $L_2(\Omega) \times H^{-1}(\Omega)$-energy estimate, in a much larger degree of generality of the version which is actually invoked in Step (ii) in the proof of Lemma 7.1, Eqn. (7.14). Thus, vastly generalizing Eqn. (7.12), here we consider the following hyperbolic equations

$$\gamma y_{tt} = \Delta y + Fy + f \qquad \text{in } Q = (0,T] \times \Omega, \tag{E.1}$$

where γ is a positive constant as in (7.12), under the following *assumptions*:

(a) F is a first-order differential operator

$$Fy = c_1(t,x)y_t + c_2(t,x) \cdot \nabla y + c_3(t,x)y, \tag{E.2}$$

with, say $C^1(Q)$-coefficients $c_1, |c_2|, c_3$ [the regularity of the latter could be relaxed];

(b) the forcing term f in (E.1) is of the following form

$$f = \dot{f}_1 + f_2, \quad \text{where} \quad f_1 \in L_1(0,T; L_2(\Omega)) \cap C([0,T]; H^{-1}(\Omega)) \tag{E.3a}$$

$$f_2 \in L_1(0,T; H^{-1}(\Omega)), \tag{E.3b}$$

where $\dot{}$ denotes time derivative.

Eqn. (7.12) of our interest in the proof of Lemma 7.1 is a specialization of the above model with $c_1(t,x) \equiv -\alpha^2(x) \in C^2(\bar{\Omega})$ and $|c_2| = c_3 \equiv 0$ in Q. Moreover, the actual forcing term f, which arises in Eqn. (7.12), is much more regular than in assumption (b), with regularity given by (7.13): thus, we may take $f_2 \equiv 0$ and

$$f_1 = -\psi_t + \frac{3}{2}\nabla\psi \cdot \nabla(\alpha^2) + (\alpha\Delta\alpha + |\nabla\alpha|^2)\psi \in C([0,T]; H^1(\Omega)),$$

by (7.4), in our application of the present setting to Eqn. (7.12), as done in Eqn. (7.14).

(c) In addition, we shall assume that the boundayr Γ of the open bounded domain $\Omega \in R^n$ is sufficiently smooth and is divided in two non-empty, open parts Γ_1, Γ_0 with $\Gamma = \Gamma_0 \cup \Gamma_1$, satisfying

$$(x - x_0) \cdot \nu(x) \le 0, \qquad \text{for } x \in \Gamma_0, \tag{E.4}$$

where $\nu(x)$ is the unit outward normal vector.

The purpose of this Appendix E is to provide the following Theorem E.1, which refers to solutions y of the hyperbolic equation (E.1), within the following class:

$$\begin{cases} \{y, y_t\} \in C([0,T]; L_2(\Omega) \times H^{-1}(\Omega)); & \text{(E.5a)} \\ y|_\Sigma \in L_2(0,T; L_2(\Gamma)) \equiv L_2(\Sigma); & \text{(E.5b)} \\ \left.\dfrac{\partial y}{\partial \nu}\right|_\Sigma \in H^{-1}(\Sigma. & \text{(E.5c)} \end{cases}$$

THEOREM E.1. *Let $T > 0$. Let $\{y, y_t\}$ be a solution of Eqn. (E.1), subject to assumptions (a), (b), and (c) above, within the class (E.5). Then, the following estimate holds true: there exists a constant $C_T > 0$, such that*

$$\begin{aligned} &\|\{y, y_t\}\|_{C([0,T];L_2(\Omega)\times H^{-1}(\Omega))} \\ &\quad \le C_T\left\{\|y|_\Sigma\|_{L_2(\Sigma)} + \left\|\left.\frac{\partial y}{\partial \nu}\right|_\Sigma\right\|_{H^{-1}(\Sigma_1)} + \|y\|_{H^{-1}(Q)}\right. \\ &\quad\quad \left. + \|f_1\|_{L_1(0,T;L_2(\Omega))} + \|f_1\|_{C([0,T];H^{-1}(\Omega))} + \|f_2\|_{L_1(0,T;H^{-1}(\Omega))}\right\}. \end{aligned} \tag{E.6}$$

REMARK E.1. *The above $L_2(\Omega) \times H^{-1}(\Omega)$-estimate (E.6) is closely related to estimates already available in the literature, such as those in [L-T.2], [L-T.5], [L-T-Y.1] [which arise in the context of uniform stabilization in the state space $L_2(\Omega) \times H^{-1}(\Omega)$ of conservative wave equations, from constant to space variable coefficient in the principal part], as well as those in [Ta.1], [Ta.2] which are established at various energy levels for general evolution operators. Yet, it does not* *appear possible to just quote the desired estimate (E.6) from existing literature such as the one above, as well as [B-L-R] and its successor.*

The estimate in [Ta.1] includes terms involving anisotropic norms of f : L_2 in the normal direction, H^{-1} in the tangential direction, introduced by Hormander. Thus, this version of the estimate requires additional regularity of f near the boundary in the normal direction, which may not be available. This extra requirement may create technical obstacles at the level of applications. This difficulty accounts for the decomposition of y into u and z, as in (E.7), (E.8) below, where u is a homogeneous problem. Another reason behind this decomposition is to avoid also boundary terms which would result in expressing the energy ($H^1 \times L_2$-norm) $E_y(t)$ of y at time t, and the energy $E_y(s)$ at time s [the inequality (E.11) below avoids these terms, since it refers to a homogeneous *problem u, such as in (E.8), with no boundary or interior non-homogeneous terms.]*

As to the energy estimate at the required lower energy level $L_2(\Omega) \times H^{-1}(\Omega)$ (as needed in Appendix E) given in [L-T-Y.1], this was written explicitly in the context of establishing uniform stabilization results for a conservative wave equation with variable (in space) coefficients in the principal part and no lower-order terms (unlike (E.1), or Eqn. (7.12) to which we apply Appendix E, which include first-order terms. *The strategy pursued in [L-T-Y.1] in the one in [L-T.5], whereby one shifts down the estimates by one unit, by multiplying the original equation by a suitably constructed tangential pseudo-differential operator of order -1 (or by an operator change of variable as in [L-T.2]), and then running the 'usual' $H^1 \times L_2$ energy estimates for the transformed problem. (This step is done in [L-T-Y.1] by using a Riemann geometric approach to reduce the variable (in space) principal part to a constant coefficient principal part, thus improving upon [L-T.5], which assumed, instead, constant coefficient principal part). Whether this technique carries over also in the presence of first-order (energy) level terms, such as F_y in (E.1) [or $\alpha^2 z_t$ in (7.12), our ultimate beneficiary] remains to be checked at present, due to the presence of an additional commutator.*

REMARK E.2. *In the proof below, we shall actually write explicitly some formulas (such as (E.16), (E.17), (E.27)) when the coefficients c_i of F are time-independent. This is only for expediency. In fact, the developments can be extended to the case where the coefficients c_i of F are also time-dependent. In this respect, we remark, once and for all, that the regularity results of [L-L-T.1], while written explicitly for the case of $F = 0$, readily extend to the case of (the lower-order term) $F \neq 0$ under suitable smoothness of the coefficients (depending on the result). This remark will not be repeated.*

Proof. **Step 1.** We decompose the solution y as

$$y = u + z, \tag{E.7}$$

where u and z solve the following problems:

$$\begin{cases} \gamma u_{tt} - \Delta u = Fu \\ u|_\Sigma = 0 \\ u(0, \cdot) = y(0, \cdot); u_t(0, \cdot) = y_t(0, \cdot); \end{cases} \qquad \begin{cases} \gamma z_{tt} - \Delta z = Fz + f & \text{(E.8a)} \\ z|_\Sigma = y & \text{(E.8b)} \\ z(0, \cdot) = z_t(0, \cdot) = 0. & \text{(E.8c)} \end{cases}$$

Step 2. (estimate for u) With reference to the u-problem in (E.8), the following estimate holds true, under assumptions (a), (c):

$$E_u(t) \equiv \|\{u(t), u_t(t)\}\|^2_{L_2(\Omega)\times H^{-1}(\Omega)} \leq C_T \left\{ \left\| \frac{\partial u}{\partial \nu}\Big|_\Sigma \right\|^2_{H^{-1}(\Sigma_1)} + \|u\|^2_{H^{-1}(Q)} \right\}. \tag{E.9}$$

The proof of (E.9) for the u-problem (E.8) rests critically on the following preliminary result: there exists a positive interval $[t_0, t_1] \subset [0, T]$, such that

$$\int_{t_0}^{t_1} E_u(t)dt \leq C_T \left\{ \left\| \frac{\partial u}{\partial \nu}\Big|_\Sigma \right\|^2_{H^{-1}(\Sigma_1)} + \|u\|^2_{H^{-1}(Q)} \right\}. \tag{E.10}$$

Moreover, since the u-problem is homogeneous (no forcing term, either in the interior, or on the boundary), then the following estimate (obtained by Gronwall's inequality) holds true, e.g., [L-T.6, Eqn. (2.3.1), p. 234]:

$$E_u(t) \leq E_u(s)e^{CT}, \quad \forall\ s, t \in [0, T]; \text{ hence } \int_{t_0}^{t_1} E_u(s)ds \geq e^{-CT}(t_1 - t_0)E_u(t_0). \tag{E.11}$$

Then, using (E.11) (right *and* left sides) in (E.10), yields (E.9), as desired.

Step 2. (estimate for z) With reference to the z-problem in (E.8), the following estimate holds true, under assumptions (a), (b):

$$E_z(t) \equiv \|\{z(t), z_t(t)\}\|^2_{L_2(\Omega)\times H^{-1}(\Omega)} \leq \|\{z, z_t\}\|^2_{C([0,T];L_2(\Omega)\times H^{-1}(\Omega))} \tag{E.12}$$

$$\leq C_T \left\{ \|y|_\Sigma\|^2_{L_2(\Sigma)} + \|f_1\|^2_{L_1(0,T;L_2(\Omega))} + \|f_1\|^2_{C([0,T];H^{-1}(\Omega))} + \|f_2\|^2_{L_1(0,T;H^{-1}(\Omega))} \right\}. \tag{E.13}$$

The validity of (E.13) follows essentially from results in [L-L-T.1]. They include: [L-L-T.1, Theorem 2.3, p. 153; or Theorem 3.3, p. 176] for the contribution to $\{z, z_t\} \in C([0,T]; L_2(\Omega) \times H^{-1}(\Omega))$ due to $y|_\Sigma \in L_2(\Sigma)$ and $f_2 \in L_1(0,T; H^{-1}(\Omega))$; and [L-L-T.1, Remark 2.8, p. 163] for $z \in C([0,T]; L_2(\Omega))$ due to $f = \dot{f}_1 + f_2$, with $f_1 \in L_1(0,T; L_2(\Omega))$ and $f_2 \in L_1(0,T; H^{-1}(\Omega))$. For completeness, it remains here to consider the problem

$$\begin{cases} \gamma z_{2,tt} - \Delta z_2 = F z_2 + \dot{f}_1 & \text{in } Q; \quad \text{(E.14a)} \\ z_2|_\Sigma = 0 & \text{in } \Sigma; \quad \text{(E.14b)} \\ z_2(0, \cdot) = z_{2,t}(0, \cdot) = 0 & \text{in } \Omega; \quad \text{(E.14c)} \end{cases}$$

$$f_1 \in L_1(0,T; L_2(\Omega)) \cap C([0,T]; H^{-1}(\Omega)), \tag{E.14d}$$

and show that

$$\|\{z_2, z_{2t}\}\|_{C([0,T];L_2(\Omega)\times H^{-1}(\Omega))} \leq C_T \left\{ \|f_1\|_{L_1(0,T;L_2(\Omega))} + \|f_1\|_{C([0,T];H^{-1}(\Omega))} \right\}. \tag{E.15}$$

Estimate (E.15) may be proved in a few ways: (i) either by energy methods (we multiply the z_2-equation (E.14a) by $\mathcal{A}^{-1}z_{2t}$, where $\mathcal{A}$ is the operator in (2.1a), and integrate by parts), or (ii) else by sine/cosine operator theory as in [L-L-T.1, Section 3] (see Remark E.2): if $\tilde{A}$ denotes the generator of the cosine operator $C(t)$, with sine operator $S(t)$, describing the solution of the abstract equation corresponding to (E.14), we have:

$$z_2(t) = \int_0^t S(t-\tau)\dot{f}_1(\tau)d\tau = -S(t)f_1(0) + \int_0^t C(t-\tau)f_1(\tau)d\tau; \tag{E.16}$$

$$z_{2,t}(t) = \int_0^t C(t-\tau)\dot{f}_1(\tau)d\tau = f_1(t) - C(t)f_1(0) + \tilde{A}\int_0^t S(t-\tau)f_1(\tau)d\tau, \tag{E.17}$$

where we have integrated by parts. Then the assumptions on f_1 in (E.14) guarantee the conclusion (E.15) from (E.16), (E.17), sincd $C(t) : H^{-1}(\Omega) \to C([0,T]; H^{-1}(\Omega))$, while $S(t) : H^{-1}(\Omega) \to C([0,T]; L_2(\Omega))$, where $H_0^1(\Omega) = \mathcal{D}((-\tilde{A})^{\frac{1}{2}})$, thus $H^{-1}(\Omega) = [\mathcal{D}((-\tilde{A}^{\frac{1}{2}}))]'$.

Step 3. Returning to the decomposition (E.7) and combining estimate (E.9) for u with estimate (E.13) for z, we arrive at the following estimate of the original y-problem (E.1):

$$E_y(t) \equiv \|\{y(t), y_t(t)\|^2_{L_2(\Omega)\times H^{-1}(\Omega)} \le \|\{y, y_t\}\|^2_{C([0,T];L_2(\Omega)\times H^{-1}(\Omega))} \tag{E.18}$$

$$\le C_T\left\{ \|y|_\Sigma\|^2_{L_2(\Sigma)} + \left\|\frac{\partial u}{\partial \nu}\Big|_\Sigma\right\|^2_{H^{-1}(\Sigma_1)} + \|u\|^2_{H^{-1}(Q)} + \|f_1\|^2_{L_1(0,T;L_2(\Omega))} + \|f_1\|^2_{C([0,T];H^{-1}(\Omega))} + \|f_2\|^2_{L_1(0,T;H^{-1}(\Omega))}\right\}. \tag{E.19}$$

Step 4. In this step, with reference to the z-problem in (E.8), we want to establish that

$$\left\|\frac{\partial z}{\partial \nu}\Big|_\Sigma\right\|_{H^{-1}(\Sigma_1)} \le C_T\left\{\|y|_\Sigma\|_{L_2(\Sigma)} + \|f_1\|_{L_1(0,T;L_2(\Omega))} + \|f_1\|_{C([0,T];H^{-1}(\Omega))} + \|f_2\|_{L_1(0,T;H^{-1}(\Omega))}\right\} \tag{E.20}$$

so that, recalling that $u = y - z$ from (E.7), we then readily obtain the following estimate

$$\left\|\frac{\partial u}{\partial \nu}\Big|_\Sigma\right\|_{H^{-1}(\Sigma_1)} \le \left\|\frac{\partial y}{\partial \nu}\Big|_\Sigma\right\|_{H^{-1}(\Sigma_1)} + \left\|\frac{\partial z}{\partial \nu}\Big|_\Sigma\right\|_{H^{-1}(\Sigma_1)} \tag{E.21}$$

$$\le C_T\left\{\|y|_\Sigma\|_{L_2(\Sigma)} + \left\|\frac{\partial y}{\partial \nu}\Big|_\Sigma\right\|_{H^{-1}(\Sigma_1)} + \|f_1\|_{L_1(0,T;L_2(\Omega))} + \|f_1\|_{C([0,T];H^{-1}(\Omega))} + \|f_2\|_{L_1(0,T;H^{-1}(\Omega))}\right\}. \tag{E.22}$$

Proof of (E.20). The validity of (E.20), for the z-problem (E.8), due to $z|_\Sigma = y|_\Sigma \in L_2(\Sigma)$ and $f_2 \in L_1(0,T;H^{-1}(\Omega))$, is a result of [L-L-T.1, Theorem 2.3, p. 153, or Theorem 3.7, p. 178]. Thus, it remains to return to the z_2-problem (E.14), and establish that

$$\left\| \frac{\partial z_2}{\partial \nu}\Big|_\Sigma \right\|_{H^{-1}(\Sigma)} \le C_T \left\{ \|f_1\|_{L_1(0,T;L_2(\Omega))} + \|f_1\|_{C(0,T];H^{-1}(\Omega))} \right\}. \tag{E.23}$$

To prove this, we consider the problem

$$\begin{cases} \gamma q_{tt} - \Delta q = Fq + f_1 & \text{in } Q; & \text{(E.24a)} \\ q|_\Sigma = 0 & \text{in } \Sigma; & \text{(E.24b)} \\ q(0,\,\cdot\,) = q_t(0,\,\cdot\,) = 0 & \text{in } \Omega. & \text{(E.24c)} \end{cases}$$

From [L-L-T.1, Theorem 2.1, p. 151; Theorem 3.5, p. 178] (notice that the required Compatibility Relation is satisfied for the z_2-problem), we obtain continuously:

$$f_1 \in L_1(0,T;L_2(\Omega)) \Rightarrow \frac{\partial q}{\partial \nu} \in L_2(\Sigma); \text{ hence } \frac{\partial q_t}{\partial \nu} \in H^{-1}(0,T;L_2(\Gamma)). \tag{E.25}$$

Next, differentiate the q-problem (E.24) to obtain

$$\begin{cases} \gamma(q_t)_{tt} - \Delta q_t = Fq_t + \dot{f}_1 & \text{in } Q; & \text{(E.26a)} \\ q_t|_\Sigma = 0 & \text{in } \Sigma; & \text{(E.26b)} \\ q(0,\,\cdot\,) = 0, (q_t)_t(0) = q_{tt}(0) = \cancel{\Delta q(0)} + \cancel{Fq(0)} + f_1(0) = f_1(0) \in H^{-1}(\Omega). & & \text{(E.26c)} \end{cases}$$

Thus problem (E.26) in q_t coincide with problem (E.14) for z_2, except (possibly) for the initial velocity $q_{tt}(0) = f_1(0) \in H^{-1}(\Omega)$. Thus, we obtain from this comparison

$$q_t(t) = z_2(t) + S(t)f_1(0), \tag{E.27}$$

where, since $f_1(0) \in H^{-1}(\Omega)$, a specialization of [L-L-T.1, Theorem 2.3, p. 153, or Theorem 3.7, p. 178] yields

$$\frac{\partial S(t)f_1(0)}{\partial \nu} \in H^{-1}(\Sigma). \tag{E.28}$$

Thus, from (E.27), (E.28), and (E.25) for q_t, we obtain

$$\frac{\partial z_2}{\partial \nu} = \frac{\partial q_t}{\partial \nu} - \frac{\partial S(t)f_1(0)}{\partial \nu} \in H^{-1}(\Sigma) \tag{E.29}$$

continuously in f_1, and (E.23) is established. Hence (E.20) and (E.22) are proved.

Step 5. Using estimate (E.22) in the right-hand side of (E.19), then yields

$$\begin{aligned} E_y(t) &\le \|\{y, y_t\}\|^2_{C([0,T];L_2(\Omega)\times H^{-1}(\Omega))} \\ &\le C_T \Bigg\{ \|y|_\Sigma\|_{L_2(\Sigma)} + \left\| \frac{\partial y}{\partial \nu}\Big|_{\Sigma_1} \right\|_{H^{-1}(\Sigma_1)} + \|u\|^2_{H^{-1}(Q)} \\ &\quad + \|f_1\|^2_{L_1(0,T;L_2(\Omega))} + \|f_1\|^2_{C([0,T];H^{-1}(\Omega))} + \|f_2\|^2_{L_1(0,T;H^{-1}(\Omega))} \Bigg\}. \end{aligned} \tag{E.30}$$

Finally, as already recalled [L-L-T.1, Remark 2.8, p. 163] yields the second part of the following inequality for the z-problem in (E.8),

$$\begin{aligned} \|z\|_{H^{-1}(Q)} &\leq \|z\|_{C([0,T];L_2(\Omega))} \\ &\leq C_T\left\{\|y|_\Sigma\|_{L_2(\Sigma)} + \|f_1\|_{L_1(0,T;L_2(\Omega))} + \|f_2\|_{L_1(0,T;H^{-1}(\Omega))}\right\}, \end{aligned} \tag{E.31}$$

while the first inequality on the left-hand side of (E.1) is obvious. (See also (E.13) and its derivation.)

Step 6. Finally, recalling $u = y - z$ from (E.7), inserting (E.31) in (E.30) yields

$$E_y(t) \equiv \|\{y(t), y_t(t)\}\|^2_{L_2(\Omega)\times H^{-1}(\Omega)} \leq \|\{y, y_t\}\|^2_{C([0,T];L_2(\Omega)\times H^{-1}(\Omega))} \tag{E.32}$$

$$\begin{aligned} &\leq C_T\left\{\|y|_\Sigma\|_{L_2(\Sigma)} + \left\|\frac{\partial y}{\partial \nu}\bigg|_{\Sigma_1}\right\|^2_{H^{-1}(\Sigma_1)} + \|y\|^2_{H^{-1}(Q)}\right. \\ &\quad \left. + \|f_1\|^2_{L_1(0,T;L_2(\Omega))} + \|f_1\|^2_{C([0,T];H^{-1}(\Omega))} + \|f_2\|^2_{L_1(0,T;H^{-1}(\Omega))}\right\}. \end{aligned} \tag{E.33}$$

Thus, (E.33) is the desired estimate (E.6). Theorem E.1 is proved. □

Appendix F: Another Result on Boosting the A-Priori Regularity of an Over-Determined Thermo-Elastic Problem

Problem. Let $\{w, w_t, \theta\}$ be a solution of the following over-determined *homogeneous* problem, where $\Sigma = (0,T] \times \Gamma$, $\Sigma_2 = (0,T] \times \Gamma_2$:

$$\left\{\begin{aligned} w_{tt} - \gamma\Delta w_{tt} + \Delta^2 w + \operatorname{div}(\alpha(x)\nabla\theta) &\equiv 0 \quad \text{in } Q = (0,T] \times \Omega; && \text{(F.1a)} \\ \theta_t - \Delta\theta - \operatorname{div}(\alpha(x)\nabla w_t) &\equiv 0 \quad \text{in } Q; && \text{(F.1b)} \\ \theta \equiv 0;\ w \equiv 0;\ \tfrac{\partial w}{\partial \nu} \equiv 0;\ \Delta w \equiv 0 \text{ on } \Sigma;\ \tfrac{\partial \Delta w}{\partial \nu} &\equiv 0 \text{ on } \Sigma_2, && \text{(F.1c)} \end{aligned}\right.$$

with initial conditions

$$\{w_0, w_1, \theta_0\} \in Y_\gamma \quad (= H^2(\Omega) \times H^1(\Omega) \times L_2(\Omega) \text{ plus B.C.}), \tag{F.1d}$$

where Y_γ is either as in (2.3) (hinged B.C.), or else as in (2.5) (clamped B.C.). Thus, by the semigroup generation of Lemma 2.1, the following *a-priori* regularity is available:

$$\left\{\begin{aligned} w &\in C([0,T]; \mathcal{D}(\mathcal{A}) = H^2(\Omega) \cap H^1_0(\Omega)); && \text{(F.2a)} \\ w_t &\in C([0,T]; \mathcal{D}(\mathcal{A}^{\frac{1}{2}}) = H^1_0(\Omega)); && \text{(F.2b)} \\ \theta &\in C([0,T]; L_2(\Omega)). && \text{(F.2c)} \end{aligned}\right.$$

We shall assume the geometrical condition (1.4): there exists a point $x_0 \in \mathbb{R}^2$, such that

$$(x - x_0) \cdot \nu(x) \leq 0 \quad \text{on } \Gamma \setminus \Gamma_2. \tag{F.3}$$

We note that problem (F.1) is *not* the dual homogeneous problem of Section 9—either (9.1)–(9.4), or (9.5)–(9.7)—that arises by duality in the observability issue, and on two counts: (i) first, the B.C. on $\frac{\partial \Delta w}{\partial \nu}$ in (F.1c) refers to w, while in (9.1d) or (9.5d) it refers to w_t; (ii) second, more importantly, this B.C. is given in (F.1c) on the finite portion Γ_2 of the boundary Γ, while in (9.1d) and (9.5d) of Section 9 it was given on the ***arbitrarily small portion*** Γ_1 of the boundary. In the notation of this paper, (F.1) requires taking $\Gamma_1 = \Gamma_2$ finite. The goal of this section is to present the following boost of the *a-priori* regularity, in the same spirit as Theorem 9.1. The proof is still, of course, by a boot-strap argument. However, the technicalities of the present proof, which refers now to problem (F.1), are very different and, in fact, much simpler than those of Section 9. Thus, even if the price to pay is a finite portion $\Gamma_1 = \Gamma_2$ rather than an infinitesimal portion Γ_1 for the support of the first control u_1, it is worth exposing it.

The issue is the one of Section 9: in order to apply the uniqueness result of Corollary 4.2.3, the above regularity (F.2) is not enough: rather, we seek to boost the regularity of a solution $\{w, w_t, \theta\}$ of problems (F.1)–(F.4) to the level $H^3(Q)$ for w_t and $H^1(Q)$ for θ_t, as required by Corollary 4.2.3. Indeed, it suffices to apply Corollary 4.2.3 on a time interval $[t_0, T]$, $t_0 > 0$, to obtain, accordingly, $w(T_1, \cdot\,) = w_t(T_1, \cdot\,) = \theta(T_1, \cdot\,) = 0$ with $T_1 = \frac{T}{2}$. After this, an application of the backward uniqueness result of [L-R-T.1] will yield $w_0 = 0$, $w_1 = 0$, $\theta_0 = 0$, and thus $w \equiv 0$, $w_t \equiv 0$, $\theta \equiv 0$ in $Q = [0, T] \times \Omega$. Thus, henceforth, we take $t_0 > 0$ arbitrarily small, and define $Q_{t_0} = [t_0, T] \times \Omega$, $\Sigma_{t_0} = [t_0, T] \times \Gamma$. We seek to boost the *a-priori* regularity (F.2) to the following level:

$$w_t \in H^3(Q_{t_0}) : \text{ i.e., } w_t \in L_2(t_0, T; H^3(\Omega));\ w_{tttt} \in L_2(t_0, T; L_2(\Omega)); \tag{F.4}$$

$$\theta_t \in H^1(Q_{t_0}) : \text{ i.e., } \theta_t \in L_2(t_0, T; H^1(\Omega));\ \theta_{tt} \in L_2(t_0, T; L_2(\Omega)). \tag{F.5}$$

That this is possible is shown by the following main result of this section.

THEOREM F.1. *Let $\{w, w_t, \theta\}$ be a solution of the over-determined problem (F.1), with* a-priori *regularity as in (F.2) and subject to the geometrical condition (F.3). Then, in fact, it satisfies the higher regularity (F.4), (F.5). Accordingly, Corollary 4.2.3 applies and yields $w_0 = 0$, $w_1 = 0$, $\theta_0 = 0$, as desired.*

Orientation. The present issue of boosting the *a-priori* regularity for the thermo-elastic problem is more complicated than the corresponding issue of boosting the *a-priori* regularity for the Kirchoff problem, as obtained in Corollary 8.2. While the basic idea is the same in both cases, in Corollary 8.2 the required estimates (6.2) (hinged case) and (7.2) (clamped case) could be applied directly. In the present thermo-elastic case, the boosting of regularity from (F.2) to (F.4), (F.5) will proceed through four steps. As already noted above, the present proof, however, is technically simpler than that in Section 9. First, the preliminary Step 1 to boost initially the regularity of the thermal variable θ, critically of θ_t, by 1 Sobolev unit is the same as Lemma 9.2 of Section 9. This is then followed by a main Step 2 for a first boost, again by 1 Sobolev unit, of the regularity of the mechanical variable w including w_{tt} (see Proposition F.3). Next, the boost of regularity of w_{tt} in Step 2 will, in turn, translate into a corresponding boost of regularity of the thermal variable, leading to the required level of θ_{tt} in Step 3 (see Lemma F.4). Finally, another main Step 4 will, in turn, boost the regularity of the mechanical variable w to the required level (F.4) (see Proposition F.5). Both main steps—in Propositions F.3 and F.4—will require the *a-priori* continuous observability estimate [L-T.6, Theorem 2.1.2(ii), Eqn. (2.1.10b)] already invoked in Eqn. (6.12) for a suitable wave equation, as well as in (9.38) of Proposition 9.4 and in Proposition 9.7. As in the last two propositions, the procedure requires, in both instances, identifying a suitable wave equation with a inhomogeneous right-hand side term (based on θ) just in $L_2(Q_{t_0})$, and with the wave equation variable just in $L_2(Q_{t_0})$ as *a-priori* regularity. Due to the vanishing of the boundary terms for the over-determined problem, plus the geometric assumption (F.3), the counterpart version of estimate (6.2) then yields a boost of regularity by 1 Sobolev unit in both position and velocity of the suitable wave equation variable.

Proof. (of Theorem F.1.) Throughout this proof, we fix $\epsilon > 0$ arbitrarily small.

Step 1.

LEMMA F.2. *With reference to problems (F.1)–(F.2), we have*

$$\theta \in L_2(0,T;H_0^1(\Omega)) \cap C([\epsilon,T];H_0^1(\Omega)); \quad \theta_t \in L_2(\epsilon,T;L_2(\Omega)),\ \epsilon > 0 \quad (\epsilon = 0,\ \textit{if } \theta_0 = 0); \tag{F.6}$$

$$w_{tt} \in L_2(0,t;L_2(\Omega)). \quad \square \tag{F.7}$$

This is Lemma 9.2 rewritten.

Step 2.

PROPOSITION F.2. *With reference to problems (F.1)–(F.2), and $\epsilon > 0$ preassigned in Step 1, we have*

$$w \in L_2(\epsilon,T;H^3(\Omega) \cap H_0^2(\Omega));\ w_t \in L_2(\epsilon,T;\mathcal{D}(\mathcal{A}) = H^2(\Omega) \cap H_0^1(\Omega)); \tag{F.8}$$

$$w_{tt} \in L_2(\epsilon,T;\mathcal{D}(\mathcal{A}^{\frac{1}{2}}) \equiv H_0^1(\Omega)) \tag{F.9}$$

($\epsilon = 0$ in (F.8), (F.9), if $\theta_0 = 0$).

Proof. Writing $\operatorname{div}(\alpha\nabla\theta) = \alpha\Delta\theta + \nabla\alpha \cdot \nabla\theta$ in the first equation (F.1a), and substituting $\Delta\theta$ from the second Eqn. (F.1b), we obtain, as in (3.1):

$$w_{tt} - \gamma\Delta w_{tt} + \Delta^2 w - \alpha\operatorname{div}(\alpha\nabla w_t) = -\alpha\theta_t - \nabla\alpha \cdot \nabla\theta. \tag{F.10}$$

This Eqn. (F.10), along with (a subset of) the B.C. in (F.1c), can be rewritten abstractly as

$$w_{tt} = -\mathcal{A}_\gamma^{-1}\mathcal{A}^2 w + \mathcal{A}_\gamma^{-1}F_\alpha w_t - \mathcal{A}_\gamma^{-1}(\alpha\theta_t) + \ell.o.t., \tag{F.11}$$

after recalling $\mathcal{A}_\gamma$ from (2.1a) and $F_\alpha w_t = \alpha \text{div}(\alpha \nabla w_t) = \alpha^2 \Delta w_t + \alpha \nabla \alpha \cdot \nabla w_t$ (see (3.3)). Thus, using this identity and applying $\mathcal{A}$ throughout (F.11), we obtain

$$(\mathcal{A}w)_{tt} = -\mathcal{A}_\gamma^{-1} \mathcal{A}^2 (\mathcal{A}w) - \mathcal{A}_\gamma^{-1} \mathcal{A}(\alpha^2 (\mathcal{A}w)_t) + \mathcal{A}_\gamma^{-1} \mathcal{A}(\alpha \nabla \alpha \cdot \nabla w_t) - \mathcal{A}_\gamma^{-1} \mathcal{A}(\alpha \theta_t). \quad \text{(F.12)}$$

Finally, recalling the identities (6.4), we rewrite (F.12) still abstractly as

$$(\mathcal{A}w)_{tt} = -\frac{\mathcal{A}(\mathcal{A}w)}{\gamma} - \frac{\alpha^2 (\mathcal{A}w)_t}{\gamma} - \frac{\alpha \theta_t}{\gamma} + f_1; \quad \text{(F.13)}$$

or, after setting a new variable,

$$q \equiv \mathcal{A}w = -\Delta w \in C([0,T]; L_2(\Omega)) \quad \text{(F.14)}$$

from (F.2a), we rewrite it as the following wave equation problem where $\Sigma_\epsilon = (\epsilon, T] \times \Gamma$:

$$\begin{cases} q_{tt} = \frac{\Delta q}{\gamma} - \frac{\alpha^2 q_t}{\gamma} + f & \text{in } (\epsilon, T] \times \Omega \equiv Q_\epsilon; & \text{(F.15a)} \\ q|_{\Sigma_\epsilon} = 0, \quad \frac{\partial q}{\partial \nu}|_{\Sigma_{\epsilon,2}} = 0 & \text{in } (\epsilon, T] \times \Gamma_2 = \Sigma_{\epsilon,2}; & \text{(F.15b)} \\ f = -\frac{\alpha \theta_t}{\gamma} + \frac{\mathcal{A}w}{\gamma^2} + \alpha \nabla \alpha \cdot \nabla w_t + \ell.o.t. \in L_2(\epsilon, T; L_2(\Omega)), & & \text{(F.15c)} \end{cases}$$

where the regularity in (F.15c) stems from (F.6) for θ_t, and (F.2a–b) for $\{w, w_t\}$. Of course, the over-determined, homogeneous B.C. in (F.15b) are a consequence of $\Delta w|_\Sigma = 0$ and $\frac{\partial \Delta w}{\partial \nu} = 0$ on $\Sigma_{\epsilon,2}$ in (F.1c). Thus, with problem (F.15), we are in the same situation as with problem (6.10), except for the over-determined B.C. for q. The same occurred with problem (9.36)–(9.38). We likewise invoke [L-T.6, Theorem 2.1.2(ii), Eqn. (2.1.10b)] (which actually applies to a general wave equation with space variable coefficients in the first-order terms, thus at the energy level). We then obtain in the notation of [L-T.6],

$$\overline{(BT)}_q|_{\Sigma_{\epsilon,2}} + \frac{C_T}{\gamma^2} \int_{Q_\epsilon} f^2 dQ + TC_T \|q\|^2_{L_2(Q_\epsilon)} \geq \int_\epsilon^T \|\{q(t), q_t(t)\}\|^2_{H^1(\Omega) \times L_2(\Omega)} dt, \quad \text{(F.16)}$$

where the boundary terms $\overline{(BT)}_q|_{\Sigma_{\epsilon,2}}$ for q satisfies, as in (9.43),

$$\begin{aligned} \overline{(BT)}_q|_{\Sigma_{\epsilon,2}} &= \int_\epsilon^T \int_{\Gamma_2} e^{\tau\phi} \left(\frac{\partial q}{\partial \nu}\right)^2 h \cdot \nu \, d\Gamma \, dt \\ &+ \int_\epsilon^T \int_{\Gamma \setminus \Gamma_2} e^{\tau\phi} \left(\frac{\partial q}{\partial \nu}\right)^2 h \cdot \nu \, d\Gamma \, dt \leq 0, \end{aligned} \quad \text{(F.17)}$$

by virtue of assumption (F.3) in the last step. By (F.17) used in (F.16), as well as by (F.14) for $q \in C([0,T]; L_2(\Omega))$, and $f \in L_2(\epsilon, T; L_2(Q))$ (by (F.15c), as required), we then obtain

$$\begin{cases} \Delta w = -q \in L_2(\epsilon, T; H_0^1(\Omega)) \\ w|_{\Sigma_\epsilon} = 0, \; \Sigma_\epsilon = (\epsilon, T] \times \Gamma \end{cases} ; \quad \begin{cases} \Delta w_t = -q_t \in L_2(\epsilon, T; L_2(\Omega)) & \text{(F.18a)} \\ w_t|_{\Sigma_\epsilon} = 0 & \text{(F.18b)} \end{cases},$$

and hence, by elliptic theory,

$$w \in L_2(\epsilon, T; H^3(\Omega)); \ w_t \in L_2(\epsilon, T; H^2(\Omega)). \tag{F.19}$$

Then (F.19), along with the B.C. for w in (F.1c), yields the desired conclusion (F.8) for $\{w, w_t\}$. As to w_{tt}, we return to (F.1a), which we rewrite abstractly, via the B.C. (F.1c), as in (9.29),

$$w_{tt} = (-\mathcal{A}_\gamma^{-1}\mathcal{A})\mathcal{A}w + \mathcal{A}_\gamma^{-1}(\alpha\mathcal{A}\theta) + \ell.o.t. \tag{F.20}$$

where now $\mathcal{A}w \in L_2(\epsilon, T; H_0^1(\Omega))$ by (F.8) just proved, and $\mathcal{A}_\gamma^{-1}(\alpha\mathcal{A}\theta) \in L_2(0, T; H_0^1(\Omega))$ by (F.6), thus obtaining $w_{tt} \in L_2(\epsilon, T; H_0^1(\Omega))$, as claimed in (F.9). □

Step 3.

LEMMA F.4. *With reference to problem (F.1)–(F.2), and $\epsilon > 0$ preassigned in Step 1, we have*

$$\theta \in L_2(2\epsilon, \mathcal{D}(\mathcal{A}) = H^2(\Omega) \cap H_0^1(\Omega)); \ \theta_1 \in L_2(2\epsilon, T; \mathcal{D}(\mathcal{A}^{\frac{1}{2}}) = H_0^1(\Omega)); \tag{F.21}$$

$$w_{ttt} \in L_2(\epsilon, T; L_2(\Omega)); \ \theta_{tt} \in L_2(2\epsilon, T; L_2(\Omega)). \tag{F.22}$$

Proof. Returning to (9.21a), we obtain for $t \geq \epsilon$:

$$\theta(t) - e^{-\mathcal{A}(t-\epsilon)}\theta(\epsilon) = -\int_\epsilon^t e^{-\mathcal{A}(t-\tau)}\mathcal{A}w_t(\tau)\alpha\, d\tau + \ell.o.t. \tag{F.23a}$$

$$= -\int_0^{t-\epsilon} e^{-\mathcal{A}s}\mathcal{A}w_t(t-s)\alpha\, ds + \ell.o.t., \tag{F.23b}$$

with $\theta(\epsilon) \in L_2(\Omega)$ by (F.2c). Thus, standard results for analytic semigroup theory imply, via $\mathcal{A}w_t \in L_2(\epsilon, T; L_2(\Omega))$ by (F.8), that

$$\mathcal{A}\theta(t) = e^{-\mathcal{A}(t-2\epsilon)}\mathcal{A}e^{-\mathcal{A}\epsilon}\theta(\epsilon) - \int_\epsilon^t \mathcal{A}e^{-\mathcal{A}(t-\tau)}\mathcal{A}w_t(\tau)d\tau \in L_2(2\epsilon, T; L_2(\Omega)), \tag{F.24}$$

as claimed in (F.21). Next, returning to identity (F.20) for w_{tt}, and differentiating in time, we obtain

$$w_{ttt} = (-\mathcal{A}_\gamma^{-1}\mathcal{A})\mathcal{A}w_t + \mathcal{A}_\gamma^{-1}(\alpha\mathcal{A}\theta_t) = \ell.o.t. \in L_2(\epsilon, T; L_2(\Omega)), \tag{F.25}$$

where the regularity in (F.25) is attained by $\mathcal{A}w_t \in L_2(\epsilon, T; L_2(\Omega))$ by (F.8), and $\mathcal{A}_\gamma^{-1}(\alpha\mathcal{A}\theta_t) \in L_2(\epsilon, T; L_2(\Omega))$ by (F.6). Thus, (F.22) is proved for w_{ttt}.

Next, differentiating (F.23b) in time yields

$$\theta_t + e^{-\mathcal{A}(t-2\epsilon)}\mathcal{A}e^{-\mathcal{A}\epsilon}\theta(\epsilon) + e^{-\mathcal{A}(t-2\epsilon)}\mathcal{A}^{\frac{1}{2}}e^{-\mathcal{A}\epsilon}\mathcal{A}^{\frac{1}{2}}w_t(\epsilon)$$

$$= -\int_0^{t-\epsilon} \mathcal{A}^{\frac{1}{2}}e^{-\mathcal{A}s}\mathcal{A}^{\frac{1}{2}}w_{tt}(t-s)\alpha\, ds \tag{F.26a}$$

$$= -\int_\epsilon^t \mathcal{A}^{\frac{1}{2}}e^{-\mathcal{A}(t-\tau)}\mathcal{A}^{\frac{1}{2}}w_{tt}(\tau)\alpha\, d\tau. \tag{F.26b}$$

Then, applying $\mathcal{A}^{\frac{1}{2}}$ throughout (F.26a) readily yields $\mathcal{A}^{\frac{1}{2}}\theta_t \in L_2(2\epsilon, T; L_2(\Omega))$, as claimed in (F.21) via standard analytic semigroup theory, by use of the (critical) regularity $\mathcal{A}^{\frac{1}{2}}w_{tt} \in L_2(\epsilon, T; L_2(\Omega))$ by (F.9), and $\theta(\epsilon) \in L_2(\Omega)$, $\mathcal{A}^{\frac{1}{2}}w_t(\epsilon) \in L_2(\Omega)$ by (F.2b-c). Finally, we differentiate (F.26a) in t and obtain

$$\begin{aligned} \theta_{tt}(t) &= e^{-\mathcal{A}(t-2\epsilon)}\mathcal{A}^2 e^{-\mathcal{A}\epsilon}\theta(\epsilon) + e^{-\mathcal{A}(t-2\epsilon)}\mathcal{A}^{\frac{3}{2}}e^{-\mathcal{A}\epsilon}\mathcal{A}^{\frac{1}{2}}w_t(\epsilon) \\ &\quad -e^{-\mathcal{A}(t-2\epsilon)}\mathcal{A}e^{-\mathcal{A}\epsilon}w_{tt}(\epsilon) - \int_\epsilon^t \mathcal{A}e^{-\mathcal{A}(t-\tau)}w_{ttt}(\tau)d\tau && \text{(F.27a)} \\ &\in L_2(2\epsilon, T; L_2(\Omega)), && \text{(F.27b)} \end{aligned}$$

by standard analytic semigroup theory, with $w_{ttt} \in L_2(\epsilon, T; L_2(\Omega))$ by (F.22) just proved, and $\theta(\epsilon)$, $\mathcal{A}^{\frac{1}{2}}w_t(\epsilon) \in L_2(\Omega)$ by (F.2b-c). Lemma F.4 is proved. □

Step 4.

PROPOSITION F.5. *With reference to problems (F.1)–(F.3), and $\epsilon > 0$ preassigned in Step 1, we have*

$$w_t \in L_2(2\epsilon, T; H^3(\Omega) \cap H_0^2(\Omega)); \; w_{tt} \in L_2(2\epsilon, T; \mathcal{D}(\mathcal{A}) = H^2(\Omega) \cap H_0^1(\Omega)); \tag{F.28}$$

$$w_{ttt} \in L_2(2\epsilon, T; \mathcal{D}(\mathcal{A}^{\frac{1}{2}}) = H_0^1(\Omega)); \; w_{tttt} \in L_2(2\epsilon, T; L_2(\Omega)); \tag{F.29}$$

($\epsilon = 0$ in (9.28), (F.29), if $\theta_0 = 0$).

Proof. We return to problem (F.15), in light of the boosted regularity of θ_{tt} in (F.22) and set, via (F.14),

$$p \equiv q_t = \mathcal{A}w_t = -\Delta w_t \in L_2(\epsilon, T; L_2(\Omega)), \tag{F.30}$$

after using the established regularity for w_t in (F.8). Thus, after differentiating problem (F.15) in t, we obtain

$$\begin{cases} p_{tt} = \frac{\Delta p_t}{\gamma} - \frac{\alpha^2 p_t}{\gamma} + g & \text{in } (2\epsilon, T] \times \Omega \equiv Q_{2\epsilon}; & \text{(F.31a)} \\ p|_{\Sigma_{2\epsilon}} \equiv 0, \quad \frac{\partial p}{\partial \nu}|_{\Sigma_{2\epsilon,2}} \equiv 0 & \text{in } (2\epsilon, T] \times \Gamma_2 \equiv \Sigma_{2\epsilon,2}; & \text{(F.31b)} \\ g = f_t \equiv -\frac{\alpha\theta_{tt}}{\gamma} + \frac{\mathcal{A}w_t}{\gamma^2} + 2\alpha\nabla\alpha \cdot \nabla w_{tt} + \ell.o.t. \in L_2(2\epsilon, T; L_2(\Omega)), & & \text{(F.31c)} \end{cases}$$

by recalling the regularity $\theta_{tt} \in L_2(2\epsilon, T; L_2(\Omega))$ from (F.22); $\mathcal{A}w_t \in L_2(\epsilon, T; L_2(\Omega))$ from (F.8) and $|\nabla w_{tt}| \in L_2(\epsilon, T; L_2(\Omega))$ from (F.9). Invoking once more [L-T.6, Theorem 2.1.2(ii), Eqn. (2.1.10b)) we obtain in the notation of [L-T.6]:

$$\overline{(BT)}_p|_{\Sigma_{2\epsilon}} + \frac{C_1}{\gamma^2}\int_{Q_{2\epsilon}} g^2 dQ + TC_T\|p\|^2_{L_2(Q_{2\epsilon})} \geq \int_{2\epsilon}^T \|\{p(t), p_t(t)\}\|^2_{H^1(\Omega)\times L_2(\Omega)}dt, \tag{F.32}$$

where the boundary terms $\overline{(BT)}_p|_{\Sigma_{2\epsilon}}$ for p on $\Sigma_{2\epsilon} = (2\epsilon, T] \times \Gamma$ is *negative*, by the same argument as in (F.17), via the vanishing of $\frac{\partial p}{\partial \nu}$ on Γ_2, and assumption (F.3) on $\Gamma \setminus \Gamma_2$: $\overline{(BT)}_p|_{\Sigma_{2\epsilon}} \leq 0$. Moreover, p is as in (F.30) and $g \in L_2(2\epsilon, T; L_2(\Omega))$ by (F.31c), as required. Eqn. (F.32) is the counterpart of (F.16) or (9.28) [and, hence, of (6.12)]. We then obtain from (F.32), via (F.30),

$$\begin{cases} \Delta w_t = -p \in L_2(2\epsilon, T; H_0^1(\Omega)) \\ w_t|_{\Sigma_{2\epsilon}} \equiv 0 \end{cases} ; \quad \begin{cases} \Delta w_{tt} = -p_t \in L_2(2\epsilon, T; L_2(\Omega)) & \text{(F.33a)} \\ w_{tt}|_{\Sigma_{2\epsilon}} \equiv 0 & \text{(F.33b)} \end{cases}$$

and hence, by elliptic theory,

$$w_t \in L_2(2\epsilon, T; H^3(\Omega)); \quad w_{tt} \in L_2(2\epsilon, T; H^2(\Omega)). \tag{F.34}$$

Then, (F.34) along with the B.C. for w_t in (F.1c) yields the desired conclusion (F.28) for $\{w_t, w_{tt}\}$.

As to w_{ttt}, we return to Eqn. (F.25), where now the regularity of $\mathcal{A}w_t$ has been boosted to $\mathcal{A}w_t \in L_2(2\epsilon, T; \mathcal{D}(\mathcal{A}^{\frac{1}{2}}) = H_0^1(\Omega))$ by (F.28) and $\mathcal{A}_\gamma^{-1}(\alpha \mathcal{A}\theta_t) \in L_2(2\epsilon, T; \mathcal{D}(\mathcal{A}^{\frac{1}{2}}))$ by (F.21). Hence, (F.25) yields $w_{ttt} \in L_2(2\epsilon, T; \mathcal{D}(\mathcal{A}^{\frac{1}{2}}))$, as claimed in (F.29).

Finally, differentiating in time (F.25) yields

$$w_{tttt} = (-\mathcal{A}_\gamma^{-1}\mathcal{A})\mathcal{A}w_{tt} - \mathcal{A}_\gamma^{-1}(\alpha \mathcal{A}\theta_{tt}) + \ell.o.t. \in L_2(2\epsilon, T; L_2(\Omega)), \tag{F.35}$$

where now $\mathcal{A}w_{tt} \in L_2(2\epsilon, T; L_2(\Omega))$ by (F.28) and $\mathcal{A}_\gamma^{-1}(\alpha \mathcal{A}\theta_{tt}) \in L_2(2\epsilon, T; L_2(\Omega))$ by (F.22). Hence, (F.35) yields $w_{tttt} \in L_2(2\epsilon, T; L_2(\Omega))$, as claimed in (F.29). □

Thus, Lemma F.4 and Proposition F.5 combined prove the required regularity (F.4), (F.5) with $t_0 = 2\epsilon$. Theorem F.1 is thus proved. □

REMARK F.1. *Plainly, the above boot-strap argument can be further continued. Here, however, we limit ourselves to what is needed to apply Corollary 4.2.3.*

Bibliography

[A-T.1] P. Albano and D. Tataru, Carleman estimates and boundary observability for a coupled parabolic-hyperbolic system, *Electronic J. of Diff. Eqns.* (22) (2000), 1–15.

[A.1] J. P. Aubin, Un théorèm de compacite. *C. R. Acad. Sc.* 256 (1963), 5042–5044.

[Av.1] G. Avalos, Exact controllability of a thermo-elastic system with control in the thermal component only, *Advances in Diff. Eqns.*, to appear.

[A-L.1] G. Avalos and I. Lasiecka, Boundary controllability of thermo-elastic plates with free boundary conditions, *SIAM J. Control*, to appear.

[B-L-R.1] C. Bardos, J. Lebeau, and J. Rauch, Sharp sufficient conditions for the observation, control and stabilization of waves from the boundary, *SIAM J. Control & Optimiz.* 30 (1992), 1024–1065.

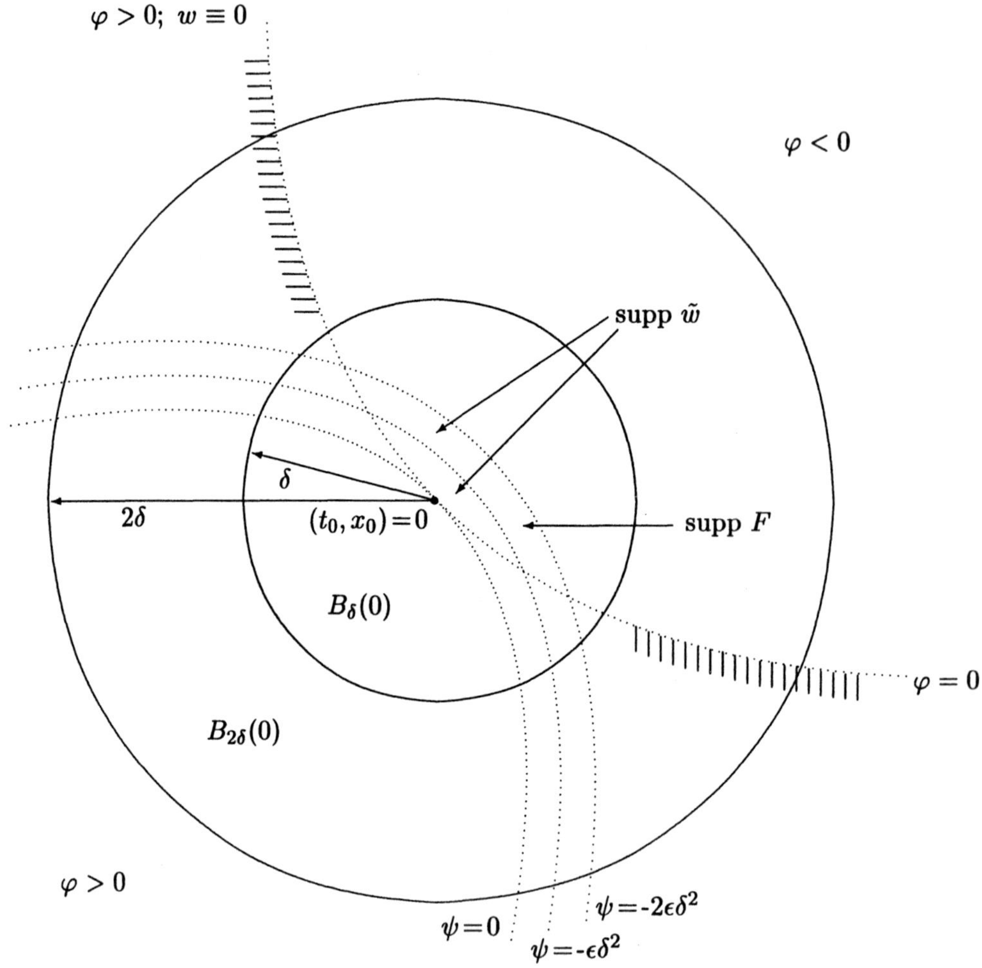

Figure 1: The geometry behind the proof of Theorem 10.3. The intersection of the curve $\psi = -2\epsilon\delta^2$ and the surface of the ball $B_\delta(0)$ is in $\varphi \geq 0$ because of the condition $\psi(t,x) \leq \varphi(t,x) - 2\epsilon(t^2 + |x|^2)$: Setting $\psi(t,x) = -2\epsilon\delta^2$ and $t^2 + |x|^2 = \delta^2$ (surface of $B_\delta(0)$) implies $\varphi(t,x) \geq 0$.

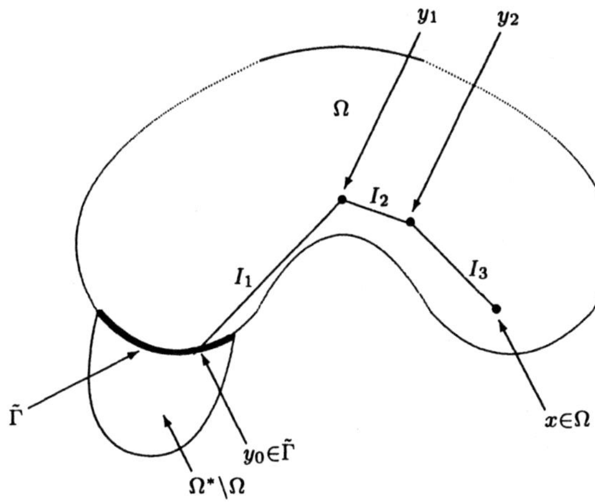

Figure 2

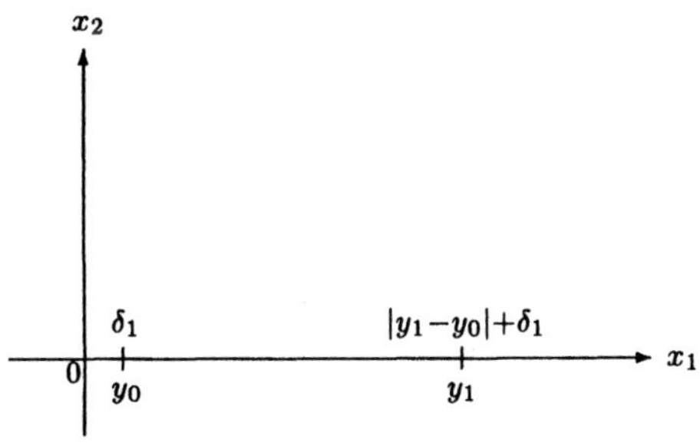

Figure 3

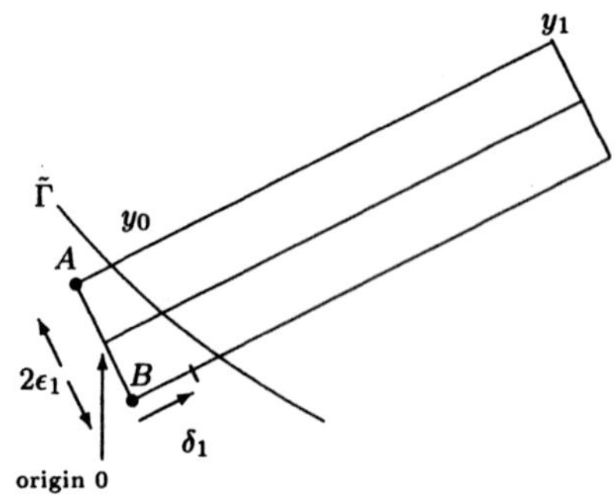

Figure 4: Extending the interval by δ_1 guarantees zero Cauchy data on the surface AB of the parallelopiped.

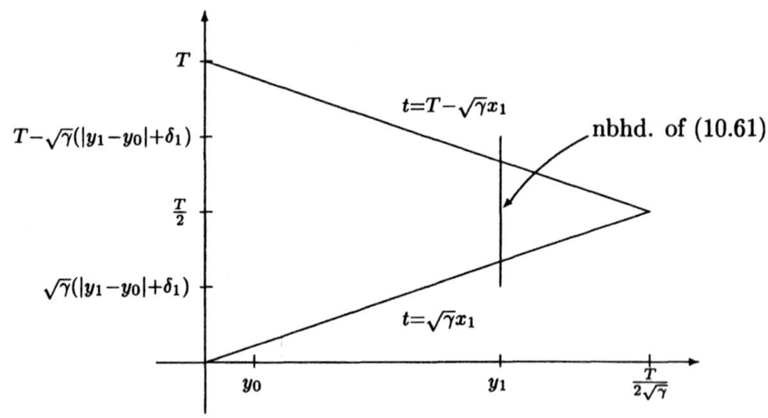

Figure 5

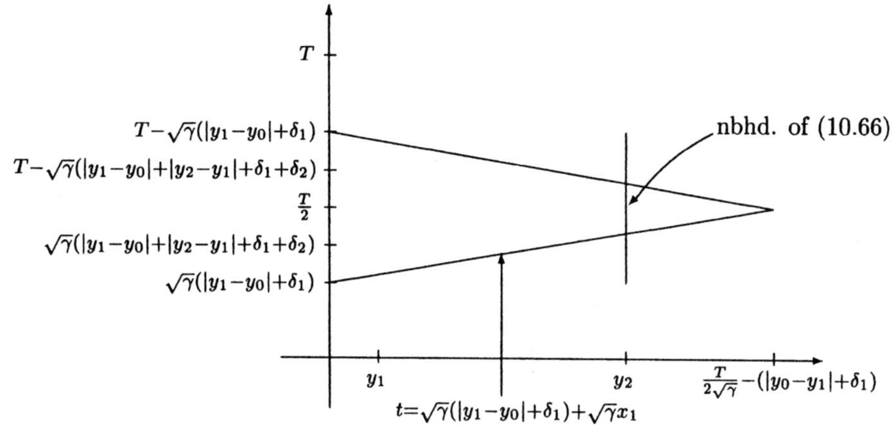

Figure 6

[D-Z.1] L. de Teresa and E. Zuazua, Controllability for the lienar system of thermo-elastic plates, *Advances in Diff. Eqns.* 1 (1996), 369–402.

[E.1] M. Eller, Uniqueness of continuation theorems, in *Direct and inverse problems of mathematical physics*, Kluver (1999)

[E.2] M.Eller, Carleman estimates with a second large parameter, preprint 1999

[E-L-T.1] M. Eller, I. Lasiecka, and R. Triggiani, Exact boundary controllability of thermoelastic plates with variable coefficients, Proceedings of Conference 'Semigroup of operators, theory and applications,' held at Los Angeles, Dec. 14–18, 1998, Birkhäuser.

[F-I.1] A. Fursikov and O. Imanuvilov, *Controllability of Evolution Equations*, Lecture Notes, Seoul National University, 1996.

[H-Z.1] S. W. Hansen and B. Zhang, Boundary control of a linear thermo-elastic beam, *J. Math. Anal. & Appl.* 210 (1997), 182–205.

[H.1] L. Hörmander, On the uniqueness of the Cauchy problem under partial analyticity assumptions, in *Geometric Optics and Related Topics*, F. Colombini & N. Lerner, eds., 1997.

[H.2] L. Hörmander *The analysis of linear partial differential operators I-IV*, Springer-Verlag, New York, 1983

[I.1] V. Isakov, On the uniqueness of continuation for a thermoelasticity system, preprint, 1998.

[I.2] V. Isakov, *Inverse problems for partial differential equations*, Springer-Verlag, New York, 1997

[I.3] V. Isakov, On uniqueness in a lateral Cauchy problem with multiple characteristics, *J. Diff. Eq.*, Vol. 134, **1**, (1997), 134-147

[Ke.1] B. Kellogg, Properties of solutions of elliptic boundary value problems, Chapter 3 in "The Mathematical Foundations of the Finite Element Method with Applications to PDE," Ed. by A. K. Aziz, Academic Press, 1972.

[K.1] V. Komornik, *Exact controllability and stabilization*, Masson, Paris, 1994.

[La.1] J. Lagnese, *Boundary Stabilization of Thin Plates*, SIAM, Philadelphia, 1989.

[La.2] J. Lagnese, The reachability problem for thermoelastic plates, *Arch. Rational Mechanics and Analysis* 112 (1990), 223-267.

[L-L.1] J. Lagnese and J. .L. Lions, *Modeling, Analysis and Control of Thin Plates*, Masson, 1988.

[Las.1] I. Lasiecka, Controllability of a viscoelastic Kirchoff plate, *Inter. Series Numer. Math.* 91 (1989), 237–247, Birkhäuser Verlag Basel.

[Li.1] J. L. Lions, *Controllabilite exacte, stabilization de systemes distribues*, Vol. 1 and 2, Masson, Paris, 1988.

[L-L-T.1] I. Lasiecka, J. L. Lions, and R. Triggiani, Nonhomogeneous boundary value problems for second order hyperbolic operators, *J. Math. Pures et Appl.* 65 (1986), 149–192.

[L-R-T.1] I. Lasiecka, M. Renardy, and R. Triggiani, Backward uniqueness for thermoelastic plates with rotational forces, *Semigroup Forum*, to appear.

[L-T.1] I. Lasiecka and R. Triggiani, Exact boundary controllability for the wave equations with Neumann boundar control, *Appl. Math. & Optimiz.* 19 (1989), 243–290. Preliminary version *Springer Verlag LNCIS* 100 (1987), 316–371.

[L-T.2] I. Lasiecka and R. Triggiani, Uniform boundary stabilization of the wave equation in a bounded region with L_2-boundary feedback in the Dirichlet B.C., *J. Diff. Eqns.* 66 (1987), 340–390.

[L-T.3] I. Lasiecka and R. Triggiani, Exact controllability of semilinear abstract systems with application to waves and plates boundary control problems, *Appl. Math. & Optimiz.* 3 (1991), 109–154.

[L-T.4] I. Lasiecka and R. Triggiani, Exact controllability and uniform stabilization of Kirchoff plates with boundary control only on $\Delta w|_\Sigma$, *J. Diff. Eqns.* 93 (1991), 62–101.

[L-T.5] I. Lasiecka and R. Triggiani, Uniform boundary stabilization of the wave equation with Dirichlet and Neumann feedback control without geometric conditions, *Appl. Math. Optimiz.* 25 (1992), 189–224.

[L-T.6] I. Lasiecka and R. Triggiani, Carleman's estimates and boundary controllability for a system of coupled, nonconservative second-order hyperbolic equations, *Lecture Notes in Pure and Applied Mathematics*, Marcel Dekker, Vol. 188, pp. 215–245, 1997.

[L-T.7] I. Lasiecka and R. Triggiani, Structural decomposition of thermoelastic semigroups with rotational forces, *Semigroup Forum* 60(1) (2000), 16–66.

[L-T.8] I. Lasiecka and R. Triggiani, Analyticity, and lack thereof, of thermoelastic semigroups, *Europ. Soc. Appl. Math., ESAIM* 4 (1999), 199–222.

[L-T.9] I. Lasiecka and R. Triggiani, *Differential and Algebraic Riccati Equations with Applications to Boundary/Point Control Problems: Continuous Theory and Approximation Theory*, LNCIS 164, Springer Verlag, 1991, pp. 160.

[L-T.10] I. Lasiecka and R. Triggiani, *Control for Partial Differential Equations: Continuous and Approximation Theories*, Vols. I and II, Cambridge University Press, Encyclopedia of Mathematics and its Applications, January 2000.

[L-T.11] I. Lasiecka and R. Triggiani, A sharp trace result of a thermo-elastic plate equation with hinged/Neumann coupled B.C., *Discrete & Continuous Dynamical Systems* 3 (1999), 585–598.

[L-T.12] I. Lasiecka and R. Triggiani, Exact null controllability of structurally damped and thermoelastic parabolic models, *Atti della Accademia Nazionale dei Lincei, Mathematica e Applicazioni*, Anno CCCXCV (1998), 43–69, Rome, Italy.

[L-T.13] I. Lasiecka and R. Triggiani, Work in progress.

[L-T-Y.1] I. Lasiecka, R. Triggiani, and P. F. Yao, An observability estimate in $L_2(\Omega) \times H^{-1}(\Omega)$ for second order hyperbolic equations with variable coefficients, in *Control of Distributed Parameter and Stochastic Systems*, edited by S. Chen, X. Li, J. Yong, and X. Zhou, Kluwer Academic Publishers 1999, Proceedings of IFIP Conference held in Hangzhou, China, June 19–22, 1998.

[L-T-Y.2] I. Lasiecka, R. Triggiani, and P. F. Yao, Inverse/observability estimates for second order hyperbolic equations with variable coefficients, *J. Math. Anal. & Appl.* 235 (1999), 13–57. Preliminary announcement in "Non-Linear Analysis: Theory, Methods and Applications," 30 (1997), 111–122.

[Lit.1] W. Littman, Remarks on global uniqueness theorems for partial differential equations, *Contemporary Mathematics*, to appear.

[Liu.1] W. Liu, Partial exact controllability and exponential stability in higher-dimensional linear thermoelasticity, ESAIM 3 (1998), 23–48. Erratum on above paper, to appear.

[R.1] J.Rauch *Partial differential equations*, Springer Verlag New York, 1991

[S.1] J. Simon, Compact sets in the space $L^p(0, T; B)$, *Ann. Math. Pura & Appl.* (4) 148 (1987), 65–96.

[Ta.1] D. Tataru, A priori estimates of Carleman's type in domains with boundaries, *J. Math. Pures et Appl.* 73 (1994), 355–387.

[Ta.2] D. Tataru, A priori psuedo-convexity energy estimates in domains with boundary: Applications to exact boundary controllability for conservative PDE's, Ph.D. dissertation, University of Virginia, May 1992.

[Ta.3] D. Tataru, Unique continuation for solutions to PDE's between Hörmander's theorem and Holmgren's theorem, *Comm. Part. Diff. Eqn.* 20 (1995), 855–884.

[Ta.4] D. Tataru, Unique continuation for operators with partially analytic coefficients, *J. Math. Pures Appl.*, 78 (1999), 505–521

[T-L.1] A. Taylor and D. Lay, *Introduction to Functional Analysis,* second edition, 1980, John Wiley.

[Tr.1] R. Triggiani, Exact boundary controllability in $L_2(\Omega) \times H^{-1}(\Omega)$ of the wave equation with Dirichlet boundary control, *Appl. Math. & Optimiz.* 18 (1988), 241–277. Preliminary version in Springer Verlag *LNCIS* 102 (1987), 291–332.

[Tr.2] R. Triggiani, Sharp regularity theory for thermo-elastic mixed problems, *Applicable Analysis* 73 (3–4) (1999), 557–572.

Matthias Eller. Georgetown University, Department of Mathematics, Washington, DC 20057-0001, USA
E-mail: MEller@tntech.edu

Irena Lasiecka. University of Virginia, Department of Mathematics, Kerchof Hall, P.O. Box 400137, Charlottesville, VA 22904-4137, USA
E-mail: il2v@virginia.edu

Roberto Triggiani. University of Virginia, Department of Mathematics, Kerchof Hall, P.O. Box 400137, Charlottesville, VA 22904-4137, USA
E-mail: rt7u@virginia.edu

Shape Derivative on a Fractured Manifold

Jamel Ferchichi and Jean-Paul Zolésio

Abstract. We present in this paper a derivative result of the Neumann tangential problem posed in a two dimension manifold packaging a regular related fracture. We appear the shape derivative of a quadratic functional cost as a limit of a jump through the crack. That is why we introduce a family of envelopes surrounding the fracture which enable us to relax certain terms and to clear up the lack of regularity which results from the presence of the fracture. We use the *min-max* derivation in order to avoid deriving the state equation and to manage the cracks singularities. We therefore write the functional as a *min-max* form on a space undertaking the hidden boundary regularity established by the tangential extractor method.

1 Introduction and Motivation

This work is in the setting of the detection techniques of a *fracture* contained in an elastic structure, usually a thin Shell. This study relies on the theory of intrinsic geometry (see [4], [6]) and the results of Laplace-Beltrami operator (see [8]). The aim of this paper is the boundary expression, (given at theorem (8.2), for the Shape gradient of the cost functional

$$J((\omega\backslash\sigma)) = \frac{1}{2}\int_{(\omega\backslash\sigma)} (\Phi - \Phi_d)^2$$

Where ω is a bounded open subset of a C^2 manifold Γ with the relative boundary $\partial_\Gamma\omega$ denoted by $\partial\omega$, σ is a connected fracture contained in ω with ends s_1 and s_2. The solution of the tangential Neumann problem on the open set $(\omega\backslash\sigma)$ is denoted Φ and Φ_d is a given heat measure.

According to the identity $\Phi'_\Gamma = \dot{\Phi}|_\Gamma - \nabla_\Gamma\Phi V(0)$ (see [12]), the function Φ'_Γ looks like it is less regular than the material derivative $\dot{\Phi}$ of the state function Φ which requires technicalities.

We begin by characterizing the shape boundary derivative Φ'_Γ of the State solution of the tangential Neumann problem on a piecewise smooth open subset ω which is contained in the surface Γ of $\mathbb{R}^N$ associated to the Laplace-Beltrami operator. Then we give a continuity result for the Tangential Neumann problem with respect to a family of envelopes surrounding the fracture σ. That allows to get rid of the lack of regularity due to the fracture. In order to show the necessary optimality condition of the initial domain and present the shape gradient of the cost functional $J((\omega\backslash\sigma))$, we will use the *min-max* theory through a hidden boundary regularity of the State provided by the *tangential extractor*. Finally, the Eulerian derivative

$dJ(\omega\backslash\sigma, V)$ turns out to be characterizing by a distributed gradient supported on the closure of the fracture $\overline{\sigma}$, its expression is given as a sum of a distributed term on γ, a jump distributed term in $L^1(\sigma)$ plus a Dirac measures at the two extremities s_i. The used technical permit to reach the situation in which the fracture σ need not to be smooth.

2 Preliminaries

2.1 Velocity Method

Let D be a smooth bounded domain of $\mathbb{R}^N$. We consider a regular open subset Ω from D. Its relative boundary will be denoted by Γ which is an oriented compact manifold. Let X be a given point of $\overline{D}$ and $t \in [0,\ \delta[$, where δ is a positif real. We define the point $x(t) = T_t(X)$ which moves on the trajectory $x \longrightarrow x(t)$ with the velocity $||\partial_t x(t)||$ equals to $||\partial_t T_t(X)||$,

$$T_t \in C^1([0,\ \delta[, W^{1,\infty}(D;\mathbb{R}^N)) \tag{2.1}$$

Let

$$V(t,x) = \frac{\partial T_t}{\partial t} \circ T_t^{-1}(x) \tag{2.2}$$

It follows

$$V \in C^0([0,\delta[, W^{1,\infty}(D;\mathbb{R}^N)) \tag{2.3}$$

Conversely, it is possible to associate transformations T_t to some vectors fields V satisfying (2.3). Let $\mathcal{E}$ be the set of vector fields satisfying (2.3), with $< V(x,t), n_{\partial D}(x) >= 0$ for $x \in \partial D$ almost everywhere and $V(x,t) = 0$ for all singular point x of ∂D. The transformations T_t is called the flow mapping associated to V. We refer to [12] for the proof of the subsequent theorem.

THEOREM 2.1. *We have the two following assertions:*

i) Let V be a vector field of $\mathcal{E}$. Transformations $T_t \in C^1([0,\ \delta[, W^{1,\infty}(D;\mathbb{R}^N))$ may be associated to V, moreover(2.2) holds.

ii) Let T_t be transformations satisfying (2.1) then there exists $V \in \mathcal{E}$ verifying (2.2). The transformations T_t is solution of the ordinary differential equation

$$\partial_t x(X,t) = V(x(X,t),t);\ \ x(X,0) = X$$

In the sequel, we point out that, in the general problem, an important issue is to keep the surface Γ fixed in the perturbation process. Such constraint is obviously solved by choosing in a general setting the speed vector field $V(t,x)$ tangent to the surface Γ: $V(t,x).n_\Gamma(x) = 0$.

We consider an open subset ω of Γ containing a fracture denoted by σ. The boundary of the open subset is also of class C^2 by parts. We consider the parameter family of open subset ω_t generated by the family of flows $T_t(V)$ associated to the vector field V. In virtue of the condition satisfied by V, the family of boundaries γ_t of ω_t is moving in the surface Γ. We design by n the out normal field on the surface Γ and by $\nu(X)$ the normal field on γ outer of ω contained in tangent space on Γ on X.

2.2 The Oriented Distance Function

DEFINITION 2.2. *the oriented distance function is defined as follows*

$$b_\Omega(x) = \begin{cases} d_\Gamma(x) & \text{if } x \in \overline{\Omega}^c \\ -d_\Gamma(x) & \text{if } x \in \Omega \end{cases}$$

Among the intrinsic geometry properties of the oriented distance function we quote the following, (we refer to [4], [5] and [6] for more informations):

i) ∇b_Ω is an extension of the normal field n on Γ.

ii) Δb_Ω is the mean curvature H of the surface Γ (i.e. $H = \delta b_\Omega|_\Gamma$.

2.3 The Projection Mapping

Let U be a tubular neighborhood of Γ given, for h small enough, by

$$U(\Gamma) = \{x \in D; |b_\Omega(x)| < h\}$$

we can associate to the oriented distance function b_Ω a projection mapping on the compact manifold Γ.

DEFINITION 2.3. *The projection mapping p is defined in [4] by*

$$p : U \longrightarrow \Gamma; p_\Gamma(x) = x - b_\Omega(x).\nabla b_\Omega(x)$$

2.4 Laplace-Beltrami Operator

DEFINITION 2.4. *The Laplace-Beltrami operator is denoted by Δ_Γ and specified, in [1], for a such regular function φ by:*

$$\Delta_\Gamma \varphi = div_\Gamma \nabla_\Gamma \varphi$$

with

$$\nabla_\Gamma \varphi = (\nabla\phi - < \nabla\phi, \nabla b_\Omega > \nabla b_\Omega)|_\Gamma, \ \ \phi \textit{ is any extension of } \varphi \textit{to a neighborhood of } \Gamma$$

$$div_\Gamma e = (divE - < DE.\nabla b_\Omega, \nabla b_\Omega >)|_\Gamma, \ \ E \textit{ is an extension of } e \textit{ to a neighborhood of } \Gamma$$

where b_Ω is the oriented distance function.

2.5 The Neumann Problem

Let F be an element given in $H^{\frac{1}{2}+\delta}(D)$ such that $F|\omega = f$ and $F|\omega_t = f_t$. We consider the tangential Neumann problem formulated in the fractured subset $\omega\backslash\sigma$:

$$\mathcal{N}T \begin{cases} -\Delta_\Gamma \Phi &= f \text{ in } \omega\backslash\sigma \\ \frac{\partial\Phi}{\partial\nu} &= 0 \text{ on } \partial(\omega\backslash\sigma) \end{cases} \tag{2.4}$$

LEMMA 2.5. *We notice that*

i) The previous problem has a unique solution in the following Hilbert space $H^1_*(\omega\backslash\sigma) = \{v \in H^1(\omega\backslash\sigma); < v, 1 >= 0\}$ *where* $< v, 1 >= 0$ *stands for* $\int_{(\omega\backslash\sigma)} v = 0$.

ii) The optimal regularity of the solution of the problem belongs to the space $H^{\frac{3}{2}-\mu}(\omega\backslash\sigma)$ *with* $\mu > 0$, *and to* $H^2(\omega')$ *for every open subset* ω' *contained in* $(\omega\backslash\sigma)$ *with empty intersection with* $\overline{\sigma}$, *we refer to [11].*

The principal aim of this paper is to exhibit the shape gradient of the cost function J. However, because of the existence of the fracture the open subset $\omega\backslash\sigma$ is not Lipschitzian, this lack of regularity involves many technical problems; especially the Green's formula is not hold. As a first step, we will investigate the tangential Neumann problem subsequently in a Lipschitzian and a piecewise smooth domain.

3 Case of a Lipschitzian Manifold

DEFINITION 3.1. *We define a Lipschitzian manifold* ω, *whom relative boundary* $\partial_\Gamma\omega = \gamma$, *as follows: It exists an injective mapping* $\lambda : [0,1] \longrightarrow \mathbb{R}^3$ *such that* $\lambda \in Lip(]0,1[;\mathbb{R}^3), \lambda([0,1]) = \gamma$ *and* $\lambda(0) = \lambda(1)$. λ *is a parametrage of* γ.

Throughout this section we assume ω to be a Lipschitz open subset which is locally on one side of its boundary. The manifold ω is containing in the surface Γ. We consider the tangential Neumann problem.

$$\mathcal{P}\begin{cases} -\Delta_\Gamma\Phi &= f \text{ in } \omega \\ \frac{\partial\Phi}{\partial\nu} &= 0 \text{ on } \partial\omega. \end{cases} \tag{3.1}$$

REMARK 3.1. *Since the boundary* $\partial\omega$ *is Lipschitzian so Green's formula holds.*

3.1 Shape Analysis

Moving Problem

¿From a heuristic point of view, looking for the shape sensitivity consists in observing the perturbation effect on the solution defined in $T_t(V)(\omega) = \omega_t$ when $t \longrightarrow 0$. For this we perturb the domain ω by the transformation T_t, it follows that

$$\mathcal{P}_t\begin{cases} -\Delta_\Gamma\Phi_t &= f_t \text{ in } \omega_t \\ \frac{\partial\Phi_t}{\partial\nu_t} &= 0 \text{ on } \partial\omega_t \end{cases} \tag{3.2}$$

REMARK 3.2. *It is clear that,* $\forall t \in [0, \delta[$, *it exists a unique solution* Φ_t *of the perturbed problem with* $\int_{\omega_t} f_t = 0$, *which motivates the definition of* f_t *in lemma (3.5).*

LEMMA 3.2. *We refer to [8] to introduce the Green's formula associated to a manifold* ω *having a boundary* γ. *Let* E *and* φ *be a regular functions, it follows*

$$-\int_\omega div_\Gamma E.\varphi = \int_\omega \{E.\nabla_\Gamma\varphi - H\varphi < E, n >_{\mathbb{R}^N}\} - \int_\gamma \varphi < E, \nu >$$

Where $H = \Delta b_\Omega|_\Gamma$ *is the mean curvature of the manifold* Γ.

Existence of the Material Derivative of the State Φ

We interested in providing the existence of the Material derivative $\dot{\Phi}$. We intend to deal with the differentiability of the map $t \to \Phi_t \circ T_t$ at the right of zero.

THEOREM 3.3. *The map $t \to \Phi_t \circ T_t$ is differentiable to the right at zero and its derivative*

$$\dot{\Phi} = \lim_{t \to 0} \frac{\Phi_t \circ T_t - \Phi}{t}$$

in $H^1_*(\omega)$, verifies the equation

$$\begin{array}{rcl} \int_{\omega_\varepsilon} \nabla_\Gamma \dot{\Phi} \nabla_\Gamma \psi & = & \int_\omega < 2\varepsilon(V) \nabla_\Gamma \Phi, \nabla_\Gamma \psi > \\ & - & \int_\omega < div_\Gamma V(0) \nabla_\Gamma \Phi, \nabla_\Gamma \psi > \\ & + & \int_\omega div_\Gamma (f.V(0)) \psi \end{array} \tag{3.3}$$

The proof of this theorem will be split in two steps. First, via Green's formula introduced in lemma 3.2, the weak formulation of the perturbed problem is given by

$$\int_{\omega_{\varepsilon t}} \nabla_\Gamma \Phi_{\varepsilon t} \nabla_\Gamma \varphi_t = \int_{\omega_t} f_t \varphi_t \forall \varphi_t \in H^1_*(\omega_t) \tag{3.4}$$

Then, we came back into the fixed domain in the first term

$$\int_{\omega_{\varepsilon t}} \nabla_\Gamma \Phi_{\varepsilon t} \nabla_\Gamma \varphi_t = \int_\omega (\nabla_\Gamma \Phi_t) \circ T_t . (\nabla_\Gamma \varphi_t) \circ T_t j(t) \tag{3.5}$$

As we have

$$(\nabla_\Gamma \Phi_t) \circ T_t = \nabla(\Phi_t \circ p) \circ T_t|_\Gamma =^* (DT_t)^{-1} \nabla(\Phi_t \circ p \circ T_t)|_\Gamma \tag{3.6}$$

We notice, via a suitable choice of test functions φ_t, that only the tangential component of the vector $\nabla(\varphi_t \circ T_t)$ vanishes. Actually, let $\varphi_t = \psi \circ T_t^{-1}$ where ψ belongs to $H^{\frac{3}{2}}(D)$, so its trace on Γ is in $H^1(\Gamma)$, with $\frac{\partial \psi}{\partial n} =< \nabla \psi,\ n >= 0$ We note that

$$(\nabla \varphi_t) \circ T_t|_\Gamma =^* (DT_t)^{-1} \nabla \psi|_\Gamma =^* (DT_t)^{-1} \nabla_\Gamma \psi$$

then due to $< \nabla(\Phi_t \circ p),\ n >= 0$, we prove that

$$< (DT_t)^{-1} .^* (DT_t)^{-1} \nabla(\Phi_{\varepsilon t} \circ p \circ T_t),\ n >= 0$$

this means that the vector $(DT_t)^{-1} .^* (DT_t)^{-1} \nabla(\Phi_t \circ p \circ T_t)$ is tangential. First, we note $\theta^t = j(t) \Phi_t \circ p \circ T_t$ and derive

$$A(t) = \int_{\omega_{\varepsilon t}} \nabla_\Gamma \Phi_t \nabla_\Gamma \varphi_t = \int_\omega < D(t) . \nabla(\theta^t), \nabla_\Gamma(\psi) > \tag{3.7}$$

where $j(t) = det(DT_t) \| ^*(DT_t)^{-1} . n \|$ and $D(t) = (DT_t)^{-1} .^* (DT_t)^{-1}$. The same thing applies for the second term, we get

$$B(t) = \int_{\omega_t} f_t \varphi_t = \int_\omega f^t \psi j(t) \tag{3.8}$$

with $f^t = f_t \circ T_t$

DEFINITION 3.4. *If the limit*

$$\lim_{t\to 0} \frac{\theta^t - \theta}{t}$$

exists strongly in $H^1_*(\omega)$ *(denoted* $\dot{\theta}$*) we say that* θ *has a material derivative in the direction of the vector field* V.

We are interested in the sequel to prove the existence of the material derivative $\dot{\theta}$. The second step of the proof will now be developed.

Weak Material derivative Let

$$z^t = \frac{\theta^t_\varepsilon - \theta_\varepsilon}{t} \in H^1_*(\omega)$$

satisfies for all $\psi \in H^1(\omega)$

$$\int_{\omega_{\varepsilon t}} \nabla_\Gamma z^t \nabla_\Gamma \psi = -\int_\omega \left\langle \frac{D(t) - I}{t} \nabla(\theta^t), \nabla_\Gamma(\psi) \right\rangle + \int_\omega \frac{f^t j(t) - f}{t} \psi \tag{3.9}$$

LEMMA 3.5. *Let* $F \in H^{\frac{1}{2}+\delta}(D)$ *such that* $F_{|\omega} = f$ *and*

$$f_t = F_{|\omega_t} - \frac{1}{|\omega|} \int_{\omega_t} F$$

so the mapping $t \to f^t$ *is weakly differentiable in* $H^{\delta-1}(\omega)$. *Furthermore:* $\frac{f^t - f}{t} \to \nabla_\Gamma F.V(0)$ *weakly in* $H^{-1}(\omega)$, *for the proof it's enough to consult [12].*

REMARK 3.3. *A direct way to get the existence and the characterization of* $\dot{\Phi} = \frac{\partial}{\partial t}(\Phi_t \circ T_t)|_{t=0}$ *is to apply the implicit function theorem. This way, we would directly get the result concerning the material derivative if the right hand side* $f_{|\Gamma}$ *of the equation is supposed more regular than* $L^2(\Gamma)$. *Here,* $f_{|\Gamma}$ *belongs to* $L^2(\Gamma)$ *does not imply the strong convergence in* $H^{-1}(\Gamma)$ *of the quotient* $(f^t - f)/t$, *as it is required to apply the implicit function theorem. This lack of regularity needs a delicate proof for the existence of* $\dot{\Phi}$. *In [12] there are counter examples for which one can not expect the mapping to be strongly differentiable in* $H^{-1}(\Gamma)$ *for any* f *in* $L^2(\Gamma)$.

By embedding the test function $\psi = z^t - l$ in (3.9) where l verifies

$$\int_\omega l = \int_\omega z^t, \quad \frac{\partial z^t}{\partial n} = \frac{\partial l}{\partial n}, \quad \text{and } \nabla_\Gamma l = \alpha \nabla_\Gamma z^t \text{with} \alpha \neq 1$$

Then we get

$$\begin{aligned} \textstyle\int_\omega \nabla_\Gamma z^t \nabla_\Gamma z^t &= \textstyle\int_{\omega_\varepsilon} \nabla_\Gamma z^t \nabla_\Gamma l - \int_\omega < \frac{D(t)-I}{t} \nabla(\theta^t), \nabla_\Gamma(z^t - l) > \\ &+ \textstyle\int_\omega \frac{f^t j(t) - f}{t}(z^t - l) \end{aligned} \tag{3.10}$$

Which enable us to point out that:

$$\|\nabla_\Gamma z^t\|_{L^2(\omega)} \le c$$

It transpires that z^t is bounded in $H^1_*(\omega)$, then there exists a subsequence which converges weakly in the same space. Let $\dot{\theta}$ be this weak limit, $\dot{\theta}$ satisfies the under equation

$$\int_\omega \nabla_\Gamma \dot{\theta} \nabla_\Gamma \psi = -\int_\omega < D'(0)\nabla_\Gamma \theta, \nabla_\Gamma \psi > + \int_\omega [fj'(0) + \nabla F V(0)]\psi \tag{3.11}$$

Obviously $\dot{\theta}$ is unique so all the sequence z^t is weakly convergent to $\dot{\theta}$ in the space $H^1_*(\omega)$.

Strong Material derivative Via the same choice of test function we prove the norms convergence, in fact

$$\begin{aligned} \lim_{t\to 0} \|\nabla_\Gamma z^t\|^2_{L^2(\omega)} &= \textstyle\int_{\omega_\epsilon} \nabla_\Gamma \dot{\theta} \nabla_\Gamma l - \int_\omega < D'(0)\nabla(\theta), \nabla_\Gamma(\dot{\theta} - l) > \\ &+ \textstyle\int_\omega [fj'(0) + \nabla F.V(0)](\dot{\theta} - l) \\ &= \|\nabla_\Gamma \dot{\theta}\|^2_{L^2(\omega)} \end{aligned} \tag{3.12}$$

We conclude that $\frac{\theta^t_\epsilon - \theta_\epsilon}{t} \to \dot{\theta}$ strongly in $H^1_*(\omega)$. Which enable us to show the existence of $\dot{\Phi}$ strongly in $H^1_*(\omega)$ since it transpires that

$$\dot{\theta} = j'(0)\Phi \circ p + j(0)\dot{\Phi} \circ p.$$

Then by introducing the last identity in equation (3.11) we deduce that $\dot{\Phi}$ satisfies the following equation:

$$\begin{aligned} \textstyle\int_{\omega_\epsilon} \nabla_\Gamma \dot{\Phi} \nabla_\Gamma \psi &= -\textstyle\int_\omega < D'(0)\nabla_\Gamma \Phi, \nabla_\Gamma \psi > \\ &- \textstyle\int_\omega < j'(0)\nabla_\Gamma \Phi, \nabla_\Gamma \psi > \\ &+ \textstyle\int_\omega [fj'(0) + \nabla_\Gamma F.V(0)]\psi \forall \psi \in H^1_*(\omega) \end{aligned} \tag{3.13}$$

Thus, the equation verified by $\dot{\Phi}$ is given by:

$$a(\dot{\Phi}, \psi) = l(\psi) \forall \psi \in H^1_*(\omega)$$

Where a is the coercive bilinear form given as follows:

$$a(\dot{\Phi}, \psi) = \int_\omega \nabla_\Gamma \dot{\Phi} \nabla_\Gamma \psi \tag{3.14}$$

and l is the following linear form:

$$\begin{aligned} l(\psi) &= \textstyle\int_\omega -D'(0)\nabla_\Gamma \Phi \nabla_\Gamma \psi - \int_\omega j'(0)\nabla_\Gamma \Phi \nabla_\Gamma \psi \\ &+ \textstyle\int_\omega [fj'(0) + \nabla_\Gamma F.V(0)]\psi \end{aligned} \tag{3.15}$$

Where the expressions of $D'(0)$ and $j'(0)$ are given by the followings lemma

LEMMA 3.6. *Let $\delta > 0$ be a given real number. The application $t \longrightarrow DT_t^{-1}$ is differentiable on $]0, \ \delta[$ and we have $\forall t \in]0, \ \delta[, \ \exists \alpha \in]0,1[$ such that $DT_t^{-1} = Id - tDV(\alpha t)$*

LEMMA 3.7. *Let $\delta > 0$ be a given real number. The application $t \in [0, \ \delta[\to j(t) \in C^{k-1}, \ k \geq 1$ is differentiable and $j'(0) = divV(0) - (DV(0)n, n) = div_\Gamma V(0)$*

Proof. The boundary Jacobian $j(t) = det(DT_t)\|{}^*DT_t^{-1}.n\|$ is differentiable for transformations $T_t(V)$ in $C^1([0,\delta[, C^2(\overline{D}, \mathbb{R}^N))$ and we have

$$\begin{aligned} \frac{\partial\|{}^*DT_t^{-1}.n\|}{\partial t}|_{t=0} &= -(DV(0)n, n). \\ \frac{\partial det(DT_t)}{\partial t}|_{t=0} &= divV(0). \end{aligned}$$

It follows that

$$-D'(0) = \{DV(0) +^* DV(0)\} = 2\varepsilon(V)$$

$$\nabla_\Gamma f.V(0) + f\,\mathrm{div}_\Gamma\,V(0) = \mathrm{div}_\Gamma(fV(0))$$

Moreover, it exists $\alpha_1, \alpha_2; 0 < \alpha_1 < \alpha_2$ such that $\alpha_1 I < \alpha(V) < \alpha_2 I$. □

This achieves the proof of the announced theorem (2.1) in page 232.

3.2 Shape Gradient Distributed Expression

According to the existence of the material derivative, we are able to give the Shape Gradient $dJ(\omega,\ V)$.

PROPOSITION 3.8. *The distributed expression of the Shape Gradient is given by:*

$$\begin{aligned} dJ(\omega, V) &= \textstyle\int_\omega[\frac{1}{2}(\Phi - \Phi_d)^2 - \nabla_\Gamma\Phi\nabla_\Gamma P]div_\Gamma V(0) - \int_\omega(\Phi - \Phi_d)\nabla_\Gamma\Phi_d.V(0) \\ &+ \textstyle\int_\omega 2\varepsilon(V)\nabla_\Gamma\Phi\nabla_\Gamma P + \int_\omega div_\Gamma(fV(0))P \end{aligned} \tag{3.16}$$

where P is the adjoint state and $\varepsilon(V)$ is the symmetrized of DV.

Proof. A mere change of variable in the cost functional formulated in ω_t leads us to

$$J(\omega_t) = \frac{1}{2}\int_{\omega_t}(\Phi_t - \Phi_d)^2 = \frac{1}{2}\int_\omega(\Phi_t o T_t - \Phi_d)^2 j(t)$$

Which yields:

$$dJ(\omega, V) = \int_\omega(\Phi - \Phi_d)(\dot{\Phi} - \nabla_\Gamma\Phi_d.V(0)) + \frac{1}{2}\int_\omega(\Phi - \Phi_d)^2 div_\Gamma V(0) \tag{3.17}$$

In order to eliminate the material derivative $\dot{\Phi}$ from the last expression, we use the following adjoint problem:

$$\begin{aligned} -\Delta_\Gamma P &= (\Phi - \Phi_d)\mathrm{in}\omega \\ \tfrac{\partial P}{\partial\nu} &= 0\mathrm{sur}\partial\omega \end{aligned} \tag{3.18}$$

Where $P \in H^1_*(\omega)$ is the adjoint of Φ. Then thanks to the conjugate form and Green's formula, it follows that:

$$\begin{aligned} dJ(\omega, V) &= a(\dot{\Phi}, P) + \tfrac{1}{2}\textstyle\int_\omega(\Phi - \Phi_d)^2 div_\Gamma V(0) - \int_\omega(\Phi - \Phi_d)\nabla_\Gamma\Phi_d.V(0) \\ &= l(P) + \tfrac{1}{2}\textstyle\int_\omega(\Phi - \Phi_d)^2 div_\Gamma V(0) - \int_\omega(\Phi - \Phi_d)\nabla_\Gamma\Phi_d.V(0) \end{aligned} \tag{3.19}$$

Thus, via the expression of the linear form l we deduce the announced proposition. □

4 Case of a Piecewise Smooth Manifold

Let ω be a piecewise smooth open subset of the manifold Γ containing m singularities s_i. We consider the same problem $\mathcal{P}$ and also the moving one.

4.1 Shape Boundary Derivative

Existence of the Shape Boundary Derivative Φ'_Γ

DEFINITION 4.1. *The shape boundary derivative Φ'_Γ is the element $(\frac{\partial}{\partial t}Y(0))|_\Gamma$ where Y is any smooth extension of Φ verifying:*

$$\begin{aligned} & Y \in C^1([0,\delta[,\ H^{\frac{3}{2}}(D) \cap H^1_*(D)) \\ & Y(0,.)|_\Gamma = \Phi(\Gamma) \\ & \frac{\partial}{\partial n}Y(0) = 0 \quad on\ \Gamma \end{aligned}$$

PROPOSITION 4.2. *The shape boundary derivative Φ'_Γ, if it exists, is given in [12] by this relation*

$$\Phi'_\Gamma \ = \ \dot{\Phi}|_\Gamma - \nabla_\Gamma\Phi.V(0) \tag{4.1}$$

where $\dot{\Phi}_\Gamma$ is the restriction of the Material derivative onto Γ.

By embedding the later relation in equation (3.3) we obtain the following result for any vector field V in $C^0([0,\ \delta[, W^{1,\infty}(D;IR^N))$. We recall from [10] the characterization:

THEOREM 4.3. *Assume that Γ is a C^2 manifold and γ is a piecewise C^2 smooth curve, then the Eulerian derivative is the solution of the following problem*

$$\left\{ \begin{aligned} -\Delta_\Gamma\Phi'_\Gamma \ & = -div_\Gamma[(2D^2.b - H)\nabla_\Gamma\Phi < V(0),n >] \\ & +(\tfrac{\partial F}{\partial n} + Hf) < V(0),n > \quad in\ \omega \\ \tfrac{\partial\Phi'_\Gamma}{\partial\nu} \ & = (f - div_\gamma\nabla_\gamma\Phi) < V(0),\ \nu > +(\nabla_\gamma\Phi.\tau) < \nabla_\gamma V(0).\tau,\ \nu > \\ & +k < \nu,\ \nu_F > (\nabla_\gamma\Phi.\tau) < V(0),\ \tau > \ on\ \partial\omega \end{aligned} \right. \tag{4.2}$$

Where k is the curvature of the curve $\partial\omega$, ν_F is the unitary normal field of the Frenet trihedral and τ is tangent vector to $\partial\omega$ which forms with ν and n a local trihedral. The proof is done in [10] for the case where γ is a C^2 smooth curve.

REMARK 4.1. *Indeed, in this case, because of the Shape boundary derivative's lack of regularity, a priori the Neumann boundary condition on $\partial\omega$ makes no sense.*

In order to prove the last result we shall establish a regularity result for the $\nabla_\Gamma\Phi$ on γ.

5 Hidden Boundary Regularity

In this section, we are interested to come over the previous difficulty and so to relax the term in cause we use the *extractor* method. Therefore, we derive a boundary hidden regularity of the State Φ. Then, we introduce the *min-max* theory in order to avoid deriving the State equation, it consists to establish the saddle-points of the *Lagrangian* related to the State-Adjoint coupled problem (whose Φ and P are solutions). We will need a hypothesis of Dirichlet condition, let $(\mathcal{H}^1 : \Phi \ = 0$ on γ_0 with $\gamma_0 \subset \gamma)$.

5.1 Extractor Method

We begin by announcing the fundamental result.

THEOREM 5.1. *We assume that Γ is a C^2 manifold and γ a piecewise smooth curve. The state Φ has a hidden boundary regularity on $\partial\omega$. Indeed $\nabla_\Gamma \Phi \in L^2(\partial\omega)$.*

We start with this technical lemma

LEMMA 5.2. *The set $C^2(\overline{\omega})$ is dense in $H^1_\Delta(\omega)$ where*

$$H^1_\Delta(\omega) = \{\varphi \in H^1(\omega); \textit{such that } \Delta_\Gamma\varphi \in L^2(\omega)\}$$

Proof. (of theorem 5.1). Let W be a vector field belongs to $C^1(\Gamma, \mathbb{R}^2)$ satisfying the hypothesis of the theorem, such that $W.n = 0$, we associate to W the flow $T_s(W)$. Thus, the tangential extractor related to W of functions sequence $\psi_n \in C^1(\overline{\omega})$ such that $(\psi_n, \ -\Delta_\Gamma\psi_n) \longrightarrow (\Phi, \ f))$ strongly in $H^1(\omega) \times L^2(\omega)$, is given by

$$\begin{aligned}
\mathcal{E}_W(\psi_n) &= \tfrac{d}{ds}\left(\int_{T_s\omega} |\nabla_\Gamma(\psi_n \circ T_s^{-1})|^2\right)_{|s=0} \\
&= \int_\omega < [D'(0) + j'(0)I]\nabla_\Gamma\psi_n, \ \nabla_\Gamma\psi_n > \\
&= \int_{\partial\omega} |\nabla_\Gamma\psi_n|^2 < W, \ \nu > \ - \ 2\int_\omega \Delta_\Gamma\psi_n \nabla_\Gamma\psi_n.W
\end{aligned}$$

and so

$$\begin{aligned}
\int_{\partial\omega} |\nabla_\Gamma\psi_n|^2 < W, \ \nu > \ &= \ \int_\omega < [D'(0) + j'(0)I]\nabla_\Gamma\psi_n, \ \nabla_\Gamma\psi_n > \\
&+ \ 2\int_\omega \Delta_\Gamma\psi_n \nabla_\Gamma\psi_n.W
\end{aligned}$$

We may choose W such that $0 < \alpha < W.\nu < \beta$ on $\partial\omega$ so also this mapping

$$\xi \longrightarrow \left(\int_{\partial\omega} |\nabla_\Gamma\xi|^2 < W, \ \nu >\right)^{\frac{1}{2}}$$

is a norm equivalent to the usual norm of $L^2(\partial\omega)$. The mapping $\xi \longrightarrow \int_{\partial\omega} |\nabla_\Gamma\xi|^2$ is weakly *lsc* on $L^2(\partial\omega)$. Since it exists $M > 0$ $\int_{\partial\omega} |\nabla_\Gamma\psi_n|^2 < M$ from the weak compacity of the closed ball in $L^2(\partial\omega)$, it exists a subsequence $\vec{\xi}_{n_k} = \nabla_\Gamma\psi_{n_k}$ converging weakly to $\vec{\xi}$ in $L^2(\partial\omega)$. It's required to prove that $\vec{\xi}$ is exactly $\nabla_\Gamma\Phi$. It's enough to use a part integration result existing in [10], in deed let $\pi \in \mathbb{D}(\partial\omega)$ with $\{s_1, s_2\} \subset (\text{supp}\pi)^c$, so

$$\int_{\partial\omega} \nabla_\Gamma\psi_{n_k}\nabla_\Gamma\pi = \int_{\partial\omega} \psi_{n_k}\mathcal{F}(\pi)$$

Where $\mathcal{F}(\pi)$ is the adequate expression existing in [10]. So also we compute the limit with k and under part argument, one easily checks that:

$$\int_{\partial\omega} \vec{\xi}.\nabla_\Gamma\pi = \int_{\partial\omega} \nabla_\Gamma\Phi\nabla_\Gamma\pi \forall\pi \in \mathbb{D}(\partial\omega)$$

which yields

$$\vec{\xi} = \nabla_\Gamma\Phi$$

thus, we conclude the existence of a boundary hidden regularity of the state Φ on $\partial\omega$. Let

$$\nabla_\Gamma\Phi \in L^2(\partial\omega).$$

and achieve the proof. □

5.2 The Min-Max Theory

We will use this technical lemma.

LEMMA 5.3. *The vector field V is given, let B_n a reunion of m-neighborhoods B_n^i of the singularities s_i such that $\chi_{B_n^i} \to \delta_{s_i}$ in L^1, where δ is the Dirac function on s_i for $i = 1, ..., m$. There exists a family of vector fields $V_n \in W^{1,\infty}(\Gamma)$ such that $V_n = 0$ on B_n and V_n is given by:*

$$V_n(x) = V(x) - (\sum_i y_n^i V)(x)$$

Where y_n^i is a function with support $B^i_{\sqrt{n}}$ containing B_n^i such that $y_n^i \geq \chi_{B_n^i}$. We will give hereafter explicitly $y_n^i \circ (\xi^i)^{-1}$ (see appendix), where ξ^i is the associated chart diffeomorphism $\xi^i : B^i_{\sqrt{n}} \to T_{s_i}\Gamma$ with $T_{s_i}\Gamma$ is the tangent space to the manifold Γ on s_i.

LEMMA 5.4. *We have the following convergence results:*

i) The function y_n^i is star weakly convergent to zero in $L^\infty(\omega)$

ii) The vector field V_n is star weakly convergent to V in $L^\infty(\omega)$

iii) The vector field V_n converges all most everywhere to V in $W^{1,\infty}$

Denote by $(\omega)_{t,n} = T_t(V_\mu^n)(\omega)$ the family of open subset generated by the flow $T_t(V_\mu^n)$ associated to the vector field V_μ^n. In the sequel, we introduce the Lagrangian saddle-points related to the coupled problem State-Adjoint in order to arise a min-max form for the Shape Gradient.

Lagrangian and Saddle-Points

We will characterize the saddle-points by the following result.

PROPOSITION 5.5. *It is known that (φ, ψ) is a saddle-point of the Lagrangian if and only if (φ, ψ) is a solution of the coupled problem State- Adjoint, we refer to [9].*

LEMMA 5.6. *Let us denote by τ_t the tangent vector to $(\partial\omega)$ then we set*

$$K^t = \{\varphi_t \in H^1(\Gamma) \cap H^2_{loc}(\Gamma \backslash B_n),\ \nabla_\Gamma\varphi_t.\tau_t \in L^2(\partial(\omega)_t)\ \varphi_t = 0 \ \text{on}\ \gamma_0\}$$

so also the functional $J(\omega_{t,n}, V_n)$ is the solution of the min-max problem:

$$\begin{aligned} J(\omega_{t,n}) &= \tfrac{1}{2}\int_{\omega_{t,n}} (\Phi_{t,n} - \Phi_d)^2 \\ &= \min_{\varphi_t \in K^t} \max_{\psi_t \in K^t} \mathcal{L}^t(\varphi_t, \psi_t). \end{aligned}$$

with $\mathcal{L}^t$ is the associated Lagrangian given by:

$$\mathcal{L}^t(\varphi_t, \psi_t) = \frac{1}{2}\int_{\omega_{t,n}} (\varphi_t - \Phi_d)^2 + \int_{\omega_{t,n}} [\nabla_\Gamma\varphi_t \nabla_\Gamma\psi_t - f\psi_t].$$

In order to come to over a fixed space, we effect a classical change of functions. Let $\varphi = \varphi_t \circ T_t$ and $\psi = \psi_t \circ T_t$ it yields

LEMMA 5.7. *Let K be the fixed space, then*

$$J(\omega_{t,n}) = \min_{\varphi\in K}\max_{\psi\in K} L^t(\varphi,\psi)$$

where

$$L^t(\varphi,\psi) = \frac{1}{2}\int_{\omega_{t,n}} (\varphi\circ T_t^{-1} - \Phi_d)^2 + \int_{\omega_{t,n}} [\nabla_\Gamma(\varphi\circ T_t^{-1})\nabla_\Gamma(\psi\circ T_t^{-1}) - f\psi\circ T_t^{-1}]$$

$$K = \{\varphi \in H^1(\Gamma)\cap H^2_{loc}(\Gamma\backslash B_n),\ \nabla_\Gamma\varphi.\tau \in L^2(\partial\omega)\ \varphi = 0 \ on\ \gamma_0\}.$$

Min-Max Differentiation

In order to concept the boundary expression of the Shape Gradient of the functional J, we will use the following import theorem.

THEOREM 5.8. *By applying the min-max differentiation, the Shape Gradient has the following forme*

$$dJ(\omega, V_n) = \frac{\partial}{\partial t} L^t(\Phi_n, P_n)_{|t=0}$$

for the proof we refer to [3] and [7].

Which yields us to prove this lemma

LEMMA 5.9. *According the previous results, the mapping: $t \in [0,\ \delta[\longrightarrow \varphi\circ T_t^{-1}(V_n) \in H^1(\Gamma)\cap H^2_{loc}(\Gamma\backslash B_n)$ is continue and differentiable in $H^1(\Gamma)$. Moreover*

$$\lim_{t\to 0} \|\frac{\varphi\circ T_t^{-1}(V_n) - \varphi}{t} - (-\nabla_\Gamma\varphi.V_n(0))\|_{H^1(\Gamma)} = 0$$

Proof. Under the regularity of φ and the continuity of the flow T_t^{-1}, the continuity of the previous mapping is obvious. Concerning the differentiability, it will be deduced equally from the same argument. Notice that

$$\varphi\circ T_t^{-1}(x) = \varphi(x) + \int_0^1 \nabla_\Gamma\varphi(x + s(T_t^{-1}(x) - x)).(T_t^{-1}(x) - x)ds$$

it results

$$\begin{aligned}\tfrac{1}{t}(\varphi\circ T_t^{-1}(x) - \varphi(x)) + (\nabla_\Gamma\varphi.V_n(0)) &= \textstyle\int_0^1[\nabla_\Gamma\varphi(x + s(T_t^{-1}(x) - x))].\frac{(T_t^{-1}(x)-x)}{t}ds \\ &- \textstyle\int_0^1 \nabla_\Gamma\varphi(x).\frac{(T_t^{-1}(x)-x)}{t}ds \\ &+ \nabla_\Gamma\varphi.[\tfrac{(T_t^{-1}(x)-x)}{t} + V_n(0,x)]\end{aligned}$$

We come to investigate the following limit when t tends to 0.

$$\lim_{t\to 0}(I_t = \|\frac{1}{t}(\varphi\circ T_t^{-1}(x) - \varphi(x)) + (\nabla_\Gamma\varphi.V_n(0))\|_{H^1(\Gamma)})$$

First, let us begin by the $L^2(\Gamma)$ norm. We denote it by I_t^1. It follows

$$\begin{aligned}(I_t^1)^2 &\le 2\textstyle\int_\Gamma |\{\int_0^1[\nabla_\Gamma\varphi(x + s(T_t^{-1}(x) - x)) - \nabla_\Gamma\varphi(x)].\frac{(T_t^{-1}(x)-x)}{t}ds\}|^2 d\Gamma \\ &+ 2\textstyle\int_\Gamma |\nabla_\Gamma\varphi.[\frac{(T_t^{-1}(x)-x)}{t} + V_n(0,x)]|^2 d\Gamma\end{aligned}$$

It's well clear, according the previous hypothesis and from the continuity of the mapping $t \longrightarrow \nabla_\Gamma\varphi.[\frac{(T_t^{-1}(x)-x)}{t}+V_n(0,x)]$ in $H^1(\Gamma)$, that the second term tends to 0 with t. Then by applying Hôlder's inequality and Lebesgue's theorem, it arises that the first term is overestimated by $c\int_0^1\{\int_\Gamma |[\nabla_\Gamma\varphi_\varepsilon(x+s(T_t^{-1}(x)-x))-\nabla_\Gamma\varphi(x)].\frac{(T_t^{-1}(x)-x)}{t}|^2 d\Gamma\}ds$ Let $h(s,t)=\int_\Gamma |[\nabla_\Gamma\varphi(x+s(T_t^{-1}(x)-x))-\nabla_\Gamma\varphi(x)].\frac{(T_t^{-1}(x)-x)}{t}|^2 d\Gamma$, we remark that $h(s,t)\le h(1,t)$ $\forall s$. Therefore $\lim_{t\to 0} I_t^1 = 0$. About the semi-norm $|.|_{1,\Gamma}$ denoted by I_t^2 converges to zźero with t. in fact, it is clear that

$$\begin{aligned}(I_t^2)^2 &\le 2\int_\Gamma |\nabla_\Gamma\{\int_0^1[\nabla_\Gamma\varphi_\varepsilon(x+s(T_t^{-1}(x)-x))-\nabla_\Gamma\varphi(x)].\tfrac{(T_t^{-1}(x)-x)}{t}ds\}|^2 d\Gamma\\ &+ 2\int_\Gamma |\nabla_\Gamma\{\nabla_\Gamma\varphi_\varepsilon.[\tfrac{(T_t^{-1}(x)-x)}{t}+V_n(0,x)]\}|^2 d\Gamma\end{aligned}$$

REMARK 5.1. *It can seems that we integrate entirely into* Γ, *in deed, since* $\frac{(T_t^{-1}(x)-x)}{t}\simeq -V_n(0)$ *when t tends to zero and due to the regularity of* φ *outside the singularities of the open subset e, the integration domain is reduced to* $(\Gamma\backslash B_n)$. *which validates the previous expressions.*

By using the same arguments as previously, it arises that I_t^2 converges to zero with t. It implies the convergence of I_t to zero with t. □

5.3 Shape Gradient Boundary Expression

First expression

By using the *min-max* differentiation result, We come to

LEMMA 5.10.

$$\begin{aligned}dJ(\omega,V_n) &= \int_{\partial\omega}[\tfrac{1}{2}(\Phi-\Phi_d)^2+\nabla_\Gamma\Phi\nabla_\Gamma P \;-\; fP] < V_n(0),\ \nu >\\ &+ \int_\omega(\Phi-\Phi_d)(-\nabla_\Gamma\Phi V_n(0))+\nabla_\Gamma P\nabla_\Gamma(-\nabla_\Gamma\Phi V_n(0))\\ &+ \int_\omega \nabla_\Gamma\Phi\nabla_\Gamma(-\nabla_\Gamma P V_n(0))-f(-\nabla_\Gamma P V_n(0)).\end{aligned}$$

Our aim is to get the Shape Gradient boundary expression. That's why the functions $(-\nabla_\Gamma P V_n(0))$ and $(-\nabla_\Gamma\Phi V_n(0))$ have to be in the space K. Which needs more condition verified by the vector field $V_n(0)$. Let $\mathcal{H}^2 : V_n(0)=0$ on γ_0. It follows

LEMMA 5.11.

$$dJ(\omega,V_n)=\int_{\partial\omega}[\frac{1}{2}(\Phi-\Phi_d)^2+\nabla_\Gamma\Phi\nabla_\Gamma P \;-\; fP] < V_n(0),\ \nu >$$

We have the following convergence result

LEMMA 5.12. *According to the hidden boundary regularity provided by the tangential extractor the function* $\nabla_\Gamma\Phi\nabla_\Gamma P$ *is in* $L^1(\partial\omega)$. *Since* $V_n(0)\longrightarrow V(0)$ *in* $L^\infty(\partial\omega)$ *weak star topology. Moreover* $V_n(0)$*and* $V(0)$ *are in* $L^\infty(\partial\omega)$. *Hence, we deduce this result.*

$$\int_{\partial\omega}\nabla_\Gamma\Phi\nabla_\Gamma P < V_n(0),\ \nu > \longrightarrow \int_{\partial\omega}\nabla_\Gamma\Phi\nabla_\Gamma P < V(0),\ \nu >, n\uparrow\infty.$$

Therefore

LEMMA 5.13. *Since we have*

$$dJ(\omega, V) = dJ(\omega, V_n) + dJ(\omega, \sum_i y_n^i V)$$

then

$$dJ(\omega, V) = \lim_{n\uparrow\infty} dJ(\omega, V_n) + \lim_{n\uparrow\infty} dJ(\omega, \sum_i y_n^i V)$$

so also

$$dJ(\omega, \sum_i y_n^i V) = dJ(\omega, \sum_i y_n^i [V - V(s_i)]) + dJ(\omega, \sum_i y_n^i V(s_i))$$

with

$$\lim_{n\uparrow\infty} dJ(\omega, \sum_i y_n^i [V - V(s_i)]) = 0, n\uparrow\infty$$

LEMMA 5.14. *We deduce easily that the sequence of punctual terms*

$$dJ(\omega, \sum_i y_n^i V(S_i) = \sum_i < G_\omega^{i,n}, V(s_i) >$$

has a limit, when $n \uparrow \infty$, which is independent on the choice of the sequence y_n^i. Where $G_\omega^{i,n}$ is a vector given by the Shape Gradient distributed expression (3.16).

$$\begin{aligned} G_\omega^{i,n} &= \int_\omega [\tfrac{1}{2}(\Phi - \Phi_d)^2 + fP - \nabla_\Gamma\Phi\nabla_\Gamma P]\nabla_\Gamma y_n^i - \int_\omega [(\Phi - \Phi_d)\nabla_\Gamma\Phi_d + \nabla f]y_n^i \\ &+ \int_\omega < \nabla_\Gamma y_n^i, \nabla_\Gamma P > \nabla_\Gamma\Phi + \int_\omega < \nabla_\Gamma y_n^i, \nabla_\Gamma\Phi > \nabla_\Gamma P \end{aligned} \tag{5.1}$$

As a consequence of these lemmas, it is easy to check that the Shape Gradient boundary expression is splitting in two terms: a continue term and a punctual one.

PROPOSITION 5.15. *We have*

$$dJ(\omega, V) = \int_{\partial\omega} [\frac{1}{2}(\Phi - \Phi_d)^2 + \nabla_\Gamma\Phi\nabla_\Gamma P - fP] < V(0), \nu > + \sum_i < G_\omega^i, V(s_i) >_{\mathbf{R}^N}$$

where

$$G_\omega^i = \lim_{n\uparrow\infty} G_\omega^{i,n}$$

$$\begin{aligned} G_\omega^i = \lim{}_{n\uparrow\infty} \quad & \{\int_\omega [\tfrac{1}{2}(\Phi - \Phi_d)^2 + fP - \nabla_\Gamma\Phi\nabla_\Gamma P]\nabla_\Gamma y_n^i \\ + & \int_\omega < \nabla_\Gamma y_n^i, \nabla_\Gamma P > \nabla_\Gamma\Phi + \int_\omega < \nabla_\Gamma y_n^i, \nabla_\Gamma\Phi > \nabla_\Gamma P\} \end{aligned} \tag{5.2}$$

PROPOSITION 5.16. *It is relatively easy to establish two possible cases concerning the punctual term.*

i) If the singularity order of the solution Φ and P in a neighborhood of s_i is equal to $\frac{1}{2}$ then G_ω^i is not vanished. (It corresponds to the flat case, [11])

ii) If the previous order is different of $\frac{1}{2}$ then G_ω^i vanishes.

Second Expression

In the following, it consists to show off another expression of the boundary Shape Gradient of the functional cost $J(\omega)$. Let us begin by giving this technical lemma

LEMMA 5.17. *We refer to [8]. We consider a mapping* $t \in [0,\ \delta[\longrightarrow u(t) = u_t \in H^1(\omega)$ *We suppose that* $u(.)$ *is differentiable in* $H^1(\omega)$. *Then*

$$\begin{aligned} \frac{\partial}{\partial t}\Big(\int_{(\omega)_t} u_t\Big)_{|t=0} &= \int_\omega u'_\Gamma(\omega, V) + \int_\omega Hu < V(0), n > \\ &+ \int_{\partial\omega} u < V(0),\ \nu > . \end{aligned}$$

Where u'_Γ is the Shape boundary derivative. Then, we deduce the following result

PROPOSITION 5.18. *The Shape Gradient boundary expression is given by*

$$dJ(\omega, V) = \frac{1}{2}\int_{\partial\omega} (\Phi - \Phi_d)^2 < V(0),\ \nu > - \int_{\partial\omega} P < \nabla_\Gamma \Phi'_\Gamma,\ \nu > . \tag{5.3}$$

REMARK 5.2. *A priori the boundary term* $\int_{\partial\omega} P < \nabla_\Gamma \Phi'_\Gamma,\ \nu >$ *makes no sense because of the regularities lack of the Shape boundary derivative* Φ'_Γ *in the singularities neighbourhood.*

Proof. (of proposition 5.18). Under the previous lemma and by Green's formula, we come to

$$\begin{aligned} dJ(\omega, V) &= \int_\omega (\Phi - \Phi_d)\Phi'_\Gamma + \tfrac{1}{2}\int_\omega H(\Phi - \Phi_d)^2 < V(0),\ n > \\ &+ \tfrac{1}{2}\int_{\partial\omega} (\Phi - \Phi_d)^2 < V(0),\ \nu > \end{aligned} \tag{5.4}$$

We use as before the method of adjoint state in order to eliminate the boundary shape derivative Φ'_Γ from the last expression

$$\begin{aligned} dJ(\omega, V) &= \int_\omega \Phi'_\Gamma \Delta_\Gamma P + \tfrac{1}{2}\int_\omega H(\Phi - \Phi_d)^2 < V(0), n > \\ &+ \tfrac{1}{2}\int_{\partial\omega} (\Phi - \Phi_d)^2 < V(0), \nu > . \end{aligned} \tag{5.5}$$

By using formally Green's formula, we come to

$$\begin{aligned} dJ(\omega, V) &= \int_\omega P\Delta_\Gamma \Phi'_\Gamma + \tfrac{1}{2}\int_\omega H(\Phi - \Phi_d)^2 < V(0),\ n > \\ &+ \tfrac{1}{2}\int_{\partial\omega} (\Phi - \Phi_d)^2 < V(0),\ \nu > \\ &- \int_{\partial\omega} P < \nabla_\Gamma \Phi'_\Gamma,\ \nu > \\ &+ \int_{\partial\omega} \Phi'_\Gamma < \nabla_\Gamma P,\ \nu > \end{aligned}$$

Or since $< \nabla_\Gamma P,\ \nu >= 0$ then

$$\begin{aligned} dJ(\omega, V) &= \int_\omega \Delta_\Gamma \Phi'_\Gamma P + \tfrac{1}{2}\int_\omega H(\Phi - \Phi_d)^2 < V(0),\ n > \\ &+ \tfrac{1}{2}\int_{\partial\omega} (\Phi - \Phi_d)^2 < V(0),\ \nu > \ - \ \int_{\partial\omega} P < \nabla_\Gamma \Phi'_\Gamma,\ \nu > . \end{aligned} \tag{5.6}$$

and as $(V, n) = 0$. □

Proof of Theorem 4.3

Thanks to the last results, one can easily check the following relaxation of the boundary condition of the Shape boundary derivative.

LEMMA 5.19. *The Shape boundary derivative Φ'_Γ is regular and is given as follows:*

$$\int_{\partial\omega} P < \nabla_\Gamma \Phi'_\Gamma,\ \nu > \; = \int_{\partial\omega} [-\nabla_\Gamma \Phi \nabla_\Gamma P + fP] < V(0),\ \nu > - < G, V(s_i) >_{\mathbb{R}^N}$$

So also,

$$\frac{\partial \Phi'_\Gamma}{\partial \nu} \in H^{\frac{-1}{2}}(\partial(\omega))$$

Thus, according the previous results we come to the proof of theorem 4.3 in page 239. In deed by embedding the relation (4.1) in the equation (3.3) and via the above lemma it will be enough to refer to [10], and so the proof is achieved.

6 Fractured Manifold

The lack of regularity of the open set $(\omega\backslash\sigma)$ and so of the solution Φ, forbidden us to have an optimal formulation for the shape functional $J(\omega\backslash\sigma)$, notably the Shape Gradient boundary expression. This suggest to introduce a regularization in order to estimate the non Lipschitzian open set $(\omega\backslash\sigma)$ by a family with parameter of piecewise smooth (and so Lipschitzian) open subsets.

6.1 Regularized Problem

We *regularize* the domain $(\omega\backslash\sigma)$ by using a family, with parameter, of singular envelopes e_ε on s_1 and s_2 surrounding the fracture σ, it will be defined subsequently. We denote by $(\omega\backslash\sigma)_\varepsilon$ the obtained regular open subset (the complementary of e_ε on $(\omega\backslash\sigma)$) in which we formulate the following homogenous tangential problem.

$$(\mathcal{N}T)_\varepsilon \left\{ \begin{array}{rcl} -\Delta_\Gamma \Phi_\varepsilon & = & f \text{ in } (\omega\backslash\sigma)_\varepsilon \\ \frac{\partial \Phi_\varepsilon}{\partial \nu_\varepsilon} & = & 0 \text{ on } \partial(\omega\backslash\sigma)_\varepsilon. \end{array} \right. \tag{6.1}$$

REMARK 6.1. *The family, with parameter ε, of open subsets $(\omega\backslash\sigma)_\varepsilon$ is Lipschitzian. As a consequence Green's formula holds.*

6.2 Shape Analysis

Since we are over a piecewise smooth subset, all the previous results concerning the Shape analysis given in section 4 hold.

Material and Eulerian Derivatives

Let us consider the moving problem posed in $(\omega\backslash\sigma)_{\varepsilon,t}$ at each fixed ε.

$$(\mathcal{N}T)_{\varepsilon,t}\begin{cases} -\Delta_\Gamma\Phi_{\varepsilon,t} &= f_t \text{in}(\omega\backslash\sigma)_{\varepsilon,t} \\ \frac{\partial\Phi_{\varepsilon,t}}{\partial\nu_{\varepsilon,t}} &= 0 \text{on} \partial(\omega\backslash\sigma)_{\varepsilon,t}. \end{cases} \tag{6.2}$$

we apply the theorem (3.3) we derive the following results.

PROPOSITION 6.1. *At each fixed ε The map: $t \to \Phi_{\varepsilon,t} \circ T_t$ is derivable in the right of zero and its derivative $\dot{\Phi}_\varepsilon = \lim_{t\to 0}\frac{\Phi_{\varepsilon,t}\circ T_t - \Phi_\varepsilon}{t}$, in $H^1_*((\omega\backslash\sigma)_\varepsilon)$, verifies the equation*

$$\begin{aligned} \int_{(\omega\backslash\sigma)_\varepsilon}\nabla_\Gamma\dot{\Phi}_\varepsilon\nabla_\Gamma\psi &= \int_{(\omega\backslash\sigma)_\varepsilon} < 2\varepsilon(V)\nabla_\Gamma\Phi_\varepsilon, \nabla_\Gamma\psi > \\ &- \int_{(\omega\backslash\sigma)_\varepsilon} < div_\Gamma V(0)\nabla_\Gamma\Phi_\varepsilon, \nabla_\Gamma\psi > \\ &+ \int_{(\omega\backslash\sigma)_\varepsilon}[fj'(0) + \nabla_\Gamma F V(0)]\psi \end{aligned} \tag{6.3}$$

and also we apply the theorem (4.3), we get at each fixed ε

PROPOSITION 6.2. *The Eulerian derivative is the solution of the following problem*

$$\begin{cases} -\Delta_\Gamma\Phi'_{\varepsilon\Gamma} &= -div_\Gamma[(2D^2.b - H)\nabla_\Gamma\Phi_\varepsilon < V(0), n >] \\ &+(\frac{\partial F}{\partial n} + Hf) < V(0), n > \quad in\ (\omega\backslash\sigma)_\varepsilon \\ \frac{\partial\Phi'_{\varepsilon\Gamma}}{\partial\nu_\varepsilon} &= (f - div_\gamma\nabla_\gamma\Phi_\varepsilon) < V(0),\ \nu_\varepsilon > +(\nabla_\gamma\Phi_\varepsilon.\tau_\varepsilon) < \nabla_\gamma V(0).\tau_\varepsilon,\ \nu_\varepsilon > \\ &+k < \nu_\varepsilon,\ \nu_{\varepsilon F} > (\nabla_\gamma\Phi_\varepsilon.\tau_\varepsilon) < V(0),\ \tau_\varepsilon > \quad on\ \partial(\omega\backslash\sigma)_\varepsilon \end{cases} \tag{6.4}$$

Shape Gradient of $J((\omega\backslash\sigma)_\varepsilon)$

PROPOSITION 6.3. *The distributed expression of the Shape Gradient is given at each fixed ε by:*

$$\begin{aligned} &dJ((\omega\backslash\sigma)_\varepsilon, V) \\ = &\int_{(\omega\backslash\sigma)_\varepsilon}[\frac{1}{2}(\Phi_\varepsilon - \Phi_d)^2 - \nabla_\Gamma\Phi_\varepsilon\nabla_\Gamma P_\varepsilon]div_\Gamma V(0) - \int_{(\omega\backslash\sigma)_\varepsilon}(\Phi_\varepsilon - \Phi_d)\nabla_\Gamma\Phi_d.V(0) \\ + &\int_{(\omega\backslash\sigma)_\varepsilon} 2\varepsilon(V)\nabla_\Gamma\Phi_\varepsilon\nabla_\Gamma P_\varepsilon + \int_{(\omega\backslash\sigma)_\varepsilon} div_\Gamma(fV(0))P_\varepsilon \end{aligned} \tag{6.5}$$

PROPOSITION 6.4. *We have, at each fixed ε*

$$\begin{aligned} dJ((\omega\backslash\sigma)_\varepsilon, V) &= \int_{\partial(\omega\backslash\sigma)_\varepsilon}[\tfrac{1}{2}(\Phi_\varepsilon - \Phi_d)^2 + \nabla_\Gamma\Phi_\varepsilon\nabla_\Gamma P_\varepsilon \quad - \quad fP_\varepsilon] < V(0),\ \nu_\varepsilon > \\ &+ \sum_i < G^i_{(\omega\backslash\sigma)_\varepsilon}, V(s_i) >_{\mathbb{R}^N} \end{aligned} \tag{6.6}$$

where $G^i_{(\omega\backslash\sigma)_\varepsilon}$ is a vector having the same expression of G^i_ω but posed in $(\omega\backslash\sigma)_\varepsilon$.

PROPOSITION 6.5.

$$dJ((\omega\backslash\sigma)_\varepsilon, V) = \frac{1}{2}\int_{\partial(\omega\backslash\sigma)_\varepsilon}(\Phi_\varepsilon - \Phi_d)^2 < V(0),\ \nu_\varepsilon > - \int_{\partial(\omega\backslash\sigma)_\varepsilon} P_\varepsilon < \nabla_\Gamma\Phi'_{\varepsilon\Gamma},\ \nu_\varepsilon > . \tag{6.7}$$

6.3 Continuity of the Neumann Problem

In this section, we study the behaviour of the Shape Gradient with respect to ε. It's obvious that we will have to prove a strong convergence result. In order to get the previous result, we will need to arise a continuity of the Neumann tangential problem with respect to the open subset e_ε. By using Green's formula we establish the weak formulation associated to the regular problem, so also

$$\int_{(\omega\backslash\sigma)_\varepsilon} \nabla_\Gamma \Phi_\varepsilon \nabla_\Gamma \varphi = \int_{(\omega\backslash\sigma)_\varepsilon} f\varphi \forall\varphi \in H^1_*((\omega\backslash\sigma)_\varepsilon)$$

it yields

$$\int_{\omega\backslash\sigma} |1_{(\omega\backslash\sigma)_\varepsilon} \nabla_\Gamma \Phi_\varepsilon|^2 \leq \|f\|_{L^2(\omega\backslash\sigma)} \|\Phi_\varepsilon\|_{L^2((\omega\backslash\sigma)_\varepsilon)}$$

Thanks to Poincare's inequality given by the space $H^1_*((\omega\backslash\sigma)_\varepsilon)$, we come to

$$\|1_{(\omega\backslash\sigma)_\varepsilon} \nabla_\Gamma \Phi_\varepsilon\|_{L^2(\omega\backslash\sigma)} \leq \lambda_\varepsilon^{\frac{-1}{2}} \|f\|_{L^2(\omega\backslash\sigma)}$$

$$\|1_{(\omega\backslash\sigma)_\varepsilon} \Phi_\varepsilon\|_{L^2(\omega\backslash\sigma)} \leq \lambda_\varepsilon^{\frac{-3}{2}} \|f\|_{L^2(\omega\backslash\sigma)}$$

Because of the dependence of the second term on ε we are not able to arise an uniformly estimation. A particular choice of the envelope e_ε enable us to overcome this difficulty.

Choice of the Envelope e_ε

The envelope e_ε will be the open subset whose boundary is the convict of the fracture σ by the T_t à $t = \varepsilon$ associated to the non autonomous vector field $E_\sigma = (E^+_\sigma,\ E^-_\sigma)$. The field E_σ satisfies the later conditions: $E^+_\sigma \in C^k(\overline{\omega\backslash\sigma_+}), E^-_\sigma \in C^k(\overline{\omega\backslash\sigma_-}) E_\sigma.n = 0, [E_\sigma] \neq 0$ on σ et $E^+_\sigma.\nu > 0$ therefore $E^-_\sigma.\nu < 0$ on σ. Et $E_\sigma(s_i) = 0$, where s_i are the extremities of the fracture σ. So also $\overline{(\partial e_\varepsilon)}_+ = T_\varepsilon(E^+_\sigma)(\sigma_+)$, et $\overline{(\partial e_\varepsilon)}_- = T_\varepsilon(E^-_\sigma)(\sigma_-)$. We denote by $\partial e_\varepsilon = \overline{(\partial e_\varepsilon)}_+ \cup \overline{(\partial e_\varepsilon)}_- = T_\varepsilon(E_\sigma)(\sigma)$. It is so clear that $\partial(e_\varepsilon)_+$ and $\partial(e_\varepsilon)_-$ are two C^∞ manifolds. Which enable us to control the first eigenvalue of the Laplace-Beltrami operator.

Boundeness of the First Eigenvalue

We begin by giving this result

PROPOSITION 6.6. *Let*

$$\lambda_\varepsilon = inf\{c_\varepsilon; c_\varepsilon \int_{(\omega\backslash\sigma)_\varepsilon} v_\varepsilon^2 \leq \int_{(\omega\backslash\sigma)_\varepsilon} |\nabla_\Gamma v_\varepsilon|^2 \forall v_\varepsilon \in H^1_*((\omega\backslash\sigma)_\varepsilon)\}$$

So i) it exists $\varphi_\varepsilon \in H^1_*((\omega\backslash\sigma)_\varepsilon), \int_{(\omega\backslash\sigma)_\varepsilon} (\varphi_\varepsilon)^2 = 1$ *such that*

$$\lambda_\varepsilon = \int_{(\omega\backslash\sigma)_\varepsilon} |\nabla_\Gamma \varphi_\varepsilon|^2$$

ii) Under hypothesis $\mathcal{H}^1$, λ_ε *is underestimated by* $c\lambda$ $\forall\varepsilon$ *et* $\exists\psi \in H^1_*(\omega\backslash\sigma)$*such that* $\int_{\omega\backslash\sigma} (\psi)^2 = 1$ *with*

$$\lambda = \int_{\omega\backslash\sigma} |\nabla_\Gamma \psi|^2$$

and λ *realise the same problem whose* λ_ε *is solution but formulated in* $(\omega\backslash\sigma)$.

In order to prove the above result we have to specify the domain's topology.

REMARK 6.2. *The open subset $(\omega\backslash\sigma)_\varepsilon$ converges, with ε, to $(\omega\backslash\sigma)$ for the Hausdorff complementary topology endowed by the metric , [1]*

$$d_{H^c}(\omega_1,\ \omega_2) = d_H(\overline{\omega}\backslash\omega_1,\ \overline{\omega}\backslash\omega_2),$$

where

$$d_H(K_1,\ K_2) = \max\{\sup_{x\in K_1}\inf_{y\in K_2}|x-y|,\ \sup_{y\in K_2}\inf_{x\in K_1}|x-y|\}.$$

is the Hausdorff distance between two closed subset of the open set ω. We say that $(\omega\backslash\sigma)_\varepsilon$ converges in the measure sense to $(\omega\backslash\sigma)$ if the corresponding characteristic functions converge strongly in $L^1(\omega)$.

Proof. (of Proposition 6.6).

i) it's sufficient to consult [12].

ii) The proof is deduced directly from hypothesis $\mathcal{H}^1$ and the following hereafter equivalence

$$\varphi_\varepsilon\ \in H^1_{\gamma_0}((\omega\backslash\sigma)_\varepsilon) \Longleftrightarrow \varphi_\varepsilon\circ T_\varepsilon\ \in H^1_{\gamma_0}(\omega\backslash\sigma)$$

with $j(\varepsilon)\longrightarrow 1$ when $\varepsilon\downarrow 0$.

□

LEMMA 6.7. *The sequences $1_{(\omega\backslash\sigma)_\varepsilon}\Phi_\varepsilon$ and $1_{(\omega\backslash\sigma)_\varepsilon}\nabla_\Gamma\Phi_\varepsilon$ are uniformly bounded in $L^2(\omega\backslash\sigma)$ with respect to ε.*

Strong Convergence

The last proposition enables us to obtain the following result.

PROPOSITION 6.8. *We come to*

i) There exist two subsequences denoted still further $1_{(\omega\backslash\sigma)_\varepsilon}\Phi_\varepsilon$ and $1_{(\omega\backslash\sigma)_\varepsilon}\nabla_\Gamma\Phi_\varepsilon$ converging in $L^2(\omega\backslash\sigma)$ respectively to μ and $\overrightarrow{\theta}$. Moreover $\overrightarrow{\theta} = \nabla_\Gamma\mu$.

ii) The sequences $1_{(\omega\backslash\sigma)_\varepsilon}\Phi_\varepsilon$ and $1_{(\omega\backslash\sigma)_\varepsilon}\nabla_\Gamma\Phi_\varepsilon$ converge strongly respectively to μ and $\overrightarrow{\theta}$.

Proof. The weak convergence is a direct consequence of the compacity argument. It remains to prove that $\overrightarrow{\theta} = \nabla_\Gamma\mu$. We will adopt the compactivor property which consists that the open set $(\omega\backslash\sigma)_\varepsilon$ soaks up all compact of the open set $(\omega\backslash\sigma)$. In deed for any compact $K\subset\omega\backslash\sigma$, $\exists$ n_k such that $\forall n\geq n_K$ we come to $K\subset(\omega\backslash\sigma)_{\varepsilon_n}$.

Let $\overrightarrow{\varphi}\in\mathbb{D}(\omega\backslash\sigma)$ whose support is K then $\exists n_K$ such that for any $n\geq n_K$, $\varphi\in ID((\omega\backslash\sigma)_{\varepsilon n})$. which yields

$$\begin{aligned}\int_{\omega\backslash\sigma}\overrightarrow{\theta}.\overrightarrow{\varphi} &= \lim_{\varepsilon\to 0}\int_{\omega\backslash\sigma}1_{(\omega\backslash\sigma)_\varepsilon}\nabla_\Gamma\Phi_\varepsilon\overrightarrow{\varphi} = \lim_{\varepsilon\to 0}\int_{(\omega\backslash\sigma)_\varepsilon}\nabla_\Gamma\Phi_\varepsilon\overrightarrow{\varphi}\\ &= \lim_{\varepsilon\to 0}-\int_{(\omega\backslash\sigma)_\varepsilon}\Phi_{\varepsilon_n}div_\Gamma\overrightarrow{\varphi} = \lim_{\varepsilon\to 0}-\int_{\omega\backslash\sigma}1_{(\omega\backslash\sigma)_\varepsilon}\Phi_\varepsilon div_\Gamma\overrightarrow{\varphi}\\ &= -\int_{\omega\backslash\sigma}\mu\, div_\Gamma\overrightarrow{\varphi} = \int_{\omega\backslash\sigma}\nabla_\Gamma\mu.\overrightarrow{\varphi}\end{aligned}\tag{6.8}$$

We result that

$$\overrightarrow{\theta} = \nabla_\Gamma \mu$$

ii) In order to get the proof. we should appear the limit uniqueness. It arises by passing to the limit in the weak formulation whose Φ_ε is solution. So also:

$$\int_{\omega\backslash\sigma} \nabla_\Gamma \mu . \nabla_\Gamma \varphi = \int_{\omega\backslash\sigma} f\varphi \forall \varphi \in H^1_*(\omega\backslash\sigma)$$

It follows that μ is the solution of the homogenous Neumann problem posed in $\omega\backslash\sigma$. Therefore μ is equal to Φ. Thus, all the sequence Φ_ε converges weakly to μ, so also to Φ. ii)As for the strong convergence, it will be obtained equally from the weak formulation.In deed, let Φ_ε be the test function. Hence

$$\int_{(\omega\backslash\sigma)_\varepsilon} |\nabla_\Gamma \Phi_\varepsilon|^2 = \int_{(\omega\backslash\sigma)_\varepsilon} f\Phi_\varepsilon$$

Or the right hand side converges to $\int_{\omega\backslash\sigma} f\mu$, which is equal to $\int_{\omega\backslash\sigma} |\nabla_\Gamma \mu|^2$. So also we derive the convergence in norms in $H^1_*(\omega\backslash\sigma)$ and so the strong convergence. □

REMARK 6.3. *The Neumann problem is continuous with respect to the perturbation $T_\varepsilon(E_\sigma)$.*

COROLLARY 6.9. *We have the same convergence result for the Adjoint problem whose P_ε is solution. We denote by P the limit of P_ε.*

Which enable us to compute the Shape Gradient limit with respect to ε.

7 Shape Gradient Convergence

In this section, under the previous results, we are interested to compute the limit of the Shape Gradient $dJ((\omega\backslash\sigma)_\varepsilon, V)$ when ε tends to zero. This result will arise from the continuity of the tangential Neumann problem.

PROPOSITION 7.1. *The distributed Gradient expression converges and its limit is given by:*

$$\begin{aligned}\lim_{\varepsilon\to 0} dJ((\omega\backslash\sigma)_\varepsilon, V) &= \int_{\omega\backslash\sigma} 2\varepsilon(V)\nabla_\Gamma\Phi\nabla_\Gamma P + \int_{\omega\backslash\sigma} div_\Gamma V(0)\nabla_\Gamma\Phi\nabla_\Gamma P \\ &+ \int_{\omega\backslash\sigma}(\Phi - \Phi_d)^2 div_\Gamma V(0) - \int_{\omega\backslash\sigma}(\Phi - \Phi_d)\nabla_\Gamma\Phi_d.V(0) \\ &+ \int_{\omega\backslash\sigma} div_\Gamma(fV)P\end{aligned}$$

and $dJ(\omega\backslash\sigma, V) = \lim_{\varepsilon\to 0} dJ((\omega\backslash\sigma)_\varepsilon, V)$.

Proof. Indeed, due to the Hausdorff convergence of the open subset $(\omega\backslash\sigma)_\varepsilon$ to $\omega\backslash\sigma$ with ε. It's enough to notice that:

$\lim_{\varepsilon\to 0} l(P_\varepsilon) = \int_{\omega\backslash\sigma} div_\Gamma V(0)\nabla_\Gamma\Phi\nabla_\Gamma P + \int_{\omega\backslash\sigma} 2\varepsilon(V)\nabla_\Gamma\Phi\nabla_\Gamma P + \int_{\omega\backslash\sigma} div_\Gamma(fV(0))P$

$\lim_{\varepsilon\to 0}\int_{(\omega\backslash\sigma)_\varepsilon}(\Phi_\varepsilon - \Phi_d)^2 div_\Gamma V(0) = \int_{\omega\backslash\sigma}(\Phi - \Phi_d)^2 div_\Gamma V(0)$

$\lim_{\varepsilon\to 0}\int_{(\omega\backslash\sigma)_\varepsilon}(\Phi_\varepsilon - \Phi_d)\nabla_\Gamma\Phi_d.V(0) = \int_{\omega\backslash\sigma}(\Phi - \Phi_d)\nabla_\Gamma\Phi_d.V(0)$

Due to the the homogenous boundary Neumann condition on $\partial(\omega\backslash\sigma)$ the material derivative of the state Φ exists and so also we doduce the continuity result for the Shape Gradient with respect to ε. □

8 Jump Through the Crack

We have this repartition $\partial(\omega\backslash\sigma)_\varepsilon = \gamma \cup \overline{(\partial e_\varepsilon)_+} \cup \overline{(\partial e_\varepsilon)_-}$ then

PROPOSITION 8.1. *By replacing $\int_{\partial(\omega\backslash\sigma)_\varepsilon} P_\varepsilon < \nabla_\Gamma \Phi'_{\varepsilon\Gamma},\ \nu_\varepsilon > d\gamma_\varepsilon$ by its expression and passing to the limit in the Shape Gradient $dJ((\omega\backslash\sigma)_\varepsilon, V)$ with ε, we come to*

$$\begin{aligned} dJ(\omega\backslash\sigma, V) &= \lim_{\varepsilon\to 0} \int_\sigma [\nabla_\Gamma \Phi^\varepsilon \nabla_\Gamma P^\varepsilon] < V(0),\ \nu > d\sigma \ - \ \int_\sigma f[P] < V(0),\ \nu > d\sigma \\ &+ \ \tfrac{1}{2}\int_\gamma (\Phi - \Phi_d)^2 < V(0),\ \nu > d\gamma + \tfrac{1}{2}\int_\sigma [(\Phi-\Phi_d)^2] < V(0),\ \nu > d\sigma \\ &+ \ \int_\gamma fP < V(0),\ \nu > d\gamma + \textstyle\sum_i < G^i_{(\omega\backslash\sigma)},\ V(s_i) >_{\mathbb{R}^N} \end{aligned} \tag{8.1}$$

with

$$[\nabla_\Gamma \Phi^\varepsilon \nabla_\Gamma P^\varepsilon]_\sigma = \nabla_\Gamma \Phi^\varepsilon_+ \nabla_\Gamma P^\varepsilon_+ - \nabla_\Gamma \Phi^\varepsilon_- \nabla_\Gamma P^\varepsilon_-$$

where $\Phi^\varepsilon_\pm = \Phi_\varepsilon \circ T_\varepsilon(E_\pm)$ et $P^\varepsilon_\pm = P_\varepsilon \circ T_\varepsilon(E_\pm)$ and $G^i_{(\omega\backslash\sigma)} = lim_{\varepsilon\downarrow 0} G^i_{(\omega\backslash\sigma)_\varepsilon}$.

Proof. It's sufficient to notice that: $\nu_\varepsilon \circ T_\varepsilon = \|{}^*D_\Gamma T_\varepsilon^{-1}.\nu\|^{-1}\, {}^*D_\Gamma T_\varepsilon^{-1}.\nu$ and to remark, when ε tends to 0, that:

i) $j(\varepsilon) = \det(DT_\varepsilon)\|{}^*D_\Gamma T_\varepsilon^{-1}.\nu\| \longrightarrow 1$ in $L^\infty(\sigma)$

ii) $\nu_\varepsilon \circ T_\varepsilon(E^+, E^-) \longrightarrow (\nu^+, \nu^-)$ in $L^\infty(\sigma)$

iii) $A(\varepsilon) \longrightarrow Id$ in $L^\infty(\sigma)$

□

REMARK 8.1. *The proposition(9.1) provide the Shape Gradient's independence of the choice of the vector field $E = (E_+, E_-)$ building the envelope e_ε. In deed when $V(0)$ is vanish in a neighborhood of γ, the expression (8.1) may be given by*

$$dJ(\omega\backslash\sigma, V) = \lim_{\varepsilon\to 0} \int_\sigma g_\varepsilon < V(0),\ \nu > d\sigma$$

as $< V(0),\ \nu >$ belongs to $C^0(\sigma)$ then g_ε converges weakly to g in the Measure space with $\overline{\sigma}$ as a support, hence

THEOREM 8.2. *The Shape boundary expression is given by:*

$$dJ(\omega\backslash\sigma, V) =< G,\ V(0) >_{\mathcal{D}'(\Gamma,T\Gamma)\times\mathcal{D}(\Gamma,T\Gamma)}$$

with

$$G = \gamma^*_\gamma(h\nu) \ + \ \gamma^*_\sigma(g\nu) + \sum_i G^i_{(\omega\backslash\sigma)}\delta_{s_i}$$

where

$$h = fP \ + \ \frac{1}{2}(\Phi - \Phi_d)^2$$

and γ^* is the adjoint of the trace operator on the corresponding boundary, let S being a boundary included in Γ then

$$\gamma_{|S} : \mathcal{D}(\Gamma, T\Gamma) \longrightarrow \mathcal{D}(S, T\Gamma)$$

Thanks to the continuity of the Neumann tangential problem with respect to the envelope e_ε and the last proposition, we have an optimal relaxation of the gradient tangential normal component of the Shape boundary derivative $\Phi'_{\varepsilon\Gamma}$. We have the following result

THEOREM 8.3.

$$\begin{aligned}\lim_{\varepsilon\to 0}\int_\sigma [P_\varepsilon \circ T_\varepsilon < \nabla_\Gamma(\Phi'_{\varepsilon\Gamma}\circ T_\varepsilon),\ \nu_\varepsilon \circ T_\varepsilon >]j(\varepsilon)d\sigma \quad &= \quad \tfrac{1}{2}\int_\gamma (\Phi-\Phi_d)^2 < V(0),\ \nu > d\gamma \\ &- \quad \int_\sigma g < V(0),\ \nu > d\sigma \\ &+ \quad \textstyle\sum_i < G^i_{(\omega\backslash\sigma)},\ V(s_i) >_{\mathbf{R}^N}\end{aligned} \tag{8.2}$$

and so $\nabla_\Gamma \Phi'_{\varepsilon\Gamma}$ *converges to a such function* h' *in* $H^{\frac{-1}{2}}(\sigma)$.

The proof is a direct consequence from (8.1).

Appendix

We will try to construct the function $y^{i,n}_\mu \circ (\xi^i)^{-1}$ denoted by $z^{i,n}_\mu$. In deed $z^{i,n}_\mu$ is define in the ball $\xi^i(B^i_{\sqrt{n}})$ with centre s_i, its radius is $\frac{1}{\sqrt{n}}$ and let $\xi^i(B^i_n)$ be the unit ball with centre s_i and radius $\frac{1}{n}$.

REMARK A.1. *ξ^i is the following projection:*

$$\xi^i : B^i_{\sqrt{n}} \longrightarrow T_{s_i}\Gamma$$

where $T_{s_i}\Gamma$ is the tangent plan to the compact manifold Γ on s_i.

We come to appear explicitly the solution of the hereafter minimization problem:

$$\min\{\int_{\xi^i(B^i_{\sqrt{n}})} |\nabla_\Gamma\psi|^2, \psi = 1 \text{ on } \xi^i(B^i_n),\ \psi = 0 \text{ on } \partial(\xi^i(B^i_{\sqrt{n}})) \text{ et } \Delta_\Gamma\psi = 0 \text{ in } \xi^i(B^i_{\sqrt{n}})\}.$$

In deed, let

$$\mathcal{J}(\psi) = \int_{\xi^i(U_i)} |\nabla_\Gamma\psi|^2$$

it leads, by passing to the polar coordinates via a C^∞ function φ with compact support centred on s_i contained in $\xi^i(B^i_{\sqrt{n}})$:

$$d\mathcal{J}(z^i_n,\ \varphi) = 4\pi\int_{\frac{1}{n}}^{\frac{1}{\sqrt{n}}} |(z^i_n)'|\varphi'\rho d\rho = 0$$

it follows:

$$(\rho|(z^i_n)'|)' = 0$$

$$|(z^i_n)'| \ + \ \rho(|(z^i_n)'|^2)^{\frac{-1}{2}}(z^i_n)'(z^i_n)'' \ = \ 0$$

$$(z^i_n)' \ + \ \rho(z^i_n)'' = 0$$

$$z_n^i(\rho) = \frac{-2}{log(n)} log(\rho) - 1$$

Thus we define the function y_n^i as follows:

$$y_n^i = \begin{cases} 1 & \text{in} B_n^i \\ z_n^i \circ \xi^i & \text{in} B_{\sqrt{n}}^i \end{cases}$$

Bibliography

[1] T. Aubin. Nonlinear Analysis on manifolds. Monge-Ampère Equations

[2] H. Brezis. Analyse fonctionnelle, Masson, 1987.

[3] M. Cuer–J.P. Zolésio. Control of Singular problem via differentation of a min-max, [J] Syst. Control Lett. 11, No.2, 151-158 (1988).

[4] M.C. Delfour–J.P. Zolésio. Shape analysis via oriented distance functions, vol. 119, pp. 426-449, July 1995, Differentiel Equations.

[5] M.C. Delfour–J.P. Zolésio. Shape analysis via distance functions, J. Funct. Anal. 123 (1994), 129-201.

[6] M.C. Delfour–J.P. Zolésio. Distance functions, curvature and shell theory; in Proc. Motion by mean curvature and related topics, A. Damlamian, J. Spruck and A. Visintin, eds, Mathematical Sciences and Applications, Gakkotosho, Tokyo.

[7] M.C. Delfour–J.P. Zolésio. Shape sensitivity analysis via min max differentiability. SIAM J. Control Optim. 26 (1988), 834-862.

[8] F. Desaint. Dérivées par rapport au domaine en géometrie intrinsèque: application aux équations de coques, thesis dissertation, university of Nice-Sophia Antipolis, dec. 1995

[9] I. Ekeland–R. Temam. Analyse convexe et problèmes variationnels. Etudes Mathématiques, Bordas, 1974.

[10] J. Ferchichi–J.P. Zolésio. Dérivation de forme dans le problème de Neumann tangentiel, (to appear).

[11] G. Grisvard. Elliptic problems in non smooth domains, Pitman, 1985

[12] J. Sokolowski– J.P. Zolésio. Shape sensitivity analysis. Springer Verlag, (1992).

Jamel Ferchichi. CMA, Ecole des Mines de Paris, and LAMSIN, Université of Tunis, 2004 route des Lucioles, BP. 93, 06902 Sophia Antipolis Cedex, France
E-mail: Jamel.Ferchichi@sophia.inria.fr

Jean-Paul Zolésio. Research Director at CNRS, Centre de Mathématiques Appliquées, Ecole des Mines de Paris, 2004 route des Lucioles, BP. 93, 06902 Sophia Antipolis Cedex, France
E-mail: Jean-Paul.Zolesio@sophia.inria.fr

Shape Sensitivity Analysis of Problems with Singularities

Gilles Fremiot and Jan Sokolowski

Abstract. In the present paper a representation theorem is proved for shape derivatives of functionals depending on solutions of boundary value problems defined in sets with geometrical singularities. The theorem is applied to a control problem with the state equation in the form of Dirichlet-Neumann boundary value problem. In the second part the shape derivative of the first eigenvalue of Laplacian in a domain with the crack is evaluated.

1 Introduction

In the paper the Hadamard formula [1] is derived for a class of shape functionals depending on solutions of boundary value problems with geometrical singularities. In section 3, the formula is applied to shape sensitivity analysis of the optimal value of cost functional for an optimal control problem. The state equation of the control problem takes the form of Dirichlet-Neumann boundary value problem. The so-called Griffith formula is derived for the shape sensitivity analysis of the control problem with respect to the perturbations of interfaces on the boundary.

In section 4, the shape sensitivity analysis of the first eigenvalue of the Laplace operator in a domain with the crack is performed. The Auchmuty variational principle is used to characterize the first eigenvalue. The representation formula of the shape gradient established in [2] is used.

The results of the present paper seem to be new, to our knowledge. We refer the reader to [1] for the general theory of material derivative method in shape sensitivity analysis which method is used in the present paper. The form of the Hadamard formula for a mixed problem is derived in [3]. The related results on shape sensitivity analysis of energy type functional with unilateral contact conditions prescribed on the crack faces are given in [4]. Finally, the topological derivative of the optimal value of the cost functional for a class of control problems is obtained in [5].

2 Structure Theorem

The structure theorem of shape derivatives is useful for applications in shape optimization, because it allows to obtain the structure of the shape gradient of a specific shape functional by means of simple verifications of hypothesis, usually in the fixed domain setting, by an application of the material derivative method [1].

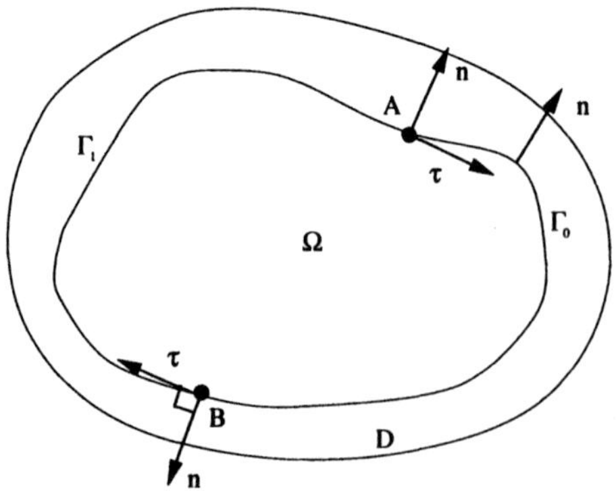

Figure 1: Domains D and Ω

Let $D \subset \mathbb{R}^2$ be a bounded domain with smooth boundary. The set D plays the role of *hold-all.* Let Ω be an open subset of D with smooth boundary $\Gamma = \partial\Omega$. We write $\Gamma = \Gamma_0 \cup \Gamma_1 \cup \{A\} \cup \{B\}$ with $\overline{\Gamma_0} = \Gamma_0 \cup \{A\} \cup \{B\}$ and $\overline{\Gamma_1} = \Gamma_1 \cup \{A\} \cup \{B\}$.

Let J be a shape functional which is shape differentiable at Ω. We refer the reader to [1] for the definition of the shape differentiability.

The shape derivative of $J(\Omega)$ in the direction of a field V is denoted by $dJ(\Omega; V)$. The velocity field V is used to construct a family of domains $\Omega_t = T_t(V)(\Omega)$ using the technique described in [1]. Without losing the generality, we can consider the problem with autonomous vector fields. Let $\mathcal{D}^k(D; \mathbb{R}^2) = C_0^k(D; \mathbb{R}^2)$ be the space of k-times continuously differentiable transformations of $\mathbb{R}^2$ with compact support in D.

We assume that $dJ(\Omega; V) = 0$ for any vector field $V \in \mathcal{D}^k(D; \mathbb{R}^2)$ such that

$$\left\{ \begin{array}{rcl} (V.\tau)(A) & = & 0, \\ (V.\tau)(B) & = & 0, \\ V.n & = & 0 \quad \text{on } \Gamma, \end{array} \right. \tag{2.1}$$

where $V.\tau$ and $V.n$ denote the tangential and normal components of the field V on Γ, respectively.

Then we have the following result on the structure of the Eulerian semiderivative $dJ(\Omega; V)$ (see [1] for the definition and properties):

THEOREM 2.1 (STRUCTURE THEOREM). *Let k be a nonnegative integer. Assume that the mapping $\mathcal{D}^k(D; \mathbb{R}^2) \ni V \mapsto dJ(\Omega; V) \in \mathbb{R}$ is linear and continuous. Moreover, we assume that if $V \in \mathcal{D}^k(D; \mathbb{R}^2)$ satisfies the conditions (2.1) then $dJ(\Omega; V) = 0$.*

Then there exist two real numbers α_A and α_B, and a linear form ϕ which is continuous on $C^k(\Gamma)$ $(\phi \in (C^k(\Gamma))')$ such that:

$$dJ(\Omega; V) = \alpha_A (V.\tau)(A) + \alpha_B (V.\tau)(B) + \phi(V.n), \quad \forall V \in \mathcal{D}^k(D, \mathbb{R}^2).$$

Proof. It is natural to consider the following set of vector fields

$$F(\Omega) = \{V \in \mathcal{D}^k(D;\mathbb{R}^2) \mid V.n = 0 \text{ on } \Gamma,\ (V.\tau)(A) = (V.\tau)(B) = 0\}. \tag{2.2}$$

According to the hypothesis that the mapping

$$V \mapsto dJ(\Omega; V)$$

is linear continuous from $\mathcal{D}^k(D;\mathbb{R}^2)$ in $\mathbb{R}$, the set $F(\Omega)$ defined by (2.2) is included in its kernel. Consequently, we can prove the following lemma.

LEMMA 2.2. *The mapping*

$$\begin{array}{rcl} \psi : \mathcal{D}^k(D;\mathbb{R}^2)/F(\Omega) & \to & C^k(\Gamma)\times\mathbb{R}\times\mathbb{R} \\ \{V\} & \mapsto & (V.n, (V.\tau)(A), (V.\tau)(B)) \end{array}$$

is an isomorphism.

Proof. (of lemma 2.2) The linear mapping $\psi : \{V\} \mapsto (V.n, (V.\tau)(A), (V.\tau)(B))$ is well defined since if $V_1 - V_2 \in F(\Omega)$ then

$$(V_1 - V_2).n = 0 \quad \text{on } \Gamma,\ ((V_1 - V_2).\tau)(A) = ((V_1 - V_2).\tau)(B) = 0.$$

Let $\{V\} \in \mathcal{D}^k(D;\mathbb{R}^2)/F(\Omega)$ be such that $\psi(\{V\}) = 0$ i.e.

$$V.n = 0 \quad \text{on } \Gamma,\ (V.\tau)(A) = (V.\tau)(B) = 0,$$

which means that $V \in F(\Omega)$ and then $\{V\} = \{0\}$. Consequently ψ is one-to-one.

Now let us show that ψ is onto. Let $(v, v_1, v_2) \in C^k(\Gamma)\times\mathbb{R}\times\mathbb{R}$. We want to find $V \in \mathcal{D}^k(D;\mathbb{R}^2)$ such that $\psi(\{V\}) = (v, v_1, v_2)$.

Ω is a smooth bounded open set and in particular is k-regular. This property means that there exists a finite number of bounded open sets $\mathcal{O}_i \subset \mathbb{R}^2$, $0 \le i \le I$, such that $\overline{\mathcal{O}_0} \subset \Omega$, $\{\mathcal{O}_i\}_{i=0}^I$ is an open covering of $\overline{\Omega}$; for all $i = 1, \ldots, I$, there exists an invertible mapping $\varphi_i : x \mapsto y = \varphi_i(x)$ which is k times continuously differentiable from $\overline{\mathcal{O}_i}$ in $\overline{B}$, $\overline{B}$ is the closed ball of $\mathbb{R}^2$ with radius 1, the inverse mapping φ_i^{-1} being k times continuously differentiable from $\overline{B}$ in $\overline{\mathcal{O}_i}$ and such that

$$\varphi_i(\mathcal{O}_i \cap \Omega) = B \cap \mathbb{R}^2_+ = \{y = (y_1, y_2) \in \mathbb{R}^2 \mid \|y\| < 1,\ y_2 > 0\}, \tag{2.3}$$

$$\varphi_i(\overline{\mathcal{O}_i} \cap \Gamma) = \{y = (y_1, y_2) \in \mathbb{R}^2 \mid |y_1| \le 1,\ y_2 = 0\} = \Sigma. \tag{2.4}$$

$\{\mathcal{O}_i, \varphi_i\}_{i=1}^I$ is a system of local maps which defines Γ.

Let us consider a partition of unity $\{\alpha_i\}_{i=1}^I$ subordoned to the covering $\{\mathcal{O}_i\}_{i=1}^I$ of Γ, that's to say a family of functions $\{\alpha_i\}_{i=1}^I$ verifying $\alpha_i \in \mathcal{D}(\mathcal{O}_i)$, $1 \le i \le I$,

$$\sum_{i=1}^{I} \alpha_i \equiv 1 \quad \text{on } \Gamma. \tag{2.5}$$

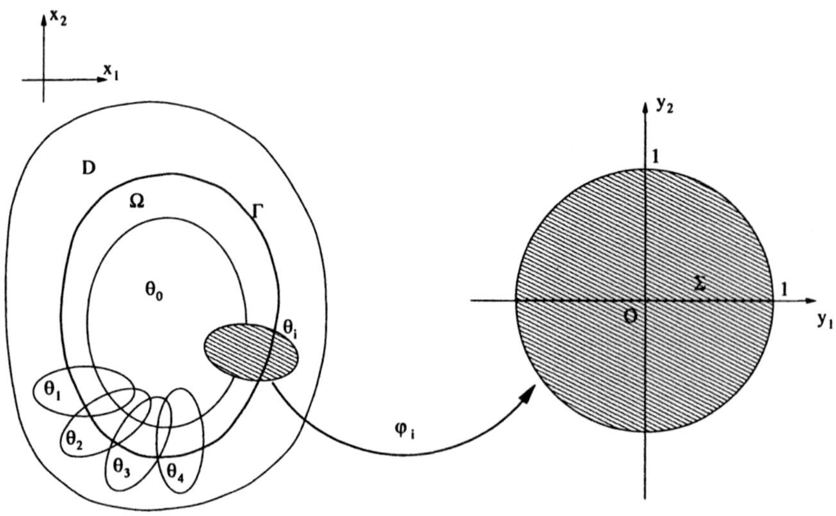

Figure 2: Mapping φ_i

We introduce in standard way a linear extension operator P from $C^k(\Gamma)$ in $\mathcal{D}^k(D;\mathbb{R})$.
It is convenient to write

$$v = \sum_{i=1}^{I} \alpha_i v, \tag{2.6}$$

in order to use the covering of Ω. For all $i = 1, \ldots, I$, we are going to define $P(\alpha_i v)$; the function Pv will be determined from the equality

$$Pv = \sum_{i=1}^{I} P(\alpha_i v). \tag{2.7}$$

For $i = 1, \ldots, I$, let us consider the function given by

$$w_i = v \circ ({\varphi_i^{-1}}_{|\Sigma}), \tag{2.8}$$

where ${\varphi_i^{-1}}_{|\Sigma}$ denotes the restriction of φ_i^{-1} to Σ, and Σ is defined by (2.4).
It is not difficult to see, according to the hypothesis, that $w_i \in C^k(\Sigma)$.
Since $w_i \in C^k(\Sigma) \simeq C^k([-1,1])$ and by definition of the space $C^k([-1,1])$, there exists $\widetilde{w_i} \in C^k(\mathbb{R} \times \{0\}) \simeq C^k(\mathbb{R})$ such that

$$\widetilde{w_i}_{|\Sigma} = w_i. \tag{2.9}$$

Let $\widetilde{W_i}$ be the function defined by the following relation

$$\widetilde{W_i}(y_1, y_2) = \widetilde{w_i}(y_1), \quad \forall y_1,\ y_2 \in \mathbb{R}. \tag{2.10}$$

It is easily seen that $\widetilde{W_i} \in C^k(\mathbb{R}^2;\mathbb{R})$ and moreover, according to (2.10) we have

$$\widetilde{W}_{i|\mathbb{R}\times\{0\}} = \widetilde{w_i}. \tag{2.11}$$

Let us introduce

$$P(\alpha_i v) = \alpha_i \widetilde{W_i} \circ \varphi_i. \tag{2.12}$$

It is easy to check that $v \mapsto Pv = \sum_{i=1}^{I} P(\alpha_i v)$ is the required extension operator.
So we have constructed $Pv \in \mathcal{D}^k(D;\mathbb{R})$ with the following property

$$Pv_{|\Gamma} = v. \tag{2.13}$$

Moreover, it is a simple matter to find a function $\phi_{v_1,v_2} \in \mathcal{D}^k(D;\mathbb{R})$ which verifies

$$\begin{cases} \phi_{v_1,v_2}(A) &= v_1, \\ \phi_{v_1,v_2}(B) &= v_2. \end{cases} \tag{2.14}$$

Indeed, let $R_1,\ R_2 \in \mathbb{R}_+^*$ be sufficiently small in order to have

$$\begin{cases} \overline{B(A,R_1)} \subset D, \\ \overline{B(A,R_2)} \subset D, \\ \overline{B(A,R_1)} \cap \overline{B(A,R_2)} = \emptyset. \end{cases} \tag{2.15}$$

Under these assumptions, there exist two functions $\varphi_A,\ \varphi_B \in C^\infty(\mathbb{R}^2;\mathbb{R})$ such that

$$\text{supp}\{\varphi_A\} \subset \overline{B(A,R_1)} \quad \text{and} \quad \varphi_A \equiv 1 \quad \text{on} \quad \overline{B(A,R_1/2)},$$

$$\text{supp}\{\varphi_B\} \subset \overline{B(B,R_2)} \quad \text{and} \quad \varphi_B \equiv 1 \quad \text{on} \quad \overline{B(B,R_2/2)}.$$

Therefore, the following function

$$\phi_{v_1,v_2} = v_1\varphi_A + v_2\varphi_B, \tag{2.16}$$

has the required properties $\phi_{v_1,v_2} \in \mathcal{D}(D;\mathbb{R})$, $\phi_{v_1,v_2}(A) = v_1$ and $\phi_{v_1,v_2}(B) = v_2$.
Let $\widetilde{n}$, $\widetilde{\tau}$ denote the extensions to an open neighborhood of Γ of normal field n and tangential field τ defined on $\Gamma = \partial D$. So we consider the vector field $V \in \mathcal{D}^k(D;\mathbb{R}^2)$ defined by the relation

$$V = (Pv)\widetilde{n} + \phi_{v_1,v_2}\widetilde{\tau}, \tag{2.17}$$

which satisfies the following conditions

$$\begin{cases} (V.\tau)(A) &= v_1, \\ (V.\tau)(B) &= v_2, \\ V.n &= v \quad \text{on } \Gamma. \end{cases}$$

This completes the proof of lemma 2.2. □

LEMMA 2.3. *There exists a linear, continuous mapping* Φ

$$\Phi : C^k(\Gamma) \times \mathbb{R} \times \mathbb{R} \to \mathbb{R}$$

such that for any vector field $V \in \mathcal{D}^k(D; \mathbb{R}^2)$,

$$dJ(\Omega; V) = \Phi\left(V.n, (V.\tau)(A), (V.\tau)(B)\right).$$

Proof. (of lemma 2.3) We define Φ by the following formula

$$\Phi(\{V\}) = dJ(\Omega; V). \tag{2.18}$$

Indeed, if $\{V'\} = \{V\}$, i.e. if $V' \in \{V\}$, we have $V' - V \in F(\Omega)$, since $F(\Omega)$ is included in the kernel of $dJ(\Omega; \cdot)$, it follows that

$$dJ(\Omega; V - V') = 0.$$

The Eulerian semiderivative $dJ(\Omega; \cdot)$ is linear by our assumption, therefore

$$dJ(\Omega; V) = dJ(\Omega; V'). \tag{2.19}$$

The relation (2.19) enables us to define Φ.
Using lemma 2.2 and the relation $\mathcal{D}^k(D; \mathbb{R}^2)/F(\Omega) \simeq C^k(\Gamma) \times \mathbb{R} \times \mathbb{R}$, it follows that

$$\{V\} = (V.n, (V.\tau)(A), (V.\tau)(B)) \tag{2.20}$$

thus

$$dJ(\Omega; V) = \Phi\left(\{V\}\right) = \Phi\left(V.n, (V.\tau)(A), (V.\tau)(B)\right). \tag{2.21}$$

Furthermore, $dJ(\Omega; \cdot)$ is linear and continuous which implies that Φ is linear and continuous.

Now, we can complete the proof of the structure theorem. Indeed, there exists a linear mapping Φ, which is continuous from $C^k(\Gamma) \times \mathbb{R} \times \mathbb{R}$ in $\mathbb{R}$, such that

$$\forall V \in \mathcal{D}^k(D; \mathbb{R}^2), \quad dJ(\Omega; V) = \Phi\left(V.n, (V.\tau)(A), (V.\tau)(B)\right)$$

with

$$\begin{aligned} \Phi \in \left(C^k(\Gamma) \times \mathbb{R} \times \mathbb{R}\right)' &= \left(C^k(\Gamma)\right)' \times \mathbb{R}' \times \mathbb{R}' \\ &= \left(C^k(\Gamma)\right)' \times \mathbb{R} \times \mathbb{R}. \end{aligned}$$

Therefore, there exist two real numbers α_A and α_B, and a linear form ϕ which is continuous on $C^k(\Gamma)$ such that the above formula can be rewritten as follows

$$dJ(\Omega; V) = \phi(V.n) + \alpha_A (V.\tau)(A) + \alpha_B (V.\tau)(B), \quad \forall V \in \mathcal{D}^k(D; \mathbb{R}^2) \tag{2.22}$$

□

This achieves the proof of the structure theorem. □

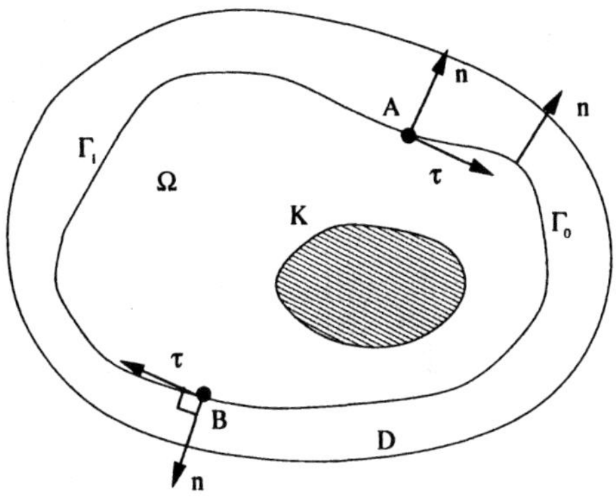

Figure 3: Domains D and $\Omega \subset D$

3 Optimal Control Problem

We apply the structure theorem in the case of cost functionals for control problems. The optimal control problem considered in this section is defined for the elliptic equation modeling the deflection of an elastic membrane.

Let us consider the domain D in $\mathbb{R}^2$ with smooth boundary. Let Ω be an open subset of D with smooth boundary $\Gamma = \partial\Omega$. We write $\Gamma = \Gamma_0 \cup \Gamma_1 \cup \{A\} \cup \{B\}$ with $\overline{\Gamma_0} = \Gamma_0 \cup \{A\} \cup \{B\}$ and $\overline{\Gamma_1} = \Gamma_1 \cup \{A\} \cup \{B\}$. Let K be an open subset of Ω, with the smooth boundary ∂K, and moreover we assume that $\overline{K} \cap \Gamma = \emptyset$. The state equation for the control problem is of the form

$$\left\{ \begin{array}{rclll} -\Delta q & = & u\chi_K & \text{in} & \Omega, \\ q & = & 0 & \text{on} & \Gamma_0, \\ \dfrac{\partial q}{\partial n} & = & 0 & \text{on} & \Gamma_1, \end{array} \right. \tag{3.1}$$

where χ_K denotes the characteristic function of K.
For given $u \in L^2(K)$, $q = q(u)$ represents the deflection of an elastic membrane loaded by the vertical force u concentrated on K. The membrane is clamped on Γ_0 and free on Γ_1. For the system (3.1), we define the cost functional

$$I(u) = \frac{1}{2} \int_K [(q - q_d)^2 + \alpha u^2] d\Omega, \tag{3.2}$$

which is minimized over the space of controls $u \in L^2(K)$, $\alpha > 0$, and $q_d \in L^2(K)$ is a given function. The minimization of the functional (3.2) with respect to u means approximation of a given function q_d in the region K by the deflection of an elastic membrane, using the smallest possible load u applied in K. The minimal value of the cost functional for this control problem defines the shape functional, depending on the geometrical domain Ω,

$$J(\Omega) = \min_{u \in L^2(K)} I(u).$$

Variation of the state $q'(v)$, corresponding to the variation v of the control u, in view of (3.1),

$$q(u+sv) = q(u) + sq'(v), \tag{3.3}$$

satisfies the equation

$$\begin{cases} -\Delta q' &= v\chi_K & \text{in } \Omega, \\ q' &= 0 & \text{on } \Gamma_0, \\ \dfrac{\partial q'}{\partial n} &= 0 & \text{on } \Gamma_1, \end{cases} \tag{3.4}$$

and, according to (3.4), the variation $dI(u;v)$ of the cost functional is given by

$$dI(u;v) = \int_K [(q(u) - q_d)q'(v) + \alpha uv] d\Omega. \tag{3.5}$$

Let us introduce the adjoint state p, which is defined by the following equation

$$\begin{cases} -\Delta p &= (q - q_d)\chi_K & \text{in } \Omega, \\ p &= 0 & \text{on } \Gamma_0, \\ \dfrac{\partial p}{\partial n} &= 0 & \text{on } \Gamma_1. \end{cases} \tag{3.6}$$

Using the adjoint state p given by (3.6) allows us to obtain the directional derivative of the cost functional (3.5) in the form

$$\begin{aligned} dI(u;v) &= \int_K [(q(u) - q_d)q'(v) + \alpha uv] d\Omega \\ &= \int_\Omega \langle \nabla p, \nabla q' \rangle d\Omega + \int_K \alpha uv d\Omega, \end{aligned}$$

and in view of the equation (3.4) satisfied by q', it follows that

$$dI(u;v) = \int_K [p + \alpha u] v d\Omega.$$

Thus the stationarity condition

$$dI(u;v) = 0, \quad \forall v \in L^2(K)$$

leads to the following equality

$$u(y) = -\frac{1}{\alpha} p(y), \text{ a.e. in } K. \tag{3.7}$$

And consequently, by using (3.7), the minimal value of the cost functional for the control problem takes the form

$$J(\Omega) = \frac{1}{2} \int_K [(q - q_d)^2 + \frac{1}{\alpha} p^2] d\Omega, \tag{3.8}$$

where p, q are given as a solution of the coupled system of equations:

$$\begin{cases} -\Delta q &= -\frac{1}{\alpha} p\chi_K & \text{in } \Omega, \\ -\Delta p &= (q-q_d)\chi_K & \text{in } \Omega, \\ q &= 0 & \text{on } \Gamma_0, \\ p &= 0 & \text{on } \Gamma_0, \\ \frac{\partial q}{\partial n} &= 0 & \text{on } \Gamma_1, \\ \frac{\partial p}{\partial n} &= 0 & \text{on } \Gamma_1. \end{cases} \tag{3.9}$$

Let us consider the perturbations of Ω by the vector field $V = (\theta_1, \theta_2)$ with $\theta_1,\ \theta_2 \in C_0^\infty(D) = \mathcal{D}(D)$ and moreover we assume that $\overline{K} \cap \text{supp}\{V\} = \emptyset$. Then we consider the transformation (see [4]) defined by:

$$\begin{cases} y_1 = x_1 - \delta\theta_1(x_1, x_2) \\ y_2 = x_2 - \delta\theta_2(x_1, x_2) \quad (\delta > 0). \end{cases} \tag{3.10}$$

The coordinates of a given point in open sets Ω, Ω_δ, are denoted by $(y_1, y_2) \in \Omega$, $(x_1, x_2) \in \Omega_\delta$, respectively.

The Jacobian of (3.10) equals to:

$$\begin{aligned} q_\delta &= 1 - \delta(\theta_{1,x_1} + \theta_{2,x_2}) + \delta^2(\theta_{1,x_1}\theta_{2,x_2} - \theta_{1,x_2}\theta_{2,x_1}) \\ &= 1 - \delta \text{div} V + \delta^2 \det(DV). \end{aligned}$$

For $\delta > 0$, δ small enough, $q_\delta > 0$, so the transformation (3.10) is one-to-one and we denote $y = y(x, \delta)$, $x = x(y, \delta)$. Let Ω_δ be the image of Ω for the transformation (3.10).

The minimization problem is defined in Ω_δ, with the cost functional

$$J(\Omega_\delta) = \frac{1}{2}\int_K [(q_\delta - q_d)^2 + \frac{1}{\alpha} p_\delta^2] d\Omega, \tag{3.11}$$

where p_δ, q_δ are the solutions of the following coupled equations (after the change of variables in order to transport the problem to the fixed domain Ω)

$$\begin{cases} \int_\Omega \langle C_\delta \cdot \nabla q_\delta, \nabla\varphi\rangle d\Omega &= -\frac{1}{\alpha}\int_K p_\delta \varphi d\Omega, \quad \forall\varphi \in H^1_{\Gamma_0}(\Omega), \\ \int_\Omega \langle C_\delta \cdot \nabla p_\delta, \nabla\psi\rangle d\Omega &= \int_K (q_\delta - q_d)\psi d\Omega, \quad \forall\psi \in H^1_{\Gamma_0}(\Omega), \end{cases} \tag{3.12}$$

where $C_\delta = \frac{1}{q_\delta} A_\delta^T \cdot A_\delta$ and A_δ takes the following form

$$A_\delta = \begin{pmatrix} 1 - \delta\theta_{1,y_1} & -\delta\theta_{2,y_1} \\ -\delta\theta_{1,y_2} & 1 - \delta\theta_{2,y_2} \end{pmatrix} = I + \delta B,$$

with

$$B = \begin{pmatrix} -\theta_{1,y_1} & -\theta_{2,y_1} \\ -\theta_{1,y_2} & -\theta_{2,y_2} \end{pmatrix}.$$

Then, we have the following result:

THEOREM 3.1. *We have the following Griffith formula*

$$\frac{dJ(\Omega_\delta)}{d\delta}|_{\delta=0} = \frac{\pi}{4}(c_p c_\eta + c_q c_\xi),$$

where c_p, c_η, c_q, c_ξ are the coefficients of singularity of solutions p, η, q, ξ, to the systems (3.9),(3.16), respectively.

Proof. Applying the implicit functions theorem gives us the existence of the material derivatives $\dot{p}$, $\dot{q} \in H^1_{\Gamma_0}(\Omega)$. Moreover, we obtain the integral identities satisfied by $\dot{p}$ and $\dot{q}$:

$$\int_\Omega \langle C' \cdot \nabla q, \nabla\varphi\rangle d\Omega + \int_\Omega \langle \nabla\dot{q}, \nabla\varphi\rangle d\Omega = -\frac{1}{\alpha}\int_K \dot{p}\varphi d\Omega, \quad \forall\varphi \in H^1_{\Gamma_0}(\Omega), \tag{3.13}$$

$$\int_\Omega \langle C' \cdot \nabla p, \nabla\psi\rangle d\Omega + \int_\Omega \langle \nabla\dot{p}, \nabla\psi\rangle d\Omega = \int_K \dot{q}\psi d\Omega, \quad \forall\psi \in H^1_{\Gamma_0}(\Omega). \tag{3.14}$$

On the other hand, in view of (3.8) the cost functional $J(\Omega)$ is shape differentiable with the shape derivative

$$dJ(\Omega; V) = \int_K [(q - q_d)\dot{q} + \frac{1}{\alpha}p\dot{p}]d\Omega. \tag{3.15}$$

In consequence, we can apply the structure theorem, which leads to the following formula for the derivative (3.15)

$$dJ(\Omega; V) = \alpha_A(V.\tau)(A) + \alpha_B(V.\tau)(B) + \phi(V.n),$$

where $\phi \in (C^1(\Gamma))'$, α_A, $\alpha_B \in \mathbb{R}$.

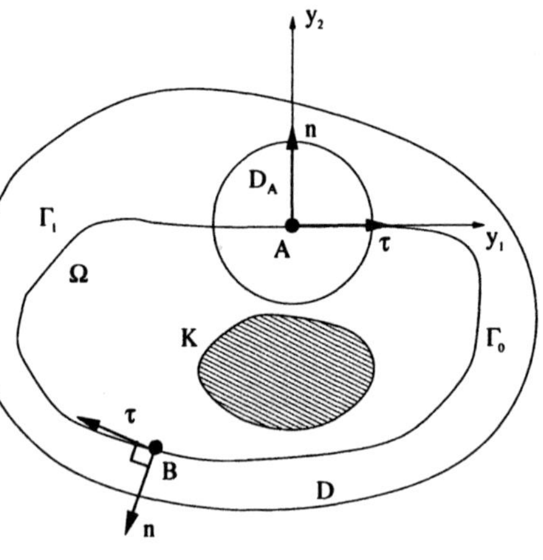

Figure 4: Domain Ω after changing variables

We show how to determine e.g. the coefficient α_A. In the remainder of this section, performing a change of variables if necessary, we may assume that there exists a neighborhood D_A

of A, $D_A \subset D$, $\overline{D_A} \cap \overline{K} = \emptyset$, such that $\Gamma \cap D_A$ is rectilinear. Moreover, we use an appropriate coordinate system with origin A. In order to identify the coefficient α_A, we consider the deformations of Ω in a small neighborhood of $\Gamma \cap D_A$. The deformations are assumed to be such that $\Gamma \cap D_A$ moves without changing direction that's why the vector field takes the form

$$V(y) = (\theta_1(y), 0),$$

where θ_1 is a smooth function supported in D_A, and $\theta_1(y) = -1$ in the vicinity of the origin A. In order to simplify the form of the Eulerian semiderivative $dJ(\Omega; V)$ and eliminate the material derivatives $\dot{p}$ and $\dot{q}$, we introduce the second level adjoint variables $\xi,\ \eta \in H^1_{\Gamma_0}(\Omega)$ (see [5]), defined by the following equations:

$$\begin{cases} -\Delta\xi - \eta\chi_K &= (q - q_d)\chi_K & \text{in } \Omega, \\ -\Delta\eta + \dfrac{1}{\alpha}\xi\chi_K &= \dfrac{1}{\alpha}p\chi_K & \text{in } \Omega, \\ \xi &= 0 & \text{on } \Gamma_0, \\ \eta &= 0 & \text{on } \Gamma_0, \\ \dfrac{\partial\xi}{\partial n} &= 0 & \text{on } \Gamma_1 \\ \dfrac{\partial\eta}{\partial n} &= 0 & \text{on } \Gamma_1, \end{cases} \tag{3.16}$$

or, in the weak form:

$$\int_\Omega \langle\nabla\xi, \nabla\varphi\rangle d\Omega - \int_K \eta\varphi d\Omega = \int_K (q - q_d)\varphi d\Omega, \quad \forall\varphi \in H^1_{\Gamma_0}(\Omega), \tag{3.17}$$

$$\int_\Omega \langle\nabla\eta, \nabla\psi\rangle d\Omega + \frac{1}{\alpha}\int_K \xi\psi d\Omega = \frac{1}{\alpha}\int_K p\psi d\Omega, \quad \forall\psi \in H^1_{\Gamma_0}(\Omega). \tag{3.18}$$

The equation (3.17) for $\varphi = \dot{q}$ leads to

$$\int_\Omega \langle\nabla\xi, \nabla\dot{q}\rangle d\Omega - \int_K \eta\dot{q} d\Omega = \int_K (q - q_d)\dot{q} d\Omega \tag{3.19}$$

and taking $\psi = \dot{p}$ in (3.18) gives us

$$\int_\Omega \langle\nabla\eta, \nabla\dot{p}\rangle d\Omega + \frac{1}{\alpha}\int_K \xi\dot{p} d\Omega = \frac{1}{\alpha}\int_K p\dot{p} d\Omega. \tag{3.20}$$

Moreover, we can substitute $\varphi = \xi$ and $\psi = \eta$ in the variational equalities (3.13),(3.14), respectively, and we obtain

$$\int_\Omega \langle C' \cdot \nabla q, \nabla\xi\rangle d\Omega + \int_\Omega \langle\nabla\dot{q}, \nabla\xi\rangle d\Omega = -\frac{1}{\alpha}\int_K \dot{p}\xi d\Omega, \tag{3.21}$$

and

$$\int_\Omega \langle C' \cdot \nabla p, \nabla\eta\rangle d\Omega + \int_\Omega \langle\nabla\dot{p}, \nabla\eta\rangle d\Omega = \int_K \dot{q}\eta d\Omega. \tag{3.22}$$

Thus, using the equalities (3.19),(3.20),(3.21),(3.22), it follows that

$$\begin{aligned} dJ(\Omega;V) &= \int_\Omega \langle \nabla\xi, \nabla\dot{q}\rangle d\Omega - \int_K \eta\dot{q} d\Omega + \int_\Omega \langle \nabla\eta, \nabla\dot{p}\rangle d\Omega + \frac{1}{\alpha}\int_K \xi\dot{p} d\Omega \\ &= -\int_\Omega \langle C'\cdot\nabla q, \nabla\xi\rangle d\Omega - \int_\Omega \langle C'\cdot\nabla p, \nabla\eta\rangle d\Omega \\ &= -\lim_{\varepsilon\to 0+}\left(\int_{\Omega_\varepsilon} \langle C'\cdot\nabla q, \nabla\xi\rangle d\Omega + \int_{\Omega_\varepsilon} \langle C'\cdot\nabla p, \nabla\eta\rangle d\Omega\right) \end{aligned}$$

where Ω_ε is the subset of Ω defined by $r > \varepsilon$ (see Fig.5). Let γ_ε be the curve given by $r = \varepsilon$ and $\pi < \theta < 2\pi$ (for ε small enough).
Let us introduce the notation

$$B_\varepsilon = \int_{\Omega_\varepsilon} \langle C'\cdot\nabla q, \nabla\xi\rangle d\Omega + \int_{\Omega_\varepsilon} \langle C'\cdot\nabla p, \nabla\eta\rangle d\Omega.$$

Moreover, we have

$$A_\delta = \begin{pmatrix} 1-\delta\theta_{1,y_1} & 0 \\ -\delta\theta_{1,y_2} & 1 \end{pmatrix}, \quad q_\delta = 1-\delta\theta_{1,y_1}$$

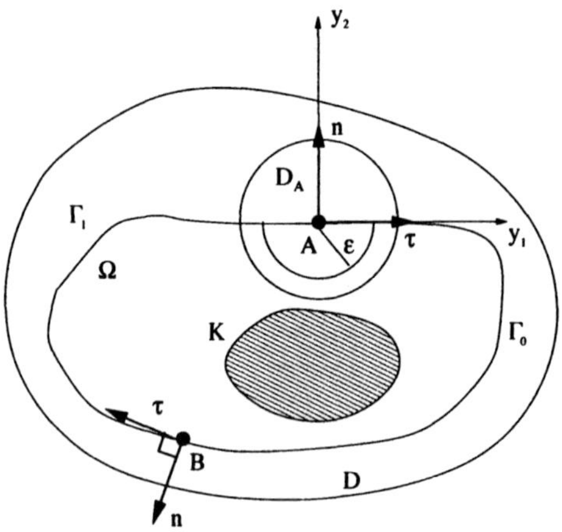

Figure 5: Domain Ω_ε

which leads to

$$C_\delta = \frac{1}{1-\delta\theta_{1,y_1}}\begin{pmatrix} (1-\delta\theta_{1,y_1})^2 + \delta^2\theta_{1,y_2}^2 & -\delta\theta_{1,y_2} \\ -\delta\theta_{1,y_2} & 1 \end{pmatrix},$$

and finally we obtain

$$C' = \frac{dC_\delta}{d\delta}|_{\delta=0} = \begin{pmatrix} -\theta_{1,y_1} & -\theta_{1,y_2} \\ -\theta_{1,y_2} & \theta_{1,y_1} \end{pmatrix}.$$

By using this relation, it follows that

$$\begin{aligned} B_\varepsilon &= \int_{\Omega_\varepsilon} \theta_{1,y_1}(-q_{y_1}\xi_{y_1} + q_{y_2}\xi_{y_2} - p_{y_1}\eta_{y_1} + p_{y_2}\eta_{y_2})d\Omega \\ &+ \int_{\Omega_\varepsilon} \theta_{1,y_2}(-q_{y_2}\xi_{y_1} - q_{y_1}\xi_{y_2} - p_{y_2}\eta_{y_1} - p_{y_1}\eta_{y_2})d\Omega. \end{aligned} \tag{3.23}$$

By integrating by parts in (3.23), we have

$$\begin{aligned} B_\varepsilon &= \int_{\gamma_\varepsilon} \theta_1(-q_{y_1}\xi_{y_1} + q_{y_2}\xi_{y_2} - p_{y_1}\eta_{y_1} + p_{y_2}\eta_{y_2})\nu_1 d\sigma \\ &+ \int_{\gamma_\varepsilon} \theta_1(-q_{y_2}\xi_{y_1} - q_{y_1}\xi_{y_2} - p_{y_2}\eta_{y_1} - p_{y_1}\eta_{y_2})\nu_2 d\sigma \\ &+ \int_{\Omega_\varepsilon} \theta_1(\xi_{y_1}\Delta q + q_{y_1}\Delta\xi + \eta_{y_1}\Delta p + p_{y_1}\Delta\eta)d\Omega. \end{aligned}$$

But, for $\varepsilon > 0$, ε small enough, $K \subset \Omega_\varepsilon$ and moreover $\overline{K} \cap \operatorname{supp}\{\theta_1\} = \emptyset$, hence

$$\theta_1\Delta q = \theta_1\Delta\xi = \theta_1\Delta p = \theta_1\Delta\eta = 0 \quad \text{on } \Omega_\varepsilon$$

and in consequence

$$\begin{aligned} B_\varepsilon &= \int_{\gamma_\varepsilon} \theta_1(-q_{y_1}\xi_{y_1} + q_{y_2}\xi_{y_2} - p_{y_1}\eta_{y_1} + p_{y_2}\eta_{y_2})\nu_1 d\sigma \\ &+ \int_{\gamma_\varepsilon} \theta_1(-q_{y_2}\xi_{y_1} - q_{y_1}\xi_{y_2} - p_{y_2}\eta_{y_1} - p_{y_1}\eta_{y_2})\nu_2 d\sigma. \end{aligned} \tag{3.24}$$

For ε small enough, $\theta_1 \equiv -1$ on γ_ε and in view of (3.24)

$$\begin{aligned} B_\varepsilon &= \int_{\gamma_\varepsilon} (q_{y_1}\xi_{y_1} - q_{y_2}\xi_{y_2} + p_{y_1}\eta_{y_1} - p_{y_2}\eta_{y_2})\nu_1 d\sigma \\ &+ \int_{\gamma_\varepsilon} (q_{y_2}\xi_{y_1} + q_{y_1}\xi_{y_2} + p_{y_2}\eta_{y_1} + p_{y_1}\eta_{y_2})\nu_2 d\sigma. \end{aligned} \tag{3.25}$$

Moreover we know [6] that

$$\begin{cases} p = p^R + c_p S, \\ q = q^R + c_q S, \\ \eta = \eta^R + c_\eta S, \\ \xi = \xi^R + c_\xi S, \end{cases} \tag{3.26}$$

where $S = \sqrt{r}\sin\left(\dfrac{\theta}{2}\right)$ is the singular function, p^R, q^R, η^R, $\xi^R \in H^2(D_A)$ and c_p, c_q, c_η, c_ξ denote the coefficients of singularity of solutions p, q, η, ξ, to (3.9), (3.16), respectively. Taking into account the decomposition (3.26) and developing in (3.25) we have

$$B_\varepsilon = B_\varepsilon^{(1)} + B_\varepsilon^{(2)} + B_\varepsilon^{(3)},$$

where $B_\varepsilon^{(1)}, B_\varepsilon^{(2)}, B_\varepsilon^{(3)}$ are defined by

$$B_\varepsilon^{(1)} = (c_p c_\eta + c_q c_\xi) \int_{\gamma_\varepsilon} [((S_{y_1})^2 - (S_{y_2})^2)\nu_1 + 2\nu_2 S_{y_1} S_{y_2}] d\sigma,$$

$$\begin{aligned} B_\varepsilon^{(2)} &= \int_{\gamma_\varepsilon} (c_\xi q_{y_1}^R + c_q \xi_{y_1}^R + c_\eta p_{y_1}^R + c_p \eta_{y_1}^R)(\nu_1 S_{y_1} + \nu_2 S_{y_2}) d\sigma \\ &+ \int_{\gamma_\varepsilon} (c_\xi q_{y_2}^R + c_q \xi_{y_2}^R + c_\eta p_{y_2}^R + c_p \eta_{y_2}^R)(\nu_2 S_{y_1} - \nu_1 S_{y_2}) d\sigma, \end{aligned}$$

$$\begin{aligned} B_\varepsilon^{(3)} &= \int_{\gamma_\varepsilon} \nu_1 (q_{y_1}^R \xi_{y_1}^R - q_{y_2}^R \xi_{y_2}^R + p_{y_1}^R \eta_{y_1}^R - p_{y_2}^R \eta_{y_2}^R) d\sigma \\ &+ \int_{\gamma_\varepsilon} \nu_2 (q_{y_2}^R \xi_{y_1}^R + q_{y_1}^R \xi_{y_2}^R + p_{y_2}^R \eta_{y_1}^R + p_{y_1}^R \eta_{y_2}^R) d\sigma. \end{aligned}$$

Moreover, the form of singular function is known in this case,

$$S_{y_1} = -\frac{1}{2\sqrt{r}} \sin\left(\frac{\theta}{2}\right) \text{ and } S_{y_2} = \frac{1}{2\sqrt{r}} \cos\left(\frac{\theta}{2}\right)$$

so using the polar coordinates, we have

$$\begin{aligned} B_\varepsilon^{(1)} &= (c_p c_\eta + c_q c_\xi) \int_\pi^{2\pi} \left(\frac{1}{4} \sin^2\left(\frac{\theta}{2}\right) - \frac{1}{4} \cos^2\left(\frac{\theta}{2}\right) \right) \cos\theta d\theta \\ &\quad -2(c_p c_\eta + c_q c_\xi) \int_\pi^{2\pi} \frac{1}{4} \cos\left(\frac{\theta}{2}\right) \sin\left(\frac{\theta}{2}\right) \sin\theta d\theta \\ &= -\frac{1}{4}(c_p c_\eta + c_q c_\xi) \int_\pi^{2\pi} (\cos^2\theta + \sin^2\theta) d\theta \\ &= -\frac{\pi}{4}(c_p c_\eta + c_q c_\xi). \end{aligned}$$

It is not difficult to see that $B_\varepsilon^{(2)} \to 0$ and $B_\varepsilon^{(3)} \to 0$ as $\varepsilon \to 0^+$, since in fact we have the estimations $B_\varepsilon^{(2)} = O(\sqrt{\varepsilon})$, $B_\varepsilon^{(3)} = O(\varepsilon)$, therefore

$$dJ(\Omega; V) = -\lim_{\varepsilon \to 0+} B_\varepsilon = \frac{\pi}{4}(c_p c_\eta + c_q c_\xi).$$

Finally, we have identified the coefficient $\alpha_A = -\frac{\pi}{4}(c_p c_\eta + c_q c_\xi)$ in the expression of the Eulerian semiderivative $dJ(\Omega; V)$ given by the structure theorem. □

4 Shape Derivative of the First Eigenvalue of the Laplacian in Nonsmooth Domain

4.1 Domain With Crack

We can show (see [2] for the proof) that the similar structure theorem, as given in section 2, holds in the case of domains with cracks. First of all, we have to specify what we mean by a domain with the crack.

Let $D \subset \mathbb{R}^2$ be a bounded domain with smooth boundary Γ, and Σ be a part of a smooth curve. We assume that $\overline{\Sigma}$ belongs to the domain D. Therefore, we consider the domain $\Omega = D \backslash \overline{\Sigma}$ with crack Σ (see Fig.6). Let us denote by A and B the tips of $\overline{\Sigma}$. Moreover, we assume that J is a domain functional which is shape differentiable at Ω.

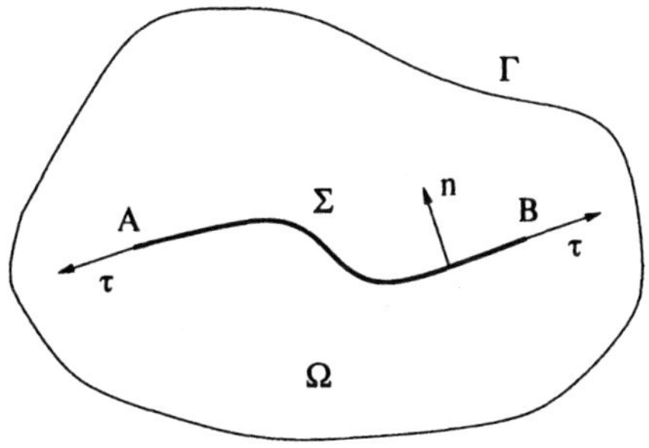

Figure 6: Domain Ω with the curved crack Σ

So, we have the following structure of the Eulerian semiderivative $dJ(\Omega; V)$:

$$dJ(\Omega; V) = \alpha_A(V.\tau)(A) + \alpha_B(V.\tau)(B) + \phi(V.n), \quad \forall V \in \mathcal{D}^k(D; \mathbb{R}^2),$$

where $k \in \mathbb{N}^*$, α_A, $\alpha_B \in \mathbb{R}$, $\phi \in (C^k(\overline{\Sigma}))'$, $V.\tau$ and $V.n$ denote the tangential and normal components of field V on $\overline{\Sigma}$, respectively.
In the following section, an application of the structure theorem for cracked domains is given.

4.2 Shape Derivative of the First Eigenvalue of the Laplace Operator in Nonsmooth Domain

Let $D \subset \mathbb{R}^2$ be a bounded domain with a smooth boundary Γ. Let Σ_l be the set defined by $\{(y_1, y_2) \mid 0 < y_1 < l, \ y_2 = 0\}$, A and B denote its tips. We assume Σ_l belongs to the domain D for $l > 0$ small enough. The domain with the crack Σ_l is denoted by $\Omega = D \setminus \overline{\Sigma_l}$.

We are going to derive the form of directional derivative of the first eigenvalue with respect to the perturbations of the crack Σ_l. According to Auchmuty's principle (see [7]), which enables to characterize the eigenvalues of the Laplace operator by a variational principle, the smallest eigenvalue $\lambda(\Omega)$ of the Laplacian in Ω is given by

$$-\frac{1}{2\lambda(\Omega)} = \min_{\varphi \in H^1_\Gamma(\Omega)} G(\varphi),$$

where G is the functional defined by

$$G(\varphi) = \frac{1}{2}\int_\Omega |\nabla\varphi|^2 dy - \sqrt{\int_\Omega \varphi^2 dy}, \quad \forall \varphi \in H^1_\Gamma(\Omega).$$

The above variational principle differs from the classical Rayleigh's ratio. Let us introduce $\mu(\Omega)$ defined by

$$\mu(\Omega) = -\frac{1}{2\lambda(\Omega)}, \tag{4.1}$$

and consequently

$$\mu(\Omega) = \min_{\varphi \in H^1_\Gamma(\Omega)} G(\varphi).$$

Let us consider the perturbations of Ω by the vector field $V = (\theta_1, \theta_2)$ with $\theta_1,\ \theta_2 \in C_0^\infty(D) = \mathcal{D}(D)$. We use the same notation as in section 3.
For $\delta > 0$, the minimization problem is defined in Ω_δ by

$$-\frac{1}{2\lambda(\Omega_\delta)} = \mu(\Omega_\delta) = \min_{\psi \in H^1_\Gamma(\Omega_\delta)} G_\delta(\psi), \tag{4.2}$$

where $\lambda(\Omega_\delta)$ denotes the smallest eigenvalue of the Laplacian in Ω_δ and

$$G_\delta(\psi) = \frac{1}{2}\int_{\Omega_\delta} |\nabla\psi|^2 dx - \sqrt{\int_{\Omega_\delta} \psi^2 dx}, \quad \forall \psi \in H^1_\Gamma(\Omega_\delta).$$

Moreover, we use the following notation

$$\lambda_\delta = \lambda(\Omega_\delta),\ \mu_\delta = \mu(\Omega_\delta),$$

and in particular, for $\delta = 0$,

$$\lambda_0 = \lambda(\Omega_0) = \lambda(\Omega),\ \mu_0 = \mu(\Omega_0) = \mu(\Omega).$$

We determine the directional derivative

$$d\mu(\Omega; V) = \lim_{\delta \downarrow 0} \frac{\mu(\Omega_\delta) - \mu(\Omega)}{\delta} = \lim_{\delta \downarrow 0} \frac{\mu_\delta - \mu_0}{\delta}.$$

The directional differentiability of μ_δ at $\delta = 0$ is equivalent to the directional differentiability of λ_δ at $\delta = 0$ because $\lambda_\delta > 0$.
By changing the variables in (4.2), in order to transport the problem to the fixed domain Ω, we obtain

$$\mu_\delta = \min_{\psi \in H^1_\Gamma(\Omega_\delta)} G_\delta(\psi) = \min_{\varphi \in H^1_\Gamma(\Omega)} G(\delta, \varphi),$$

with

$$G(\delta, \varphi) = \frac{1}{2}\int_\Omega |A_\delta \cdot \nabla\varphi|^2 \frac{dy}{q_\delta} - \sqrt{\int_\Omega \varphi^2 \frac{dy}{q_\delta}}, \quad \forall \varphi \in H^1_\Gamma(\Omega),$$

where $A_\delta = I + \delta B$, B is the matrix defined in section 3 and $q_\delta = 1 - \delta \mathrm{div} V + \delta^2 \det(DV)$.
Denote

$$M_\delta = M(\Omega_\delta) = \left\{\varphi \in H^1_\Gamma(\Omega) \mid \mu_\delta = G(\delta, \varphi)\right\},$$

the set of minimizers of $G(\delta, \cdot)$ over $H^1_\Gamma(\Omega)$, in particular we denote $M_0 = M(\Omega)$.
In order to show the differentiability of μ_δ at $\delta = 0^+$, we need the following lemma.

LEMMA 4.1. *Let $\{\delta_k\}_{k=1}^{+\infty}$ be a sequence such that $\delta_k \downarrow 0$ for $k \to +\infty$. Then for any sequence $\{z_k\}$ with $z_k \in M_{\delta_k}$, there exists a subsequence, also denoted by $\{z_k\}$, and an element $z^* \in M_0$, such that*

$$z_k \to z^* \textit{ strongly in } H^1_\Gamma(\Omega) \quad \textit{as } k \to +\infty.$$

Proof. It is divided into small steps.

Step 1: $\{z_k\}$ is bounded in $H^1_\Gamma(\Omega)$.
$z_k \in M_{\delta_k}$ i.e. z_k minimizes $G(\delta_k, \cdot)$ over $H^1_\Gamma(\Omega)$ and in consequence

$$G(\delta_k, z_k) \leq G(\delta_k, 0) = 0$$

which leads to

$$\frac{1}{2}\int_\Omega |A_{\delta_k} \cdot \nabla z_k|^2 \frac{dy}{q_{\delta_k}} - \sqrt{\int_\Omega z_k^2 \frac{dy}{q_{\delta_k}}} \leq 0 \tag{4.3}$$

and taking into account that for k large enough, we have the following estimations

$$\frac{1}{2} \leq q_{\delta_k} \leq \frac{3}{2} \quad \text{and} \quad |A_{\delta_k} \cdot \nabla z_k|^2 \geq \frac{1}{4}|\nabla z_k|^2,$$

the inequality (4.3) leads to

$$\frac{1}{12}\int_\Omega |\nabla z_k|^2 dy - \sqrt{2}\sqrt{\int_\Omega z_k^2 dy} \leq 0,$$

and by Poincaré's inequality there exists a constant C such that

$$\int_\Omega |\nabla z_k|^2 dy \leq C.$$

Since $\{z_k\}$ is a bounded sequence in $H^1_\Gamma(\Omega)$ and, moreover, the embedding $H^1_\Gamma(\Omega) \hookrightarrow L^2(\Omega)$ is compact, there exists a subsequence, still denoted by $\{z_k\}$, such that

$$\begin{aligned} z_k &\rightharpoonup z^* \text{ weakly in } H^1_\Gamma(\Omega), \\ z_k &\to z^* \text{ strongly in } L^2(\Omega) \quad \text{as } k \to +\infty. \end{aligned}$$

Step 2: $z^* \in M_0$.
z_k minimizes $G(\delta_k, \cdot)$ over $H^1_\Gamma(\Omega)$, therefore,

$$G(\delta_k, z_k) \leq G(\delta_k, \varphi), \quad \forall \varphi \in H^1_\Gamma(\Omega). \tag{4.4}$$

First, it is not difficult to see that for any fixed function φ,

$$G(\delta_k, \varphi) \to G(0, \varphi) \quad \text{as } k \to +\infty.$$

Moreover, by the sequential lower semi-continuity of the functional $\varphi \longmapsto \int_\Omega |\nabla\varphi|^2 dy$ with respect to the weak topology of $H^1_\Gamma(\Omega)$ and in view of the strong convergence of the sequence $z_k \to z^*$ in $L^2(\Omega)$, it follows that

$$\liminf_{k\to+\infty} G(\delta_k, z_k) \geq G(0, z^*).$$

Using the inequality (4.4), we have

$$G(0,z^*) \leq \liminf_{k\to+\infty} G(\delta_k, z_k) \leq \liminf_{k\to+\infty} G(\delta_k, \varphi) = \lim_{k\to+\infty} G(\delta_k,\varphi) = G(0,\varphi)$$

which implies that

$$G(0,z^*) \leq G(0,\varphi), \quad \forall\varphi \in H^1_\Gamma(\Omega). \tag{4.5}$$

The inequality (4.5) means that $z^* \in M_0$.

Step 3: strong convergence $z_k \to z^*$ in $H^1_\Gamma(\Omega)$.
Showing this property is based on the stationarity conditions satisfied by z_k and z^*, respectively. Indeed, let $z_k \in M_{\delta_k}$ and $z* \in M_0$ which means that the following equations are satisfied

$$\int_\Omega \langle A_{\delta_k} \cdot \nabla z_k, A_{\delta_k} \cdot \nabla\varphi\rangle \frac{dy}{q_{\delta_k}} = \frac{\displaystyle\int_\Omega z_k\varphi \frac{dy}{q_{\delta_k}}}{\sqrt{\displaystyle\int_\Omega z_k^2 \frac{dy}{q_{\delta_k}}}}, \quad \forall\varphi \in H^1_\Gamma(\Omega), \tag{4.6}$$

and

$$\int_\Omega \langle\nabla z^*, \nabla\psi\rangle\, dy = \frac{\displaystyle\int_\Omega z^*\psi dy}{\sqrt{\displaystyle\int_\Omega z^{*2} dy}}, \quad \forall\psi \in H^1_\Gamma(\Omega). \tag{4.7}$$

By substituting $\varphi = z_k$ in (4.6) and $\psi = z^*$ in (4.7), we obtain

$$\int_\Omega |A_{\delta_k} \cdot \nabla z_k|^2 \frac{dy}{q_{\delta_k}} = \sqrt{\int_\Omega z_k^2 \frac{dy}{q_{\delta_k}}}, \tag{4.8}$$

$$\int_\Omega |\nabla z^*|^2 dy = \sqrt{\int_\Omega z^{*2} dy}. \tag{4.9}$$

Consequently, by using (4.8),(4.9),

$$\begin{aligned}\int_\Omega |\nabla z_k|^2 dy &= \int_\Omega |\nabla z_k|^2 dy - \int_\Omega |A_{\delta_k} \cdot \nabla z_k|^2 \frac{dy}{q_{\delta_k}} + \int_\Omega |A_{\delta_k} \cdot \nabla z_k|^2 \frac{dy}{q_{\delta_k}} \\ &= \alpha_k + \int_\Omega |A_{\delta_k} \cdot \nabla z_k|^2 \frac{dy}{q_{\delta_k}}\end{aligned}$$

$$
\begin{aligned}
&= \alpha_k + \sqrt{\int_\Omega z_k^2 \frac{dy}{q_{\delta_k}}} \\
&= \alpha_k + \sqrt{\int_\Omega z_k^2 \frac{dy}{q_{\delta_k}}} - \sqrt{\int_\Omega z^{*2} dy} + \sqrt{\int_\Omega z^{*2} dy} \\
&= \alpha_k + \beta_k + \sqrt{\int_\Omega z^{*2} dy} \\
&= \alpha_k + \beta_k + \int_\Omega |\nabla z^*|^2 dy,
\end{aligned}
$$

where

$$
\begin{aligned}
\alpha_k &= \int_\Omega |\nabla z_k|^2 dy - \int_\Omega |A_{\delta_k} \cdot \nabla z_k|^2 \frac{dy}{q_{\delta_k}}, \\
\beta_k &= \sqrt{\int_\Omega z_k^2 \frac{dy}{q_{\delta_k}}} - \sqrt{\int_\Omega z^{*2} dy}
\end{aligned}
$$

and

$$\alpha_k,\ \beta_k \to 0 \quad \text{as } k \to +\infty.$$

We have shown that

$$\int_\Omega |\nabla z_k|^2 dy \to \int_\Omega |\nabla z^*|^2 dy \quad \text{as } k \to +\infty. \tag{4.10}$$

On the other hand, the convergence

$$z_k \rightharpoonup z^* \text{ weakly in } H^1_\Gamma(\Omega) \quad \text{as } k \to +\infty$$

means in particular, that

$$\nabla z_k \rightharpoonup \nabla z^* \text{ weakly in } (L^2(\Omega))^2 \quad \text{as } k \to +\infty. \tag{4.11}$$

And finally, using (4.10) and (4.11), we can conclude that

$$z_k \to z^* \text{ in } H^1_\Gamma(\Omega) \quad \text{as } k \to +\infty$$

□

REMARK 4.1. *$G(\cdot,\varphi)$ is right-differentiable at $\delta = 0$ for any $\varphi \in H^1_\Gamma(\Omega)$. Moreover, we have the following formula*

$$
\begin{aligned}
\frac{\partial G}{\partial \delta}(0,\varphi) &= \lim_{s \downarrow 0} \frac{G(s,\varphi) - G(0,\varphi)}{s} \\
&= \frac{1}{2}\int_\Omega |\nabla\varphi|^2 \operatorname{div} V dy + \int_\Omega \langle \nabla\varphi, B \cdot \nabla\varphi \rangle dy - \frac{\int_\Omega \varphi^2 \operatorname{div} V dy}{2\|\varphi\|_{L^2(\Omega)}}.
\end{aligned}
$$

Now, we can prove the right-differentiability of μ_δ at $\delta = 0$. To this end, first we estimate the upper limit. We have

$$\begin{aligned}\mu_\delta - \mu_0 &= G(\delta, \varphi_\delta) - G(0, \varphi_0), \quad \forall \varphi_\delta \in M_\delta,\ \forall \varphi_0 \in M_0 \\ &\leq G(\delta, \varphi_0) - G(0, \varphi_0), \quad \forall \varphi_0 \in M_0,\end{aligned}$$

thus for $\delta > 0$,

$$\frac{\mu_\delta - \mu_0}{\delta} \leq \frac{G(\delta, \varphi_0) - G(0, \varphi_0)}{\delta}$$

which implies that

$$\begin{aligned}\limsup_{\delta \downarrow 0} \frac{\mu_\delta - \mu_0}{\delta} &\leq \limsup_{\delta \downarrow 0} \frac{G(\delta, \varphi_0) - G(0, \varphi_0)}{\delta} \\ &\leq \lim_{\delta \downarrow 0} \frac{G(\delta, \varphi_0) - G(0, \varphi_0)}{\delta} \\ &\leq \frac{\partial G}{\partial \delta}(0, \varphi_0)\end{aligned}$$

according to remark 4.1.
We have obtained the inequality

$$\limsup_{\delta \downarrow 0} \frac{\mu_\delta - \mu_0}{\delta} \leq \frac{\partial G}{\partial \delta}(0, \varphi_0), \quad \forall \varphi_0 \in M_0. \tag{4.12}$$

On the other hand

$$\begin{aligned}\mu_\delta - \mu_0 &= G(\delta, \varphi_\delta) - G(0, \varphi_0), \quad \forall \varphi_\delta \in M_\delta,\ \forall \varphi_0 \in M_0 \\ &\geq G(\delta, \varphi_\delta) - G(0, \varphi_\delta), \quad \forall \varphi_\delta \in M_\delta.\end{aligned}$$

By using Taylor's expansion, there exists s, $0 \leq s \leq \delta$, such that

$$\frac{G(\delta, \varphi_\delta) - G(0, \varphi_\delta)}{\delta} = \frac{\partial G}{\partial \delta}(s, \varphi_\delta),$$

and it follows that

$$\frac{\mu_\delta - \mu_0}{\delta} \geq \frac{\partial G}{\partial \delta}(s, \varphi_\delta). \tag{4.13}$$

For $\delta \downarrow 0$, $s \downarrow 0$ and in view of lemma 4.1, there exists $\varphi^* \in M_0$ such that

$$\varphi_\delta \to \varphi^* \text{ in } H^1_\Gamma(\Omega) \quad \text{with } \delta \to 0.$$

In consequence

$$\frac{\partial G}{\partial \delta}(s, \varphi_\delta) \to \frac{\partial G}{\partial \delta}(0, \varphi^*) \quad \text{as } \delta \to 0,$$

and passage to the limit in (4.13) gives us

$$\frac{\partial G}{\partial \delta}(0, \varphi^*) \leq \liminf_{\delta \downarrow 0} \frac{\mu_\delta - \mu_0}{\delta}. \tag{4.14}$$

Finally, combining (4.12),(4.14) leads to

$$\frac{\partial G}{\partial \delta}(0,\varphi^*) \le \liminf_{\delta \downarrow 0} \frac{\mu_\delta - \mu_0}{\delta} \le \limsup_{\delta \downarrow 0} \frac{\mu_\delta - \mu_0}{\delta} \le \frac{\partial G}{\partial \delta}(0,\varphi_0), \quad \forall \varphi_0 \in M_0,$$

where $\varphi^* \in M_0$.
We can conclude that μ_δ is differentiable at $\delta = 0^+$ with the following formula for the directional derivative

$$\begin{aligned} d\mu(\Omega; V) &= \min \left\{ \frac{\partial G}{\partial \delta}(0,\varphi) \mid \varphi \in M_0 \right\} \\ &= \frac{\partial G}{\partial \delta}(0,\varphi^*), \end{aligned}$$

where

$$\frac{\partial G}{\partial \delta}(0,\varphi) = \frac{1}{2}\int_\Omega |\nabla\varphi|^2 \mathrm{div} V dy + \int_\Omega \langle \nabla\varphi, B \cdot \nabla\varphi \rangle dy - \frac{\int_\Omega \varphi^2 \mathrm{div} V dy}{2\|\varphi\|_{L^2(\Omega)}}. \tag{4.15}$$

The differentiability of μ_δ at $\delta = 0$ leads to the differentiability of λ_δ at $\delta = 0$. In view of (4.1),(4.2), we obtain the following formula

$$\begin{aligned} d\lambda(\Omega; V) &= 2\lambda^2(\Omega) d\mu(\Omega; V) \\ &= 2\lambda^2(\Omega) \min \left\{ \frac{\partial G}{\partial \delta}(0,\varphi) \mid \varphi \in M_0 \right\} \\ &= 2\lambda^2(\Omega) \frac{\partial G}{\partial \delta}(0,\varphi^*). \end{aligned}$$

By applying the structure theorem for the shape functional $\lambda(\Omega)$ in domains with cracks, it follows that

$$d\lambda(\Omega; V) = \alpha_A (V.\tau)(A) + \alpha_B (V.\tau)(B) + \phi(V.n), \quad \forall V \in \mathcal{D}^k(D; \mathbb{R}^2)$$

where $\phi \in (C^k(\overline{\Sigma_l}))'$, α_A, $\alpha_B \in \mathbb{R}$ with $k = 1$.
Moreover, we can derive the explicit form e.g. of α_A.
Indeed, if $V(y) = (\theta_1(y), 0)$ where θ_1 has the support in D and $\theta_1(y) = -1$ in the vicinity of the origin A, by using the same method as in previous section, for optimal control problem, i.e. integrating on Ω_ε (where Ω_ε is the subset of Ω defined, in polar coordinates, by $r > \varepsilon$) and passing to the limit as $\varepsilon \to 0^+$, we obtain

$$\begin{aligned} \alpha_A &= 2\lambda^2(\Omega) \frac{\pi c_{\varphi^*}^2}{4} \\ &= 2\lambda^2(\Omega) \min \left\{ \frac{\pi c_\varphi^2}{4} \mid \varphi \in M_0 \right\}, \end{aligned}$$

where c_{φ^*} and c_φ respectively denote the coefficients of singularity of functions φ^* and φ.
Finally

$$c_{\varphi^*}^2 = \min_{\varphi \in M_0} c_\varphi^2.$$

REMARK 4.2. *In general, the set $M_0 = M(\Omega)$ is strictly included in the space of eigenfunctions.*

REMARK 4.3. *If Ω is a regular domain i.e. with smooth boundary Γ, integrating by parts in the relation (4.15) leads to the well known result*

$$\frac{\partial G}{\partial \delta}(0,\varphi) = -\frac{1}{2}\int_{\Gamma} |\nabla\varphi|^2 V.nd\Gamma$$

Bibliography

[1] J. SOKOŁOWSKI, J.-P. ZOLÉSIO. *Introduction to shape optimization: shape sensitivity analysis.* Springer Series in Computational Mathematics. 16. Berlin etc.: Springer-Verlag, (ISBN 3-540-54177-2). 250 p. (1992).

[2] G. FREMIOT, J. SOKOŁOWSKI. *The structure theorem for the Eulerian derivative of shape functionals defined in domains with cracks*, to appear in Siberian Mathematical Journal.

[3] B. PALMERIO, A. DERVIEUX. *Hadamard's variational formula for a mixed problem and an application to a problem related to a Signorini-like variational inequality*, Numer. Funct. Anal. Optimiz. **1**, 113-144, 1979.

[4] A.-M. KHLUDNEV, J. SOKOŁOWSKI. *Griffith formula for elasticity system with unilateral conditions in domains with cracks*, European Journal of Elasticity/Solids. vol.19, No.1, 2000, 105-120.

[5] J. SOKOŁOWSKI, A. ZOCHOWSKI. *Topological derivative for optimal control problems*, Les prépublications de l'Institut Élie Cartan, 13(1999). To appear in Control and Cybernetics.

[6] P. GRISVARD. *Singularities in boundary value problems.* Recherches en Mathématiques Appliquées. 22. Paris: Masson, (ISBN 2-225-82770-2). Berlin: Springer-Verlag, (ISBN 3-540-55450-5). xiv, 198 p. (1992).

[7] G. AUCHMUTY. *Duality for non-convex variational principles.* J. Differential Equations **50** (1983), 80-145.

Gilles Fremiot. Institut Elie Cartan, Laboratoire de Mathématiques, Université Henri Poincaré Nancy I, B.P. 239, 54506 Vandoeuvre lès Nancy Cedex, France
and INRIA-Lorraine, Projet Numath, France
E-mail: fremiot@iecn.u-nancy.fr

Jan Sokolowski. Institut Elie Cartan, Laboratoire de Mathématiques, Université Henri Poincaré Nancy I, B.P. 239, 54506 Vandoeuvre lès Nancy Cedex, France
and INRIA-Lorraine, Projet Numath, France
E-mail: sokolows@iecn.u-nancy.fr

Mapping Method in Optimal Shape Design Problems Governed by Hemivariational Inequalities

Leszek Gasiński[1]

Abstract. In this paper we consider nonlinear hemivariational inequalities of elliptic and parabolic type as well as optimal shape design problems described by these inequalities. The existence results for hemivariational inequalities are given. Using the example of parabolic case the existence result for the optimal shape problem is shown.

1 Introduction

Hemivariational inequalities were introduced in the 80-ies by P.D. Panagiotopoulos as a natural description of physical problems governed by nonmonotone and possibly multivalued laws (see Panagiotopoulos [17], [18], Moreau, Panagiotopoulos and Strang [13]). The mathematical models for such problems deal with potentials given by nonconvex, possibly nondifferentiable functions. In [19], Panagiotopoulos introduced the notion of a nonconvex superpotential, being a generalization of the convex superpotential introduced by Moreau (see [12]). This generalization led to a new type of variational inequalities, called hemivariational inequalities, which cover boundary value problems for PDE's with nonmonotone, nonconvex and possibly multivalued laws.

The aim of this paper is to present some existence results for an optimal shape design problem for systems described by hemivariational inequalities of elliptic and parabolic type. These problems may be formulated as control problems in which hemivariational inequalities appear as state equations and the role of controls is played by sets from a family of admissible shapes. The cost functional to be minimized is of general (not necessary integral) form. Such control problems governed by variational inequalities, of both elliptic and parabolic type, were studied by Liu and Rubio in [7] and [8], where the applications of these problems to the so called electrochemical machining are given.

Proofs of the existence of optimal shapes are based on the direct method of the calculus of variations. We use the mapping method introduced by Micheletti [9] (see also Murat and

[1]The author was partially supported by the State Committee for Scientific Research of Poland (KBN) under Research Grant 2 P03A 040 15.

Simon [14] or Sokołowski and Zolesio [23]), which provides both a class of admissible shapes and a topology in this class of domains. The admissible shapes are obtained as the images of a fixed open bounded subset of $\mathbb{R}^N$ by regular bijections in $\mathbb{R}^N$ (see Section 2.2 for details).

The plan of the paper is as follows. In Section 2 we recall the notation and properties of the Clarke subdifferential (Section 2.1) and of the mapping method (Section 2.2). Section 3 is devoted to the elliptic case, where we give different formulations of elliptic hemivariational inequalities (Section 3.1) as well as formulate the shape design problem governed by an elliptic hemivariational inequality (Section 3.2). Analogously in Section 4 we study the parabolic case. We formulate parabolic hemivariational inequality (Section 4.1) and give the existence result for an optimal problem described by this inequality (Section 4.2).

2 Preliminaries

2.1 Clarke Subdifferential

In this section we recall the notion of the Clarke subdifferential as well as some its properties.

Let X be a Banach space and X' its topological dual. By $\langle\cdot,\cdot\rangle_{X'\times X}$ we will denote the duality brackets between X' and X. A function $f : X \longmapsto \mathbb{R}$ is said to be locally Lipschitz if for every $x \in X$ we can find a neighborhood U of x and a constant $k_x > 0$ depending on U, such that $|f(y) - f(z)| \leq k_x \|y - z\|_X$ for all $y, z \in U$. It is well known from convex analysis that a proper, convex and lower semicontinuous function $g : X \longmapsto \overline{\mathbb{R}} = \mathbb{R} \cup \{+\infty\}$ is locally Lipschitz in the interior of its effective domain $\mathrm{dom} g \stackrel{df}{=} \{x \in X : g(x) < +\infty\}$. For a locally Lipschitz function $f : X \longmapsto \mathbb{R}$, every $x \in X$ and $h \in X$, we define the Clarke directional derivative of f at x in the direction h by

$$f^0(x;h) \stackrel{df}{=} \limsup_{\substack{y \to 0 \text{ in } X \\ t \searrow 0 \text{ in } \mathbb{R}}} \frac{f(x+y+th) - f(x+y)}{t}.$$

It is easy to check that the function $X \ni h \longmapsto f^0(x;h) \in \mathbb{R}$ is sublinear and continuous (in fact $|f^0(x;h)| \leq k_x \|h\|_X$ and hence $f^0(x;\cdot)$ is Lipschitz). So by the Hahn-Banach theorem $f^0(x;\cdot)$ is the support function of a nonempty, convex and w^*-compact set $\partial f(x)$, defined by

$$\partial f(x) \stackrel{df}{=} \{x^* \in X' : \ f^0(x;h) \geq \langle x^*, h\rangle_{X'\times X} \text{ for all } h \in X\}$$

(see Clarke [3], Proposition 2.1.2, p.27). The w^*-compactness of $\partial f(x)$ follows from the fact that it is bounded. The set $\partial f(x)$ is called the Clarke subdifferential of f at x. For every $x \in X$ there exists $k_x > 0$ such that for every $x^* \in \partial f(x)$ we have $\|x^*\|_{X'} \leq k_x$. Also, if $f, g : X \longmapsto \mathbb{R}$ are locally Lipschitz functions, then $\partial(f+g)(x) \subseteq \partial f(x) + \partial g(x)$ and $\partial(\alpha f)(x) = \alpha \partial f(x)$ for all $\alpha \in \mathbb{R}$. Moreover, if $f : X \longmapsto \mathbb{R}$ is convex (so locally Lipschitz as well), then the Clarke subdifferential defined above and subdifferential in the sense of convex analysis coincide and $f^0(x;h) = f'(x;h) \stackrel{df}{=} \lim_{t\searrow 0} \dfrac{f(x+th) - f(x)}{t}$ (the usual directional derivative of convex functions). Also, if f is strictly differentiable at x (in particular if f is continuously Gâteaux differentiable at x), then $\partial f(x) = \{f'(x)\}$.

For a given $\beta \in L^{\infty}_{loc}(\mathbb{R})$ by $\widehat{\beta} : \mathbb{R} \longmapsto 2^{\mathbb{R}}$ we denote a multifunction obtained from β by "filling in the gaps" at its discontinuity points, i.e.

$$\widehat{\beta}(\xi) \overset{df}{=} [\underline{\beta}(\xi), \overline{\beta}(\xi)],$$

where

$$\underline{\beta}(\xi) \overset{df}{=} \lim_{\delta \to 0^+} \operatorname{ess\,inf}_{|t-\xi| \leq \delta} \beta(t), \quad \overline{\beta}(\xi) \overset{df}{=} \lim_{\delta \to 0^+} \operatorname{ess\,sup}_{|t-\xi| \leq \delta} \beta(t)$$

and $[\cdot, \cdot]$ denotes the interval. It is well known (cf. Chang [2]) that a locally Lipschitz function $j : \mathbb{R} \longmapsto \mathbb{R}$ can be determined up to an additive constant, by the relation

$$j(\xi) = \int_0^{\xi} \beta(s)\, ds$$

and that $\partial j(\xi) \subset \widehat{\beta}(\xi)$. Moreover, if limits $\beta(\xi \pm 0)$ exist for every $\xi \in \mathbb{R}$, then $\partial j(\xi) = \widehat{\beta}(\xi)$.

The smooth critical point theory uses a compactness-type condition known as the "Palais-Smale condition". In the present nonsmooth setting, this condition takes the following form. We say that the locally Lipschitz function $f : X \longmapsto \mathbb{R}$ satisfies the generalized Palais-Smale condition if any sequence $\{x_n\}_{n \geq 1} \subseteq X$ along which the sequence $\{f(x_n)\}_{n \geq 1}$ is bounded and $m(x_n) \overset{df}{=} \min\{\|x^*\|_{X'} : x^* \in \partial f(x_n)\} \longrightarrow 0$ as $n \to +\infty$, has a strongly convergent subsequence. When $f \in C^1(X)$, then since $\partial f(x_n) = \{f'(x_n)\}$, we see that the above definition of the nonsmooth Palais-Smale condition coincides with classical (smooth) one (see Rabinowitz [21]). The Palais-Smale condition and the nonsmooth Palais-Smale condition are fairly strong and in particular if f is also bounded from below, then they imply that f is coercive.

2.2 Mapping Method

In this section we recall the notion and basic properties of the mapping method (cf. Murat and Simon [14]). Roughly speaking, this method consists in finding the optimal shapes in a class of admissible domains obtained as images of a fixed set. An appropriate topology in the class will allow us to obtain an existence result for the optimal shape design problem.

Let C be a bounded open subset of $\mathbb{R}^N$ with a boundary ∂C of class $W^{i,\infty}$, $i \geq 1$ and such that $\operatorname{int}\overline{C} = C$. Then, following Murat and Simon [14] and Gasiński [5], we introduce, for $k \geq 1$, the following spaces

$$W^{k,\infty}(\mathbb{R}^N; \mathbb{R}^N) \overset{df}{=} \{\varphi \,|\, D^{\alpha}\varphi \in L^{\infty}(\mathbb{R}^N; \mathbb{R}^N) \text{ for all } \alpha, 0 \leq |\alpha| \leq k\},$$

where derivatives $D^{\alpha}\varphi$ are understood in the distributional sense. By $\mathcal{O}^{k,\infty}$ we will denote the space of bounded open sets of $\mathbb{R}^N$, which are isomorphic with C, i.e.

$$\mathcal{O}^{k,\infty} \overset{df}{=} \{\Omega \mid \Omega = T(C),\ T \in \mathcal{F}^{k,\infty}\},$$

where $\mathcal{F}^{k,\infty}$ is the space of regular bijections in $\mathbb{R}^N$, defined by

$$\mathcal{F}^{k,\infty} \overset{df}{=} \{T : \mathbb{R}^N \longmapsto \mathbb{R}^N \mid T \text{ is bijective and } T, T^{-1} \in \mathcal{V}^{k,\infty}\},$$

where

$$\mathcal{V}^{k,\infty} \stackrel{df}{=} \{T : \mathbb{R}^N \longmapsto \mathbb{R}^N \mid T - I \in W^{k,\infty}(\mathbb{R}^N; \mathbb{R}^N)\}.$$

In other words $\mathcal{F}^{k,\infty}$ represents the set of essentially bounded perturbations (with essentially bounded derivatives) of identity in $\mathbb{R}^N$. It can be seen that if C has a $W^{k,\infty}$ boundary, then every set $\Omega \in \mathcal{O}^{k,\infty}$ also has the boundary of class $W^{k,\infty}$. Endowing the space $W^{k,\infty}(\mathbb{R}^N; \mathbb{R}^N)$ with the norm

$$\|\varphi\|_{k,\infty} \stackrel{df}{=} \text{ess sup}_{x\in\mathbb{R}^N} \left(\sum_{0\leq|\alpha|\leq k} |D^\alpha \varphi|^2_{\mathbb{R}^N} \right)^{\frac{1}{2}},$$

we define on $\mathcal{O}^{k,\infty} \times \mathcal{O}^{k,\infty}$ a function

$$\delta_{k,\infty}(\Omega_1, \Omega_2) \stackrel{df}{=} \inf_{T\in\mathcal{F}^{k,\infty}, T(\Omega_1)=\Omega_2} \left(\|T - I\|_{k,\infty} + \|T^{-1} - I\|_{k,\infty}\right).$$

The mapping $\delta_{k,\infty}$ is a pseudo-distance on $\mathcal{O}^{k,\infty}$ since it does not satisfy the triangle inequality (see Section 2.4 of Murat and Simon [14]) but it can be easily modified into a distance function. Namely, there exists a positive constant μ_k such that $d_{k,\infty}$ defined by $d_{k,\infty} = \sqrt{\min(\delta_{k,\infty}, \mu_k)}$ is a metric on $\mathcal{O}^{k,\infty}$. Moreover the space $(\mathcal{O}^{k,\infty}, d_{k,\infty})$ is a complete metric space. If $k \geq 2$, then the embedding of $\mathcal{O}^{k,\infty}$ into $\mathcal{O}^{k-1,\infty}$ is compact. More precisely, if $k \geq 2$ and $\mathcal{B}$ is a bounded (in $\delta_{k,\infty}$), closed subset of $\mathcal{O}^{k,\infty}$, then for any sequence $\{\Omega_n\}_{n\geq1} \subset \mathcal{B}$, there exist a subsequence $\{\Omega_{n_\nu}\}_{\nu\geq1}$ of $\{\Omega_n\}_{n\geq1}$ and a set $\Omega \in \mathcal{B}$ such that $\Omega_{n_\nu} \longrightarrow \Omega$ in $\mathcal{O}^{k-1,\infty}$ (see Proposition 2.3, Theorem 2.2 and Theorem 2.4 of Murat and Simon [14]).

It is also known (cf. Section 2 in Murat and Simon [14]) that $\Omega_n \longrightarrow \Omega$ in $\mathcal{O}^{k,\infty}$ if and only if there exist T_n and T in $\mathcal{F}^{k,\infty}$ such that $T_n(C) = \Omega_n$, $T(C) = \Omega$ and $T_n \longrightarrow T$, $T_n^{-1} \longrightarrow T^{-1}$ in $W^{k,\infty}(\mathbb{R}^N; \mathbb{R}^N)$.

Some other facts on the mapping method, are summarized in the following lemma.

LEMMA 2.1. *Let $k \geq 1$.*
a) If $T \in \mathcal{F}^{1,\infty}$ and $\Omega = T(C)$, then $u \in L^2(\Omega)$ if and only if $u \circ T \in L^2(C)$ and $u \in H^1(\Omega)$ if and only if $u \circ T \in H^1(C)$. Moreover, if $u_n \longrightarrow u$ in $H^1(\Omega)$ (or in $H^1(C)$) and $T \in \mathcal{F}^{k,\infty}$, then $u_n \circ T \longrightarrow u \circ T$ in $H^1(C)$ (or $u_n \circ T^{-1} \longrightarrow u \circ T^{-1}$ in $H^1(\Omega)$).
b) Let $u \in H^l(\mathbb{R}^N)$ with $l = 0$ or 1. Then the mapping $T \mapsto u \circ T$ is continuous from $\mathcal{V}^{k,\infty}$ to $H^l(\mathbb{R}^N)$ at every point $T \in \mathcal{F}^{k,\infty}$.
c) The following mappings are continuous

$$T \mapsto J_T^{-1} \text{ from } \mathcal{V}^{k,\infty} \text{ to } W^{k-1,\infty}(\mathbb{R}^N; \mathbb{R}^{N^2}),$$

$$T \mapsto \det J_T \text{ from } \mathcal{V}^{k,\infty} \text{ to } W^{k-1,\infty}(\mathbb{R}^N; \mathbb{R})$$

at every point $T \in \mathcal{F}^{k,\infty}$ (J_T denotes here the standard Jacobian matrix of T).

For the proofs of (a) - (c) of the above lemma, we refer, respectively to Lemma 4.1, Lemma 4.4(i) and Lemma 4.3 and 4.2 of Murat and Simon [14] (see also Liu and Rubio [7]).

It is interesting to observe some relationships between the convergence in $\mathcal{O}^{k,\infty}$ and other types of convergence of sets.

Let D be an open subset of $\mathbb{R}^N$. If by 1_D we denote the characteristic function of D, then we have the following relation:

$$\text{if } \Omega_n \longrightarrow \Omega_0 \text{ in } \mathcal{O}^{k,\infty}, \text{ then } 1_{\Omega_n} \longrightarrow 1_{\Omega_0} \text{ in } L^2(\mathbb{R}^N).$$

If by H^c we denote the Hausdorff topology introduced by Hausdorff complementary metric:

$$d(\Omega_1,\Omega_2) \stackrel{df}{=} \max\left(\sup_{x\in\mathbb{R}^N\setminus\Omega_1} \inf_{y\in\mathbb{R}^N\setminus\Omega_2} \|x-y\|_{\mathbb{R}^N}, \sup_{x\in\mathbb{R}^N\setminus\Omega_2} \inf_{y\in\mathbb{R}^N\setminus\Omega_1} \|x-y\|_{\mathbb{R}^N} \right)$$

then we have the following relation:

$$\text{if } \Omega_n \longrightarrow \Omega_0 \text{ in } \mathcal{O}^{k,\infty} \text{ and } \operatorname{int}\overline{C} = C, \text{ then } \Omega_n \xrightarrow{H^c} \Omega_0.$$

H^c-convergence has an important property of "covering" of the compacts, namely,

$$\text{if } \Omega_n \xrightarrow{H^c} \Omega_0, \text{ then } \forall G \subset\subset \Omega_0\ \exists n_G \in \mathbb{N}\ \forall n \geq n_G:\ G \subseteq \Omega_n.$$

3 Elliptic Case

3.1 Elliptic Hemivariational Inequality

In this section we present some possible formulations of the elliptic hemivariational inequalities, relations between them as well as different ways of obtaining the existence results for these formulations.

Let Ω be an open and bounded subset of $\mathbb{R}^N$ $(N \geq 1)$ and let $V = V(\Omega) = H^1(\Omega)$ be the Sobolev space. By V' we will denote its dual space and by $\langle\cdot,\cdot\rangle_{V'\times V}$ the duality brackets between V and V'. Let $H = H(\Omega) = L^2(\Omega)$ be the Hilbert space with scalar product $(\cdot,\cdot)_H$. Let $j:\mathbb{R} \longmapsto \mathbb{R}$ be a locally Lipschitz function, $\mathrm{L}: V \longmapsto V'$ a linear operator and $f \in V'$.

First possible formulation of elliptic hemivariational inequality is the following:

$$(EHI_1) \qquad \begin{cases} \text{Find } u \in V \text{ such that} \\ \langle \mathrm{L}\, u, v-u\rangle_{V'\times V} + \int_\Omega j^0(u; v-u)\, dx \\ \qquad\qquad \geq \langle f, v-u\rangle_{V'\times V} \quad \forall v \in V. \end{cases}$$

We will transform this problem into the inclusion form. Let us define a functional $\mathrm{F}: V \longmapsto \mathbb{R}$ by

$$\mathrm{F}(u) \stackrel{df}{=} \frac{1}{2}\langle \mathrm{L}\, u, u\rangle + \mathrm{J}(u) - \langle f, u\rangle_{V'\times V} \qquad \forall u \in V,$$

where $\mathrm{J}: V \longmapsto \mathbb{R}$ is given by $\mathrm{J}(u) \stackrel{df}{=} \int_\Omega j(u(x))\, dx$. Then we can rewrite (EHI_1) as an inclusion

$$(EHI_2) \qquad \begin{cases} \text{Find } u \in V \text{ such that} \\ 0 \in \partial\, \mathrm{F}(u), \end{cases}$$

or equivalently

$$(EHI_2')\qquad \begin{cases} \text{Find } u \in V \text{ such that} \\ f \in \mathrm{L}\,u + \partial\,\mathrm{J}(u). \end{cases}$$

As we see every solution of (EHI_1) is also a solution of (EHI_2) (so also (EHI_2')). The opposite implication is also true under an additional assumption (the so called growth condition) on ∂j, namely

$$\forall \xi \in \mathbb{R}\, \forall \eta \in \partial j(\xi): \quad |\eta| \le c(1 + |\xi|),$$

with some constant $c > 0$ (for details see Panagiotopoulos [19]).

Another formulation of elliptic hemivariational inequality contains the selection χ of the Clarke subdifferential ∂j. Namely, we consider the following problem:

$$(EHI_3)\qquad \begin{cases} \text{Find } u \in V \text{ such that} \\ \langle \mathrm{L}\,u, v\rangle_{V'\times V} + (\chi, v)_H = \langle f, v\rangle_{V'\times V} \quad \forall v \in V \\ \chi(x) \in \partial j(u(x)) \quad \text{for a.e. } x \in \Omega \\ \chi \in H. \end{cases}$$

If $j : \mathbb{R} \longmapsto \mathbb{R}$ is of the form $j(\xi) = \int_0^\xi \beta(s)\, ds$ (see Section 2.1) and limits $\beta(\xi \pm 0)$ exist, then any solution of (EHI_3) is also a solution of (EHI_1).

The existence results for all above formulations of elliptic hemivariational inequalities can be obtained using different methods. Each of them requires of course a different set of assumptions on L, f and j. In [15], Naniewicz proved an existence result for (EHI_2) using the surjectivity results for pseudomonotone and coercive operators (see Browder and Hess [1], Proposition 9 and Theorem 3).

Another way of receiving an existence result for (EHI_2') is to check that functional F defined above satisfies generalized Palais-Smile condition (see Section 2.1) and apply so called deformation lemma (compare Chang [2], Theorems 3.1 and 3.5).

Final possibility is to use the Galerkin method and receive the sequence of problems which solutions converge to the solution of (EHI_3) (compare Miettinen [10], Section 1.3.3).

Now we would like to concentrate on one possible set of assumptions, which guarantee the existence of solutions of (EHI_3) (so also (EHI_1) and (EHI_2)) and which will be used in formulating the shape design problem in the next section. For the operator L and the function β we put the following assumptions:

$\underline{H(\mathrm{L})_\Omega}$ $\mathrm{L} : V \longmapsto V'$ is a linear operator given by

$$\langle \mathrm{L}\,u, v\rangle_{V'\times V} \overset{df}{=} \int_\Omega \left[\left(\overline{A}(x)\nabla u(x), \nabla v(x)\right)_{\mathbb{R}^N} + a_0(x)u(x)v(x)\right] dx$$

which is continuous (i.e. $\exists M > 0\ \forall u \in V: \ \|\mathrm{L}\,u\|_{V'} \le M\|u\|_V$), symmetric and coercive on V (i.e. $\exists \alpha > 0\ \forall v \in V: \ \langle \mathrm{L}\,u, u\rangle_{V'\times V} \ge \alpha\|u\|_V^2$), where $\overline{A} \in [C(\mathbb{R}^N) \cap L^\infty(\mathbb{R}^N)]^{N^2}$ is a matrix and $a_0 \in C(\mathbb{R}^N) \cap L^\infty(\mathbb{R}^N)$, $a_0(x) \ge \widetilde{a} > 0$ a.e. in $\mathbb{R}^N$.

$\underline{H(\beta)}$ $\beta \in L^\infty_{loc}(\mathbb{R})$ is such that

(i) limits $\beta(\xi \pm 0)$ exists for each $\xi \in \mathbb{R}$;

(ii) the graph of β increases ultimately i.e. there exist $\overline{\xi} \in \mathbb{R}$ such that

$$\operatorname{ess\,sup}_{(-\infty,-\overline{\xi})} \beta(\xi) \leq 0 \leq \operatorname{ess\,inf}_{(\overline{\xi},+\infty)} \beta(\xi);$$

(iii) there exists $c_0 > 0$ such that $|\beta(\xi)| \leq c_0(1 + |\xi|)$ for $\xi \in \mathbb{R}$.

If by $S_E(\Omega)$ we denote the set of all solutions of (EHI_3), then we can formulate an existence theorem as follows.

THEOREM 3.1. *If hypotheses $H(\mathrm{L})_\Omega$, $H(\beta)$ hold and $f \in V'$, then problem (EHI_3) admits a solution, i.e. $S_E(\Omega) \neq \emptyset$.*

For the proof of this theorem we refer to Naniewicz and Panagiotopoulos [16], Theorem 3.4.

3.2 Optimal Shape Design Problem for Elliptic Hemivariational inequality

In this section we consider the control problem governed by elliptic hemivariational inequality.

Let us assume the following assumptions on a family $\mathcal{B}$ of admissible shapes and on functional J:

$\underline{H(C,\mathcal{B})}$ C is a bounded open set in $\mathbb{R}^N$ with boundary of class $W^{i,\infty}$, $i \geq 1$ such that $\operatorname{int}\overline{C} = C$ and $\mathcal{B}$ is a bounded closed subset of $\mathcal{O}^{k,\infty}$, with $k \geq 3$ and $1 \leq i \leq k$.

$\underline{H_E(\mathrm{J})}$ $\mathrm{J} : D_E(\mathrm{J}) \stackrel{df}{=} \bigcup_{\Omega\in\mathcal{B}} (\{\Omega\} \times S_E(\Omega)) \longmapsto \mathbb{R}$ is a functional which is lower semicontinuous with respect to the following convergence in $D_E(\mathrm{J})$:

$(\Omega_n, u_n) \longrightarrow (\Omega_0, u_0)$ in $D_E(\mathrm{J})$ iff $\Omega_n \longrightarrow \Omega_0$ in $\mathcal{O}^{k-1,\infty}$ and $\underline{u_n} \longrightarrow \underline{u_0}$ in $H(\mathbb{R}^N)$, where by $\underline{u}$ we denote the extension by zero of the function $u \in V(\Omega)$, namely

$$\underline{u}(x) \stackrel{df}{=} \begin{cases} u(x) & \text{if } x \in \Omega \\ 0 & \text{if } x \in \mathbb{R}^N \setminus \Omega. \end{cases}$$

The assumption of lower semicontinuity of functional J with respect to the above defined convergence is slightly weaker than the lower semicontinuity with respect to the local convergence (compare Denkowski and Migórski [4], assumption $H(\mathrm{J})$ and Definition 3).

The optimal shape design problem consists in solving the following control problem:

$$\text{(OPEI)} \qquad \begin{cases} \text{Find } (\Omega^*, u^*) \in D_E(\mathrm{J}) \text{ such that} \\ \mathrm{J}(\Omega^*, u^*) = \min\limits_{\Omega\in\mathcal{B}} \min\limits_{u\in S_E(\Omega)} \mathrm{J}(\Omega, u), \end{cases}$$

in which the controls are the sets Ω changing in the family $\mathcal{B} \subseteq \mathcal{O}^{k,\infty}$.

In our assumptions we do not need to specify the form of the cost functional J. Nevertheless, in practice, it is usually of integral form, namely

$$\mathrm{J}(\Omega, u) = \int_\Omega l(x, u, \nabla u)\, dx.$$

The sufficient conditions for lower semicontinuity of the functional J with respect to the local convergence were given by Serrin in [22], for instance, the integrand $l(x,u,p)$ should be nonnegative, continuous in (x,u,p) and strictly convex in p.

Now we can formulate the existence theorem for $(OPEI)$.

THEOREM 3.2. *If*
(i) hypotheses $H(C,\mathcal{B})$, $H(\beta)$, $H_E(\mathrm{J})$ *hold;*
(ii) for every $\Omega \in \mathcal{B}$, *hypothesis* $H(\mathrm{L})_\Omega$ *holds with parameters* M, α, $\overline{A}$, a_0 *and* $\widetilde{a}$ *not depending on sets* Ω*;*
(iii) $f \in H(\mathbb{R}^N)$,
then problem $(OPEI)$ *admits at least one solution.*

The proof is a slight modification of the one given by Denkowski and Migórski in [4], Theorem 3.

4 Parabolic Case

4.1 Parabolic Hemivariational Inequality

In parabolic case we can also consider analogous formulations of hemivariational inequalities as in elliptic one. We concentrate only on one of them, for which we formulate the existence result and give an application to optimal shape design problem.

Let Ω be an open, bounded subset of $\mathbb{R}^N$ and $I > 0$. Let us introduce the following spaces: $V = V(\Omega) = H^1(\Omega)$, $H = H(\Omega) = L^2(\Omega)$, $\mathcal{V} = \mathcal{V}(\Omega) = L^2(0,I;V)$, $\mathcal{V}' = \mathcal{V}'(\Omega) = L^2(0,I;V')$, $\mathcal{H} = \mathcal{H}(\Omega) = L^2(0,I;H)$, $\mathcal{W} = \mathcal{W}(\Omega) = W(0,I;V) = \{v : v \in \mathcal{V},\ v' \in \mathcal{V}'\}$. The inclusion $V \subset H \subset V'$ implies $L^2(0,I;V) \subset L^2(0,I;H) \subset L^2(0,I;V')$, as $[L^2(0,I;V)]' \simeq L^2(0,I;V')$ (see Lions and Magenes [6]). By parabolic hemivariational inequality we mean the following problem:

$$(\mathrm{PHI}) \qquad \begin{cases} \text{Find } u \in \mathcal{W} \text{ such that} \\ \langle u'(t), v\rangle_{\mathcal{V}'\times\mathcal{V}} + \langle \mathrm{L}\, u(t), v\rangle_{\mathcal{V}'\times\mathcal{V}} + (\chi(t), v)_{\mathcal{H}} \\ \qquad\qquad = \langle f(t), v\rangle, \quad \forall\, v \in V, \quad \text{a.e. } t \in (0,I), \\ u(0) = \psi \quad \text{in } \Omega, \\ \chi(t,x) \in \partial j(u(t,x)) \quad \text{a.e. } (t,x) \in (0,I)\times\Omega \\ \chi \in \mathcal{H}(\Omega). \end{cases}$$

If by $S_P(\Omega)$ we denote the set of all solutions of (PHI), then we can formulate an existence theorem as follows.

THEOREM 4.1. *If hypotheses* $H(\mathrm{L})_\Omega$, $H(\beta)$ *hold and* $f \in \mathcal{V}'$, $\psi \in H(\mathbb{R}^N)$, *then problem* (PHI) *admits a solution, i.e.* $S_P(\Omega) \neq \emptyset$.

For the proof we refer to Miettinen [11].

4.2 Optimal Shape Design Problem for Parabolic Hemivariational Inequality

In this section we consider the control problem governed by parabolic hemivariational inequality.

Let us assume the following assumptions on cost functional J:

$\underline{H_P(\mathrm{J})}$ $\mathrm{J} : D_P(\mathrm{J}) \overset{df}{=} \bigcup_{\Omega \in \mathcal{B}} (\{\Omega\} \times S_P(\Omega)) \longmapsto \mathbb{R}$ is a functional which is lower semicontinuous with respect to the following convergence in $D_P(\mathrm{J})$:

$(\Omega_n, u_n) \longrightarrow (\Omega_0, u_0)$ in $D_P(\mathrm{J})$ iff $\Omega_n \longrightarrow \Omega_0$ in $\mathcal{O}^{k-1,\infty}$ and $\underline{u_n} \longrightarrow \underline{u_0}$ in $\mathcal{H}(\mathbb{R}^N)$, where by $\underline{u}$ we denote the extension by zero of the function $u \in \mathcal{V}(\Omega)$, namely

$$\underline{u}(t,x) \overset{df}{=} \begin{cases} u(t,x) & \text{if } x \in \Omega \\ 0 & \text{if } x \in \mathbb{R}^N \setminus \Omega. \end{cases}$$

By the optimal shape design problem governed by (PHI) we mean the following problem:

$$(\mathrm{OPPI}) \qquad \begin{cases} \text{Find } (\Omega^*, u^*) \in D_P(\mathrm{J}) \text{ such that} \\ \mathrm{J}(\Omega^*, u^*) = \min\limits_{\Omega \in \mathcal{B}} \min\limits_{u \in S_P(\Omega)} \mathrm{J}(\Omega, u). \end{cases}$$

It is a control problem where the controls are the sets Ω changing in the family $\mathcal{B} \subseteq \mathcal{O}^{k,\infty}$.

The existence theorem for $(OPPI)$ is following.

THEOREM 4.2. *If*
(i) hypotheses $H(C,\mathcal{B})$, $H(\beta)$, $H_P(\mathrm{J})$ hold;
(ii) for every $\Omega \in \mathcal{B}$, hypothesis $H(\mathrm{L})_\Omega$ holds with parameters M, α, $\overline{A}$, a_0 and $\widetilde{a}$ not depending on sets Ω;
(iii) $f \in \mathcal{H}(\mathbb{R}^N)$, $\psi \in H(\mathbb{R}^N)$,
then problem $(OPPI)$ admits at least one solution.

In the proof of the last theorem the crucial role plays the fact that the map $\mathcal{B} \ni \Omega \longmapsto S_P(\Omega) \subseteq \mathcal{W}(\Omega)$ has a graph closed in the sense of the following lemma.

LEMMA 4.3. *Let us assume that the hypotheses of Theorem 4.2 hold. Let $\{\Omega_n\}_{n\geq 1} \subseteq \mathcal{B}$, $\Omega_0 \in \mathcal{B}$, $\{T_n\}_{n\geq 1} \subseteq \mathcal{F}^{k,\infty}$, $T_0 \in \mathcal{F}^{k,\infty}$ be such that $\Omega_n = T_n(C)$ for all $n \geq 1$ and $\Omega_0 = T_0(C)$. Let $u_n \in S_P(\Omega_n)$, $\widehat{u}_n(t,X) \overset{df}{=} u_n(t,T_n(X))$ for all $n \geq 1$ and $u^* \in \mathcal{W}(C)$. If $\Omega_n \longrightarrow \Omega_0$ in $\mathcal{O}^{k,\infty}$, $\widehat{u}_n \longrightarrow u^*$ weakly in $\mathcal{W}(C)$, then there exists $u_0 \in S_P(\Omega_0)$ such that $u^*(t,X) = u_0(t,T_0(X))$.*

For the proof see Proposition 1 of Gasiński [5]. Next lemma gives a priori estimates on a solution of (PHI).

LEMMA 4.4. *Let us assume that hypotheses $H(\mathrm{L})_\Omega$, $H(\beta)$ hold, $f \in \mathcal{V}'$ and $\psi \in H$. If $u \in S_P(\Omega)$, then the following estimate holds:*

$$\|u\|_{\mathcal{W}} \leq b\,(1+|\Omega|)\, e^{d(1+|\Omega|)} \left(1 + |\Omega| + \|\psi\|_{H(\Omega)} + \|f\|_{\mathcal{V}'(\Omega)}\right),$$

with constants $b, d > 0$ depending only on α, M, a_0, $\widetilde{a}$, I, c_0, and not depending on Ω (by $|\Omega|$ we denote the Lebesgue measure of the set Ω).

For the proof see Lemma 5 of Gasiński [5]. Now we can prove our existence theorem.

Proof. (of Theorem 4.2) We apply the direct method of the calculus of variations. Let

$$\{(\Omega_n, u_n)\}_{n\geq 1} \subseteq D_P(\mathrm{J})$$

be a minimizing sequence for $(OPPI)$. As the embedding of $\mathcal{O}^{k,\infty}$ into $\mathcal{O}^{k-1,\infty}$ is compact (see Section 2.2) so $\mathcal{B}$ is compact in $\mathcal{O}^{k-1,\infty}$ and we can choose a subsequence of $\{\Omega_n\}_{n\geq 1}$ (still indexed by n) and a set $\Omega_0 \in \mathcal{B}$ such that $\Omega_n \longrightarrow \Omega_0$ in $\mathcal{O}^{k-1,\infty}$. This means that there exist $\{T_n\}_{n\geq 1} \subseteq \mathcal{F}^{k-1,\infty}$ and $T_0 \in \mathcal{F}^{k-1,\infty}$ such that $\Omega_n = T_n(C)$, $\Omega_0 = T_0(C)$ and $T_n - T_0 \longrightarrow 0$, $T_n^{-1} - T_0^{-1} \longrightarrow 0$ in $W^{k-1,\infty}(\mathbb{R}^N;\mathbb{R}^N)$.

From the relationship between $\mathcal{O}^{k,\infty}$-convergence and the convergence of characteristic functions of open sets (see Section 2.2), we obtain that $1_{\Omega_n} \longrightarrow 1_{\Omega_0}$ in $H(\mathbb{R}^N)$ which gives, in particular, that the sequence $\{|\Omega_n|\}_{n\geq 1}$ is bounded, so also sequences $\{\|\psi\|_{H(\Omega_n)}\}_{n\geq 1}$ and $\{\|f\|_{\mathcal{H}(\Omega_n)}\}_{n\geq 1}$ are bounded. Since $u_n \in S_P(\Omega_n)$, so using Lemma 4.4, we can obtain that the sequence $\{\|u_n\|_{\mathcal{W}(\Omega_n)}\}_{n\geq 1}$ is bounded. Putting $\widehat{u}_n(t,X) \stackrel{df}{=} u_n(t, T_n(X))$ and using Lemma 2.1 of Liu and Rubio [8] (see also the remark below Lemma 2.1 in [8]), we obtain that the sequence $\{\|\widehat{u}_n\|_{\mathcal{W}(C)}\}_{n\geq 1}$ is bounded. Thus, taking a next subsequence if necessary, we have

$$\widehat{u}_n \longrightarrow u^* \quad \text{weakly in } \mathcal{W}(C),$$

with some $u^* \in \mathcal{W}(C)$. From the compactness of the embedding $\mathcal{W}(C) \subset \mathcal{H}(C)$, we get

$$\widehat{u}_n \longrightarrow u^* \quad \text{in } \mathcal{H}(C).$$

From Lemma 4.3, we have that $u^*(t,X) = u_0(t, T_0(X))$ with some $u_0 \in S_P(\Omega_0)$. So the pair (Ω_0, u_0) is admissible for $(OPPI)$.

Let $\underline{\widehat{u}_n}$ and $\underline{u^*}$ denote the functions in $\mathcal{H}(\mathbb{R}^N)$ obtained from $\widehat{u}_n$ and u^*, respectively, by extending them by zero outside C. So, we have

$$\underline{\widehat{u}_n} \longrightarrow \underline{u^*} \quad \text{in } \mathcal{H}(\mathbb{R}^N).$$

From Lemma 2.2 of Liu and Rubio [8], we also have

$$\underline{u_n} \longrightarrow \underline{u_0} \quad \text{in } \mathcal{H}(\mathbb{R}^N),$$

where

$$\underline{u_n}(t,x) = \begin{cases} u_n(t,x) & \text{if } x \in \Omega_n \\ 0 & \text{if } x \in \mathbb{R}^N \setminus \Omega_n, \end{cases}$$

$$\underline{u_0}(t,x) = \begin{cases} u_0(t,x) & \text{if } x \in \Omega_0 \\ 0 & \text{if } x \in \mathbb{R}^N \setminus \Omega_0. \end{cases}$$

Hence, due to the hypothesis $H(\mathrm{J})$, we conclude that (Ω_0, u_0) solves the problem $(OPPI)$ and the proof of the theorem is complete.

□

Bibliography

[1] Browder F.E., Hess P., *Nonlinear Mappings of Monotone Type in Banach Spaces*, J. of Funct. Anal., 11 (1972), 251-294.

[2] Chang K.C., *Variational methods for nondifferentiable functionals and applications to partial differential equations*, J. Math. Anal. Appl., 80 (1981), 102-129.

[3] Clarke F.H., *Optimization and Nonsmooth Analysis*, John Wiley & Sons, New York, 1983.

[4] Denkowski Z., Migórski S., *Optimal Shape Design Problems for a Class of Systems Described by Hemivariational Inequality*, J. Global. Opt., 12 (1998), 37-59.

[5] Gasiński L., *Optimal Shape Design Problems for a Class of Systems Described by Parabolic Hemivariational Inequality*, J. Global. Opt., 12 (1998), 299-317.

[6] Lions J.L., Magenes E., *Non-Homogeneous Boundary Value Problems and Applications*, Springer-Verlag, Berlin, 1972.

[7] Liu W.B., Rubio J.E., *Optimal Shape Design for Systems Governed by Variational Inequalities, Part 1: Existence Theory for the Elliptic Case*, J. Optim. Th. Appl., 69 (1991), 351-371.

[8] Liu W.B., Rubio J.E., *Optimal Shape Design for Systems Governed by Variational Inequalities,, Part 2: Existence Theory for Evolution Case*, J. Optim. Th. Appl., 69 (1991), 373-396.

[9] Micheletti A.M., *Metrica per famiglie di domini limitati e proprietà generiche degli autovalori*, Annali della Scuola Normale Superiore di Pisa, 28 (1972), 683-693.

[10] Miettinen M., *Approximation of Hemivariational Inequalities and Optimal Control Problems*, Thesis, University of Jyväskylä, Finland, Report 59, 1993.

[11] Miettinen M., *A Parabolic Hemivariational Inequality*, Nonlinear Analysis, 26 (1996), 725-734.

[12] Moreau J.J., *Le Notions de Sur-potential et les Liaisons Unilatérales en Élastostatique*, C.R. Acad. Sc. Paris, 267A (1968), 954-957.

[13] Moreau J.J., Panagiotopoulos P.D., Strang G., *Topics in Nonsmooth Mechanics*, Birkhäuser, Basel, 1988.

[14] Murat F., Simon J., *Sur le Controle par un Domaine Geometrique*, Preprint no. 76015, University of Paris, 6 (1976), 725-734.

[15] Naniewicz Z., *On the Pseudo-Monotonicity of Generalized Gradients of Nonconvex Functions*, Applicable Analysis, 47 (1992), 151-172.

[16] Naniewicz Z., Panagiotopoulos P.D., *Mathematical Theory of Hemivariational Inequalities and Applications*, Dekker, New York, 1995.

[17] Panagiotopoulos P.D., *Nonconvex Superpotentials in the Sense of F.H. Clarke and Applications*, Mech. Res. Comm., 8 (1981), 335-340.

[18] Panagiotopoulos P.D., *Nonconvex Problems of Semipermeable Media and Related Topics*, Z. Angew. Math. Mech., 65 (1985), 29-36.

[19] Panagiotopoulos P.D., *Inequality Problems in Mechanics and Applications. Convex and Nonconvex Energy Functions*, Birkhäuser, Basel, 1985.

[20] Panagiotopoulos P.D., *Modeling of Nonconvex Nonsmooth Energy Problems. Dynamic Hemivariational Inequalities with Impact Effects*, J. Comp. Appl. Math., 63 (1995), 123-138.

[21] Rabinowitz P.H., *Minimax Methods in Critical Point Theory with Applications to Differential Equations*, CBMS, Regional Conf. Series in Math, Vol. 65, AMS, Providence, R.I. (1986).

[22] Serrin J., *On the Definition and Properties of Certain Variational Integrals*, Trans. Amer. Math. Soc., 101 (1961), 139-167.

[23] Sokołowski J., Zolesio J.P., *Introduction to Shape Optimization. Shape Sensitivity Analysis*, Springer Verlag, 1992.

Leszek Gasiński. Institute of Computer Science, Jagiellonian University, ul. Nawojki 11, 30-072 Kraków, Poland
E-mail: gasinski@softlab.ii.uj.edu.pl

Existence of Free-Boundary for a Two Non-Newtonian Fluids Problem

Nicolas Gomez and Jean-Paul Zolésio

Abstract. The flow of two immiscible fluids separated by an interface is considered. When the two fluids undergo the action of volume forces and are non-Newtonian of Norton-Hoff type with different consistencies, an existence result for the free interface is derived. Due to the slow motion of viscous fluids, a quasi-steady model for the evolution on a time interval I is convenient. Provided the total domain D occupied by both fluids is constant, our model is a cylindrical evolution problem: we have to find the flow u on $I \times D$, and the domain $\Omega(t)$ occupied by one of the fluid. Using Norton-Hoff law, we come to the following boundary value problem

$$\begin{aligned} -div\,(K(t,x)|\varepsilon(u(t,x))|^{p-2}\varepsilon(u(t,x))) - \nabla P(t,x) &= f \text{ on } I \times D \\ div\, u(t,x) &= 0 \text{ on } I \times D \\ u(t,x) &= 0 \text{ on } I \times \partial D \end{aligned}$$

where $\varepsilon(u)$ is the symmetrized of Du and $1 < p \leq 2$. The coefficient $K(t,x)$ is linked to the viscosity of the fluid at position x and time t, and thus is a discontinuous function related to $\Omega(t)$. Due to the lack of regularity of Norton-Hoff operator, classical techniques for two-fluids problems fail. Nevertheless, using some techniques from Shape Analysis, we derive the existence of a solution $(u(t,x), \Omega(t))$. The idea is to consider fictious evolutions of the domain $\tilde{\Omega}(t)$ and find the correspondent $\tilde{u}$. The use of a fixed-point argument yield the result. This requires sharp techniques for non-smooth evolution of the domain and sensitivity result with respect to the coefficients in Norton-Hoff equation, which we both recently derived.

1 Introduction

Norton-Hoff law was introduced in [8] to study the creeping of steel at high-temperature. Generalised by Friaâ in [3], it describes quasi-static evolution of viscous flows with a power law as stress-strain relation: $\sigma = K|\varepsilon(u)|^{p-2}\varepsilon(u)$, where σ is the Cauchy stress tensor and $\varepsilon(u) = \frac{1}{2}(Du + Du^*)$ is the linearised strain velocity tensor. The coefficient K is the consistency of the material and p is the exponent. They are rheological characteristics of the material. When $p = 2$, Stokes' problem for newtonian fluid case is recovered. The fluid is named dilatant when $p > 2$ and is a Norton-Hoff media for $1 < p < 2$, which will be the only case considered in this paper. In [10], Temam proved that Prandtl-Reuss law for plasticity is recovered from Norton-Hoff model when p tends to 1.

In this paper, we prove an existence result in a two-fluids problem. We consider two fluids of Norton-Hoff type with the same exponent occupying a fixed region D of $\mathbb{R}^N$ and undergoing the action of time-dependant volume forces f. We assume the fluids are non-miscible and denote α and β their consistencies. The evolution of this system on a time interval $I = [0,1]$ is modelized as the solution of the following non-linear transmission problem

$$\mathcal{P} \begin{cases} \sigma = K|\varepsilon(u)|^{p-2}\varepsilon(u) & \text{on } I \times D \\ -div\ (\sigma) + \nabla P = f & \text{on } I \times D \\ div\ u(t,\cdot) = 0 & \text{on } I \times D \\ u = 0 & \text{on } I \times \partial D \\ \\ \partial_t K + \nabla K \cdot u = 0 & \text{on } I \times D \\ K(0,\cdot) = \alpha\chi_\Omega + \beta(1-\chi_\Omega) & \end{cases}$$

where Ω is the subset of D occupied by the fluid of consistency α at the initial time. Our point of view is to look for the solution as an evolution of the initial domain Ω (and thus an evolution of the interface between the two fluids).

As a first step, we will investigate Norton-Hoff problem in the domain D with non-smooth coefficient K. An existence result is given for this singular vector-valued elliptic operator. Then we will look for the solution of $\mathcal{P}$ as a fixed point of an operator $V \mapsto u^V$, where V is a virtual evolution of the interface between the two fluids. Since non smooth evolution of the domains have to be considered, we will use some recent developments of the Speed Method for Shape Analysis, that were initiated in [13] and [12]. Some continuity property of the solution of Norton-Hoff problem are also obtained.

Throughout this paper we assume Ω is a Lipschitz open subset of D which is locally on one side of its boundary.

2 Norton-Hoff Problem with Discontinuous Coefficient

In the following, D will denote a smooth bounded open subset of $\mathbb{R}^N$. For $g \in L^{p'}(D)$ and $K \in L^\infty(D)$, we denote $\wp(K,g)$ the boundary value problem

$$\wp(K,g) \begin{cases} \sigma(u) = |\varepsilon(u)|^{p-2}\varepsilon(u) \\ -div\ (K\sigma(u)) - \nabla P = g \text{ on } D \\ div\ u = 0 \text{ on } D \\ u = 0 \text{ on } \partial D \end{cases} \tag{2.1}$$

We denote

$$\mathcal{W} = W_0^{1,p}(D,\mathbb{R}^N) \text{ and } \mathcal{W}_{\text{div}} = \{v \in \mathcal{W} \mid div\ v = 0\} \tag{2.2}$$

which are the natural spaces involved in the study of $\wp(K,g)$ (see also [4]). The following proposition provides a Banach space structure both on $\mathcal{W}$ and $\mathcal{W}_{div}$.

PROPOSITION 2.1. *The mapping $\|\cdot\|$ defined from $\mathcal{W}$ to $\mathbb{R}$ by*

$$\|v\| = \left(\int_D |\varepsilon(v)|^p\right)^{\frac{1}{p}} \tag{2.3}$$

is a norm on $\mathcal{W}$, equivalent to the one induced by the canonical one of $W^{1,p}(D,\mathbb{R}^N)$.

For this norm, $\mathcal{W}_{\text{div}}$ is closed in $\mathcal{W}$, so it is a Banach space for the induced norm, still denoted $\|\cdot\|$ for simplicity.

Proof. Since $\|u\| = \|\varepsilon(u)\|_{L^p(D,\mathbb{R}^{N^2})}$, it is clear that

$$\forall u, w \in \mathcal{W}, \ \|u+w\| \leq \|u\| + \|w\|$$

Moreover, if $u \in \mathcal{W}$ is such that $\|u\| = 0$ then $\varepsilon(u) = 0$ on D. Accordingly u is a rigid body motion: $u(x) = a + b \cdot x$ where b is a anti-symmetric matrix (see [11] for instance). Thus its kernel is an hyperplane which is not compatible with $u_{\partial D} = 0$ unless $u = 0$.

We now prove the equivalence between $\|\cdot\|$ and $\|\cdot\|_{1,p}$, the norm induced on $\mathcal{W}$ by the usual norm of $W^{1,p}(D,\mathbb{R}^N)$.

For any $u \in \mathcal{W}$ we have

$$\|u\| = \frac{1}{2}\|Du + Du^*\|_{L^p(D,\mathbb{R}^{N^2})} \leq \|Du\|_{L^p(D,\mathbb{R}^{N^2})}$$

$$\|u\| \leq \|Du\|_{L^p(D,\mathbb{R}^{N^2})} + \|u\|_{\mathbb{R}mL^p(D,\mathbb{R}^N)} = \|u\|_{1,p} \tag{2.4}$$

Both Poincaré's and Korn's inequalities hold in $\mathcal{W}$ (*cf* [11, 6, 4]): there exists constants $c_P(D)$ and $c_K(D)$ such that, for any u in $\mathcal{W}$,

$$\|u\|_{L^p(D,\mathbb{R}^N)} \leq c^P(D)\|Du\|_{L^p(D,\mathbb{R}^{N^2})} \tag{2.5}$$

$$\|Du\|_{L^p(D,\mathbb{R}^{N^2})} \leq c^K(D)\left(\|\varepsilon(u)\|_{L^p(D,\mathbb{R}^{N^2})} + \|u\|_{L^p(D,\mathbb{R}^N)}\right) \tag{2.6}$$

We assume there exists a sequence (u_n) in $\mathcal{W}$ and a sequence (k_n) in $\mathbb{R}^+$ which tends to infinity such that $\|u_n\|_{1,p} > k_n\|u_n\|$. Without loss of generality, we suppose $\|u_n\|_{L^p(D,\mathbb{R}^N)} = 1$. By Korn's inequality (2.6) we have $c^K(D)\|u_n\| + c^K(D) > k_n\|u_n\|$ for any n. Hence

$$\frac{c^K(D)+1}{k_n - c^K(D)} > \|u_n\| \ \text{and} \ \frac{c^K(D)(c^K(D)+1)}{k_n - c^K(D)} + c^K(D) \geq \|Du_n\|_{L^p(D,\mathbb{R}^{N^2})}$$

Thus the sequence (u_n) is bounded in $\mathcal{W}$ endowed with the classical topology of $W_0^{1,p}(D,\mathbb{R}^N)$. So it converges weakly towards u_* in $\mathcal{W}$. Since $\|\cdot\|$ is lowerly semi-continuous,

$$\|u_*\| \leq \liminf_{n\to\infty} \|u_n\| = 0$$

So $u_* = 0$ which is contradictory with $\|u_n\|_{L^p(D,\mathbb{R}^N)} = 1$. Accordingly, there exists c such that

$$\forall u \in \mathcal{W}, \quad \|u\|_{1,p} \leq c\|u\|$$

and the equivalence is proven. □

2.1 Continuity of the Operator

For K in $\mathcal{K}$, we consider the operator

$$\begin{aligned} A_K : \mathcal{W} &\to \mathcal{W}' \\ v &\mapsto -div\,(K|\varepsilon(v)|^{p-2}\varepsilon(v)) \end{aligned}$$

This section will prove the following result.

PROPOSITION 2.2. *The operator A_K belongs to $\mathcal{C}^{0,p-1}(\mathcal{W},\mathcal{W}')$.*

We start proving that

$$\begin{aligned} a : \mathcal{W} &\to L^{p'}(D,\mathbb{R}^{N^2}) \\ v &\mapsto |\varepsilon(v)|^{p-2}\varepsilon(v) \end{aligned}$$

satisfies the usual monotonicity assumptions, using the following technical lemma.

LEMMA 2.3. *Let $1 < p \leq 2$ and m an integer. Then*

$$\forall (x,y) \in \mathbb{R}^m \times \mathbb{R}^m$$

$$\left||x|^{p-2}x - |y|^{p-2}y\right| \leq 2^{2-p}|x-y|^{p-1} \tag{2.7}$$

$$(|x|^{p-2}x - |y|^{p-2}\,,\, x-y) \geq (|x|+|y|)^{p-2}\,|x-y|^2 \tag{2.8}$$

Proof. It is sufficient to prove these inequalities when both x and $|y|$ are fixed. Let us notice that we can suppose they both are non-zero: otherwise the inequalities are obvious.

We fix a non-zero $x \in \mathbb{R}^m$ and $r > 0$. The mapping $y \mapsto \left||x|^{p-2}x - |y|^{p-2}y\right|$, restricted to S_r, the sphere of radius r in $\mathbb{R}^m$, reaches its maximum on when $y = -\lambda x$ with $\lambda = \dfrac{r}{|x|}$. Accordingly, for any $y \in S_r$,

$$\frac{\left||x|^{p-2}x - |y|^{p-2}y\right|}{|x-y|^{p-1}} \leq \frac{\left||x|^{p-2}x + |\lambda|^{p-1}|x|^{p-2}x\right|}{|x-\lambda|^{p-1}} = \frac{1+\lambda^{p-1}}{(1+\lambda)^{p-1}}$$

The right-hand side is maximum when $\lambda = 1$ so

$$\frac{\left||x|^{p-2}x - |y|^{p-2}y\right|}{|x-y|^{p-1}} \leq 2^{2-p}$$

□

Accordingly a simple calculus yield

LEMMA 2.4. *For any u and v in $\mathcal{W}$, a.e. in D,*

$$|a(u)-a(v)| \leq 2^{2-p}|\varepsilon(u-v)|^{p-1} \tag{2.9}$$

$$\langle A(u)-A(v), u-v\rangle_{\mathcal{W}'\times\mathcal{W}} \geq (|\varepsilon(u)|+|\varepsilon(v)|)^{p-2}|\varepsilon(u-v)|^2$$

is a norm on $\mathcal{W}$, equivalent to $\|\cdot\|$. Moreover n_K^p is convex and l.s.c.

2.2 A Duality Method for a Solution of $\wp$

In this section, we fix $K \in \mathcal{K}$ and use classical duality argument to prove the existence of a solution to $\wp(K,g)$. Let us consider the following Lagrangian ℓ_K:

$$\begin{array}{rcl} W_0^{1,p}(D,\mathbb{R}^N)\times L^{p'}(D) & \longrightarrow & \mathbb{R} \\ (u,q) & \longmapsto & \displaystyle\int_D \frac{K}{p}|\varepsilon(u)|^p - gu - q(div\ u) \end{array}$$

PROPOSITION 2.5. *The function ℓ_K has at least a saddle point (u_K, P_K).*

Proof. The following conditions are sufficient for the existence of a saddle point (we refer to [2]):

i) For any q in $L^{p'}(D)$, $\ell_K(.,q)\ :\ v \mapsto \ell_K(v,q)$ is convex and l.s.c. on $W_0^{1,p}(D,\mathbb{R}^N)$.

ii) For any u in $W_0^{1,p}(D,\mathbb{R}^N)$, $\ell_K(u,.)\ :\ q \mapsto \ell_K(u,q)$ is concave and l.s.c. on $L^{p'}(D)$.

iii) There exists $q_0 \in L^{p'}(D)$ such that $\lim\limits_{\|u\|\to+\infty} \ell_K(u,q_0) = +\infty$

iv) $\lim\limits_{\|q\|\to+\infty}\ \inf\limits_{v\in W_0^{1,p}(D,\mathbb{R}^N)} \ell_K(v,q) = -\infty$

Condition (i) is satisfied. Indeed for any q in $L^{p'}(D)$, $\ell_K(.,q)$ is the sum of n_K^p and a linear functional. Accordingly $\ell_K(.,q)$ is convex and l.s.c.

Since $\ell_K(u,q)$ is linear with respect to q, condition (ii) obviously holds.

The choice $q_0 = 0$ is convenient for condition (iii) since $\ell_K(.,0) = n_K^p$ is coercive.

We eventually achieve the proof by noticing that

$$\lim_{\|q\|\to+\infty}\ \inf_{v\in W_0^{1,p}(D,\mathbb{R}^N)} \ell_K(v,q)$$

$$\leq \lim_{\|q\|\to+\infty}\ \Big(\inf_{v\in W_0^{1,p}(D,\mathbb{R}^N)} n_K(v)^p - < g,v > +\ \inf_{v\in W_0^{1,p}(D,\mathbb{R}^N)} -\langle q, div\ v\rangle\Big)$$

$$\leq \inf_{v\in W_0^{1,p}(D,\mathbb{R}^N)} (n_K(v)^p - \langle g,v\rangle) + \lim_{\|q\|\to+\infty}\ \inf_{v\in W_0^{1,p}(D,\mathbb{R}^N)} \langle q, div\ v\rangle)$$

Since $\inf\limits_{v\in W_0^{1,p}(D,\mathbb{R}^N)} n_K(v)^p - \langle g,v\rangle \leq 0$,

$$\inf_{v\in W_0^{1,p}(D,\mathbb{R}^N)} (n_K(v)^p - \langle g,v\rangle) + \lim_{\|q\|\to+\infty}\ \inf_{v\in W_0^{1,p}(D,\mathbb{R}^N)} \langle q, div\ v\rangle = -\infty$$

which achieve the proof of the existence for a saddle point at ℓ_K. □

PROPOSITION 2.6. *For any K in $\mathcal{K}$, the problem $\wp(K,g)$ has a unique solution (u_K, P_K) in $\mathcal{W} \times L^{p'}(D)/\mathbb{R}$.*

Proof. We are going to prove that a saddle point of ℓ_K is a solution of $\wp(K,g)$.

If (u_K, P_K) is a saddle point of ℓ_K, one has

$$\forall (v,q) \in W_0^{1,p}(D,\mathbb{R}^N)\times L^{p'}(D)\ ,\ \ell_K(u_K,q) \leq \ell_K(u_K,P_K) \leq \ell(v,P_K) \tag{2.10}$$

The first inequality yield

$$\int_D \frac{K}{p}|\varepsilon(u_K)|^p - g.u_K - q div\ u_K \leq \int_D \frac{K}{p}|\varepsilon(u_K)|^p - g.u_K - P_K div\ v$$

which holds for any q in $L^{p'}(D)$. Hence, $div\ u_K = 0$ so $u_K \in \mathcal{W}$.

According [2], first order optimality conditions holds. As a consequence, the second inequality of (2.10) implies, for any v in $W_0^{1,p}(D, \mathbb{R}^N)$ $\langle \partial_1 \ell(u_K, P_K), v \rangle = 0$. Hence

$$\int_D K|\varepsilon(u_K)|^{p-2}\varepsilon(u_K)..\varepsilon(v) - gv - P_K div\ v = 0$$

Using Green's formulae and by part integration we come to

$$\langle -K div\ \left(K|\varepsilon(u_K)|^{p-2}\varepsilon(u_K)\right) - g + \nabla P_K, v\rangle_{W^{-1,p'} \times W_0^{1,p}} = 0$$

which holds for any v in $W_0^{1,p}(D, \mathbb{R}^N)$.

Hence,

$$\begin{cases} -div\ [K|\varepsilon(u_K)|^{p-2}\varepsilon(u_K)] + \nabla P_K &= g \quad \text{on } D \\ u_K &= 0 \quad \text{on } \partial D \\ div\ u_K &= 0 \quad \text{on } D \end{cases}$$

and we have a solution of $\wp(K, g)$.

A mere application of Green's formulae prove that a solution of $\wp(K, g)$ is a saddle point of ℓ_K.

Accordingly, if (u', P') is another solution of $\wp(K, g)$, it is another saddle point of ℓ_K. Since the set of saddle points is a Cartesian product (see *e.g.* [2]), (u', P_K) is a third saddle point. Moreover, both u_K and u' are solutions of the problem

$$\begin{cases} -div\ [K|\varepsilon(u)|^{p-2}\varepsilon(u)] &= g \quad \text{on } D \\ u &= 0 \quad \text{on } \partial D \end{cases}$$

with $\tilde{g} = g - \nabla P_K$. But this problem has an unique solution, which is the minimum of the continuous, strictly convex and coercive functional

$$\begin{array}{ccc} \mathcal{W} & \to & \mathbb{R} \\ v & \mapsto & \displaystyle\int_D \frac{K}{p}|\varepsilon(v)|^p - \tilde{g}v \end{array}$$

Hence $u_K = u'$.

If we assume that (u_K, P_K) and (u_K, P') both are solutions of $\wp(K, g)$, the we come to $\nabla(P_K - P') = 0$ on D. Hence P_K and P' are equals up to a constant *i.e.* they are equals in $L^{p'}/\mathbb{R}$. □

3 Two Phases Model

In order to prove an existence result for the transmission problem $\mathcal{P}$, we split it in two problems: an evolution *via* the flow of a smooth vector-field of the domain Ω occupied by the fluid of consistency α and a quasi-steady Norton-Hoff problem with discontinuous coefficient. We use the so-called Speed Method for the shape evolution problem.

3.1 Smooth Evolution of the Domain

Let $k > 0$ be an integer. We denote

$$\mathcal{V}_k = \{V \in \mathcal{C}^0(I, \mathcal{C}^k(D, \mathbb{R}^N)) \mid \forall t \in I, div\, V(t,\cdot) = 0,\ V.n = 0 \text{ on } \partial D\}$$

where n is the outward normal unit vector of ∂D. It is a classical result of O.D.E. theory that any V in $\mathcal{V}_k$ has a flow mapping T^V which is solution of the initial value problem $\partial T = V \circ T$ with $T(0,\cdot) = Id_D$. Moreover, $T^V \in \mathcal{C}^1(I, \mathcal{C}^k(D, \mathbb{R}^N))$ is invertible and $(T^V)^{-1} \in \mathcal{C}^0(I, \mathcal{C}^k(D, \mathbb{R}^N)) \cap \mathcal{C}^1(I, \mathcal{C}^{k-1}(D, \mathbb{R}^N))$. Those properties are the bases of the so-called Speed Method and we refer to [9] for details.

For a domain $\Omega \subset D$, one can consider the family

$$(\Omega_t(V))_{t\in I} = (T^V(t, \Omega))_{t\in I} \tag{3.1}$$

This process preserves the regularity: if Ω is of class $\mathcal{C}^r$ (resp. Lipschitz regular) all the $\Omega_t(V)$ are of class $\mathcal{C}^{\max\{k,r\}}$ (resp. Lipschitz regular) because of the regularity of T^V. Due to the assumption $div\ V = 0$, the volume is preserved during the deformation: $meas\ \Omega_t(V) = meas\ \Omega$. The assumption $V.n = 0$ on ∂D yields that $\Omega_t(V) \subset D$: this parameterization of the deformation allows to take into account some geometrical constraints. In the Speed Method framework, Shape Analysis is done through the study of the variations of a shape-dependent criterion generated by these deformations of the domain.

LEMMA 3.1. *The mapping* $\chi^V = t \mapsto \chi_{\Omega_t(V)} = \chi_\Omega \circ T^V(t,\cdot)$ *is solution to the initial value problem*

$$\left\{\begin{array}{l} \partial_t \chi + \nabla\chi.V = 0 \\ \chi(0,\cdot) = \chi_\Omega \end{array}\right. \tag{3.2}$$

Consequently, the following problems are equivalent

$$\left\{\begin{array}{l} -div\ (K|\varepsilon(u)|^{p-2}\varepsilon(u)) + \nabla P = f \\ u \in L^\infty(I, \mathcal{W}_{\rm div}) \\ K(t,\cdot) = \alpha\chi^V(t) + \beta(1-\chi^V(t)) \end{array}\right. \quad \text{and} \quad \mathcal{P}^V \left\{\begin{array}{l} -div\ (K|\varepsilon(u)|^{p-2}\varepsilon(u)) + \nabla P = f \\ u \in L^\infty(I, \mathcal{W}_{\rm div}) \\ \partial_t K + \nabla K.V = 0 \\ K(0,\cdot) = \alpha\chi_\Omega + \beta(1-\chi_\Omega) \end{array}\right. \tag{3.3}$$

If it exists, the solution to one of this problem is the flow of the two materials of consistency α and β in the domain D while Ω, the volume occupied by one of the fluid, is moved with a speed-field V. This problem has no physical meaning since the evolution of the interface is given.

PROPOSITION 3.2. *If f belongs to $L^\infty(I, W^{-1,p'}(D, \mathbb{R}^N))$, the problem $\mathcal{P}^V$ has an unique solution (u^V, P^V) such that $u^V \in L^\infty(I, \mathcal{W}_{div})$.*

Proof. For almost any t in I, $K(t,\cdot) = \alpha\chi_{\Omega_t(V)} + \beta(1-\chi_{\Omega_t(V)}) \in \mathcal{K}$. Accordingly, using Proposition 2.6, the problem $\wp(K(t,\cdot), f(t))$ has an unique solution (u_t, P_t) in $\mathcal{W}_{\rm div} \times L^{p'}(D)$. Since the solution (u_t, P_t) of $\wp(K(t,\cdot), f(t))$ is a saddle point of $\ell_{K(t,\cdot),f(t)}$, u_t is the minimum of the functional defined on $\mathcal{W}_{\text{div}}$ by $v \mapsto \int_D \frac{1}{p} K(t,\cdot)|\varepsilon(v)|^p - f(t).v$. Thus, $\int_D \frac{1}{p} K(t,\cdot)|\varepsilon(u_t)|^p - f(t).u_t \leq 0$. This yields, for almost any t in I,

$$\frac{\alpha}{p}\|u_t\|^{p-1} \leq \|f\|_{L^\infty(I,W^{-1,p'}(D,\mathbb{R}^N))}$$

which proves $u^V : t \mapsto u_t$ belongs to $L^\infty(I,\mathcal{W}_{\text{div}})$. □

If it exists, a fixed point of the mapping $V \mapsto u^V$ would be a solution to the transmission problem $\mathcal{P}$. In this framework, such a fixed point can not be found since, as far as we know, we do not have $u_t \in \mathcal{C}^k(D,\mathbb{R}^N)$. In fact, the question of (spatial) regularity for non-linear vector-valued degenerated operators of Norton-Hoff type, is still opened, though it has been settled for the scalar cases. We refer to [5] for details or further considerations.

In the following, we use a weakened version of the evolution of domains which will provide a fixed point. This method has been recently developed and further details or applications to Euler equation in fluid mechanics can be found in [13, 12, 1].

3.2 Non-Smooth Evolution of Domains

From now on we assume

$$p > \frac{2N}{2+N} \tag{3.4}$$

in order to have $\mathcal{W}_{\text{div}} \subset L^2(D,\mathbb{R}^N)$. We denote

$$\mathcal{V} = \left\{V \in L^2(I, L^2(D,\mathbb{R}^N)) \mid div\ V = 0 \text{ a.e. on } I \times D,\ < V, n >= 0 \text{ on } I \times \partial D\right\}$$

and

$$\mathcal{L} = L^2(I, L^2(D)) \cap H^1(I, H^{-1}(D)) \quad \subset \mathcal{C}^0(I, H^{-\frac{1}{2}}(D))$$

It has been proven in [1] that for any $V \in \mathcal{V}$ and any measurable subset Ω of D, there exists a solution $\chi^V \in \mathcal{L}$ to the transport problem

$$\mathcal{T}(V) \begin{cases} \dot{\chi}(t) + \nabla\chi(t)V(t) = 0 & \text{in } D, \text{ a.e. } t \text{ in } I \\ \chi(0) = \chi_\Omega & \text{in } D \end{cases} \tag{3.5}$$

which values are characteristic functions. With this relaxation of the evolution of domains, we can extend the mapping $V \mapsto u^V$ to $\mathcal{V}$ so that it takes values in $L^\infty(I,\mathcal{W}_{\text{div}}) \subset \mathcal{V}$. The existence of a fixed point now depends upon continuity and compactness of this extended mapping. This will derive from the following continuity property.

THEOREM 3.3. *Let $(V_n)_{n\in\mathbb{N}}$ be a sequence in $L^2(I,\mathcal{W}_{div})$ which converges weakly towards V in $L^2(I,\mathcal{W}_{div})$.*

Then the solutions χ^{V_n} of $\mathcal{T}(V_n)$ converges towards χ^V in $L^2(I\times D)$.

The proof is similar to the one given for proposition 2.2 page 417.

Proof. For shortness we write χ_n instead of χ^{V_n} for $n \in \mathbb{N}$ and χ_* instead of χ^V. Moreover, since (χ_n) is bounded in $L^2(I\times D)$ we can assume that $\chi_n \rightharpoonup \lambda$ in $L^2(I\times D)$.

□

3.3 Shape Sensitivity

In order to prove the continuity of $V \mapsto u^V$ we are interested in the dependence of the solution of a Norton-Hoff problem with respect to the domain. This problem was investigated in [7] for the control of shape-dependant functional using the so-called Shape Differential Equation making regularity assumptions on the domain. Here, we derive a new result for a non-smooth case which is needed when dealing with non-smooth evolution of the domain.

In this section, we denote $\mathcal{O}_{\text{lip}}$ the set of Lipshitz domains included in D. For $\Omega \in \mathcal{O}_{\text{lip}}$ an $g \in W^{-1,p'}(D, \mathbb{R}^N)$, we will denote $K_\Omega = \alpha\chi_\Omega + \beta(1 - \chi_\Omega)$ and by $u_{\Omega,g}$ the solution of $\wp(K_\Omega, g)$.

PROPOSITION 3.4. *Let $(\Omega_n)_{n\geq 0}$ be a sequence in $\mathcal{O}_{lip}$ such that $(\chi_{\Omega_n})_{n\geq 0}$ converges in $L^2(D)$ towards a characteristic function $\chi_* = \chi_{\Omega_*}$ with $\Omega_* \in \mathcal{O}_{lip}$.*

Then $u_{\Omega_n,g}$ converges towards $u_{\Omega_*,g}$, weakly in $\mathcal{W}$.

The proposition is mainly deduced from compacity arguments based on the following boundedness property. A particular case of this lemma was yet used to prove Proposition 3.2.

LEMMA 3.5. *For any Ω in $\mathcal{O}_{lip}$,*

$$\|u_{\Omega,g}\| \leq \left[\frac{p}{\alpha}\|g\|_{W^{-1,p'}(D,\mathbb{R}^N)}\right]^{\frac{1}{p-1}}$$

Proof. Since, for any $\Omega \in \mathcal{O}_{\text{lip}}$

$$\forall v \in \mathcal{W}_{\text{div}}, \quad \int_D \frac{K_\Omega}{p}|\varepsilon(u_{\Omega,g})|^p - gu_{\Omega,g} \leq \int_D \frac{K_\Omega}{p}|\varepsilon(v)|^p - gv \tag{3.6}$$

Hence we have

$$\int_D \frac{K_\Omega}{p}|\varepsilon(u_{\Omega,g})|^p - gu_{\Omega,g} \leq 0 \tag{3.7}$$

Hence, $\displaystyle\int_D \frac{K_\Omega}{p}\left|\varepsilon\left(u_{\Omega,g}\right)\right|^p - gu_{\Omega,g} \leq 0$ and thus,

$$\int_D \frac{\alpha}{p}\left|\varepsilon\left(u_{\Omega,g}\right)\right|^p \leq \int_D gu_{\Omega,g}$$

So we have $\dfrac{\alpha}{p}\|u_{\Omega,g}\|^p \leq \|g\|_{W^{-1,p'}(D,\mathbb{R}^N)}\|u_{\Omega,g}\|$, so we come to the conclusion. □

We can now prove Proposition 3.4.

Proof. We denote

$$\chi_n = \chi_{\Omega_n,g} \quad u_n = u_{\Omega_n,g} \quad K_n = K_{\Omega_n}$$

By previous lemma, the sequence $(u_n)_{n\geq 0}$ is bounded in $\mathcal{W}_{div}$ and thus, one can extract a subsequence $(u_{n_k})_{k\geq 0}$ which converges weakly in $\mathcal{W}_{\text{div}}$. We name u_* this weak-limit, and assume, that $(u_n)_{n\geq 0}$ itself converges weakly towards u_*.

With (3.6) we have, for any n,

$$\forall v \in \mathcal{W}_{\text{div}}, \quad \int_D \frac{K_n}{p}|\varepsilon(u_n)|^p - gu_n \leq \int_D \frac{K_n}{p}|\varepsilon(v)|^p - gv$$

If $K_* = K_{\Omega_*}$, the sequence (K_n) converges (strongly) towards K_* in $L^2(D)$ but also in $L^{p'}(D)$. So

$$\liminf_{n\to\infty} \int_D \frac{K_n}{p}|\varepsilon(u_n)|^p - gu_n \leq \int_D \frac{K_*}{p}|\varepsilon(v)|^p - gv \tag{3.8}$$

But,

$$\begin{aligned}\liminf_{n\to\infty} \int_D \frac{K_n}{p}|\varepsilon(u_n)|^p - gu_n &\geq \left(\liminf_{n\to\infty} \int_D \frac{K_n}{p}|\varepsilon(u_n)|^p\right) - \int_D gu_* \\ &\geq \frac{1}{p}\|\sqrt[p]{K_n}\varepsilon(u_n)\|_{L^p(D,\mathbb{R}^{N^2})} - \int_D gu_*\end{aligned}$$

Since the norm of $L^p(D, \mathbb{R}^{N^2})$ is weakly l.s.c., we will have

$$\liminf_{n\to\infty} \int_D \frac{K_n}{p}|\varepsilon(u_n)|^p - gu_n \geq \int_D \frac{K_*}{p}|\varepsilon(u_*)|^p - gu_* \tag{3.9}$$

as soon as $\sqrt[p]{K_n}\varepsilon(u_n) \rightharpoonup \sqrt[p]{K_*}\varepsilon(u_*)$ in $L^p(D, \mathbb{R}^{N^2})$. This holds. Indeed, for any $\zeta \in L^{p'}(D, \mathbb{R}^{N^2})$, Green's formulae provides

$$\int_D \sqrt[p]{K_n}\varepsilon(u_n)..\zeta = -\langle u_n, div\ (\sqrt[p]{K_n}\zeta)\rangle_{\mathcal{W}\times\mathcal{W}'}$$

The convergence of $(\sqrt[p]{K_n}\zeta)$ towards $\sqrt[p]{K_*}\zeta$ is strong in $L^p(D, \mathbb{R}^{N^2})$ so

$$\langle u_n, div\ (\sqrt[p]{K_n}\zeta)\rangle_{\mathcal{W}\times\mathcal{W}'} \to \langle u_*, div\ (\sqrt[p]{K_*}\zeta)\rangle_{\mathcal{W}\times\mathcal{W}'}$$

Hence we actually have $\sqrt[p]{K_n}\varepsilon(u_n) \rightharpoonup \sqrt[p]{K_*}\varepsilon(u_*)$ in $L^p(D, \mathbb{R}^{N^2})$. Accordingly, equations (3.8) and (3.9) yield, for any v in $\mathcal{W}_{\text{div}}$,

$$\int_D \frac{K_*}{p}|\varepsilon(u_*)|^p - gu_* \leq \int_D \frac{K_*}{p}|\varepsilon(v)|^p - gv$$

Eventually, u_* is the minimum of Φ_{Ω_*}, and by uniqueness $u_* = u(\Omega_*)$. □

It is important to notice that, as far as we know, a strong convergence in this proof (*i.e.* a strong continuity in the Proposition) can not be obtained.

3.4 Existence Result for $\mathcal{P}$

For any given V in $\mathcal{V}$, the Proposition 3.3 provides the existence of a family of domains $(\Omega_t(V))_{t\in I}$ where we can solve the quasi-static problem

$$\mathcal{P}(V) \quad \begin{cases} -div\ (K_{\Omega_t(V)}\sigma(u)) + \nabla P = f \text{ on } I \times D \\ div\ u = 0 \text{ on } D \\ u = 0 \text{ on } \partial D \end{cases}$$

which corresponds to the flow of each phase under a given time dependent distributed force f, and a prescribed deformation of the interface between the two fluids. The free interface is the tube

$$\mathfrak{T}(V) = \{\{t\} \times \Omega_t(V) \mid t \in I\}$$

build by a field V which is a fixed point of the mapping $V \mapsto u(V)$, where $u(V)$ is the solution of the problem $\mathcal{P}(V)$. The following of this section is devoted to the proof of the existence of the fixed point.

THEOREM 3.6. *The problem $\mathcal{P}$ has a solution in $L^2(I, \mathcal{W})$: there exists V in $L^2(I, \mathcal{W})$ such that $u^V = V$.*

Proof. Let V_0 be a given field in $L^2(I, \mathcal{W}_{\text{div}})$. For any integer n we denote $V_{n+1} = u^{V_n} = t \mapsto u_{\Omega_t(V_n)}$, the solution of $\mathcal{P}(V_n)$. Since

$$||V_{n+1}||_{L^2(I,\mathcal{W}_{\text{div}})} \leq ||u^{V_n}||_{L^\infty(I,\mathcal{W}_{\text{div}})} = \sup_{t\in I} ||u_{\Omega_t(V_n),f(t)}||$$

lemma 3.5 provides

$$||V_{n+1}||_{L^2(I,\mathcal{W}_{\text{div}})} \leq \left[\frac{\alpha}{p}||f||_{L^\infty(I,W^{-1,p'}(D,\mathbb{R}^N))}\right]^{\frac{1}{p-1}}$$

Hence, up to passing to a subsequence, there exists $V_* \in L^2(I, \mathcal{W}_{\text{div}})$ such that $V_n \rightharpoonup V_*$. Theorem 3.3 proves that the corresponding solutions of the non-smooth transport problem converges strongly in $L^2(I \times D)$: $\chi^{V_n} \to \chi^{V_*}$. Accordingly, for almost any t in I, $\chi^{V_n}(t) \to \chi^{V_*}(t)$. Then Proposition 3.4 provides, $V_{n+1}(t) = u_{\Omega_t(V_n),f(t)} \rightharpoonup u_{\Omega_t(V_*),f(t)}$ in $\mathcal{W}_{div}$, a. e. in I. But the sequence V_n is bounded in $L^\infty(I, \mathcal{W}_{div})$ (*cf* lemma 3.5), so with Lebesgue's Theorem we can prove V_n converges towards $u_{\Omega_t(V_*),f(t)}$ weakly in $L^2(I, \mathcal{W}_{div})$. By uniqueness of the weak limit, we eventually have $V_* = t \mapsto u_{\Omega_t(V_*)} = u^{V_*}$ which achieve the proof of the existence of a fixed point. □

Bibliography

[1] J.P. Delfour, M.; Zolésio. *Shapes and Geometries: analysis, differential calculus and optimization.* Springer-Verlag, 1999.

[2] I. Ekeland and R. Temam. *Analyse convexe et problèmes variationnels.* Dunod, Paris, 1974.

[3] A. Friaâ. La loi de Norton-Hoff généralisée en plasticité et viscoplasticité. Thèse de doctorat d'état, 1979.

[4] G. Geymonat and P. Suquet. Functionnal spaces for norton-hoff materials. *Math. Meth. in the Appl. Sci.*, 8:206–222, 1986.

[5] M. Giaquinta. *Introduction to Regularity Theory for Non-Linear Elliptic Systems.* Birkhauser, 1993.

[6] J. Gobert. Une inéquation fondamentale de la théorie de l'élasticité. *Bull. Soc. Roy. Sci. Liège*, 31(3-4), 1962.

[7] N. Gomez and J.P. Zolésio. Shape sensitivity and large deformation of the domain for norton-hoff flow. In G. Leugering, editor, *Proceedings of the IFIP-WG7.2 conference, Chemnitz*, volume 133 of *Int. Series of Num. Math.*, pages 167–176, 1999.

[8] F.H. Norton. *The Creep of Steel at High Temperature.* Mc Graw Hill, 1929.

[9] J. Sokolowski and J.P. Zolésio. *Introduction to Shape Optimization: Shape sensitivity analysis*, volume 16 of *Computational Mathematics.* Springer-Verlag, New York, Berlin, Heidelberg, 1992.

[10] R. Temam. A generalized Norton-Hoff model and the Prandtl-Reuss law of plasticity. *Arch. for Rat. Mech. and Anal.*, 95(2):137–183, 1986.

[11] R. Temam. *Problèmes Mathématiques en Plasticité.* Dunod Paris, 1986.

[12] J.P. Zolésio. *Notes of a Lecture on Shape Optimization given in Troia.* Cellina, ed., 1998.

[13] J.P. Zolésio. In G. Leugering, editor, *Proceedings of the IFIP-WG7.2 conference, Chemnitz*, volume 133 of *Int. Series of Num. Math.*, 1999.

Nicolas Gomez. Oakland University, School of Engineering and Computer Science, Rochester, MI 48309-4478, USA
E-mail: gomez@oakland.edu

Jean-Paul Zolésio. Research Director at CNRS, Centre de Mathématiques Appliquées, Ecole des Mines de Paris, 2004 route des Lucioles, BP. 93, 06902 Sophia Antipolis Cedex, France
E-mail: Jean-Paul.Zolesio@sophia.inria.fr

Some New Problems Occurring in Modeling of Oxygen Sensors

Jean-Pierre Yvon, Jacques Henry and Antoine Viel

Abstract. The aim of this paper is to present a new analysis of a model of oxygen sensors. They are used in motor vehicles to measure exhaust gas oxygen. A previous analysis based on electroneutrality did not completely agree with the numerical results. By introducing a new scaling taking into account relatively large applied potential this difficulty appears to be overcome. An asymptotic analysis within this new scaling is presented.

1 Presentation - Objectives

An oxygen sensor is basically a zirconia in which a flux of oxygen can be created by a difference of pressure at each end of the crystal. This flux is carried by the migration of charged oxygen vacancies in the crystal. The modeling of such a system implies the analysis of several phenomena: electromigration, chemical reactions at interfaces between zirconia and external gases, electrical exchanges with electrodes. This leads to a set of equations which are similar to those of semiconductor devices (also including the concentration of oxygen vacancies) coupled with kinetic equations at boundaries.

In some models, developed in a previous paper [5], it turns out that the theoretical analysis of asymptotic behavior under electroneutrality is not in total adequacy with numerical results. This observation has initialized a new analysis of the electrical part of the model which shows that another phenomenon, very similar to what happens for reverse-bias semiconductors, must be taken into account. Roughly speaking it is necessary to introduce another small parameter in the boundary conditions for potential to take into account the large value of the applied potential. The main scope of the paper is to analyze this question by providing some results which, in some aspects, are related to those obtained by [3] and to present the corresponding numerical results.

2 The Physical Framework

2.1 General Equations

The most commonly used oxygen sensor in the automotive industry involves a measurement of the open circuit-potential generated by the difference in the exhaust-gas oxygen partial pressure relative to a reference gas, which is usually air.

This sensor is based on an yttria-stabilized zirconia which allows a flow of ionic oxygen due to the presence of oxygen vacancies created by a suitable doping represented by the reaction:

$$2Y'_{Zr} + V_O^{\cdot\cdot} \rightleftharpoons 0$$

where $V_O^{\cdot\cdot}$ denotes the doubly positive charged oxygen vacancies (a dot denotes a positive charge and a quote a negative charge). This doping creates a concentration of fixed positive charges denoted by c_Y. Below we will consider one dimensional models in which concentrations of various species are function of time t and space x, with $0 \leq x \leq L$, where L is the width of the crystal. So, for instance, we assume that $c_Y = c_Y(x)$ and we may consider that the initial concentration of vacancies $\bar{c}_V$, created by the doping, is given.

At each boundary (i.e. at $x = 0$ and $x = L$) fixation of oxygen can be represented by the reaction:

$$\frac{1}{2}O_2 + V_O^{\cdot\cdot} \rightleftharpoons O_O^{\times} + 2h^{\cdot} \tag{2.1}$$

where $O_O^{\times}$ denotes a fixed oxygen ion in the lattice of vacancies, h denotes a hole. Actually the process of exchange between exhaust gas and sensor is far more complicated because it involves catalytic reactions between the various gases, but analysis of these phenomena is beyond the scope of this paper (we refer to [1] for a possible analysis of this question). This boundary mechanism induces a flux of several species through the sensor: ionic oxygen, vacancies, holes and electrons. As we may consider that the relevant information is the concentration of actual vacancies (i.e. sites not occupied by oxygen) we are led to consider three charged species: vacancies, holes and electrons, the concentration of which is denoted respectively by c_V, c_h, and c_e. The last variable to consider is the electrical potential ϕ which is also a function of x and t.

General equations governing electromigration are

$$\frac{\partial c_i}{\partial t} = -\nabla N_i + R_i \tag{2.2}$$

with $i = V, h, e$, the terms R_e or R_h correspond to reaction terms (for vacancies there is no such a term: $R_V = 0$) and N_i, the flux of species i, is given by

$$N_i = -D_i \nabla c_i - z_i F u_i c_i \nabla \phi \tag{2.3}$$

where D_i refers to the ionic diffusion coefficient, z_i is the charge number, F is the Faraday constant, u_i is the mobility. The potential is governed by the classical Poisson's equation:

$$-\varepsilon_0 \Delta \phi = F \sum_{i=e,h,V,Y} z_i c_i. \tag{2.4}$$

By analogy with the semiconductor framework, the reaction term in (2.2) corresponds to a generation-recombination term — for holes and electrons — and can be represented by

$$R_e = -R_h = k_0(n_i^2 - c_h c_e).$$

where n_i is the intrinsic carrier concentration. In the numerical simulations we will not consider such a term which does not seem to play a substantial role in the results.

2.2 Boundary Conditions

The main purpose of this paper is to focus on the boundary layers which appear if electroneutrality is nearly satisfied everywhere except at the boundaries. This singular behavior of vacancy concentration has been pointed out by several authors, see for instance [7].

Unfortunately the analysis of boundary conditions is very hard and the literature on this subject is far from coherent. In this paper we will analyze a possible formulation of the problem and its consequences, both physical and mathematical.

The first equation, which is classical in the relevant literature, is obtained by applying the mass-action law to fixation of oxygen at boundaries, represented by reaction (2.1):

$$c_h^2(\bar{c}_V - c_V) - k_e P_{O_2}^{1/2} = 0 \tag{2.5}$$

where $\bar{c}_V$ in the initial concentration of vacancies (as a result of the doping) and k_e is a suitable equilibrium constant.

A second relation, corresponding to thermodynamical equilibrium, which is also a classical assumption, is

$$c_e c_h = n_i^2 \tag{2.6}$$

Boundary conditions for potential can be taken as, for instance, Dirichlet conditions:

$$\phi(0) = \phi_0, \quad \phi(L) = \phi_1 \tag{2.7}$$

A third condition at the boundaries is required in order to have a properly posed problem (both from a mathematical and physical standpoint). This last relation must convey the nature of the electrical exchanges between the zirconia and the electrode. One possibility is to consider that the electron concentration is the same in the zirconia and in the platinum electrode (see [1]). An other possibility is to copy the continuity equation used in the semiconductor framework and to write

$$2c_V + c_h - c_e = 0 \tag{2.8}$$

which means that we assume continuity of the total concentration of *mobile* charges through the interface. This is the situation explored below.

It must be noticed that the set of boundary conditions (2.5)(2.6) and (2.8) determines uniquely the values of c_V, c_h, and c_e at $x = 0, L$.

3 A Short Analysis of the Model

3.1 Analysis of the Numerical Results

From now on we will restrict ourselves to the stationary case. After a classical scaling, a small parameter ε appears in equation (2.4) at the same position as ε_0. The main point concerns the behavior of the stationary system (2.2), (2.3), (2.4) when ε goes to 0. The numerical simulations show that some unexpected phenomena occur. The first one concerns the shape of the boundary layers for vacancy concentration. Under the given boundary conditions there is a dissymmetry in the two boundary layers and a change of concavity occurs at $x = 0$.

A second point concerns the limit behavior under electroneutrality. Actually if we follow the approach developed in [4] it is possible to determine the limit boundary conditions obtained under electroneutrality (i.e. when $\varepsilon \to 0$). For instance, if we consider the boundary conditions at $x = 0$, we introduce the real number $d_0 > 0$ solution of

$$2d^2 c_v(0) + dc_h(0) - d^{-1}c_e(0) - c_Y = 0$$

the boundary conditions for the reduced problem will be

$$\bar{c}_e(0) = d_0^{-1}c_e(0),\ \bar{c}_h(0) = d_0 c_h(0),\ \bar{c}_V(0) = d_0^2 c_V(0), \bar{\phi}(0) = \phi - \frac{RT}{F}\log d_0. \tag{3.1}$$

with analogous definitions in $x = 1$. As a matter of fact it turns out that the numerical extrapolated boundary values for the various concentrations do not coincide with the predicted values given by (3.1).

This question is closely related with other observations made by several authors concerning the scaling in semiconductor equations. In order to concentrate on this specific problem we may consider a simpler case in which there is only one species.

3.2 A Simplified Case

From now on we will consider a simpler situation which still conveys the phenomenon we want to analyze. Actually we will check in the numerical simulations that the behavior of solutions for the simplified problem will be very similar to the one of the original problem.

For this purpose we assume that there is only one species and that, after scaling, the values of various constants may be set at 1, so that we may consider that the only small parameter is ε. Then, the concentration of the sole mobile species being denoted by $u(x)$, we may consider that the original system has been reduced to:

$$\left\{ \begin{array}{lcl} \varepsilon\phi'' & = & 1-u \\ (u' + u\phi')' & = & 0 \end{array} \right. \tag{3.2}$$

where the prime denotes the spatial derivation, with boundary conditions

$$\left\{ \begin{array}{lcl} \phi(0) & = & \phi_0 \\ u(0) & = & q_0 \end{array} \right., \quad \left\{ \begin{array}{lcl} \phi(1) & = & 0 \\ u(0) & = & q_1 \end{array} \right. \tag{3.3}$$

If ϕ_0 is of order 1 and ε is small enough, this leads asymptotically to the electroneutral model. In this situation we have $u(x) = 1$, except for boundary layers with magnitude of order $O(\sqrt{\varepsilon})$

at $x = 0$ and $x = 1$ (if $q_i \neq 1$, $i = 0, 1$). These boundary layers correspond to some zones in which the density of charges does not vanishes and is in $O(1)$. In these regions the potential should not change of concavity, which is in contradiction to the obtained numerical results.

Then an idea is to consider that the applied potential ϕ_0 is "large" and depends on ε in the following way

$$\phi_0 = \frac{V}{\varepsilon^\alpha} \tag{3.4}$$

where α is given. If we make the change of variable:

$$\phi = \frac{v}{\varepsilon^\alpha}$$

system (3.2) becomes

$$\left\{ \begin{array}{rcl} \varepsilon^{1-\alpha} v'' & = & 1 - u \\ (\varepsilon^\alpha u' + uv')' & = & 0 \end{array} \right. \tag{3.5}$$

with

$$\left\{ \begin{array}{rcl} v(0) & = & V \\ u(0) & = & q_0 \end{array} \right. , \quad \left\{ \begin{array}{rcl} v(1) & = & 0 \\ u(0) & = & q_1 \end{array} \right. \tag{3.6}$$

If $\alpha = 1$ this does not lead necessarily to an electroneutral situation and it may appear some charged regions in $O(1)$. Furthermore if we assume that $q_0 = 0$ this corresponds exactly to the situation analyzed in [3] where the existence of a depletion zone ($u = 0$) (which does not correspond to what is known in physics of such sensors) is shown. This leads to consider situations in which $0 < \alpha < 1$. In this case one gets boundary layers of order $O(\varepsilon^{\frac{1-\alpha}{2}})$.

The specific case $\alpha = \frac{1}{2}$ corresponds to a least degeneration because we get two terms in $O(\sqrt{\varepsilon})$. The second argument in favor of this choice of α is the fact that we may observe numerically a kind of depletion zone of order $O(\varepsilon^{1/4})$ which is coherent, as we will see, with the analysis developed in [3]. This situation is studied below.

4 Analysis of the Simplified Model

4.1 Formulation of the Problem

If we follow the presentation of the previous section, we are led to introduce the "small" parameter:

$$\lambda = \sqrt{\varepsilon}$$

in such a way that system (3.5) can be written as

$$\lambda u''(x) + (u(x)v'(x))' = 0, \tag{4.1}$$

$$\lambda v''(x) = 1 - u(x), \tag{4.2}$$

with the boundary conditions:

$$u(0) = q_0,\ u(1) = q_1,\ v(0) = 1,\ v(1) = 0, \tag{4.3}$$

with $q_i \geq 0$. In what follows we will assume, for simplicity, that $q_0 \leq q_1$ and $q_1 \neq 0$. The first step consists in showing that this system admits a solution at least.

THEOREM 4.1. *For all $\lambda > 0$, the system (4.1) (4.2) and (4.3) has solution (u_λ, v_λ) such that*

$$q_0 \le u_\lambda(x) \le 1, \;\; x \in [0,1] \tag{4.4}$$

and such that there exists $\xi_\lambda \in [0,1]$ such that u_λ is increasing in $[0,\xi_\lambda]$ and decreasing in $[\xi_\lambda, 1]$ (i.e. u is unimodal). In the case where $q_1 = 1$, the function u_λ is monotone increasing on the whole interval $[0,1]$.

Proof.— Existence can be proven by a fixed point argument as in [2]. The fact that, for $x \in [0,1]$, we have $0 \le u(x) \le 1$ is a part of the proof. The last part of the theorem results from the fact that any solution u cannot have a minimum in the open interval $]0,1[$. First of all it is useful to express the solution of (4.1) with respect to v:

$$u(x) = e^{-\frac{v(x)}{\lambda}} \left[q_0 e^{1/\lambda} + (q_1 - q_0 e^{1/\lambda}) \frac{\psi(x)}{\psi(1)} \right] \tag{4.5}$$

where ψ is defined by

$$\psi(x) = \int_0^x e^{\frac{v(t)}{\lambda}}\, dt \tag{4.6}$$

Equation (4.5) can be written as

$$u(x) = e^{-\frac{v(x)}{\lambda}} \left[q_0 \left(1 - \frac{\psi(x)}{\psi(1)}\right) e^{1/\lambda} + q_1 \frac{\psi(x)}{\psi(1)} \right]$$

which shows that u cannot vanish in the open interval $]0,1[$ (even if $q_0 = 0$, q_1 being strictly positive, by assumption).

Now if u is minimum in a point $a \in]0,1[$ we have:

$$u'(a) = 0, \;\; u''(a) \ge 0.$$

But equations (4.1)(4.2) give

$$\lambda u''(a) + u(a)v''(a) = \lambda u''(a) + \frac{1}{\lambda} u(a)(1 - u(a)) = 0$$

which implies that $u(a) = 0$ (the other possibility $u(a) = 1$ being absurd as a corresponds to a minimum) which is impossible as we have seen before. It follows that u must have only one maximum in $[0,1]$. It is then easy to prove that u is increasing in the vicinity of $x = 0$.

4.2 Asymptotic Analysis: Study of u_λ

This section is devoted to the study of the behavior of (u_λ, v_λ) as λ goes to zero. A first result concerns u_λ, this result being independent of the fact that $q_0 = 0$ or not.

THEOREM 4.2. *For all $x \in]0,1[$ we have:*

$$\lim_{\lambda \to 0} u_\lambda(x) = 1. \tag{4.7}$$

Proof.— In order to present a simpler proof we will assume that $q_1 = 1$ which implies that u_λ is monotone increasing (a complete proof is presented in [6]).

The proof proceeds by contradiction by assuming that there exists $a \in]0,1[$ such that $u_\lambda(a)$ does not converge to 1. By extracting a suitable subsequence, still indexed by λ, we may assume that for all λ (in a subset of $\mathbb{R}_+$)

$$u_\lambda(a) \le 1 - \alpha$$

with $\alpha > 0$. The point is to get a contradiction.

First of all, as u_λ is not decreasing, we have in fact:

$$\forall x \in [0,a]\,,\ u_\lambda(x) \le 1 - \alpha$$

so that we can state that

$$\begin{cases} \alpha \le \lambda v_\lambda''(x) \le 1 & \text{for} \quad x \in [0,a] \\ 0 \le \lambda v_\lambda''(x) \le 1 & \text{for} \quad x \in]a,1] \end{cases}$$

Thus we get the following bounds for v_λ:

$$\underline{\varphi}_\lambda \le v_\lambda \le \overline{\varphi}_\lambda$$

where $\underline{\varphi}_\lambda$ is the solution of

$$\begin{cases} \lambda\varphi''(x) = 1, \ x \in [0,1] \\ \varphi(0) = 1, \varphi(1) = 0 \end{cases}$$

and $\overline{\varphi}_\lambda$ is the solution of

$$\begin{cases} \lambda\varphi''(x) = \begin{cases} \alpha & \text{for} \quad x \in [0,a] \\ 0 & \text{for} \quad x \in]a,1] \end{cases} \\ \varphi(0) = 1, \varphi(1) = 0 \end{cases}$$

Some simple calculations show that

$$\underline{\varphi}_\lambda(x) = \frac{1}{2\lambda}(x-1)(x-2\lambda) \tag{4.8}$$

and

$$\overline{\varphi}_\lambda(x) = \begin{cases} \frac{\alpha}{2\lambda}x^2 + \beta x + 1 & \text{for} \quad x \in [0,a] \\ p(x-1) & \text{for} \quad x \in]a,1] \end{cases}$$

with

$$\beta = \frac{\alpha a}{2\lambda}(a-2) - 1, \ p = \frac{\alpha a^2}{2\lambda} - 1.$$

We will show that, in these conditions, $u_\lambda(a)$ is not bounded when λ tends to 0, which is clearly impossible. The expression of $u_\lambda(a)$ is given by (4.5):

$$u_\lambda(a) = e^{-\frac{v_\lambda(a)}{\lambda}} \left\{ q_0 e^{1/\lambda} + (1 - q_0 e^{1/\lambda}) \frac{\psi_\lambda(a)}{\psi_\lambda(1)} \right\} \tag{4.9}$$

with ψ_λ still given by (4.6), where v is replaced by v_λ. First of all one has

$$e^{-\frac{v_\lambda(a)}{\lambda}} \ge e^{-\frac{\overline{\varphi}_\lambda(a)}{\lambda}}$$

with

$$\overline{\varphi}_\lambda(a) = p(a-1) = -(\frac{\alpha a^2}{2\lambda} - 1)(1-a).$$

Then, as $0 < a < 1$, we get

$$-\frac{\overline{\varphi}_\lambda(a)}{\lambda} \sim (1-a)\frac{\alpha a^2}{2\lambda^2} \to +\infty \text{ when } \lambda \to 0$$

so that

$$e^{-\frac{v_\lambda(a)}{\lambda}} \to +\infty, \text{ when } \lambda \to 0. \tag{4.10}$$

It remains to study the behavior of the second factor of (4.9) and, for this purpose let z_λ be the unique point such that

$$v_\lambda(z_\lambda) = 0, \ 2\lambda \leq z_\lambda \leq \xi_\lambda$$

where ξ_λ is the smallest zero of $\overline{\varphi}_\lambda$. Then we can write

$$\psi_\lambda(a) = \int_0^{z_\lambda} e^{\frac{v_\lambda(x)}{\lambda}} dx + \int_{z_\lambda}^{a} e^{\frac{v_\lambda(x)}{\lambda}} dx$$

but, as $v_\lambda \leq 0$ on the interval $[z_\lambda, a]$, we get

$$0 \leq \int_{z_\lambda}^{a} e^{\frac{v_\lambda(x)}{\lambda}} dx \leq a - z_\lambda \leq a.$$

In a similar manner we may write

$$\psi_\lambda(1) = \int_0^{z_\lambda} e^{\frac{v_\lambda(x)}{\lambda}} dx + \int_{z_\lambda}^{1} e^{\frac{v_\lambda(x)}{\lambda}} dx$$

so that we obtain the following expression:

$$\frac{\psi_\lambda(a)}{\psi_\lambda(1)} = \frac{\int_0^{z_\lambda} e^{\frac{v_\lambda(x)}{\lambda}} dx + A_\lambda}{\int_0^{z_\lambda} e^{\frac{v_\lambda(x)}{\lambda}} dx + B_\lambda} \tag{4.11}$$

with A_λ and B_λ bounded. One has

$$\int_0^{z_\lambda} e^{\frac{v_\lambda(x)}{\lambda}} dx \geq \int_0^{z_\lambda} e^{\frac{\varphi_\lambda(x)}{\lambda}} dx \geq \int_0^{2\lambda} e^{\frac{\varphi_\lambda(x)}{\lambda}} dx$$

and

$$\begin{aligned} \int_0^{2\lambda} e^{\frac{\varphi_\lambda(x)}{\lambda}} dx &= \int_0^{2\lambda} e^{\frac{1}{2\lambda^2}(1-x)(2\lambda-x)} dx \\ &\geq \int_0^{2\lambda} e^{\frac{1}{2\lambda^2}(1-2\lambda)(2\lambda-x)} dx \\ &= \frac{2\lambda^2}{1-2\lambda}\left\{e^{\frac{1-2\lambda}{\lambda}} - 1\right\} \end{aligned}$$

so that, when $\lambda \to 0$, we get

$$\int_0^{2\lambda} e^{\frac{\varphi_\lambda(x)}{\lambda}} dx \longrightarrow +\infty.$$

From the previous results and (4.11) it follows that

$$\frac{\psi_\lambda(a)}{\psi_\lambda(1)} \to 1 \text{ when } \lambda \to 0.$$

Then we can deduce that there exists a constant $\gamma > 0$ such that

$$qe^{1/\lambda} + (1 - qe^{1/\lambda})\frac{\psi_\lambda(a)}{\psi_\lambda(1)} = qe^{1/\lambda}\left\{1 - \frac{\psi_\lambda(a)}{\psi_\lambda(1)}\right\} + \frac{\psi_\lambda(a)}{\psi_\lambda(1)} \geq \gamma,$$

as the left hand side member is the sum of a non negative term and of a term which converges to 1. This, combined with (4.10), shows that $u_\lambda(a)$ cannot remain bounded as λ goes to zero.

4.3 Asymptotic Analysis: Study of v_λ

The behavior of v_λ depends on the fact that $q_0 = 0$ or not. In order to simplify the proofs we will assume, from now, that

$$q_1 = 1 \tag{4.12}$$

which implies that u_λ is monotone non decreasing in $[0, 1]$.

THEOREM 4.3. *If $q_0 = 0$ then, for all $x \in]0, 1]$:*

$$\lim_{\lambda \to 0} v_\lambda(x) = 0. \tag{4.13}$$

Proof.— Let us consider the negation of (4.13) by assuming that there exists $a \in]0, 1[$ such that

$$v_\lambda(a) \not\to 0 \text{ when } \lambda \to 0.$$

So we have the alternative:

(i) there exist a subsequence λ_n and a real $\beta > 0$ such that

$$v_{\lambda_n}(a) \geq \beta > 0.$$

Then we get immediately

$$|u_{\lambda_n}(a)| \leq \left| \frac{e^{-\frac{\beta}{\lambda_n}} \psi_{\lambda_n}(a)}{\psi_{\lambda_n}(1)} \right| \leq e^{-\frac{\beta}{\lambda_n}}$$

which vanishes when $\lambda \to 0$ and is a contradiction to the result of theorem 4.2.

(ii) there exist a subsequence λ_n and a real $\beta > 0$

$$v_{\lambda_n}(a) \leq -\beta < 0.$$

As in the proof of theorem 4.2 we may bound below v_{λ_n} by the function $\underline{\varphi}$ given by (4.8) and, by convexity, bound above by $\overline{\varphi}$ which is the function affine by piece, defined by:

$$\overline{\varphi}(x) = \begin{cases} -\frac{\beta+1}{a}x + 1 & \text{for} \quad 0 \leq x < a \\ \frac{\beta}{1-a}(x-1) & \text{for} \quad a \leq x \leq 1 \end{cases}$$

Then, if we drop the index n (for clarity), we have

$$u_\lambda(a) \geq \frac{e^{\frac{\beta}{\lambda}}\psi_\lambda(a)}{\psi_\lambda(1)}$$

with ψ_λ still defined by (4.6). So it is quite clear that, if

$$\frac{\psi_\lambda(a)}{\psi_\lambda(1)} \to 1 \text{ when } \lambda \to 0, \tag{4.14}$$

we will get a contradiction, because in that case we would have $u_\lambda(a)$ unbounded when λ goes to 0. Let us introduce z_λ such that $v_\lambda(z_\lambda) = 0$. The previous double inequality verified by v_λ implies that

$$2\lambda \leq z_\lambda \leq \frac{a}{1+\beta} < a.$$

By splitting the integral

$$\int_0^a e^{\frac{v_\lambda(x)}{\lambda}}\,dx = \int_0^{z_\lambda} e^{\frac{v_\lambda(x)}{\lambda}}\,dx + \int_{z_\lambda}^a e^{\frac{v_\lambda(x)}{\lambda}}\,dx$$

and, similarly

$$\int_0^1 e^{\frac{v_\lambda(x)}{\lambda}}\,dx = \int_0^{z_\lambda} e^{\frac{v_\lambda(x)}{\lambda}}\,dx + \int_{z_\lambda}^1 e^{\frac{v_\lambda(x)}{\lambda}}\,dx,$$

the proof can be completed in a way similar to theorem 4.2 by using the function $\underline{\varphi}$ which bounds below v_λ.

The case where $q_0 > 0$ is quite different as it is shown in the numerical results. In fact it is possible to show the following result:

THEOREM 4.4. *If $q_0 > 0$ and $q_1 = 1$ then, (u_λ, v_λ) solution of (4.1),(4.2) and (4.3), satisfies, for $x \in]0,1]$:*

$$\lim_{\lambda\to 0} u_\lambda(x) = 1, \tag{4.15}$$

$$\lim_{\lambda\to 0} v_\lambda(x) = 1 - x. \tag{4.16}$$

The first result consists in showing that v_λ is bounded.

PROPOSITION 4.5. *For λ small enough and $x \in [0,1]$ we have:*

$$v'_\lambda(x) \leq 0$$

and

$$0 \leq v_\lambda(x) \leq 1 - x \tag{4.17}$$

Proof.— Integration of equation (4.1) shows that there exists a constant α_λ such that

$$\lambda u_\lambda'(x) + u_\lambda(x) v_\lambda'(x) = \alpha_\lambda. \tag{4.18}$$

The expression of u_λ becomes

$$u_\lambda(x) = e^{-\frac{v_\lambda(x)}{\lambda}} \left[q e^{\frac{1}{\lambda}} + \frac{\alpha_\lambda}{\lambda} \psi_\lambda(x) \right]$$

In particular, one has

$$\frac{\alpha_\lambda}{\lambda} \psi_\lambda(1) = 1 - q e^{\frac{1}{\lambda}} \tag{4.19}$$

which shows that, for λ small enough,

$$\alpha_\lambda < 0. \tag{4.20}$$

As u_λ is not decreasing, we have $u_\lambda'(x) \geq 0$ and (4.18) implies $v_\lambda'(x) \leq 0$, for $x \in [0,1]$. The double inequality (4.17) is then obtained by using the convexity of v_λ.

PROPOSITION 4.6. *For any $a \in]0,1[$ there exists a constant K such that, for λ small enough:*

$$|v_\lambda|_{H^1[a,1]} \leq K.$$

Proof.— The proof is very similar to the one given in [4] and is based on the fact that we can, partially, eliminate u in the system — we drop the index λ for legibility — in such a way that v is solution of

$$\lambda^2 v^{(4)}(x) - (u(x) v'(x))' = 0$$

Let φ be a smooth function such that $\varphi(x) = 0$ in a neighborhood of 0 and $\varphi(x) = 1$ for $x \geq a$. By multiplying the previous equation by φ we get

$$\lambda^2 (v\varphi)^{(4)} - (u(v\varphi)')' = 2\lambda^2 \varphi' v''' + \lambda^2 \varphi'' v'' + 2\lambda^2 (\varphi' v')'' + \lambda^2 (v\varphi'')'' - uv'\varphi' - (uv\varphi')'.$$

Multiplication of this equation by $v\varphi$ and integration on $[0,1]$ lead to, after some tedious calculations

$$\begin{aligned}
\lambda^2 \int_0^1 |(v\varphi)''|^2 \, dx + \int_0^1 u|(v\varphi)'|^2 \, dx = \\
-2\lambda \int_0^1 (1-u)[(v\varphi)'\varphi' + v\varphi\varphi''] \, dx + \lambda \int_0^1 (1-u)\varphi'' v\varphi \, dx \\
-2\lambda^2 \int_0^1 [(v\varphi')' - v\varphi''](v\varphi)' \, dx + \lambda^2 \int_0^1 v\varphi''(v\varphi)'' \, dx \\
- \int_0^1 u[(v\varphi)' v\varphi' - v^2 \varphi'^2] \, dx + \lambda \int_0^1 uv\varphi'(v\varphi)' \, dx,
\end{aligned}$$

this expression resulting from some integrations by parts which use extensively the properties of φ and v at the boundaries.

The fact that $q_0 \le u(x) \le 1$ and that v_λ is bounded in L^∞ allow to bound above the various terms of this expression in order to get an inequality of the form

$$\lambda^2 \int_0^1 |(v\varphi)''|^2\,dx + \int_0^1 u(x)|(v\varphi)'|^2\,dx \le K\lambda|(v\varphi)'|_{L^2} + K\lambda^2|(v\varphi)''|_{L^2} + K, \tag{4.21}$$

which completes the proof.

From the previous estimations, we can extract subsequences, still denoted u_λ and v_λ, such that, when $\lambda \to 0$:

$$u_\lambda \to 1, \text{ uniformly in } [a,1]\ (a>0, \text{ arbitrary }) \tag{4.22}$$

$$v_\lambda \to \overline{v}, \text{ in } H^1(a,1) \text{ weakly and a.e.} \tag{4.23}$$

PROPOSITION 4.7. *The limit function $\overline{v}$ defined by (4.23) has the form*

$$\overline{v}(x) = \overline{\alpha}x + \overline{\beta} \tag{4.24}$$

with

$$\overline{\alpha} + \overline{\beta} = 0, \ 0 \le \overline{\beta} \le 1$$

Proof. — Let $a > 0$ be fixed, integration of (4.18) gives

$$\lambda u_\lambda(1) - \lambda u_\lambda(a) + \int_a^1 u_\lambda(x)v_\lambda'(x)\,dx = (1-a)\alpha_\lambda.$$

Taking into account the previous estimation it is straightforward to see that α_λ is convergent in $\mathbb{R}$ and, if $\overline{\alpha}$ denotes the limit, we get

$$\int_a^1 \overline{v}'(x)dx = (1-a)\overline{\alpha}.$$

As $a > 0$ is arbitrary it follows immediately that

$$\overline{v}'(x) = \overline{\alpha}\ ,\ \ \overline{v}(x) = \overline{\alpha}x + \overline{\beta}. \tag{4.25}$$

At last, the weak convergence in $H^1(a,1)$ implies

$$\lim_{\lambda\to 0} v_\lambda(1) = \overline{v}(1)$$

so that $\overline{\alpha} + \overline{\beta} = 0$ with $\overline{\alpha} \le 0$ according to (4.20).

PROPOSITION 4.8. *There exists a sequence of real numbers (β_λ) such that*

$$\lim_{\lambda\to 0} \beta_\lambda = \overline{\beta}$$

where $\overline{\beta}$ is defined by (4.24) and we have

$$v_\lambda'(0) = -\sqrt{\frac{2(1-\beta_\lambda)}{\lambda} + 2q_0} \tag{4.26}$$

Proof.— Starting from the equation (4.1) we have

$$\lambda u_\lambda''(x) = (u_\lambda(x)v_\lambda'(x))' = \lambda u_\lambda''(x) + [(1-\lambda v_\lambda''(x))v_\lambda'(x)]' = 0$$

which can be written as

$$\lambda u_\lambda''(x) + v_\lambda''(x) - \frac{\lambda}{2}\frac{d^2}{dx^2}|v_\lambda'(x)|^2 = 0.$$

This implies existence of real α_λ and β_λ such that

$$\lambda u_\lambda(x) + v_\lambda(x) - \frac{\lambda}{2}|v_\lambda'(x)|^2 = \alpha_\lambda x + \beta_\lambda. \tag{4.27}$$

This relation written in $x = 0$ establishes the result (4.26). Then we remark that the sequence (α_λ) is the same as the one introduced in (4.18). For any $a > 0$ we have

$$\lambda\int_a^1 u_\lambda(x)\,dx + \int_a^1 v_\lambda(x)\,dx - \frac{\lambda}{2}\int_a^1 |v_\lambda'(x)|^2\,dx = \int_a^1 (\alpha_\lambda x + \beta_\lambda)\,dx \tag{4.28}$$

in which we can take the limit when λ tends to 0. For the left hand side member of (4.28) this is a consequence of the previous results. As the sequence (α_λ) converges, it also true for the sequence (β_λ), the limit of which being denoted by $\widetilde{\beta}$. This leads to

$$\int_a^1 \overline{v}(x)\,dx = \int_a^1 (\overline{\alpha}x + \widetilde{\beta})\,dx$$

which shows that $\widetilde{\beta} = \overline{\beta}$.

Proof of the theorem 4.4

We consider the subsequence defined in (4.23) which, according to (4.24), has the limit:

$$\overline{v}(x) = \overline{\alpha}x + \overline{\beta}$$

with $0 \le \overline{\beta} \le 1$. Now the sole result to prove is that $\overline{\beta} = 1$.

For this purpose we proceed by contradiction and we assume that $0 \le \overline{\beta} < 1$, which implies that

$$\overline{\alpha} \le 0 \tag{4.29}$$

and that

$$\theta(\lambda) := \sqrt{\frac{2(1-\beta_\lambda)}{\lambda} + 2q} \to +\infty \tag{4.30}$$

when $\lambda \to 0$. The sought contradiction will result from two different estimations of $\psi_\lambda(1)$. First of all, according to (4.19), we have

$$1 = \frac{\alpha_\lambda}{\lambda}\psi_\lambda(1) + qe^{\frac{1}{\lambda}},$$

hence

$$\psi_\lambda(1) = \frac{\lambda}{\alpha_\lambda}\left(1 - qe^{\frac{1}{\lambda}}\right). \tag{4.31}$$

On the other hand we will get a bound for $\psi_\lambda(1)$ by using the Taylor expansion of v_λ in a neighborhood of 0:

$$v_\lambda(x) = 1 - \theta(\lambda)x + \frac{1-q}{2\lambda}x^2 + \frac{x^3}{6}v_\lambda'''(\xi).$$

But, as $\lambda v_\lambda'''(x) = -u_\lambda'(x) \leq 0$, we have for all $x \in [0,1]$:

$$v_\lambda(x) \leq 1 - \theta(\lambda)x + \frac{1-q}{2\lambda}x^2 \tag{4.32}$$

We start from

$$\psi_\lambda(1) = \int_0^1 e^{\frac{v(x)}{\lambda}}\,dx = \int_0^\lambda e^{\frac{v(x)}{\lambda}}\,dx + \int_\lambda^1 e^{\frac{v(x)}{\lambda}}\,dx.$$

On the interval $[0,\lambda]$ we can, using (4.32), establish the following inequalities :

$$\int_0^\lambda e^{\frac{v(x)}{\lambda}}\,dx \leq \int_0^\lambda e^{\frac{1}{\lambda}\left[1-\theta(\lambda)x+\frac{1-q}{2\lambda}x^2\right]}\,dx \leq \frac{\lambda}{\theta(\lambda)}e^{\frac{1}{\lambda}}e^{\frac{1-q}{2}}\left(1-e^{-\theta(\lambda)}\right).$$

On the interval $[\lambda,1]$, by using a convexity argument, it is possible to bound above v_λ by the affine function:

$$-\frac{r(\lambda)}{1-\lambda}(1-x), \quad \text{with} \quad r(\lambda) = 1 - \lambda\theta(\lambda) + \frac{1-q}{2}\lambda$$

(we remark that, taking into account the expression of $\theta(\lambda)$ given by (4.30), $r(\lambda) \to 1$ when $\lambda \to 0$). Thus we have

$$\int_\lambda^1 e^{\frac{v_\lambda(x)}{\lambda}}\,dx \leq \int_\lambda^1 e^{-\frac{r(\lambda)}{\lambda(1-\lambda)}(1-x)}\,dx = \frac{\lambda(1-\lambda)}{r(\lambda)}\left[e^{\frac{1}{\lambda}}e^{\frac{1-q}{2}}e^{-\frac{\theta(\lambda)}{\lambda}} - 1\right]$$

All preceding calculations can be summarized by

$$\psi_\lambda(1) \leq \lambda e^{\frac{1}{\lambda}}R(\lambda)$$

where

$$R(\lambda) = \frac{e^{\frac{1-q}{2}}}{\theta(\lambda)}\left(1-e^{-\theta(\lambda)}\right) + \frac{1-\lambda}{r(\lambda)}\left(e^{\frac{1-q}{2}}e^{-\frac{\theta(\lambda)}{\lambda}} - e^{-\frac{1}{\lambda}}\right).$$

As a consequence of (4.30), $R(\lambda) \to 0$ when $\lambda \to 0$. The relation (4.31) and the previous results lead to

$$-\frac{1}{\alpha_\lambda}\left(q - e^{-\frac{1}{\lambda}}\right) \leq R(\lambda) \tag{4.33}$$

which is the contradiction we are seeking because, according to (4.29), the left hand side member of (4.33) tends to $\frac{q}{|\alpha|} > 0$, while the right hand side member of (4.33) tends to 0.

4.4 Numerical Experiments

System (4.1), (4.2) with boundary conditions

$$u(0) = q_0, \; u(1) = 1, \; v(0) = 1, \; v(1) = 0, \tag{4.34}$$

Fig. 1 -- Concentration u [boundary conditions u(0)=0 u(1)=1]

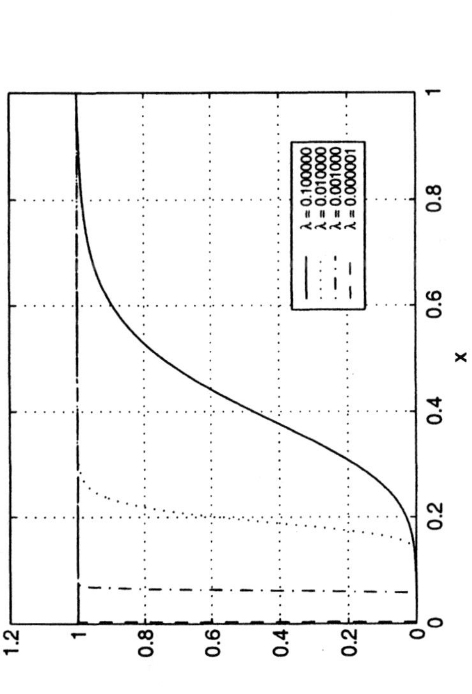

Fig.2 -- Potential v [boundary conditions v(0)=1 v(1)=0]

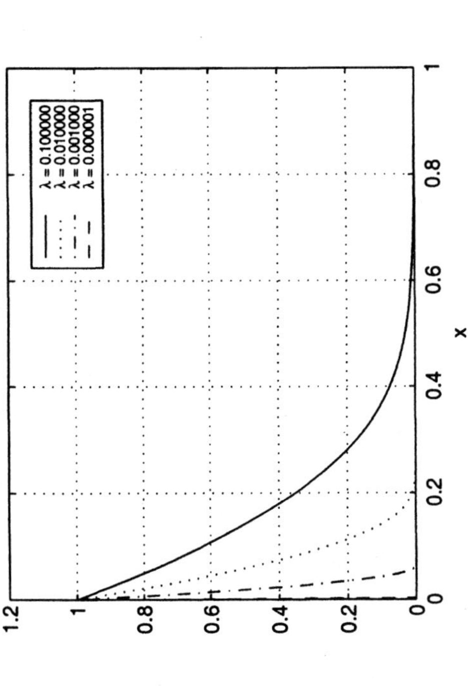

Fig. 3 -- Concentration u [boundary conditions u(0)=0.01 u(1)=1]

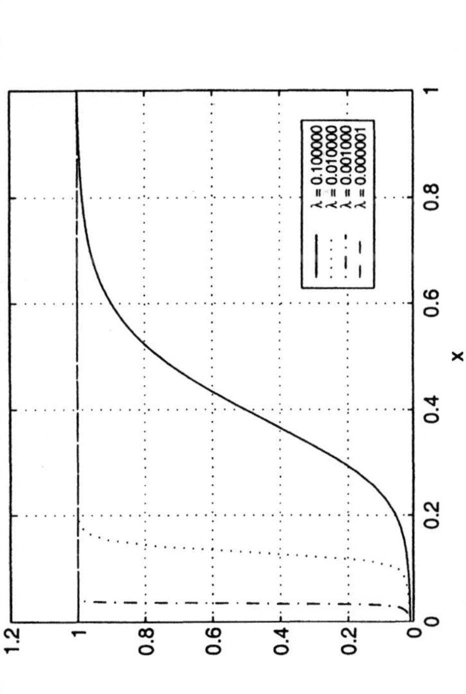

Fig.4 -- Potential v [boundary conditions v(0)=1 v(1)=0]

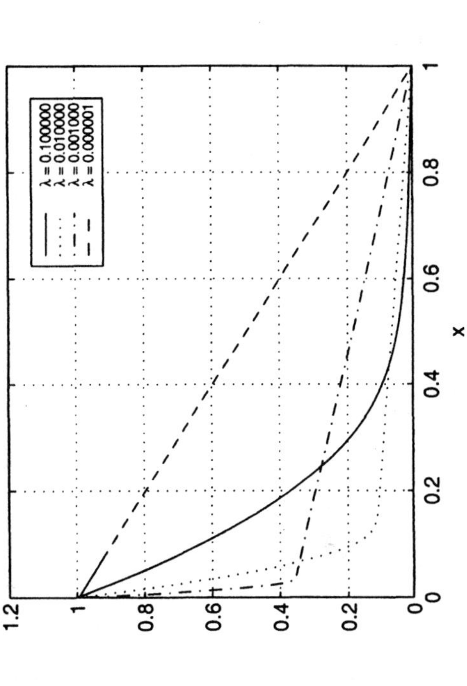

has been solved numerically for various values of λ. Figure 1 and 2, which correspond to $q_0 = 0$, show the convergence of u_λ and v_λ. The results corresponding to $q_0 = 0.01$ are shown on the figures 3 and 4.
The numerical experiments confirm a drastic difference of behavior as λ goes to zero between the cases $q_0 = 0$ and $q_0 > 0$. The main difference between the two cases lies in the behavior of the potential: in the second case, in spite of the smallness of q_0, the asymptotic value of v_λ, $(1 - x)$, is completely different from the value obtained when $q_0 = 0$ ($v_\lambda \to 0$). Actually, some further numerical results – not presented here – show that the speed of convergence decreases as q_0 goes to zero, which is coherent with the discontinuity of the limit at $q_0 = 0$.

Bibliography

[1] BRAILSFORD A.D., YUSSOUF M., LOGOTHETIS E.M., *Theory of gas sensors*, Sensors and Actuators B., 13-14, pp. 135-138 (1993)

[2] BREZZI F., CAPELO A., MARINI L.D., *Singular perturbation problems in semiconductor devices*, in Proc. II MAS Workshop on numerical analysis, J.P. Hennart Ed., Lect. Notes in Math. 1230, pp. 191-198 (1986).

[3] BREZZI F., CAPELO A., GASTALDI L., *A Singular Perturbation Analysis of Reversed-biased Semiconductor Diodes*, SIAM J. Math Anal, vol 20, No 2, pp. 372-38 (1989).

[4] HENRY J., LOURO B., *Asymptotic Analysis of reaction-diffusion-electromigration systems*, Asymptotic Analysis 10, pp. 1-24 (1995)

[5] HENRY J., VIEL A., YVON J.P., *Modelling of Oxygen Sensors*, IFIP TC-7 Conf. Detroit 1997.

[6] VIEL A., Ph. D Thesis University of Technology of Compigne (1999).

[7] WANG T., SOLTIS R.E., LOGOTHETIS E.M. *et al.*, *Static Characteristics of ZrO_2 Exhaust Gas Oxygen Sensors*, SAE Technical paper series, paper 930352, Detroit Michigan (1993)

Jean-Pierre Yvon. INSA-Rennes, CS 14315, 35043 Rennes Cedex, France
E-mail: Jean-Pierre.Yvon@insa-rennes.fr

Jacques Henry. INRIA, B.P. 105, 78153 Le Chesnay Cedex, France
E-mail: Jacques.Henry@inria.fr

Antoine Viel. UTC, Département GI, BP 649, 60206 Compiegne Cedex, France
E-mail: Antoine.Viel@utc.fr

Adaptive Control of a Wake Flow Using Proper Orthogonal Decomposition[1]

Konstantin Afanasiev and Michael Hinze

Abstract. We present an effective quasi-optimal control method for the instationary Navier-Stokes equations. It in an adaptive manner utilizes the snapshot form of the proper orthogonal decomposition to construct a low dimensional approximation of the Navier-Stokes equations which then is used as subsidiary condition in the underlying optimization problem. The numerical solution of the reduced control problem is used as control for the full equations and to compute new snapshots. The set of the new snapshots then is added to the set of the snapshots already available, the low dimensional approximation of the Navier-Stokes equations is re-computed and the reduced optimization is restarted. This process is iterated until convergence is achieved. As numerical example we present tracking-type control of the periodic flow around a cylinder. The numerical results of the quasi-optimal approach are compared to that of the optimal approach.

1 Introduction

In many applications it can be useful to control fluid flow and to optimize its characteristics. Typical applications include the reduction of drag and the lag or elimination of the transition from laminar to turbulent flow regime. The solutions of the underlying optimal control problems satisfy a system of coupled nonlinear partial differential equations involving the time-dependent Navier-Stokes equations. Solving these systems for realistic flows numerically is unworkable with most currently available computing facilities. Hence there is a need for suboptimal control strategies which incorporate the nonlinear character of the flow and which are amenable to controlling the flow using presently available computing environments.

In this work we present an iterative quasi-optimal approach to the control of the instationary Navier-Stokes equations. It is based on approximating the nonlinear dynamics of the Navier-Stokes equations by a reduced order model obtained by the snapshot form of the proper orthogonal decomposition [20] and carrying out an exact optimization for the reduced system.

[1]Supported by the Sonderforschungsbereich 557 Beeinflussung komplexer turbulenter Scherströmungen, sponsored by the Deutsche Forschungsgemeinschaft.

The numerical solution of the reduced control problem is used as control for the full equations and to compute new snapshots. The set of the new snapshots then is added to the set of the snapshots already available, the low dimensional approximation of the Navier-Stokes equations is re-computed and the reduced optimization is restarted. This process is iterated until convergence is achieved.

Related approaches to the control of the instationary Navier-Stokes equations can be found in [11, 15, 17, 21]. An excellent overview on diverse aspects of control for fluids can be obtained from [9]. For further references we also refer to [1, 4, 5, 6, 7, 8, 13, 14, 18].

The paper is organized as follows. In Section 2 we formulate the distributed control problem for the instationary Navier-Stokes equations and provide an appropriate analytical frame for its solution and for the proper formulation of Newton's method. Section 3 is devoted to the description of the reduced order model approach and its application to the instationary Navier-Stokes equations. Furthermore, in the same section the adaptive quasi-optimal control approach is introduced. Finally, in Section 4 we give some numerical examples including a comparison of optimal and quasi-optimal control computations for the two dimensional periodic flow around a cylinder.

2 The Optimal Control Problem

We consider the optimal control problem

$$\left.\begin{array}{ll} \min_{(y,u)\in W\times U} \; J(y,u) := \frac{1}{2}\int\limits_{Q_o} |y-z|^2\,dxdt + \frac{\gamma}{2}\int\limits_{Q_c} |u|^2\,dxdt & \\ \text{subject to} & \\ \frac{\partial y}{\partial t} + (y\cdot\nabla)y - \nu\Delta y + \nabla p = Bu & \text{in } Q = (0,T)\times\Omega, \\ \text{div } y = 0 & \text{in } Q, \\ y(t,\cdot) = 0 & \text{on } \Sigma = (0,T)\times\partial\Omega, \\ y(0,\cdot) = y_0 & \text{in } \Omega. \end{array}\right\} \tag{2.1}$$

where $Q_c := \Omega_c \times (0,T)$ and $Q_o := \Omega_o \times (0,T)$, with Ω_c and Ω_o subsets of the bounded spatial domain $\Omega \subset \mathbb{R}^2$ denoting control and observation volumes, respectively.

To define the spaces and operators required for the investigation of (2.1) we introduce the solenoidal spaces

$$H = \{v \in C_0^\infty(\Omega)^2 \colon \text{ div } v = 0\}^{-|\cdot|_{L^2}}, V = \{v \in C_0^\infty(\Omega)^2 \colon \text{ div } v = 0\}^{-|\cdot|_{H^1}},$$

with the superscripts denoting closures in the respective norms. Further we define

$$W = \{v \in L^2(V) \colon v_t \in L^2(V^*)\} \quad \text{and} \quad Z := L^2(V) \times H,$$

where W is endowed with the norm

$$|v|_W = (|v|^2_{L^2(V)} + |v_t|^2_{L^2(V^*)})^{1/2},$$

and set $\langle\cdot,\cdot\rangle := \langle\cdot,\cdot\rangle_{L^2(V^*),L^2(V)}$, with V^* denoting the dual space of V. Here $L^2(V)$ is an abbreviation for $L^2(0,T;V)$ and similarly $L^2(V^*) = L^2(0,T;V^*)$. Recall that up to a set of measure zero in $(0,T)$ elements $v \in W$ can be identified with elements in $C([0,T];H)$. In (2.1) further $U = L^2(Q_c)$ denotes the Hilbert space of controls which is identified with its dual U^*. The the cost functional $J\colon L^2(H) \times U \to \mathbb{R}$ is assumed to be bounded from below, weakly lower semi-continuous, twice Fréchet differentiable with locally Lipschitzian second derivative, and radially unbounded in u, i.e. $J(y,u) \to \infty$ as $|u|_U \to \infty$, for every $y \in W$. We define the nonlinear mapping

$$e\colon W \times U \to Z^*$$

by

$$e(y,u) = (\tfrac{\partial y}{\partial t} + (y \cdot \nabla)y - \nu\Delta y - Bu, y(0) - y_0)$$

where $Bu(t,\cdot)$ denotes the Leray-projection of the extension by zero of $u(t,\cdot)$ to the whole of Ω and $y_0 \in H$. In variational form the constraints in (2.1) can be equivalently expressed as:

given $u \in U$ find $y \in W$ such that $y(0) = y_0$ and

$$\langle y_t, v\rangle + \langle (y \cdot \nabla)y, v\rangle + \nu(\nabla y, \nabla v)_{L^2(L^2)} = \langle Bu, v\rangle \;\forall v \in L^2(V). \tag{2.2}$$

It is well known, see [22] that for every $u \in U$ (2.2) admits a unique solution $y(u) \in W$. Therefore, with respect to existence (2.1) can equivalently be rewritten as

$$\min \hat{J}(u) = J(y(u),u) \;\text{ subject to }\; u \in U, \tag{2.3}$$

where $y(u) \in W$ satisfies $e(y(u),u) = 0$. It is proved in [1] that (2.3) admits a solution $(y^*,u^*) \in W \times U =: X$. We shall frequently refer to the linearized Navier-Stokes system and the adjoint equations given next:

$$\begin{cases} v_t + (v \cdot \nabla)y + (y \cdot \nabla)v - \nu\Delta v = f & \text{in } \Omega \text{ a.e. on } (0,T], \\ v(0) = v_0. \end{cases} \tag{2.4}$$

and

$$\begin{cases} -w_t + (\nabla y)^t w - (y \cdot \nabla)w - \nu\Delta v = g & \text{in } \Omega \text{ a.e. on } [0,T). \\ w(T) = 0. \end{cases} \tag{2.5}$$

The following proposition is proved in [13].

PROPOSITION 2.1. *Let $x = (y,u) \in W \times U$. Then $e_y(x)\colon W \to Z^*$ is a homeomorphism. Moreover, if the inverse of its adjoint $e_y^{-*}(x)\colon W^* \to Z$ is applied to an element $g \in W^* \cap L^\alpha(V^*)$, where $\alpha \in [1,4/3]$ then setting $(w,w_0) := e_y^{-*}(x)g \in L^2(V) \times H$ we have $w_t \in L^\alpha(V^*)$, $w(0) = w_0$ and w is the variational solution to* (2.5).

The derivatives of the operator e were characterized in [13] and are cited for the convenience of the reader.

PROPOSITION 2.2. *The operator $e = (e^1, e^2)\colon X \to Z^*$ is twice continuously differentiable with Lipschitz continuous second derivative. The action of the first two derivatives of e^1 are given by*

$$\langle e_x^1(x)(w,s), \phi\rangle = \langle w_t, \phi\rangle + \langle (w \cdot \nabla)y, \phi\rangle + \langle (y \cdot \nabla)w, \phi\rangle + \nu(\nabla w, \nabla\phi)_{L^2(L^2)} - \langle Bs, \phi\rangle_{L^2(L^2)},$$

where $x = (y, u) \in X, (w, s) \in X$ *and* $\phi \in L^2(V)$, *and*

$$\langle e^1_{xx}(x)(w,s)(v,r),\phi\rangle = \langle e^1_{yy}(x)(w,v),\phi\rangle = \langle (w\cdot\nabla)v,\phi\rangle + \langle (v\cdot\nabla)w\phi, v\rangle =: \langle v, H(\phi)w\rangle_{W,W^*}, \quad (2.6)$$

where $(v, r) \in X$.

As a consequence of variations of standard estimates for the Navier-Stokes and the linearized Navier-Stokes equations and the implicit function theorem the first derivative of the mapping $u \to y(u)$ at u in direction δu is given by

$$y'(u)\delta u = -e_y^{-1}(x)e_u(x)\delta u, \quad (2.7)$$

where $x = (y(u), u)$. By the chain rule we thus obtain

$$\langle \hat{J}'(u), \delta u\rangle_U = \langle J_u(x) - e_u^*(x)e_y^{-*}(x)J_y(x), \delta u\rangle_U.$$

Introducing the variable

$$\lambda = -e_y^{-*}(x)J_y(x) \in Z \quad (2.8)$$

we obtain the Riesz representation for the first derivative of $u \to \hat{J}(u)$:

$$\hat{J}'(u) = J_u(x) + e_u^*\lambda. \quad (2.9)$$

Here $\lambda = (\lambda^1, \lambda^0) \in Z$, $\lambda^1_t \in L^{4/3}(V^*)$, $\lambda^1 \in C(H)$ and λ^1 is the variational solution of

$$\begin{cases} -\lambda^1_t + (\nabla y)^t\lambda^1 - (y\cdot\nabla)\lambda^1 - \nu\Delta\lambda^1 = -J_y(x) \\ \lambda^1(T) = 0, \end{cases} \quad (2.10)$$

where the first equation holds in $L^{4/3}(V^*) \cap W^*$.

The computation of the representation of the second derivative of $\hat{J}''$ of $\hat{J}$ is more involved. Introducing the Lagrangian $L\colon X \times Z \to \mathbb{R}$

$$L(x,\lambda) = J(x) + \langle e(x), \lambda\rangle_{Z^*,Z} \quad (2.11)$$

and the matrix operators

$$T(x) = \begin{pmatrix} -e_y^{-1}(x)e_u(x) \\ Id_U \end{pmatrix} \in \mathcal{L}(U, X) \quad (2.12)$$

we observe that the second derivative of $\hat{J}$ can be expressed as

$$\hat{J}''(u) = T^*(x)L_{xx}(x,\lambda)T(x), \quad (2.13)$$

where $x = (y(u), u)$ and the second derivative of L with respect to x is given by

$$L_{xx}(x,\lambda) = \begin{pmatrix} J_{yy}(x) + \langle e^1_{yy}(x)(\cdot,\cdot), \lambda^1\rangle & 0 \\ 0 & J_{uu}(x) \end{pmatrix} \in \mathcal{L}(X, X^*).$$

As numerical solution method for the optimal control problem (2.3) we suggest Newton's algorithm which for the sake of reference is specified next.

ALGORITHM 2.1 (NEWTON ALGORITHM).

1. *Choose* $u^0 \in N(u^*)$, *set* $k = 0$.

2. *Do until convergence*

   i) *solve* $\hat{J}''(u^k)\delta u^k = -\hat{J}'(u^k)$,

   ii) *update* $u^{k+1} = u^k + \delta u^k$,

   iii) *set* $k = k + 1$.

Here u^* denotes a local solution of problem (2.3). The numerical complexity of Algorithm 2.1 applied to problem (2.3) is investigated in [12]. In the same paper also a local convergence prove is given. Numerical experiments show that despite the implementational complexity the performance of the algorithm compared to that of gradient-type methods with respect to computation time is superior [13]. However, to anticipate parts of the discussion of the numerical results in Section 4 the running time of Newton's method for optimal control computations for the cylinder flow on a realistic grid utilizing a ORIGIN 200 takes days. Moreover, several gradient iterations have to be applied to provide a suitable initial guess to ensure local convergence of Newton's algorithm.

3 Reduced Order Modeling (ROM) Using Proper Orthogonal Decomposition (POD)

The ROM approach to optimal control problems such as (2.1) is based on approximating the nonlinear dynamics by a Galerkin technique utilizing basis functions that contain characteristics of the expected controlled flow. This is in contrast to finite element based Galerkin schemes where the basis elements are not related to the physical properties of the system that they approximate. Consequently one expects that only a few basis elements will provide a reasonably good description of the controlled flow. Various ROM techniques differ in the choice of the basis functions [11, 15]. Here we use the snapshot variant of POD introduced by Sirovich in [20] to obtain a low-dimensional approximation of the Navier-Stokes equations and to define the control space.

First we describe the model reduction. To begin with let $y^1, \ldots, y^n$ denote an ensemble of snapshots of the flow corresponding to different time instances. For the flow we make the Ansatz

$$y = \bar{y} + \sum_{i=1}^{n} \alpha_i \Phi_i \tag{3.1}$$

with modes Φ_i and mean $\bar{y}$ that are obtained as follows.

1. Compute mean $\bar{y} = \frac{1}{n} \sum_{i=1}^{n} y^i$

2. Build correlation matrix $K = k_{ij}$, $k_{ij} = \int_{\Omega} (y^i - \bar{y})(y^j - \bar{y})\, dx$

3. Compute eigenvalues $\lambda_1, \dots, \lambda_n$ and eigenvectors $v^1, \dots, v^n$ of K
4. Set $\Phi_i := \sum_{j=1}^{n} v_j^i (y^j - \bar{y})$
5. Normalize $\Phi_i = \frac{\Phi_i}{\|\Phi_i\|}$

Now choose the eigenvectors of the matrix K pairwise orthogonal (this is possible since K is symmetric). Then it follows that the modes Φ_i are pairwise orthonormal. Moreover, they are optimal with respect the L^2 scalar product in the sense that no other basis of $D := \text{span}\{y_1 - \bar{y}, \dots, y_n - \bar{y}\}$ can contain more energy in fewer elements, compare [2, 16]

In order to obtain a low-dimensional basis for the Galerkin Ansatz modes corresponding to small eigenvalues are neglected. To make this idea more precise let $D^M := \text{span}\{\Phi_1, \dots, \Phi_M\}$ $(1 \le M \le N := \text{dim}D)$ and define the relative information content of this basis by

$$\text{I}(M) := \sum_{k=1}^{M} \lambda_k \Big/ \sum_{k=1}^{N} \lambda_k.$$

If the basis is required that contains δ% of the total information contained in the space D, say the dimension M of the subspace D^M is determined by

$$M = \text{argmin}\,\{\text{I}(M);\ \text{I}(M) \ge \delta\}\,. \tag{3.2}$$

The reduced dynamical system is obtained by plugging in (3.1) into the Navier-Stokes System and using a subspace D^M containing sufficient information as test space. This results in

$$(y_t, \Phi_j) + \nu(\nabla y, \nabla \Phi_j) + ((y \cdot \nabla)y, \Phi_j) = (Bu, \Phi_j) \quad (j = 1, \dots, M).$$

which may be rewritten as

$$\dot{\alpha} + A\alpha = n(\alpha) + \mathcal{M}_1\beta + r, \quad \alpha(0) = a_0.$$

Here the initial values y_0 are approximated by $\bar{y} + \sum_{k=1}^{M} a_0^k \Phi_k$ and the i-th component a_0^i of the M-vector a_0 is given by $(y_0 - \bar{y}, \Phi_i)$. The matrix A is the POD stiffness matrix and the inhomogeneity r results from the contribution of the mean $\bar{y}$ to the Ansatz in (3.1). The matrix $\mathcal{M}_1$ has the entries $m_{ij}^1 = \int_{\Omega_c} \Phi_i \Phi_j \, dx$, and the M-vector β results from the contribution of the controls, i.e. for them we make the Ansatz

$$u = \sum_{i=1}^{M} \beta_i \Phi_i \chi_{\Omega_c}. \tag{3.3}$$

Here, χ_{Ω_c} denotes the characteristic function of the control domain Ω_c. We note that $\mathcal{M}_1$ is the identity matrix in the casr $\Omega_c = \Omega$. The reduced optimization problem corresponding to (2.1) is obtained by plugging in (3.1) and (3.3) into the cost functional and utilizing the reduced dynamical system as subsidiary condition in the optimization process. Finally, writing $\int_{\Omega_0} |y - z|^2 \, dx = (\alpha - \alpha^z)^t \mathcal{M}_2 (\alpha - \alpha^z)$, where the entries of $\mathcal{M}_2$ are given by $m_{ij}^2 = \int_{\Omega_0} \Phi_i \Phi_j \, dx$ and the desired state z is written as $z = \bar{y} + \sum_{k=1}^{M} \alpha_k^z \Phi_k$ we obtain

$$(\text{ROM}) \begin{cases} \min J(\alpha, \beta) = \frac{1}{2} \int_0^T (\alpha - \alpha^z)^t \mathcal{M}_2 (\alpha - \alpha^z) \, dt + \frac{\gamma}{2} \int_0^T \beta^t \mathcal{M}_1 \beta \, dt \\ \text{s.t.} \\ \dot{\alpha} + A\alpha = n(\alpha) + \mathcal{M}_1\beta + r, \quad \alpha(0) = a_0. \end{cases} \tag{3.4}$$

With these preparations we are in the position to formulate the quasi-optimal control strategy. We choose the value δ for (3.2) and a sequence of increasing numbers N_j.

ALGORITHM 3.1 (POD-BASED ADAPTIVE CONTROL).

1. *Let a set of snapshots* $y_i^0, i = 1, \ldots, N_0$ *be given and set j=0.*
2. *Compute* M *according to* (3.2) *with* N *replaced by* N_j.
3. *Compute POD modes and solve the reduced optimization problem* (3.4) *for* $u^j \in D_j^M$.
4. *Compute the state* y^j *corresponding to the current control* u^j *and add the snapshots* $y_i^{j+1}, i = N_j + 1, \ldots, N_{j+1}$ *to the snapshot set* $y_i^j, i = 1, \ldots, N_j$.
5. *If* $|u^{j+1} - u^j|$ *is not sufficiently small, set* $j = j+1$ *and goto 2.*

It cannot be claimed that the controls obtained with this algorithm in some sense converge to an optimal control. However, the numerical results to be presented justify the approach, especially with regard to the computing times.

4 Numerical Results

Here we present numerical computations related to the approaches presented in the previous paragraphs. The flow configuration is given by the two-dimensional channel with a circular cylinder and coincides with that used in [19] as a benchmark for Navier-Stokes solver, see Fig. 1. The control gain is to track the Stokes flow in a given observation volume Ω_o by applying

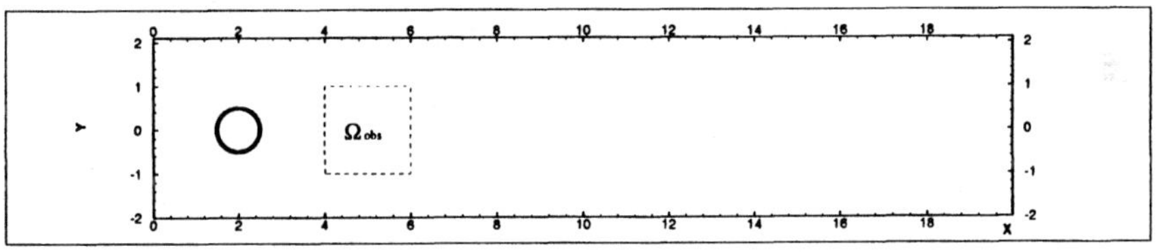

Figure 1: Flow domain and observation volume

control forces in a control volume Ω_c. The Reynolds number Re= $1/\nu$ for this configuration is determined by

$$\text{Re} = \frac{\bar{U}d}{\mu}$$

with $\bar{U}$ denoting the bulk velocity at the inlet, d the diameter of the cylinder and μ denoting the molecular viscosity of the fluid. In all numerical experiments presented we choose a parabolic inflow profile at the inlet, $d = 1$, Re=100 and as channel length $l = 19.5d$. At the outflow boundary so called do-nothing boundary conditions are used, see [10]. This means that the boundary conditions used for the numerical computations are different from them given in (2.1). For the spatial discretization the Taylor-Hood finite element on a grid with 7808 triangles,

16000 velocity and 4096 pressure nodes is used. As time interval we use $[0, T]$ with $T = 3.4$ which coincides with the length of one period of the wake flow. For the time discretization we use a fractional step Θ-scheme [3] or a semi-implicit Euler-scheme on a grid containing 500 points, which corresponds to a time step size of $\delta t = 0.0068$. The total number of variables in the optimization problem (2.1) therefore is of order 5.4×10^7 (primal, adjoint and control variables). In all numerical examples $\gamma = 2.10^{-2}$.

For the POD we add 100 snapshots to the snapshot set in every iteration of Algorithm 3.1. The relative information content of the basis formed by the modes is required to be larger than 99.99%. In Fig. 2 the vorticity and the streamlines of the uncontrolled flow (top) and the Stokes flow (bottom) are presented. For the numerical solution of the reduced optimization problems the Schur-complement SQP-algorithm is used. All computations are performed on a ORIGIN 200.

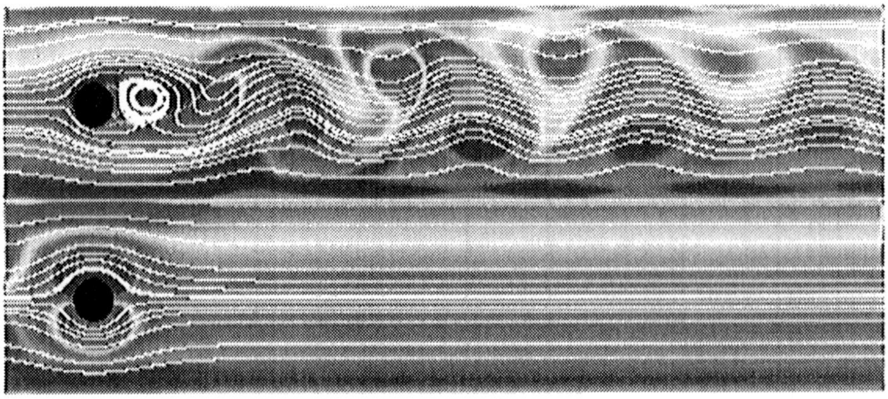

Figure 2: Uncontrolled flow (top) and Stokes flow (bottom)

Example 1 We first present the results for $\Omega_c = \Omega_o = \Omega$. Fig. 3 compares the evolution of the cost in $[0, T]$ for the optimal control and the control obtained by Algorithm 3.1. The adaptive algorithm terminates after 5 iterations and achieves a remarkable cost reduction. To recover 99.99% of the energy stored in the snapshots in the first iteration 10 modes have to be taken, 20 in the second iteration, 26 in the third, 30 in the fourth, and 36 in the final iteration. The computation of the optimal control takes approximately 140 CPU hours with the Newton method including the initialization process with a step-size controlled gradient algorithm. To obtain a relative error $|\nabla J(u^n)|/|\nabla J(u^0)|$ lower than 10^{-2} 32 gradient iterations are needed. The value of the cost functional for the uncontrolled flow is $J(u^0) = 22.658437$, after 32 iterations this value is reduced to $J(u^{32}) = 1.138325$. Here, as initial control $u^0 = 0$ is taken. Note that every gradient step amounts to solving the non-linear Navier-Stokes equations in (2.1), the adjoint equations (2.10) and a further Navier-Stokes system for the computation of the step-size in the gradient algorithm, compare [13]. Newton's algorithm then is initialized with u^{32} and 3 Newton steps further reduce the value of the cost functional to $J(u^*) = 1.090321$. The numerical amount of work for Newton's method in the outer loop coincides with that of one gradient step. The Newton system 2.i) in Algorithm 3.1 has to be solved iteratively, where every iteration amounts to solving the systems (2.4) and (2.5), respectively with appropriate right-hand-sides f and g, respectively. Algorithm 3.1 takes 5 iterations and 8 CPU hours to obtain the quasi-optimal control $\tilde{u}^*$. The final value of the cost is $J(\tilde{u}^*) = 6.440180$.

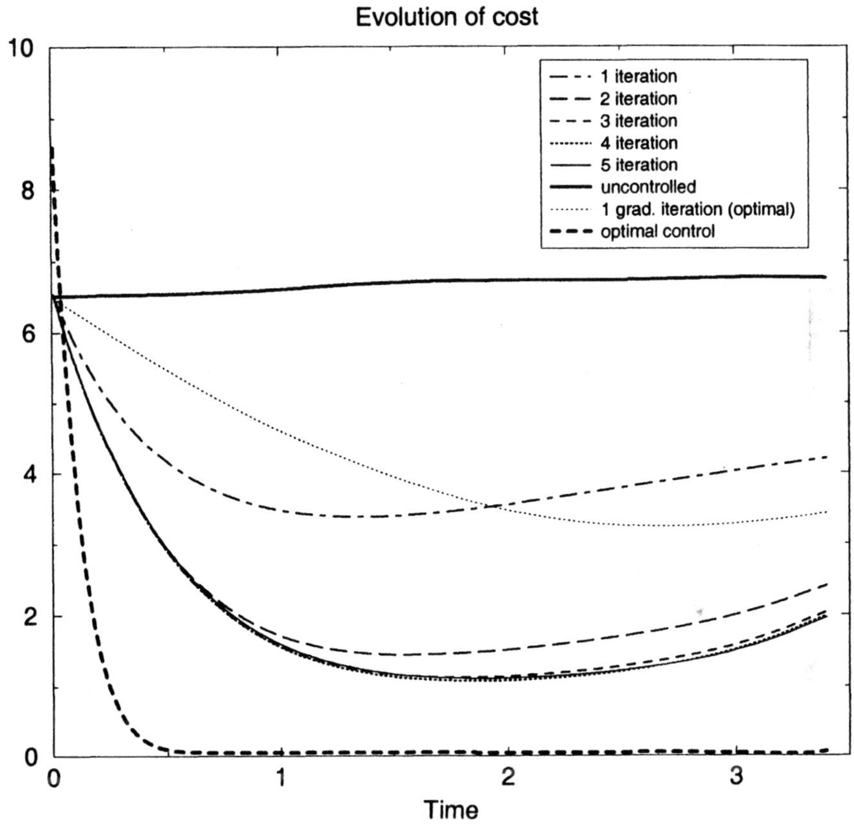

Figure 3: Example 1, Evolution of the cost $\frac{1}{2}(\alpha - \alpha^2)^t \mathcal{M}_2(\alpha - \alpha^z) + \frac{\gamma}{2}\beta^t \mathcal{M}_1 \beta$

Fig. 4 shows a zoom on the evolutions of the control cost for both approaches. The optimal cost at $t = 0$ has the value 122.71 and is approximately 10 times larger than the quasi-optimal cost at the same time. Fig. 5 shows the streamlines and the vorticity of the flow controlled by the adaptive approach at $t = 3.4$ (top) and the mean flow $\bar{y}$ (bottom), the latter formed with the snapshots of all 5 iterations. The controlled flow no longer contains vortex sheddings and is approximately stationary. Recall that the controls are sought in the space of deviations from the mean flow. This explains the remaining recirculations behind the cylinder.

In the previous example the observation volume Ω_o and the control volume Ω_c each cover the whole spatial domain. From the practical point of view this is not feasible. However, from the numerical standpoint this is a complicated situation, since the inhomogeneities in the primal

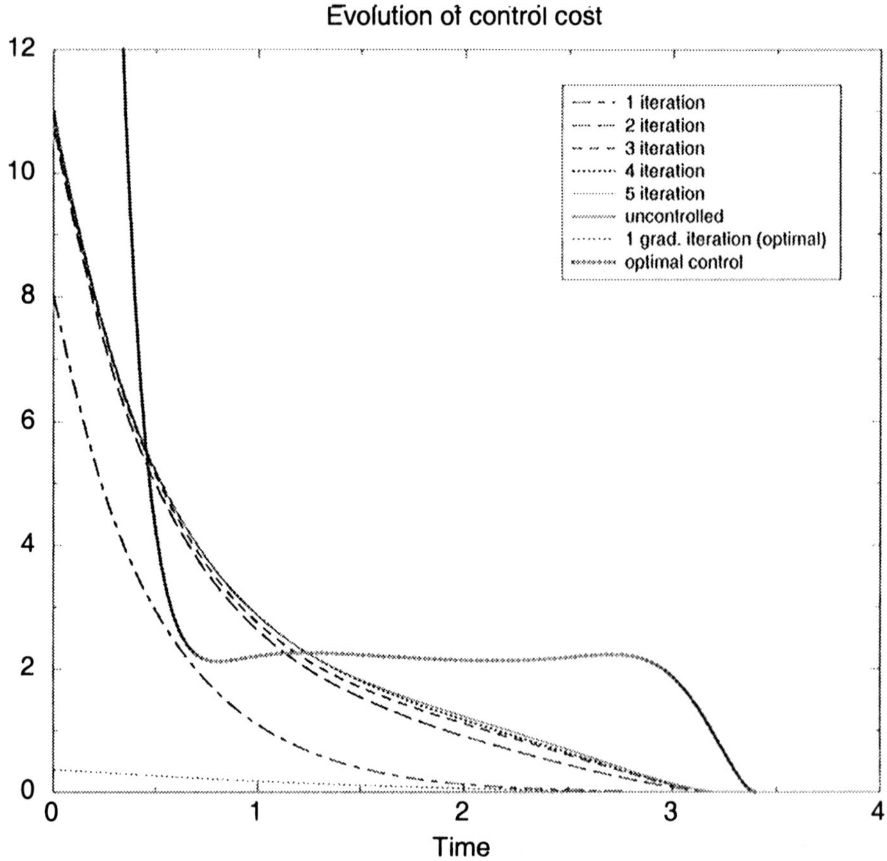

Figure 4: Example 1, Evolution of control control cost $\beta^t \mathcal{M}_1 \beta$

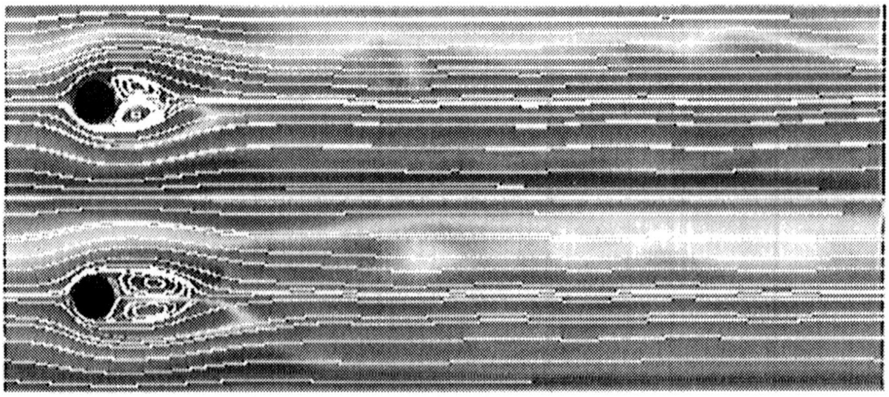

Figure 5: Example 1: Controlled flow (top) and mean flow $\bar{y}$ (bottom)

and adjoint equations are large. We next present a numerical example with different observation and control volumes. This results in smaller control and observation volumes than in Example 1, and thus the primal and adjoint equations in the optimal control algorithm are numerically simpler to solve.

Example 2 Here $\Omega_c = \Omega_o = [4,6] \times [-1,1]$. Fig. 6 compares the evolution of the cost in $[0,T]$ for the optimal control and the adaptively improved control. The adaptive algorithm in this example terminates after 6 iterations and again achieves a remarkable cost reduction. The best value of the cost in the quasi-optimal case is nearly obtained already after the second iteration of Algorithm 3.1. The computation of the optimal control takes approximately 100 CPU hours utilizing a gradient algorithm with step-size control and initial control $u^0 = 0$. Using the same stopping criterion as in Example 1 the algorithm takes 28 iterations to reduce the cost from $J(u^0) = 4.087893$ to $J(u^{28}) = 0.637754$. The quasi-optimal control $\tilde{u}^*$ is computed by 6 iterations of Algorithm 3.1 within 7 CPU hours. The optimal function value in this case is $J(\tilde{u}^*) = 0.942525$.

In Fig. 7 a zoom is shown that compares the evolutions of the control cost for both approaches. The value of the optimal cost at the beginning of the control horizon has the value 36.11 and is again approximately a magnitude larger than that of the quasi-optimal cost.

An interesting observation is made when comparing the evolution of $|y - z|^2_{L^2(\Omega)}$ in the control horizon. As is indicated in Fig. 8 the quasi-optimal approach obtains a reduction of this difference similar to that of the optimal control approach, but at significantly lower control costs. This is due to the global nature of the Ansatz and the test functions in the POD method. As a consequence also local deviations have an immediate global impact. Of course, this is different for the optimal approach since the the finite element method is used to approximate the optimal controls.

5 Conclusions

An adaptive quasi-optimal control approach for the wake flow around a cylinder is presented. The method is based on the snapshot form of the POD for the Navier-Stokes equations and iteratively computes controls by solving low-dimensional optimization problems. Compared to the optimal control approach the quasi-optimal method is much less time consuming and provides competitive controls. Moreover, due to the global nature of the POD basis functions already local measurements can be sufficient for achieving more global control goals.

Bibliography

[1] **Abergel, F. & Temam, R.** On some Control Problems in Fluid Mechanics. *Theoret. Comput. Fluid Dynamics*, 1:303–325, 1990.

[2] **Atwell, J.A. & King, B.B.** Proper Orthogonal Decomposition for reduced basis feedback controllers for parabolic equations. ICAM Report 99-01-01, 1999. Interdisciplinary Center

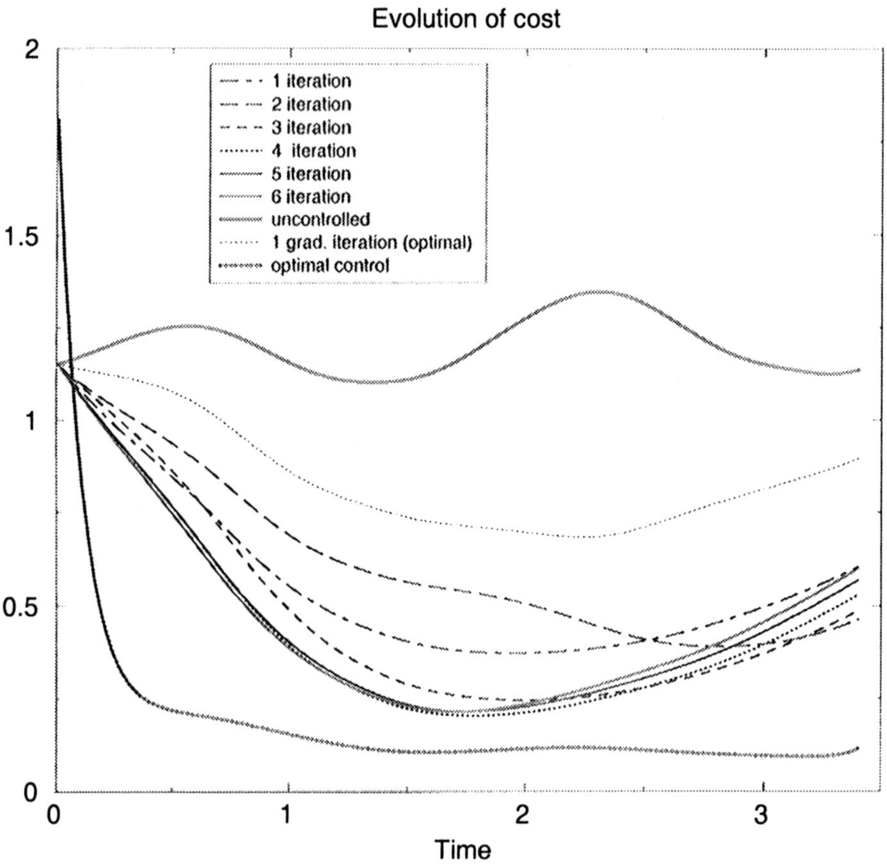

Figure 6: Example 2, Evolution of the cost $\frac{1}{2}(\alpha - \alpha^z)^t \mathcal{M}_2(\alpha - \alpha^z) + \frac{\gamma}{2}\beta^t \mathcal{M}_1 \beta$

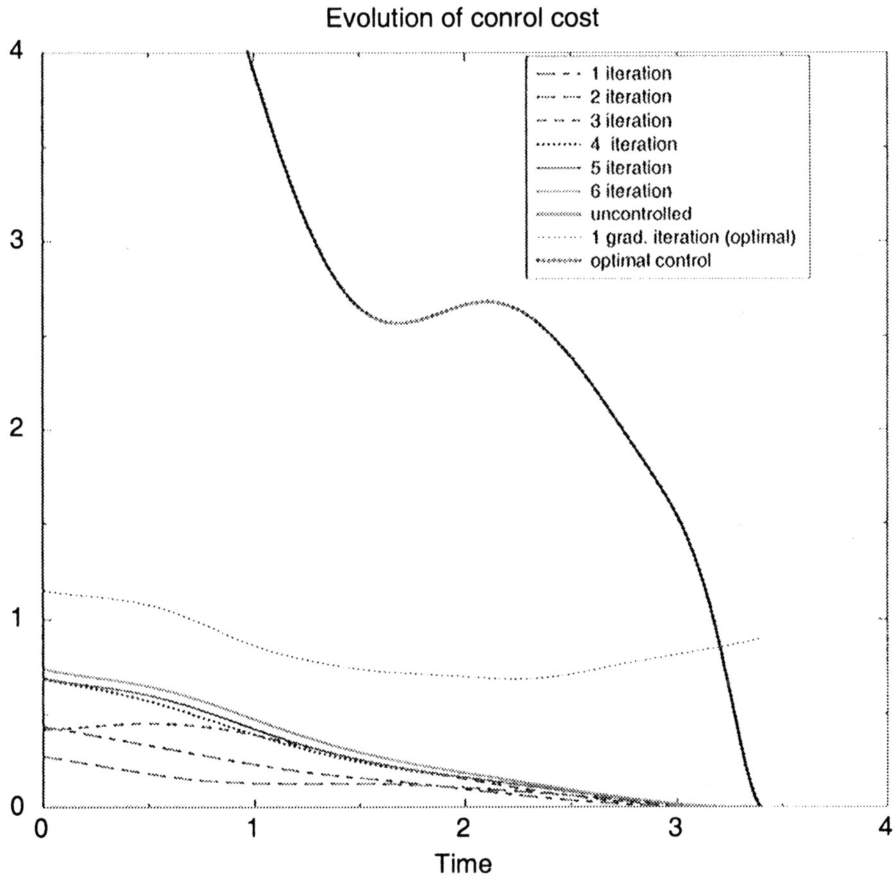

Figure 7: Example 2, Evolution of the control cost $\beta^t \mathcal{M}_1 \beta$

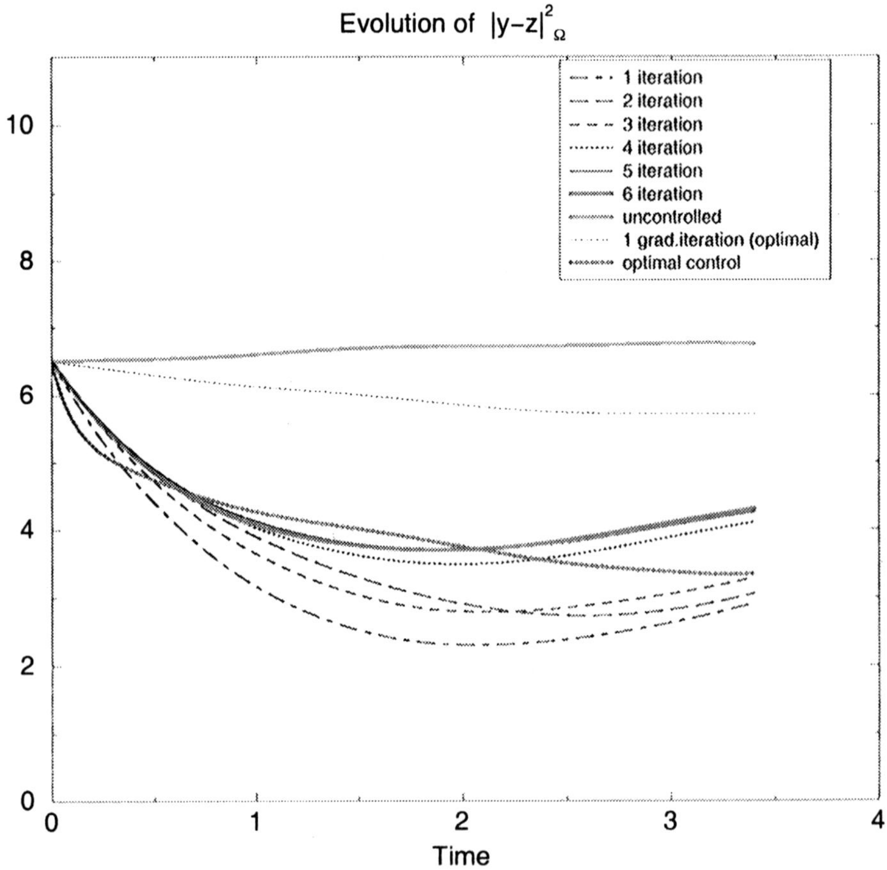

Figure 8: Example 2, Evolution of $|y - z|^2_{L^2(\Omega)}$

for Applied Mathematics, Virginia Polytechnic Institute and State University Blacksburg, Virginia.

[3] **Bänsch, E.** An adaptive Finite-Element-Strategy for the three-dimensional time-dependent Navier-Stokes-Equations. *J. Comp. Math.*, 36:3–28, 1991.

[4] **Berggren, M.** Numerical solution of a flow control problem: Vorticity reduction by dynamic boundary action. *Siam J. Sci. Comput.*, Vol. 19(No. 3), 1998.

[5] **Bewley, T.; Choi, H.; Temam, R. & Moin, P.** Optimal feedback control of turbulent channel flow. CTR Annual Research Briefs, 1993. Center for Turbulence Research, Stanford University/NASA Ames Research Center, 3-14.

[6] **Chang, Y. & Collis, S.** Active control of turbulent channel flows by large eddy simulation. In *Proceedings of the FEDSM99*. ASME, 1998.

[7] **Choi, H.** Suboptimal Control of Turbulent Flow Using Control Theory. In *Proceedings of the International Symposium. on Mathematical Modelling of Turbulent Flows*, 1995. Tokyo, Japan.

[8] **Glowinski, R.** Finite element methods for the numerical simulation of incompressible viscous flow; Introduction to the Control of the Navier-Stokes Equations. *Lectures in Applied Mathematics*, 28, 1991.

[9] **Gunzburger, M.D.** *Flow Control.* IMA. Springer, 1995.

[10] **Heywood, J.G.; Rannacher, R. & Turek, S.** Artificial Boundaries and Flux and Pressure Conditions for the Incompressible Navier-Stokes Equations. *Preprint SFB 359*, 94 - 06, 1994. Universität Heidelberg.

[11] **Hinze, M. & Kunisch, K.** On suboptimal Control Strategies for the Navier-Stokes Equations. Preprint No. 568/1997, 1997. Technische Universität Berlin, Deutschland, see also ESAIM: Proceedings, Vol. 4, 1998, 181-198, http://www.emath.fr/proc/Vol.4 (1998), France.

[12] **Hinze, M. & Kunisch, K.** Newton's method for tracking-type control of the instationary Navier-Stokes equations. Preprint, 1999, to appear in ENUMATH99, P. Neittaanmalzi et al. (Eds.)

[13] **Hinze, M. & Kunisch, K.** Second order methods for optimal control of time-dependent fluid flow. Bericht Nr. 165, Spezialforschungsbereich Optimierung und Kontrolle, 1999. Institut für Mathematik, Karl-Franzens-Universität Graz.

[14] **Hou, L.S.; Gunzburger, M.D.; Manservisi, S.; Turner, J. & Yan, Y.** Computations of optimal controls for incompressible flows. In *1997 ASME Fluids Engineering Division Summer Meeting*, 1997.

[15] **Ito, K. & Ravindran, S.S.** A reduced basis method for control problems governed by pdes. Preprint, 1997. Department of Mathematics, North Carolina State University, USA.

[16] **Kunisch, K. & Volkwein, S.** Control of Burgers equation by a reduced order approach using Proper Orthogonal Decomposition. Bericht Nr. 138, Spezialforschungsbereich Optimierung und Kontrolle, September 1998. Univsersität Graz, Österreich, to appear in J. Opt. Theory and Appl.

[17] **Ly, H.V. & Tran, H.T.** Modelling and control of physical processes using proper orthogonal decomposition. Report, 1998. CRSC-TR98-37, Center for Research in Scientific Computation, North Carolina State University.

[18] **Manservisi, S.** *Optimal boundary and distributed controls for the velocity tracking problem for Navier-Stokes flows.* PhD thesis, Faculty of the Virginia Polytechnic Institute and State University, Blacksburg, 1997.

[19] **Schäfer, M. & Turek, S.** Benchmark computations of laminar flow around a cylinder. Report, 1996. IWR, Universität Heidelberg.

[20] **Sirovich, L.** Turbulence and the dynamics of coherent structures, Part I-III. *Quarterly of Applied Mathematics*, 45:561–590, 1987.

[21] **Tang, K.Y.; Graham, W.R. & Peraire, J.** Active Flow Control using a Reduced Order Model and Optimum Control. AIAA Nr. 1996-19-46, 1996. see also Computational Aerospace Sciences Laboratory, MIT Department of Aeronautics and Astronautics.

[22] **Temam, R.** *Navier-Stokes Equations.* North-Holland, 1979.

Konstantin Afanasiev. Fachbereich Mathematik, MA 6-3, Technische Universität, Berlin, D-10623 Berlin, Germany
E-mail: afanasiev@pi.tu-berlin.de

Michael Hinze. Fachbereich Mathematik, MA 6-3, Technische Universität, Berlin, D-10623 Berlin, Germany
E-mail: hinze@math.tu-berlin.de

Nonlinear Boundary Feedback Stabilization of Dynamic Elasticity with Thermal Effects

Irena Lasiecka[1]

Abstract. An n-dimensional elastodynamic system accounting for thermal effects is considered. The main aim of this paper is to construct a ***square integrable*** nonlinear boundary feedback which would stabilize the overall dynamics. It is shown that a nonlinear monotone feedback acting on velocity of the the boundary displacement provides uniform decay rates for the energy function associated with all ***weak solutions***. The key role in these results is played by "sharp regularity" estimates for the boundary traces established for elastic systems.

1 Introduction

1.1 The Model and the Motivation

We consider a model describing nonlinear oscillations of an elastic medium occupying a bounded region $\Omega \in R^n$ and subjected to thermal effects. References related to modeling of elastic systems can be found in [19, 17, 16, 15, 11, 4, 29, 5, 3].

Here, the variable $u = (u_1, u_2, \dots u_n)$ represents displacement of a n elastic medium in n-dimensional domain Ω with sufficiently smooth boundary $\Gamma = \Gamma_0 \cup \Gamma_1$. We shall assume that $\Gamma_0 \cap \Gamma_1 = \emptyset$. The variable ϕ describes the thermal strain resultants affecting the displacement. More specifically, ϕ is the average of thermal stress (see formulas (6.8), (6.9) in [18]). The governing equations to be considered are given by:

$$\begin{aligned} u_{tt} - div[\mathcal{C}[\epsilon(u)] + \nabla\phi = 0 \text{ in } \ \Omega \times (0,\infty) \\ \phi_t = \Delta\phi - div u_t; \ \text{ in } \Omega \times (0,\infty) \end{aligned} \tag{1.1}$$

with Dirichlet boundary conditions on the portion of the boundary Γ_0

$$u = 0 \ \text{ on } \Gamma_0 \times (0,\infty)$$

[1]This research is partially supported by the NSF Grant DMS-9504822 and the Army Research Office Grant DAAH04-96-1-0059.

The boundary conditions on the boundary Γ_1 are prescribed in the form of tractions and given by

$$C\epsilon(u)\nu + ku - \phi\nu = f; \ k \geq 0 \tag{1.2}$$

Here, the vector ν denotes the outward normal direction to the boundary. Similarly, we denote by $\frac{\partial}{\partial \tau}$ the tangential direction on Γ, ie $\frac{\partial}{\partial \tau} = \nabla \cdot \tau$ where $\tau = [\tau_1 \cdots \tau_{n-1}]$ is an oriented, unit tangent vector to the boundary.

The function $f \in L_2(0,\infty; L_2(\Gamma_1))$ represents a boundary control (to be found) for the system.

The boundary conditions imposed on the temperature ϕ are of Robin type and given by

$$\frac{\partial}{\partial \nu}\phi + \lambda\phi = 0 \ \text{ on } \Gamma \times (0,\infty); \ \lambda \geq 0 \tag{1.3}$$

With (1.1) we associate the initial conditions

$$u(0) = u_0, u_t(0) = u_1, \phi(0) = \phi_0 \text{ in } \Omega \tag{1.4}$$

E represents the Young's modulus and the constant $0 < \mu < 1/2$ is Poisson's modulus. The fourth order tensor C is defined by

$$C\epsilon \equiv \frac{E}{(1-2\mu)(1+\mu)}[\mu \ trace \ \epsilon \ I + (1-2\mu)\epsilon]$$

where the strain tensor is given by $\epsilon(u) \equiv 1/2(\nabla u + \nabla^T u)$. It can be easily verified that the tensor C is symmetric and strictly positive.

The natural energy functional associated with the model (1.1), (1.2), (1.3) is given by

$$E(t) = E_k(t) + E_p(t); \ \ E_k(t) = \int_\Omega |u_t|^2 d\Omega \tag{1.5}$$

with the potential energy given by:

$$E_p(t) = \int_\Omega \left[C\epsilon(u)\epsilon(u) + |\phi|^2\right] d\Omega + k\int_{\Gamma_1} |u|^2 d\Gamma_1 \tag{1.6}$$

It is well known that $E_p(t)$ is topologically equivalent to $[H^1(\Omega)]^n \times L_2(\Omega)$ topology. In particular, the following inequalities resulting from Korn's inequality and Sobolev's embeddings will be used frequently:,

$$|u|^2_{H^1(\Omega)} \leq C[|\epsilon(u)|^2_{L_2(\Omega)} + k\int_{\Gamma_1} |u|^2 d\Gamma_1] \tag{1.7}$$

where $k > 0$ whenever $\Gamma_0 = \emptyset$.

Our goal, motivated by [27], is to find a nonlinear boundary control $f \in L_2(0,\infty; L_2(\Gamma_1))$, in the form of a feedback acting on the velocity of the displacement $f \equiv f(u_t)$, such that the resulting nonlinear feedback system is *uniformly stable*. This amounts to the fact that the energy of the system decays to zero, at an uniform rate, when $t \to \infty$.

There are considerable number of papers dealing with the boundary stabilization of elastic equations/ systems *without* the thermal effects [15, 19, 16, 31, 9, 10, 21, 1]. More recently, thermoelastic systems were considered in [26, 27] and this particular problem of boundary stabilization has been addressed in [26, 27]. However, due to technical/mathematical difficulties (more on this below), the authors of [26, 27] consider *different boundary conditions* associated with the system. These are given by

$$C\epsilon(u)\nu + ku - \mu[\frac{\partial u_j}{\partial x_j}\nu_i - \frac{\partial u_j}{\partial x_i}\nu_j] + ku = -(x - x_0)\nu g(u_t); \; x_0 \in R^n \tag{1.8}$$

For this "new" system, the authors prove uniform decay of the energy, for the *linear case* in [26], and for the nonlinear case [27]. In the nonlinear case, the authors of [27] adopt methods of [23] in order to cope with nonlinearities without a specified growth at the origin. Technical assumptions involving nonlinearities and the geometry of the domain which are imposed in [27] include:

ASSUMPTION 1.1. *1. $g(u) = h(u_i); i = 1 \dots n$; where a function h: (i) is continuous, monotone, zero at the origin, of linear growth at the infinity and moreover (ii) satisfies Lipschitz condition $|h(s_1) - h(s_2)| \leq C|s_1 - s_2|$; for $|s_1 - s_2| \geq 1$ and Holder condition $|h(s_1) - h(s_2)| \leq C|s_1 - s_2|^q$; for $|s_1 - s_2| \geq 1$ and some $q > 0$.*

2.

$$\begin{aligned} (x - x_0) \cdot \nu \geq \delta > 0 \; on \; \Gamma_1 \\ (x - x_0) \cdot \nu \leq 0 \; on \; \Gamma_0 \end{aligned} \tag{1.9}$$

The above result, when interpreted in terms of the original problem formulated by (1.1), (1.11), (1.2), (1.3) gives

$$f \equiv f_i(\frac{\partial u_i}{\partial x_j}, u_t) \equiv -(x - x_0) \cdot \nu g(u_t) - \mu[\frac{\partial u_j}{\partial x_j}\nu_i - \frac{\partial u_j}{\partial x_i}\nu_j] \tag{1.10}$$

as the stabilizing feedback for the model. However, due to the presence on the right side of (1.10) of boundary derivatives with respect to the spatial variables, this feedback is not well defined for finite energy solutions. Even more, for smooth solutions this feedback is *not square integrable* $L_2(0, \infty, L_2(\Gamma_1))$. This fact undermines applicability of such feedback in the context of control theory and raises a legitimate question: can one construct a square integrable control feedback $f(u, u_t)$ which would stabilize the system? In fact, this particular question was also raised in [27], where it was addressed as an open problem.

The main goal of the present paper is to provide an affirmative answer to the above question by showing that a goal of uniform stabilization with a *square integrable feedback control*, defined for all finite energy solutions, is achieved with a simple feedback control given by

$$f \equiv g(u_t) \tag{1.11}$$

where nonlinear vector function g is assumed continuous, monotone increasing, zero at the origin and of linear growth at infinity. Moreover, we obtain $f = g(u_t) \in L_2(0, \infty, L_2(\Gamma_1))$ for all finite energy solutions. Thus, our result provides not only an affirmative answer to a question raised in [27], but the technical assumptions imposed on the nonlinearity and the geometry of the domain are much less restrictive than the ones assumed in [27] (see Assumption 1.1 as compared to Assumption 1.2 formulated below).

We note that the main technical difficulty of the problem lies in obtaining appropriate estimates for the tangential derivatives on the boundary of finite energy solutions. (Notice that the topology induced by finite energy says anything about the regularity of the traces on the boundary). In fact, the reason for assuming in [27] more complex structure of the feedback given by (1.10) was precisely to eliminate altogether the difficulties arising due to the presence in the estimates of these tangential derivatives. A careful inspection of the proof in [27] reveals that the particular combination of the terms on the right hand side of (1.8) "cancels out", in appropriate estimates, the terms which are troublesome. Instead, in order to solve a correct problem, these boundary derivatives need to be estimated. This is done by applying microlocal estimates developed in the context of elasticity [7, 21] together with methods developed in [22] for treating thermoelasticity. In fact, the result of this paper is really "buried" in [22, 21], where more complex fully nonlinear von Karman system was considered. This particular thermo-elastodynamic system discussed in the present paper, is just a component of full von Karman system with thermoelasticity, studied in [22]. Therefore, the arguments leading to the establishments of uniform decays which are [presented in this paper are much simpler and more direct than the ones provided in [22].

1.2 Formulation of the Results

To begin our analysis, it is essential that the feedback system (1.1) with f given by (1.11) be well-posed. In fact, the following wellposedness/regularity results follow from standard monotone operator theory applied to boundary feedback problems [20, 23].

THEOREM 1.1.

- *(1)* ***Weak solutions.*** *(i) There exists a* ***unique****, global solution of finite energy corresponding to the system (1.1),(1.2),(1.3) with f given by (1.11). This is to say that for any initial data*

$$u_0, u_1, \phi_0 \in [H^1_{\Gamma_0}(\Omega)]^n \times [L_2(\Omega)]^n \times L_2(\Omega);$$

there exists a unique solution

$$(u, \phi) \in C(0,T; [H^1(\Omega)]^n \times L_2(\Omega))$$
$$u_t \in C(0,T; [L_2(\Omega)]^n); \quad \phi \in L_2(0,T; H^1(\Omega))$$

which depend continuously on the initial data. Here the spaces $H^1_{\Gamma_0}(\Omega)$ and $H^2_{\Gamma_0}(\Omega)$ denote the usual Sobolev spaces which incorporate zero boundary conditions on Γ_0.

(ii) If, in addition,

$$(g(s), s)_{R^n} \geq m|s|^2; m > 0, |s| \geq 1 \tag{1.12}$$

then

$$f = g(u_t) \in L_2(0, \infty; L_2(\Gamma_1)) \tag{1.13}$$

- ***(2) Regular solutions.*** *Assume, in addition, that $g \in C^1$. For any initial data subject to the regularity in part (1) and, in addition,*

$$u_0 \in [H^2(\Omega)]^n, \ u_1 \in [H^1_{\Gamma_0}(\Omega)]^n; \ \phi_0 \in H^2(\Omega)$$

and subject to compatibility conditions specified on the boundary, there exists a unique, global solution

$$(u, \phi) \in C(0, T; [H^2(\Omega)]^n \times H^2(\Omega)) \cap C^1(0, T; [H^1(\Omega)]^n \times L_2(\Omega))$$

where $T > 0$ is arbitrary.

Moreover, regular solutions depend continuously on the initial data in the topology of regular solutions (as defined above).

To formulate our main stability result, we introduce the function $\mathcal{H}(s)$ which is assumed concave, strictly increasing functions, zero at the origin and such that the following inequalities are satisfied for $|s| \leq 1$

$$\mathcal{H}(sg(s))) \geq s^2 + |g(s)|^2 \ ,$$

Such a function can be easily constructed in view of the monotonicity assumption imposed on g, [23].

We are ready to state the main result of this paper. To this end, we introduce the following hypothesis which will be assumed throughout the paper.

ASSUMPTION 1.2.

(1) Let $\boldsymbol{h}(x)$ be a vector field given by $\boldsymbol{h}(x) \equiv x - x_0$; $x_0 \in R^2$ We assume that

$$\boldsymbol{h} \cdot \nu \leq 0 \ on \ \Gamma_0 \tag{1.14}$$

(2) There exist positive constants $0 < m \leq M$ such that for $|s| \geq R$ with a constant R sufficiently large, we have:

$$m|s|^2 \leq (g(s), s)_{R^n} \leq M|s|^2; \tag{1.15}$$

Our main result is:

THEOREM 1.2. *Let u, ϕ be a weak solution to the original system given by (1.1), (1.2), (1.3), (1.11). Assume that the constant k is positive whenever $\Gamma_0 = \emptyset$ and $\lambda > 0$. Then, the following estimate holds*

$$E(t) \leq Cs(t/T_0 - 1)); \quad t \geq T_0 \tag{1.16}$$

where the real variable function $s(t)$ converges to zero as $t \to \infty$ and it satisfies the following ordinary differential equation

$$s_t(t) + q(s(t)) = 0, \quad s(0) = E(0) \tag{1.17}$$

The (nonlinear), monotone increasing function $q(s)$ is determined entirely from the behavior at the origin of the nonlinear functions g and it is given by the following algorithm.

$$q \equiv I - (I + p)^{-1} \tag{1.18}$$

$$p \equiv (I + \mathcal{H}_0)^{-1}(\frac{\cdot}{K}) \tag{1.19}$$

$$\mathcal{H}_0 \equiv \mathcal{H}(\cdot / mes\Sigma_1) \tag{1.20}$$

where $\Sigma_1 \equiv \Gamma_1 \times (0, T)$ and the constant $K > 0$ does not depend on $E(0)$.

REMARK 1.1. *Note that the decay rates established in Theorem 1.2 can be computed explicitly (elementary calculus) by solving the nonlinear ODE system (1.17), once the behavior of $g(s)$ at the origin is specified. If the nonlinear function g is bounded from below by a linear function, then it can be shown that the decay rates predicted by Theorem 1.2 are exponential. This is to say that there exist positive constants C, ω, possibly depending on $E(0)$, such that*

$$E(t) \leq Ce^{-\omega t}, \quad for \ t > 0.$$

If, instead, this function has a polynomial growth (resp. exponential decay) at the origin, than the decay rates are algebraic (resp. logarithmic) (see [23] where these elementary computations are explicitly provided).

REMARK 1.2. *The idea of expressing decay rates for the energy function in terms of solutions of appropriately constructed nonlinear ODE goes back to [23]. In order to achieve this, a critical role is played by a function $\mathcal{H}$ which captures the behavior of nonlinear function $g(s)$ at the origin. We mention that the same idea, including the construction of function $\mathcal{H}$, was used recently in [27] (see also [28]) where it was employed in a more restrictive setting, under additional regularity assumptions imposed on the nonlinear function g.*

REMARK 1.3. *It should be noted that our results, in addition to providing a bona fide square integrable feedback control for the system, do not require any geometric constraints ("star shaped" type) imposed on the controlled portion of the boundary (ie Γ_1). This is in contrast with [26, 27] and also with other literature dealing with a problem of boundary stabilization of elastodynamic linear systems see-[1, 13, 14] and references therein. To cope with this technical issue, we use "sharp" regularity results developed for the traces of elastodynamic solutions [7, 8]. By doing this we are able to establish the needed "sharp" regularity of boundary traces for the thermoelastic system - a result of independent PDE interest.*

REMARK 1.4. *One can also consider heat equation with Dirichlet boundary data. The analysis is even simpler in this case. We could also take Neumann boundary conditions for the wave part on the uncontrolled part of the boundary Γ_0. In such case, one would need to assume convexity of Γ_1. On the other hand, in this pure Neumann case there is no need to assume that $\overline{\Gamma_0} \cap \overline{\Gamma_1} =$ emptyset. This is in line with recent results corresponding to the wave equations[25].*

REMARK 1.5. *We note that in order to guarantee the decay rates for the energy function, the presence of the damping term u_t on the boundary is necessary. Indeed, this follows from the fact that thermal dissipation in two and higher dimensional linear system of elasticity may* *provide only strong stability and not the exponential stability- see[6, 12]. For this reason, it is necessary to introduce "dissipation " acting at least on the selenoidal part [2] of the horizontal displacement.*

The remainder of this paper is devoted to the proof of Theorem 1.2. Section 2 presents trace regularity results valid for the thermoelastic systems. These results are critical for stability analysis presented in Sections 3 and 4. The main body of the proof of Theorem 1.2 is in section 3.

2 Preliminary Results and Trace Regularity for Thermoelastic Systems

In this section we shall formulate and prove several preliminary estimates which deal with the trace regularity of solutions to the equations given by (1.1), (1.11). These results, while important in proving the main theorem, are also of independent interest.

2.1 Dissipativity Equality

A starting point is, as usual, the disspativity equality which states that the energy of the entire system is nonincreasing. This fact alone does not prove, of course, that the energy is decaying, but it is a necessary preliminary step of stability analysis.

LEMMA 2.1. *Let u, w be a finite energy solution of system (1.1), (1.2), (1.3), (1.11). Then, for any $s \leq t$*

$$E(t) + 2\int_s^t [\int_{\Gamma_1} g(u_t)u_t d\Gamma_1 + \int_\Omega |\nabla\phi|^2 d\Omega]dz + 2\int_s^t \int_\Gamma \lambda|\phi|^2 d\Gamma dz = E(s) \tag{2.1}$$

Proof. The proof is standard and it follows by classical energy type of argument (we multiply (1.1), (1.11) by u_t, ϕ integrate over $\Omega \times (s,t)$ and apply the Divergence Theorem first to smooth solutions which is then extended by density to all weak solutions. If function g does not display enough regularity and "smooth" solutions do not exist, we apply regularization procedure as in [23], to obtain the estimate for the "regularized problem" and then pass with the limit on the regularization parameter. □

2.2 Trace Regularity

This subsection provides trace regularity results which are critical for the proof of stability estimates *without* assuming the geometric conditions on Γ_1 and *without* considering the tangential derivatives of the displacements in the structure of the stabilizing feedback. These estimates are based on the corresponding trace estimates valid for the linear model of dynamic elasticity [7]. Similar estimates for the wave equation are derived in [24]. Here the main idea is to obtain the estimates for the tangential derivatives on the boundary in terms of the velocity traces and the lower order terms (defined below). To formulate these results we introduce some notation. Let $T > 0$ be fixed. In fact, from now on we shall assume that T is sufficiently large and greater than the finite speed of propagation corresponding to equation (1.1). We denote: $Q \equiv [0,T] \times \Omega$, $\Sigma_\alpha \equiv [\alpha, T-\alpha] \times \Gamma_1$ where $\alpha < T/2$. $\Sigma_1 \equiv [0,T] \times \Gamma_1$, $\Sigma_0 \equiv [0,T] \times \Gamma_0$, $\Sigma \equiv [0,T] \times \Gamma$

We shall also use the following notation for Sobolev norms:

$$|u|_{\alpha,\Omega} \equiv |u|_{H^\alpha(\Omega)}, \ |u|_{\alpha,\Gamma} \equiv |u|_{H^\alpha(\Gamma)}$$

and for the inner products

$$(u,v)_\Omega \equiv (u,v)_{L_2(\Omega)}; \ \langle u,v\rangle_\Gamma \equiv (u,v)_{L_2(\Gamma)}$$

For $\alpha < 0$, $H^\alpha(\Omega) \equiv [H^\alpha(\Omega)]'$ where the duality is with respect to $L_2(\Omega)$ inner products. By the same symbol we shall also denote norms/ inner products of two copies of L_2 or H^α spaces. This should not create any confusion, since the meaning will be clear from the context.

The constant C is a generic constant, different in various occurrences. $C(E(0))$ denotes the quantities bounded in terms of $E(0)$.

LEMMA 2.2. *Let u, ϕ be a finite energy solution corresponding to the system (1.1), (1.11), (1.2), (1.3). Then, for any $0 < \epsilon < 1/4$, there exist a constant C such that the following trace regularity is valid:*

$$\int_{\Sigma\alpha} |\nabla u|^2 d\Sigma_\alpha \le C_\alpha \int_{\Sigma_1} [|u_t|^2 + |g(u_t)|^2] d\Sigma_1$$
$$+C_\alpha \int_0^T |\phi|^2_{1,\Omega} dt + C_\alpha \int_0^T |u|^2_{1-\epsilon,\Omega} dt \tag{2.2}$$

REMARK 2.1. *Notice that the regularity of the trace of ∇u, stated in Lemma 2.2, does not follow from the standard interior regularity of finite energy solutions via the Trace Theory. These are independent regularity results which rely heavily on microlocal arguments applied to the dynamic system of elasticity.*

Proof. STEP 1. We begin with the following trace regularity result valid [7] for the linear model of dynamic elasticity. Define

$$F(x,y,t) \equiv -\nabla\phi(x,y,t) \tag{2.3}$$

where w is a finite energy solution corresponding to the system (1.1). Then the solution u satisfies the following "linear" system of dynamic elasticity

$$u_{tt} - divC\epsilon(u) = F \text{ in } Q \tag{2.4}$$

According to [7], for all $\epsilon < 1/2$, we have the estimate:

$$\int_{\Sigma_\alpha} |\nabla u_T|^2 d\Sigma_\alpha \le C_\alpha \int_0^T [|u_t|^2_{0,\Gamma_1} + |\epsilon(u)\nu|^2_{0,\Gamma_1} + |F|^2_{-1/2,\Omega} + |u|^2_{1-\epsilon,\Omega}] dt \tag{2.5}$$

where, we recall, $|u|_{-1/2,\Omega} \equiv |u|_{[H^1/2(\Omega)]'}$ with duality considered as a duality pairing with respect to $L_2(\Omega)$ topology.

Using the boundary conditions satisfied on Γ_1 yields

$$\int_{\Sigma_\alpha} |\nabla u\tau|^2 d\Sigma_\alpha \le C_\alpha \int_0^T [|u_t|^2_{0,\Gamma_1} + |g(u_t)|^2_{0,\Gamma_1} + |\phi|^2_{0,\Gamma_1} + |F|^2_{-1/2,\Omega} + |u|^2_{1-\epsilon,\Omega}]dt \tag{2.6}$$

REMARK 2.2. *The estimate in inequality (2.6), when applied to the homogeneous system of dynamic elasticity, states that the traces of the tangential derivatives of u are bounded by the traces of velocity modulo lower order terms. A result of similar nature was first obtained for the classical wave equation in[24].*

STEP 2. We shall estimate the fourth term on the right hand side of inequality in (2.6)

PROPOSITION 2.3. *Let $0 < \epsilon < 1/2$. Then the function F defined in (2.3) satisfies for all $t \ge 0$:*

$$|F(t)|_{-1/2,\Omega} \le C|\phi(t)|_{1/2+\epsilon,\Omega} \tag{2.7}$$

Proof. Let $\psi \in H^{1/2}(\Omega)$. Direct computations involving duality argument give:

$$(F, \psi)_{0,\Omega} = (-\nabla\phi, \psi)_\Omega \le C|\phi|_{1/2+\epsilon,\Omega}|\psi|_{1/2-\epsilon,\Omega} \tag{2.8}$$

which, via duality, proves the assertion in the proposition. □

STEP 3. We shall next estimate the normal derivatives of the vector u.

PROPOSITION 2.4. *For all $\epsilon < 1/4$ we have*

$$\int_{\Sigma_\alpha} |\nabla u\nu|^2 d\Sigma_1 \le C \int_{\Sigma_\alpha} [|g(u_t)|^2 + |\nabla u\tau|^2 + |u|^2 + |\phi|^2] d\Sigma_\alpha \tag{2.9}$$

Proof. Reading off the boundary conditions for the variable u we obtain the relation:

$$\epsilon(u)\nu = \vec{g} \tag{2.10}$$

where we introduce the variable

$$\vec{g} \equiv -C^{-1}[g(u_t) + ku - \phi\nu]$$

$\vec{g}$ satisfies the estimate

$$\begin{aligned} |\vec{g}|^2_{L_2(\Sigma_\alpha)} &\le C[|g(u_t)|^2_{L_2(\Sigma_\alpha)} + |u|^2_{L_2(\Sigma_\alpha)} + |\phi|^2_{L_2(\Sigma_\alpha)}] \\ &\le C[|g(u_t)|^2_{L_2(\Sigma_\alpha)} + |u|^2_{L_2(\Sigma_\alpha)} + |\phi|^2_{L_2(\Sigma_\alpha)}]dt \\ &\le C|g(u_t)|^2_{L_2(\Sigma_\alpha)} + C\int_\alpha^{T-\alpha} [|u|^2_{L_2(\Sigma_\alpha)} + |\phi|^2_{L_2(\Sigma_\alpha)}]dt \end{aligned} \tag{2.11}$$

On the other hand, denoting $\vec{d} \equiv \nabla u\tau$, and writing

$$\epsilon(u) \cdot \nu = \vec{g}; \quad \nabla u\tau = \vec{d} \tag{2.12}$$

leads to the algebraic linear system of the form $A\vec{u} = [\vec{g}, \vec{d}]^T$, where $\vec{u}$ has n^2 coordinates equal to $\frac{\partial u_i}{\partial x_j}$; $i,j = 1 \dots n$ We shall show that this system is uniquely solvable for each point $x \in \Gamma_1$. To accomplish this it is enough to show

$$(\nabla u + \nabla^T u)\nu = 0, \ \nabla u \tau = 0 \Rightarrow \nabla u \equiv 0; \quad on\ \Gamma_1 \tag{2.13}$$

In what follows we adopt the convention that the vector is a column vector.

From $\nabla u \tau = 0$ we infer that there exist a vector $k \in R^n$ such that $\nabla u_i = k_i \nu^T$, $k = [k_1, k_2, \dots k_n]$, or $\nabla u = k\nu^T$ Therefore

$$\nabla u \nu = k\nu^T \cdot \nu = k(\nu, \nu)_{R^n} = k$$

$$\nabla^T u \nu = \nu k^T \cdot \nu = \nu(k, \nu)_{R_n}$$

which imply

$$(\nabla u + \nabla^T u)\nu = k + (k \cdot \nu)\nu$$

The first condition in (2.13) then gives

$$k + (k \cdot \nu)\nu = 0$$

and taking inner product with a normal vector ν

$$k \cdot \nu + (k \cdot \nu) = 0 \Rightarrow k \cdot \nu = 0$$

Going back to

$$\nabla^T u \nu = \nu(k, \nu)_{R_n}$$

we obtain that

$$\nabla^T u \nu = 0 \Rightarrow \nabla u \nu = 0$$

which then combined with $\nabla u \tau = 0$ gives the desired conclusion $\nabla u = 0$. This allows for local (around each point in Γ_1) solvability of system in (2.12). Compactness of Γ yields the inequality:

$$\int_{\Sigma_\alpha} |\frac{\partial u_i}{\partial x_j}|^2 d\Sigma_\alpha \leq C \int_{\Sigma_\alpha} [|\vec{g}|^2 + |\vec{d}|^2] d\Sigma_\alpha \tag{2.14}$$

The above estimate together with (2.11) leads to the result in (2.9). □

STEP 4

Collecting the results of the estimates (2.6), (2.9), (2.7) we obtain:

$$\int_{\Sigma\alpha} |\nabla u|^2 d\Sigma_\alpha \leq C_\alpha \int_{\Sigma_1} [|u_t|^2 + |g(u_t)|^2 + |\phi|^2] dx dt$$
$$+ C_\alpha \int_0^T |\phi|^2_{1,\Omega} dt + C_\alpha \int_0^T [|u(t)|_{1-\epsilon}] dt \tag{2.15}$$

which estimate, via Trace Theorem leads to the desired result in Lemma 2.2 □

3 Stabilizability Estimate

The main aim in this section is to prove the following stabilizability estimate

LEMMA 3.1. *Let u, ϕ be a weak solution to (1.1), (1.11), (1.2), (1.3). Assume the geometric condition (1.14) on Γ_0. Then, there exists T large enough such that for any constant $0 < \epsilon < 1/4$ the following estimate is valid:*

$$E(T) + \int_0^T E(t)dt \leq C_T[\int_{\Sigma_1} [|u_t|^2 + |g(u_t)|^2]d_{\Sigma_1} + \int_0^T |\nabla\phi|^2_{0,\Omega}dt + C_T lot(u) \tag{3.1}$$

where we have used the notation for the "lower order terms" (lot(u))

$$lot(u) \equiv \int_0^T |u(t)|^2_{1-\epsilon,\Omega}dt \tag{3.2}$$

The estimate of the Lemma above, critical to the proof of the main stabillizability result, is an inverse type of the estimate. Indeed, it allows to reconstruct the energy of the system (1.1), modulo lower order terms, from the measurements of velocities of the horizontal displacement on the boundary and from the gradients of the temperature.

The reminder of this section is devoted to the proof of the Lemma 3.1. Here, the strategy used for the proof is to apply first the usual " differential multipliers" method (for an exposition of this method see the books [18], [15], [13] and references therein). These lead to the estimate for the energy in terms of *all* boundary traces and the lower order terms (see Lemma 3.2 below). The next, crucial, step is to eliminate the "unwanted" boundary traces, which is done by using sharp trace regularity results presented in Section 2. One should mention that the above procedure applies well when the regularity of solutions is sufficiently high in order to justify necessary PDE calculations. This is the case when the nonlinear function g is sufficiently smooth (see part II of Theorem 1.1) and the initial data are in appropriately high Sobolev's spaces. If, instead, function g is only continuous, then a special "regularization" argument as in [23] can be used, in order to construct "smooth" approximations of the original solutions. Performing the calculations on the approximate problem and the passage to the limit at the very end allow to reconstruct inequalities which remain valid for all weak solutions. In order to avoid the additional technical complications and for the sake of clarity of exposition we shall not do this here and for details we refer the reader to [23] and also [9], [10], in the context of plates.

3.1 Variational Formulation and Preliminary Identities

We begin by writing system (1.1) in a variational form. To this end let be given two test functions $\xi \in H^1(\Omega) \times H^1(\Omega)$. The thermoelastic system admits the following variational form:

$$(u_{tt}, \xi)_\Omega + (\mathcal{C}\epsilon(u), \epsilon(\xi))_\Omega + (\nabla\phi, \xi)_\Omega + \langle g(u_t) + ku - \phi\nu, \xi\rangle_{\Gamma_1} - \langle \mathcal{C}\epsilon(u)\nu, \xi\rangle_{\Gamma_0} = 0 \tag{3.3}$$

Note that we have used the boundary conditions satisfied on Γ.

We shall apply this variational form with various choices of test functions ξ and ψ. In order to facilitate verification of rather tedious computations we will provide below few elementary tensor identities.

In what follows the vector field $\boldsymbol{h}$ always denotes the radial vector field. ie $\boldsymbol{h} \equiv x - x_0$, $x_0 \in R^2$

$$\epsilon(\nabla u \boldsymbol{h} = \epsilon(u) + M \tag{3.4}$$

where the tensor M is given by;

$$M \equiv \begin{bmatrix} D^2_{x_1,x_i} u_1 \boldsymbol{h}_i & 1/2[D^2_{x_2,x_i} u_1 \boldsymbol{h}_i + D^2_{x_1,x_i} u_2 \boldsymbol{h}_i] \\ 1/2[D^2_{x_2,x_i} u_1 \boldsymbol{h}_i + D^2_{x_1,x_i} u_2 \boldsymbol{h}_i] & D^2_{x_2,x_i} u_2 \boldsymbol{h}_i \end{bmatrix} \tag{3.5}$$

and we have adopted double index notation to indicate the summation of the terms. If A is any symmetric fourth order tensor identified by its coefficients $a_{i,j}$,

$$A \equiv \{a_{i,j}\}$$

then it is straightforward to show that

$$AM = a_{k,j} D^2_{x_k,x_i} u_j \boldsymbol{h}_i \tag{3.6}$$

Let B be another symmetric tensor such that $a_{j,i} = c_{j,l} b_{l,i}$, with constant and symmetric coefficients $c_{j,i}$. Then:

$$div[(AB)\boldsymbol{h}] = 2A \cdot B + c_{i,l} D_{x_k} b_{l,j} b_{i,j} \boldsymbol{h}_k = 2A \cdot B + 2a_{j,i} D_{x_k} b_{j,i} \boldsymbol{h}_k \tag{3.7}$$

In the particular case when the tensors A and B are given by

$$A = \mathcal{C}[\epsilon(u)]; \quad B = \epsilon(u)$$

the formula above reads:

$$\begin{aligned} div[\mathcal{C}[\epsilon(u)][\epsilon(u)]\boldsymbol{h}] = \\ 2\mathcal{C}[\epsilon(u)][\epsilon(u)] + 2a_{i,j}[D^2_{x_k,x_j} u_i + w_{x_j} D^2_{x_i,x_k} w]\boldsymbol{h}_k \end{aligned} \tag{3.8}$$

Taking $A = \mathcal{C}\epsilon(u)$; $B = \epsilon(u)$, we obtain

$$div[(\mathcal{C}\epsilon(u)\epsilon(u))\boldsymbol{h}] = 2\mathcal{C}\epsilon(u)\epsilon(u) + 2a_{i,j}[D^2_{x_k,x_j} u_i]\boldsymbol{h}_k \tag{3.9}$$

3.2 Multipliers Estimate

In this subsection we shall prove a preliminary estimate which shows that the energy of the system is bounded by the boundary traces of the solutions, the gradients of the temperature , modulo the lower order terms.

LEMMA 3.2. *Let u, ϕ be a solution to (1.1),(1.11) with prescribed boundary conditions. Assume the geometric condition (1.14) to hold on Γ_0. Then, there exists T large enough, such that for any constant $0 < \epsilon < 1/4$ the following estimate is valid:*

$$E(T) + \int_0^T E(t)dt \leq C\int_{\Sigma_1} [|u_t|^2 + |g(u_t)|^2 + |\nabla u|^2]d\Sigma_1$$
$$+C[\int_0^T |\phi|^2_{1,\Omega}dt + lot(u) \tag{3.10}$$

Proof. Computations carried below in Step 1, based on " differential multipliers method", are reminiscent of these performed earlier in [17], [16] and later in [30].

STEP 1 (First multiplier- estimate for $|u_t|_{0,\Omega}$).

We use $\xi \equiv \nabla u(h$ as a test function in the variational equality(3.3).

By the virtue of (3.4),(3.6) applied with the second order tensor

$$A \equiv C\epsilon(u)$$

we obtain

$$(C\epsilon(u), \epsilon(\nabla u(h)) = (C\epsilon(u), \epsilon(u)) + (a_{i,j}, D^2_{x_k,x_j}u_i h_k) \tag{3.11}$$

Applying this relation in variational formulation (3.3) and integrating over Q yields

$$(u_t, \nabla u(h)_\Omega|_0^T - 1/2\int_\Sigma |u_t|^2 h\nu d\Sigma + \int_Q |u_t|^2 dQ$$
$$+\int_0^T [(C\epsilon(u), \epsilon(u))_\Omega + (a_{i,j}, D^2_{x_k,x_j}u_i h_k)_\Omega$$
$$+\langle g(u_t) + ku - \phi\cdot\nu, \nabla u(h\rangle_{\Gamma_1} - \langle C\epsilon(u), \epsilon(u)\nu h\rangle_{\Gamma_0} + (\nabla\phi, \nabla u(h)_\Omega]dt = 0 \tag{3.12}$$

where we have used the fact that u vanishes on Γ_0 and therefore

$$C\epsilon(u)\nu\nabla u(h = C\epsilon(u)\epsilon(u)\nu h \quad on \quad \Gamma_0 \tag{3.13}$$

To see (3.13), it suffices to notice the following identities taking place on Γ_0

$$\nabla uh = \frac{\partial u}{\partial\nu}\nu h; trace\epsilon(u) = \frac{\partial u}{\partial\nu}\nu$$
$$\epsilon(u)\nu = [\frac{\partial u_1}{\partial\nu}\nu_1^2, \frac{\partial u_2}{\partial\nu}\nu_2^2] + (1/2)\nu^T(\frac{\partial u}{\partial\nu}\nu^T)$$
$$\epsilon(u)\epsilon(u) = \frac{\partial u_i^2}{\partial\nu}\nu_i^2 + (1/2)(\frac{\partial u}{\partial\nu}\nu^T)^2 \tag{3.14}$$

Hence:

$$\epsilon(u)\nu\nabla u(h = \{[\frac{\partial u_1}{\partial\nu}\nu_1^2, \frac{\partial u_2}{\partial\nu}\nu_2^2] + (1/2)\nu^T(\frac{\partial u}{\partial\nu}\nu^T)\}\frac{\partial u}{\partial\nu}\nu h =$$
$$(\frac{\partial u_i^2}{\partial\nu}\nu_i^2 + (1/2)(\frac{\partial u}{\partial\nu}\nu^T)^2)\nu h = \epsilon(u)\epsilon(u)\nu h$$
$$trace\ (\epsilon(u))I\nu\nabla u(h = (\frac{\partial u}{\partial\nu})^2\nu h = trace\ \epsilon(u)I\epsilon(u)\nu h \tag{3.15}$$

The identity in (3.13) follows now from (3.15) and the definition of $\mathcal{C}$.

By using (3.9) and the divergence Theorem in (3.12) we obtain:

$$(u_t, \nabla u(h)_\Omega|_0^T - 1/2 \int_\Sigma |u_t|^2 h\nu d\Sigma + \int_Q |u_t|^2 dQ$$

$$- \int_0^T [+\langle g(u_t) + ku - \phi\nu, \nabla u(h)\rangle_{\Gamma_1}$$

$$-1/2\langle \mathcal{C}\epsilon(u), \epsilon(u)\nu h\rangle_{\Gamma_0} + 1/2\langle \mathcal{C}\epsilon(u), \epsilon(u)\nu h\rangle_{\Gamma_1} + (\nabla\phi, \nabla u(h)_\Omega]dt = 0 \tag{3.16}$$

$$\int_0^T |u_t|_{0,\Omega}^2 dt \le C[E(0) + E(T)] + \int_{\Sigma_0} \mathcal{C}\epsilon(u)\epsilon(u)\nu h d\Sigma_0$$

$$+1/2 \int_{\Sigma_1} [|u_t|^2 + \nabla u(h[g(u_t) + ku - \phi\nu]]d\Sigma_1$$

$$+C \int_0^T |\phi|_{1,\Omega}|u|_{1,\Omega} dt$$

$$\le C[E(0) + E(T)] + C \int_{\Sigma_1} [|u_t|^2 + |g(u_t)|^2 + |\nabla u|^2]d\Sigma_1$$

$$+ \int_0^T [\epsilon|u|_{1,\Omega}^2 + C_\epsilon|\phi|_{1,\Omega}^2]dt$$

$$\le C[E(0) + E(T)] + C \int_{\Sigma_1} [|u_t|^2 + |g(u_t)|^2 + |\nabla u|^2]d\Sigma_1$$

$$+ \int_0^T [\epsilon|u|_{1,\Omega}^2 + +|\phi|_{1,\Omega}^2]dt \tag{3.17}$$

STEP 2 (second multiplier- the difference of potential and kinetic energy)

We apply variational equality (3.3) with the following choice of test functions: $\xi = u$, $\psi = w$.

$$(u_t, u)_\Omega|_0^T + \int_0^T [(\mathcal{C}\epsilon(u), \epsilon(u))_\Omega - |u_t|_{0,\Omega}^2 + (\nabla\phi, u)_\Omega$$

$$+\langle g(u_t) + ku - \phi\nu, u\rangle_{\Gamma_1} - \langle \mathcal{C}\epsilon(u)\nu, u\rangle_{\Gamma_0}]dt = 0 \tag{3.18}$$

Here we took into account the boundary conditions on Γ_0 for the variable w.

(3.18) implies

$$\int_0^T [E_p(t) - E_k(t)]dt \le C[E(0) + E(T)]$$

$$+C \int_0^T [|\phi|_{1,\Omega}^2 + |g(u_t)|_{0,\Gamma_1}^2]dt + Clot(u) \tag{3.19}$$

Combining with the results of Step 1 (inequality (3.17) we obtain

$$\int_0^T E_k(t)dt \le \epsilon \int_0^T E_p(t)dt + C[E(0) + E(T)]$$

$$+C \int_{\Sigma_1} [|u_t|^2 + |g(u_t)|^2 + |\nabla u|^2]d\Sigma_1 + C_\epsilon \int_0^T |\phi|_{1,\Omega}^2 dt \tag{3.20}$$

Combining the above with inequality (3.19) in Step 2, and selecting suitably small ϵ yields

$$\int_0^T E(t)dt \leq C[E(0) + E(T)] + C\int_{\Sigma_1} [|u_t|^2 + |g(u_t)|^2 + |\nabla u|^2] d\Sigma_1$$
$$+C\int_0^T |\phi|^2_{1,\Omega}dt + Clot(u) \tag{3.21}$$

Taking T large enough and applying energy identity (2.1) to eliminate $E(0), E(T)$ we obtain the result in Lemma 3.2. □

Our next step is to absorb the boundary traces in the inequality in Lemma 3.2.

3.3 Absorption of Boundary Traces - Completion of the Proof of Lemma 3.1

In this subsection we shall show that the boundary traces, involving the first order spatial derivatives of u are redundant. This will be done with a help of trace regularity results formulated in Section 2.

LEMMA 3.3. *Under the assumptions of Lemma 3.2, we have:*

$$E(T) + \int_0^T E(t)dt \leq CE + C\int_{\Sigma_1} [|u_t|^2 + |g(u_t)|^2] d\Sigma_1$$
$$+C\int_0^T |\phi|^2_{1,\Omega}dt + Clot(u) \tag{3.22}$$

Proof. From the result of Lemma 3.2 applied to the interval $[\alpha, T-\alpha]$ we obtain:

$$E(T-\alpha) + \int_\alpha^{T-\alpha} E(t)dt \leq C\int_{\Sigma_\alpha} [|u_t|^2 + |g(u_t)|^2 + |\nabla u|^2] d\Sigma_\alpha$$
$$+C[\int_\alpha^{T-\alpha} |\phi|^2_{1,\Omega}dt + lot(u)] \tag{3.23}$$

Here we took the advantage of the dissipativity property in Lemma 2.1 which allows to upper-bound $E(\alpha)$ by $E(0)$. On the other hand, from regularity results stated in Lemma 2.2 we infer the estimates

$$\int_{\Sigma\alpha} |\nabla u|^2 d\Sigma_\alpha \leq C_\alpha[\int_{\Sigma_1} [|u_t|^2 + |g(u_t)|^2] d\Sigma_1 + \int_0^T |\phi|^2_{1,\Omega}dt + lot(u)] \tag{3.24}$$

Combining (3.23), (3.24) and recalling, again, the dissipativity equality (2.1) gives

$$E(T-\alpha) + \int_\alpha^{T-\alpha} E(t)dt \leq C_\alpha \int_{\Sigma_1} [|u_t|^2 + |g(u_t)|^2] d_{\Sigma_1}$$
$$+C_\alpha \int_0^T |\phi|^2_{1,\Omega}dt + Clot(u) \tag{3.25}$$

By applying (3.23) and taking ϵ suitably small, we obtain:

$$E(T-\alpha) + \int_{\alpha}^{T-\alpha} E(t)dt \le CE(0) + C_\alpha \int_{\Sigma_1} [|u_t|^2 + |g(u_t)|^2] d_{\Sigma_1}$$
$$+ C_\alpha \int_0^T [|\phi|^2_{1,\Omega}] dt + C_\alpha lot(u) \quad (3.26)$$

To complete the proof of the Lemma we need to estimate the contribution of the energy on the subintervals $[0,\alpha]$ and $[T-\alpha,T]$. To accomplish this we denote the right hand side of (3.26) by $\mathcal{F}$. From dissipativity relation (2.1) and (3.26) we have for all $t \in [0,T]$:

$$E(t) \le E(T-\alpha) + C_\alpha \int_{\Sigma_1} [|u_t|^2 + |g(u_t)|^2] d\Sigma_1 + C_\alpha \int_0^T + |\phi|^2_{1,\Omega} dt \le C_\alpha \mathcal{F} \quad (3.27)$$

Hence

$$\int_0^\alpha E(t)dt + \int_{T-\alpha}^T E(t)dt \le C_\alpha \mathcal{F} \quad (3.28)$$

which combined with (3.26) and dropping irrelevant dependence on α

$$E(T) + \int_0^T E(t)dt \le C\mathcal{F} \quad (3.29)$$

Using once more dissipativity relations (2.1)

$$E(T) + \int_0^T E(t)dt \le CE(0) + C\int_{\Sigma_1} [|u_t|^2 + |g(u_t)|^2] d_{\Sigma_1}$$
$$+ C\int_0^T |\phi|^2_{1,\Omega} dt + Clot(u) \quad (3.30)$$

which completes the prof of Lemma 3.3. □

Proof. (of Lemma 3.1) follows from Lemma 3.3 and dissipativity equality (2.1). Indeed, from (2.1) and Lemma 3.3 we obtain:

$$TE(T) \le \int_0^T E(t)dt \le CE(T) + C\int_{\Sigma_1} [|u_t|^2 + |g(u_t)|^2] d_{\Sigma_1}$$
$$+ C\int_0^T |\phi|^2_{1,\Omega} dt + Clot(u) \quad (3.31)$$

Taking T large enough so that $TE(T) \ge 2CE(T)$ we obtain

$$E(T) \le C\int_{\Sigma_1} [|u_t|^2 + |g(u_t)|^2] d_{\Sigma_1} + C[\int_0^T |\phi|^2_{1,\Omega} dt + lot(u)] \quad (3.32)$$

Combining with (3.31)

$$\int_0^T E(t)dt \leq C \int_{\Sigma_1} [|u_t|^2 + |g(u_t)|^2] d_{\Sigma_1}$$
$$+C \int_0^T |\phi|_{1,\Omega}^2 dt + Clot(u) \tag{3.33}$$

Adding the results of inequalities in (3.33) and (3.32) yields the desired conclusion in Lemma 3.1. □

4 Completion of the Proof of Theorem 1.2

4.1 Absorption of the Lower Order Terms

The main result of this section is

LEMMA 4.1. *Let (u, ϕ) be a solution to (1.1), (1.11) with prescribed boundary conditions. Let $T > 0$ be large enough. Assume that $k > 0$ or $\Gamma_0 \neq \emptyset$. Then, there exists a constant C_T such that*

$$lot(u) \leq C_T \int_0^T [|\phi|_{1,\Omega}^2 + |u_t|_{0,\Gamma_1}^2 + |g(u_t)|_{0,\Gamma_1}^2] dt \tag{4.1}$$

Proof. The argument is, as usual, by contradiction. We take a sequence of initial data

$$(u_n(0), u_{nt}(0), \phi_n(0))$$

with the corresponding solutions $(u_n(t), u_{nt}(t), \phi_n(t))$. Contradicting to (4.1) we infer

$$\frac{lot(u_n)}{\int_0^T |\phi_n|_{1,\Omega}^2 + |u_{tn}|_{0,\Gamma_1}^2 + |g(u_{tn})|_{0,\Gamma_1} dt} \to \infty \tag{4.2}$$

By Lemma 3.1 we must have

$$\int_0^T |\phi_n|_{1,\Omega}^2 + |u_{tn}|_{0,\Gamma_1}^2 + |g(u_{tn})|_{0,\Gamma_1}^2 dt \to 0 \tag{4.3}$$

$$\int_0^T |\phi_n|_{1,\Omega}^2 + |u_n|_{1,\Omega}^2 + |u_{nt}|_{0,\Omega}^2 dt \leq C \tag{4.4}$$

These bounds and the compactness of Sobolev's embeddings allow us to deduce that on a subsequence, denoted by the same symbol, we have:

$$u_n \to u \textit{ weakly}^* \textit{ in } L_\infty(0, T; H^1(\Omega)); \quad u_{nt} \to u_t \textit{ weakly}^* \textit{ in } L_\infty(0, T; L_2(\Omega))$$
$$\phi_n \to 0 \textit{ in } L_2(0, T; H^1(\Omega))$$
$$lot(u_n) \to lot(u); \quad u_{tn}, g(u_{tn}) \to 0 \textit{ in } L_2(\Sigma_1) \tag{4.5}$$

The above convergence allow us to pass with the limit on the original equation and to deduce that u, ϕ satisfy

$$\begin{aligned} u_{tt} - divC\epsilon(u) = 0; \text{ in } \Omega \times (0,T) \\ C\epsilon(u)\nu + ku = 0; u_t = 0, \ on\ \Gamma_1 \times (0,T); \\ u = 0 \ on\ \Gamma_0 \end{aligned} \tag{4.6}$$

Here we have used the fact that by (4.3)

$$C\epsilon(u_n)\nu + ku_n = \phi_n\nu - g(u_{nt}) \to 0; \ on\ \Gamma_1$$

(4.6) implies after denoting $\bar{u} \equiv u_t$

$$\begin{aligned} \bar{u}_{tt} - divC\epsilon(\bar{u}) = 0 \ in\ \Omega \times (0,T) \\ C\epsilon(\bar{u})\nu + k\bar{u} = 0, \bar{u} = 0; \ on\ \Gamma_1 \times (0,T); \bar{u} = 0 \ on\ \Gamma_0 \times (0,T) \end{aligned} \tag{4.7}$$

By Holmgren's unique continuation result applied to the system of thermoelasticity we infer that $\bar{u} = u_t \equiv 0$. Feeding this information back to the static form of equations in (4.6) yields $u \equiv 0$ and by (4.5) $lot(u) = 0$. Thus we must have (see (4.5))

$$lot(u_n) \to 0; \tag{4.8}$$

We define next

$$\tilde{u_n} \equiv \frac{u_n}{c_n}, \ \tilde{\phi_n} \equiv \frac{\phi_n}{c_n}, \ c_n^2 \equiv lot(u_n) \to 0$$

Without loss of generality we can assume that $c_n > 0$ for large n. Clearly from (4.2)

$$lot(\tilde{u}_n) = 1, \ \int_0^T [|\tilde{\phi_n}|^2_{1,\Omega} + |\tilde{u_{nt}}|^2_{0,\Gamma_1}]dt \to 0 \tag{4.9}$$

Dividing by c_n the inequality in Lemma 3.1 gives :

$$\int_0^T [|\tilde{u_n}|^2_{1,\Omega} + |\tilde{u_{nt}}|^2_{0,\Omega}]dt \le C \tag{4.10}$$

which, in turn, gives the convergence (on a subsequence)

$$\begin{aligned} \tilde{w_{nt}} \to \tilde{w_t}; \ \tilde{u_{nt}} \to \tilde{u_t} \ weakly\ in\ L_2(0,T; L_2(\Omega)) \\ \tilde{u_n} \to \tilde{u} \ weakly\ in\ L_2(0,T; H^1(\Omega)); \ \ lot(\tilde{u_n}) \to lot(\tilde{u}). \end{aligned} \tag{4.11}$$

So that

$$lot(\tilde{u}) = 1; \tag{4.12}$$

Our next step is to pass with the limit on the equation for $\tilde{u}_n$ ie

$$\begin{aligned} \tilde{u}_{ntt} - divC[\epsilon(\tilde{u_n})] + \nabla\tilde{\phi}_n = 0; \ in\ \Omega \times (0,T) \\ C[\epsilon(\tilde{u_n})] \cdot \nu + k\bar{u}_n - \tilde{\phi}_n\nu = -\frac{g(u_{nt})}{c_n}; \ on\ \Gamma_1 \times (0,T) \\ \tilde{u}_n = 0; \ on\ \Gamma_0 \times (0,T) \end{aligned} \tag{4.13}$$

By virtue of (4.9) and (4.11) and recalling that $u_n \to 0$ *in* Ω see (4.8), $\tilde{\phi_n} \to 0$ in $H^1(\Omega)$ (see 4.9), the passage with the limit on all terms in the equation (4.13) is straightforward and it gives

$$\begin{aligned} \tilde{u}_{tt} - div C\epsilon(\tilde{u}) = 0 \; in \; \Omega \times (0,T) \\ C\epsilon(\tilde{u})\nu + k\tilde{u} = 0; \tilde{u}_t = 0; \; on \; \Gamma_1 \times (0,T) \\ \tilde{u} = 0; \; on \; \Gamma_0 \times (0,T) \end{aligned} \tag{4.14}$$

As before we obtain that $\tilde{u}_t \equiv 0$ and going back to the equation above (in the static version) we infer

$$\epsilon(\tilde{u}) = 0; \; in \; \Omega; \; C\epsilon(\tilde{u})\nu + k\tilde{u} = 0 \; on \; \Gamma_1; \quad \tilde{u} = 0 \; on \; \Gamma_0 \tag{4.15}$$

So we obtain (note that $k > 0$ when $\Gamma_0 = \emptyset$), $\tilde{u} = 0$ which contradicts (4.12). □

4.2 Completion of the Proof of Theorem 1.2

By combining the results of Lemmas 4.1 and 3.1 we have obtained:

LEMMA 4.2. *Let u, ϕ be a solution to (1.1), (1.11) with prescribed boundary conditions. Assume the geometric condition (1.14) on Γ_0. Then, there exists T large enough such that the following estimate is valid:*

$$E(T) + \int_0^T E(t)dt \le C_T[\int_{\Sigma_1} [|u_t|^2 + |g(u_t)|^2] d_{\Sigma_1} + \int_0^T |\phi|^2_{1,\Omega} dt] \tag{4.16}$$

In what follows we shall denote

$$\Sigma_A \equiv \{(t,x) \in \Sigma_1 : |u_t| \le 1\}; \qquad \Sigma_B \equiv \Sigma_1 - \Sigma_A$$

recalling the definition of the function $\mathcal{H}$ we obtain:

$$\begin{aligned} \int_{\Sigma_1} [|u_t|^2 + |g(u_t)|^2] d\Sigma_1 &\le \int_{\Sigma_A} \mathcal{H}(u_t, g(u_t)) d\Sigma_A + C \int_{\Sigma_B} g(u_t) u_t d\Sigma_B \\ &\le \int_{\Sigma_1} [\mathcal{H}(u_t, g(u_t)) + C u_t g(u_t)] d\Sigma_1 \end{aligned} \tag{4.17}$$

By using Jensen's inequality we infer:

$$\int_{\Sigma_1} [|u_t|^2 + |g(u_t)|^2 \le [CI + \mathcal{H}_0] \int_{\Sigma_1} u_t g(u_t) d\Sigma_1 \tag{4.18}$$

Denoting by

$$\mathcal{F} \equiv \int_{\Sigma_1} u_t g(u_t) d\Sigma_1 + \int_{\Sigma} \lambda |\phi|^2 d\Sigma + \int_0^T |\nabla\phi|^2_{0,\Omega} + |\nabla\theta|^2_{0,\Omega} dt \tag{4.19}$$

the inequality in (4.16) (recall $\lambda > 0$) reads:

$$E(0) + E(T) + \int_0^T E(t)dt \leq C_T[\mathcal{F} + \mathcal{H}_0(\mathcal{F})] \tag{4.20}$$

Since the function $\mathcal{H}_0 + I$ is monotone increasing, we can write:

$$[I + \mathcal{H}_0]^{-1}(\frac{E(T)}{C_T}) \leq \mathcal{F} = E(0) - E(T) \tag{4.21}$$

where we have used the dissipativity equality (2.1). This in turn gives:

$$p(E(T)) + E(T) \leq E(0) \tag{4.22}$$

where the monotone function p is defined in (1.20) in Section 1 with $K = C_T$. Thus we have proved

LEMMA 4.3. *Let u, ϕ be a solution to the original equation. Then, there exist a constant $T > 0$ such that*

$$p(E(T)) + E(T) \leq E(0) \tag{4.23}$$

where the monotone function p is defined in section 1.

The final conclusion of Theorem follows now from (4.23) and Lemma 3.2, 3.3 in [23]
The estimates which are first established for regular solutions first, are extended to all weak solutions by the virtue of uniqueness of weak solutions established in Theorem 1.1. □.

Bibliography

[1] F. Alabau and V. Komornik. " Boundary observability, controllability and stabilization of linear elastodynamic systems ". *SIAM J. Control Preprint*, to appear.

[2] A. Benabdallah and I. Lasiecka. " Exponential decay rates for a full von Karman system of dynamic thermoelasticity ". *Journal Differential Equations*, to appear.

[3] A. Benabdallah and D. Teniou. " Exponential Stability of a Von karman model with Thermal efffects ". *Electronic Journal of Differential Equations*, 7:1–13, 1998.

[4] Ph. Ciarlet and P. Rabier. " *Les Equations de von Karman* ". Springer Verlag, 1982.

[5] G. Duvaut and J.L.Lions. " *Les Inequations en Mecaniques et en Physiques* ". Dunod, 1972.

[6] G.Lebeau and E. Zuazua. " Sur la decroissance non uniforme de l'energie dans les systemes de la thermoelasticite lineaire ". *C.R. Acad. Sci. Paris*, 324:409–415, 1997.

[7] M. A. Horn. " Sharp trace regularity of the traces to solutions of dynamic elasticity ". *Journal of Mathematical Systems, Estimation and Control*, 8:217–229, 1998.

[8] M. A. Horn. " Uniform stabilization of elastodynamic systems with boundary dissipation ". *Journal Mathematical Analysis ans Applications*, 1998.

[9] M. A. Horn and I. Lasiecka. "Uniform decay of weak solutions to a von Karman plate with nonlinear dissipation ". *Differential and Integral Equations*, 7:885–908, 1994.

[10] M. A. Horn and I. Lasiecka. "Global stabilization of a dynamic von Karman plate with nonlinear boundary feedback ". *Applied Mathematics and Optimization*, 31:57–84, 1995.

[11] T. Von Karman. "Festigkeitprobleme in Maschinenbau ". *Encyklopedie der Mathematischen Wissenschaften*, 4:314–385, 1910.

[12] H. Koch. " Slow decay in linear thermoelasticity ". *Preprint SFB 97-47*, 1997.

[13] V. Komornik. *"Exact controllability and stabilization, The multipliers method"*. Masson, 1994.

[14] J. Lagnese. "Boundary stabilization of linear elastodynamic systems ". *SIAM J. on Control*, 21:968- 984, 1983.

[15] J. Lagnese. *" Boundary Stabilization of Thin Plates"*. SIAM, 1989.

[16] J. Lagnese. "Modeling and stabilization of nonlinear plates ". *International Ser. Num. Math.*, 100:247–264, 1991.

[17] J. Lagnese. "Uniform asymptotic energy estimates for solutions of the equations of dynamic plane elasticity with nonlinear dissipation at the boundary ". *Nonlinear Analysis*, 16:35–54, 1991.

[18] J. Lagnese and J.L.Lions. *" Modeling, Analysis and Control of Thin Plates "*. Masson, 1988.

[19] J. Lagnese and G. Leugering. "Uniform stabilization of a nonlinear beam by nonlinear boundary feedback ". *Journal of Differential Equations*, 91:355–388, 1991.

[20] I. Lasiecka. " Existence and uniqueness of the solutions to second order abstract equations with nonlinear and nonmonotine boundary conditions ". *Nonlinear Analysis*, 23:797–823, 1994.

[21] I. Lasiecka. " Uniform stabilizability of a full von Karman system with nonlinear boundary feedback ". *SIAM J. on Control*, 36 nr 4:1376–1422, 1998.

[22] I. Lasiecka. "Uniform decay rates for full von Karman system of dynamic thermoelasticity with free boundary conditio ns and partial boundary dissipation. ". *Communications on PDE's*, 24:1801–1847, 1999.

[23] I. Lasiecka and D. Tataru. " Uniform boundary stabilization of semilinear wave equations with nonlinear boundary damping ". *Differential and Integral Equations*, 6:507–533, 1993.

[24] I. Lasiecka and R. Triggiani. " Uniform stabilization of the wave equation with Dirichlet and Neumann feedback control without geometrical conditions ". *Applied Mathematics and Optimization*, 25:189–224, 1992.

[25] I. Lasiecka, R. Triggiani, and X. Zhang. " Nonconservative wave equations with unbounded Neumann B.C.: global uniqueness and observability ". *Proceedings of AMS: Contemporary Mathematics*, to appear.

[26] W. Liu. " Partial exact controllability and exponential stability in higher-dimensional linear thermoelasticity ". *ESAIM*, 3:23–48, 1998.

[27] W.J. Liu and E. Zuazua. "Uniform stabilization of the higher dimensional system of thermoelasticity with a nonlinear boundary feedback ". *Quarterly in Applied Mathematics.*

[28] W.J. Liu and E. Zuazua. " Decay rates for dissipative wave equations ". *Ricerca di Palermo*, to appear.

[29] N. Morozov. " Non-linear vibrations of thin plates with allowance for rotational inertia ". *Sov. Math*, 8:1137–1141, 1967.

[30] J. Puel and M. Tucsnak. " Boundary stabilization for the von Karman equations ". *SIAM J. on Control*, 33:255–273, 1996.

[31] J. Puel and M. Tucsnak. "Global existence for the full von Karman system ". *Applied Mathematics and Optimization*, 34:139–161, 1996.

Irena Lasiecka. University of Virginia, Department of Mathematics, Kerchof Hall, P.O. Box 400137, Charlottesville, VA 22904-4137, USA
E-mail: il2v@virginia.edu

Domain Optimization for Unilateral Problems by an Embedding Domain Method

Andrzej Myśliński[1]

Abstract. The paper deals with the mathematical modeling and shape optimization of elastic contact problems. The fictitious domain approach and Lagrange multipliers approach are employed to solve numerically the shape optimal design problem for an elastic body in unilateral contact with a rigid foundation. This approach allows to perform the computations on a fixed domain. The finite element approximation is employed. The design sensitivity analysis for the discrete problem is performed. The Lagrangian approach is used to solve numerically the state system. The descent gradient method with projection is used as an optimization method. Numerical results are provided.

1 Introduction

The paper is concerned with the numerical solution of a shape optimal design problem of an elastic body in unilateral contact with a rigid foundation. The equilibrium state of this contact problem is described by an elliptic variational inequality of the second order. The existence, uniqueness and regularity of solutions to this variational inequality were investigated in [4, 10].

The shape optimization problem for the elastic bodies in contact consists in finding, in a contact region, such shape of the boundary of the domain occupied by the body that the normal contact stress is minimized. It is assumed that the volume of the domain occupied by the body is constant. Moreover the function describing the boundary of the domain occupied by the body and its derivative are bounded.

Shape optimization of elastic or hyperelastic contact problems were considered, among others in [10, 12, 14, 19]. The existence of optimal solutions was investigated in [10]. The necessary optimality conditions were formulated in [12, 19] using the material derivative approach as well as in [10] using the optimal control theory approach. The convergence of the finite element approximation was investigated in [10, 12]. Numerical results are reported in [10, 12, 14].

[1]Supported by the Polish State Scientific Committee under the grant 8 T 11A 008 15 as well as of the Warsaw University of Technology.

In these papers the classical approach to numerical solving of optimal shape design problems based on boundary or domain variations methods [9, 10] was used. In this approach the state problem is solved many times on a domain changing during the computation. The boundary or domain variation methods require calculation of a new triangularization of the optimized domain, updating the stiffness matrix and load vector at each iteration of the numerical algorithm. Since the optimized domain usually has complicated geometrical structure the whole computational process is tedious, time consuming and expensive. To overcome this difficulty, in response to growing number of industrial applications of the optimal shape design problems, fixed domain methods [5, 6, 7, 8, 9, 11, 15, 17, 18] for solving these problems are being developed. Fixed domain methods are based on using fictitious or embedding domain method [6]. The fictitious domain method for solving the state system described by partial differential equations consists in transforming the original state system defined in the complicated geometry domain into a new system defined in a given fixed simpler geometry domain containing the original domain with the same differential operator [6]. This method allows to use fairly structured meshes on a simple geometry domain containing the current geometry and fast solvers. The solution of the state equation in the fictitious domain is enforced to satisfy original boundary conditions. Embedding domain methods for solving elliptic equations with Dirichlet boundary conditions were investigated in [5, 6, 18]. Neumann problem was investigated in [7] where penalty approach was employed to satisfy original boundary conditions.

Fixed domain methods for solving shape optimal design problems were considered in [3, 9, 11, 15, 17]. In [3, 9] the original shape optimal design problem for the Laplace state equation with Dirichlet boundary condition was transformed into equivalent one using a new control variable on the right-hand side of the state equation defined in the fixed domain containing the original one. The convergence of the finite element approximation as well as a numerical result obtained by using nonsmooth optimization method are presented in [3, 9]. In [8] the optimal design problem for the obstacle problem was formulated using the fictitious domain approach. In [11, 15, 17] the original shape optimal design problem for Dirichlet system is replaced by a constrained control problem with either boundary control or distributed control. The state system is defined on a fixed domain larger than the original one. The existence of optimal controls and optimality conditions are investigated in [15, 17].

This paper deals with the numerical solving of a shape optimization problem for linear contact problem using the fictitious domain approach. To the author knowledge this approach for contact problems has not been investigated yet and the present paper is continuation of the author papers [12, 13, 14]. In the paper we shall formulate a 2D contact problem in variable domain. The boundary of the domain is the variable subject to optimization. The cost functional is an approximation of the normal contact stress as introduced in [12]. The fictitious domain formulation to this optimization problem is given using the boundary control technique [15] and the equivalence to the classical formulation is shown. The piecewise linear finite element approximation for displacements and constant for Lagrange multipliers is proposed and its convergence is shown. The design sensitivity analysis for the discrete shape optimization problem is performed. The Lagrangian multiplier method for solving the discretized state system is proposed [10]. The descent gradient method is proposed for solving the optimization problem. Numerical examples are provided.

2 Problem Formulation

Consider deformations of an elastic body occupying domain $\Omega = \Omega(v) \subset R^2$ depending on a Lipschitz continuous function v. The domain Ω has the following geometrical structure:

$$\Omega = \Omega(v) = \{(x_1, x_2) \in R^2 \ : \ 0 < x_1 < g_1, \ \ 0 \leq v(x_1) < x_2 < g_2\} \tag{2.1}$$

where g_1 and g_2 are given positive constants. We shall consider domains $\Omega(v)$ depending on a function v from the set U_{ad} defined by:

$$U_{ad} = \{v \in C^{0,1}([0, g_1]) \ : \ 0 \leq v(x_1) \leq c_0, \ | \frac{dv}{dx_1} | \leq c_1 \ \forall x_1 \in [0, g_1], \ \int_{\Omega(v)} dx = c_2\} \tag{2.2}$$

where c_0, c_1, c_2 are given positive constants. The set U_{ad} is assumed to be nonempty. $C^{0,1}([0, g_1])$ denotes the set of Lipschitz continuous functions on $[0, g_1]$. The boundary Γ of the domain Ω is divided into three pieces Γ_0, Γ_1 and $\Gamma_2 = \Gamma_2(v)$ such that:

$$\Gamma = \Gamma_0 \cup \Gamma_1 \cup \Gamma_2, \ \ \Gamma_i \cap \Gamma_j = \emptyset \ \ i, j = 0, 1, 2, \ i \neq j \tag{2.3}$$

We shall assume that the body is subjected to body forces $f = (f_1, f_2)$. Moreover surface tractions $p = (p_1, p_2)$ are applied along the boundary Γ_1. The body is clamped along the portion Γ_0 of the boundary Γ. It is assumed that the contact conditions are prescribed on the portion Γ_2 of the boundary Γ.

We denote by $u = (u_1, u_2)$ the displacement of the body and by $\sigma = \{\sigma_{ij}(u(x))\}$, $i, j = 1, 2$ the stress field in the body. We shall consider elastic bodies obeying Hooke's law [4]:

$$\sigma(u) = \sigma_{ij}(u(x)) = c_{ijkl}(x) e_{kl}(u(x)) \ \ i, j, k, l = 1, 2 \tag{2.4}$$

$$e_{kl} = e_{kl}(u(x)) = \frac{1}{2}(u_{k,l} + u_{l,k}) \ \ k, l = 1, 2 \tag{2.5}$$

where

$$c_{ijkl}(x) \in L^\infty(\Omega), \ \ c_{ijkl} = c_{jikl} = c_{klij} \tag{2.6}$$

$$c_{ijkl} t_{ij} t_{kl} \geq \alpha_0 t_{ij} t_{ij} \tag{2.7}$$

holds for almost all $x \in \Omega$, for all symmetric 2×2 matrices t_{ij} and some positive constant α_0. Note, that the summation convention [4] over repeated indices is used in (2.4)-(2.7) as well as throughout the paper.

In an equilibrium state a stress field σ satisfies the system of equations [4, 10]:

$$\sigma_{ij}(x)_{,j} = -f_i(x) \ \ x \in \Omega, \ \ i, j = 1, 2 \tag{2.8}$$

where $\sigma_{ij}(x)_{,j} = \partial \sigma_{ij}(x) / \partial x_j$, $i, j = 1, 2$. The following boundary conditions are given:

$$u_i = 0 \ \text{ on } \Gamma_0 \ i, j = 1, 2 \tag{2.9}$$

$$\sigma_{ij}(x)n_j = p_i(x) \text{ on } \Gamma_1 \; i,j = 1,2 \tag{2.10}$$

$$u_2 \geq -v, \;\; \sigma_N \geq 0, \;\; (u_2 + v)\sigma_N = 0 \text{ on } \Gamma_2 \tag{2.11}$$

$$\sigma_T = 0 \quad \text{on } \Gamma_2 \tag{2.12}$$

where

$$u_N = u_i n_i, \qquad u_T = u - n u_N, \qquad n = \{n_i\}$$

$$\sigma_N = \sigma_{ij} n_i n_j, \qquad \sigma_T = \{\sigma_{iT}\}, \qquad \sigma_{iT} = \sigma_{ij} n_j - \sigma_N n_i$$

and $n = \{n_i\}$ is an outward unit normal vector to the boundary Γ. We shall consider variational form of the problem (2.8) –(2.12). Let us introduce:

$$V = \{z \in [H^1(\Omega)]^2 : \; z_i = 0 \text{ on } \Gamma_0 \,, i = 1,2\}, \;\; K = \{z \in V \; : \; z_2 \geq -v \text{ on } \Gamma_2\} \tag{2.13}$$

where $H^1(\Omega)$ denotes the Sobolev space [1] . Let us denote by $a(.,.) : V \times V \to R$ the bilinear form defined by:

$$a(u,z) = \int_\Omega c_{ijkl}(x) e_{ij}(u) e_{kl}(z) dx \tag{2.14}$$

We denote by $l(.) : V \to R$ the linear form given by:

$$l(z) = \int_\Omega f_i z_i dx + \int_{\Gamma_1} p_i z_i ds \tag{2.15}$$

where $f \in [L^2(\Omega)]^2$ and $p \in [H^{-1/2}(\Gamma_1)]^2$.

The system (2.8) – (2.12) we can write in an equivalent variational form [2, 4, 10, 19]: *For a given function $v \in U_{ad}$ find an element $u \in K$ such that:*

$$a(u, z-u) \geq l(z-u) \;\; \forall z \in K \tag{2.16}$$

The existence of a unique solution to (2.16) was shown in [4]. We shall call (2.16) the primal variational formulation of problem (2.8)-(2.12).

2.1 Shape Optimization Problem Formulation

Let $\hat{\Omega} \subset R^2$ be a domain such that for all functions $v \in U_{ad} : \Omega(v) \subset \hat{\Omega}$. Let M be the set:

$$M = \{\phi \in [H^1(\hat{\Omega})]^2 : \phi \geq 0 \text{ on } \hat{\Omega}, \| \phi \|_{H^1(\hat{\Omega})} \leq 1\} \tag{2.17}$$

We shall consider the following shape optimization problem: *For a given function $\phi \in M$, find function $v \in U_{ad}$ minimizing the cost functional*

$$J_\phi(v) = \int_{\Gamma_2(v)} \sigma_N \phi_N ds \tag{2.18}$$

over the set U_{ad} given by (2.2). σ_N is the normal component of the stress field $\sigma(u)$ given by (2.16).

The cost functional (2.18) approximates the normal contact stress on the boundary $\Gamma_2(v)$, i.e., the original boundary flux cost functional [5]. It is known [10] that the contact stress attains high values in the middle of the contact area and the goal of structural engineers is to reduce this stress as much as possible. The shape optimization problem with the original boundary flux cost functional is too difficult to handle [10, 19]. We modify the original cost functional using an auxiliary function ϕ [12, 19] and the form (2.18). The goal of the optimization problem (2.18) is to find such boundary $\Gamma_2(v)$ depending on the function v that the normal contact stress is minimized. It is assumed that the volume of the body is constant as well as that the function v is bounded and has bounded first derivative. The optimization problem (2.18) has at least one solution [10, 12].

3 Fictitious Domain Formulation

Let $\hat{\Omega} = (0, g_1) \times (0, g_2)$ be a fixed domain such that $\hat{\Omega} \supset \bar{\Omega}(v)$ for all $v \in U_{ad}$. Denote by:

$$\hat{\Gamma}_0 = \{x \in R^2 : x_1 = 0 \vee g_1, x_2 \in (0, g_2)\}, \quad \hat{\Gamma}_1 = \Gamma_1$$

$$\hat{\Gamma}_2 = \{x \in R^2 : x_1 \in (0, g_1), x_2 = 0\}$$

The problem (2.8)-(2.12) takes the form:

$$\hat{\sigma}_{ij}(x)_{,j} = -\hat{f}_i(x) \ \ x \in \hat{\Omega}, \ \ i, j = 1, 2 \tag{3.1}$$

The following boundary conditions are given:

$$\hat{u}_i = 0 \ \text{ on } \hat{\Gamma}_0 \ i, j = 1, 2 \tag{3.2}$$

$$\hat{\sigma}_{ij}(x) n_j = p_i(x) \ \text{ on } \hat{\Gamma}_1 \ i, j = 1, 2 \tag{3.3}$$

$$\hat{u}_2 \geq 0, \ \ \hat{\sigma}_N \geq 0, \ \ \hat{u}_2 \hat{\sigma}_N = 0 \ \text{ on } \hat{\Gamma}_2 \tag{3.4}$$

$$\hat{u}_{2|\Gamma_2(v)} \geq -v, \ \hat{\sigma}_{N|\Gamma_2(v)} \geq 0, \ \ (\hat{u}_{2|\Gamma_2(v)} + v)\hat{\sigma}_{N|\Gamma_2(v)} = 0 \tag{3.5}$$

$$\hat{\sigma}_T = 0 \ \ \text{on } \hat{\Gamma}_2 \tag{3.6}$$

Let us define:

$$\begin{aligned} &\hat{V} = \{\hat{z} \in [H^1(\hat{\Omega})]^2 : \ \hat{z}_i = 0 \text{ on } \hat{\Gamma}_0 \, , i = 1, 2\}, \quad \hat{K} = \{\hat{z} \in \hat{V} \ : \ \hat{z}_2 \geq 0 \text{ on } \hat{\Gamma}_2\}, \\ &\Lambda_1 = \{\lambda \in H^{-1/2}(\Gamma_2) : \lambda \geq 0\}, \ \ \Lambda_2 = \{\lambda \in H^{1/2}(\Gamma_2) : \lambda \geq 0\}, \ \ \Lambda_3 = L^\infty(\Gamma_2) \end{aligned} \tag{3.7}$$

The problem (3.1)-(3.6) can be written in the variational form: *For a given function $v \in U_{ad}$ find an element $\hat{u} \in \hat{K}$ and $\lambda = (\lambda_1, \lambda_2, \lambda_3) \in \Lambda = (\Lambda_1 \times \Lambda_2 \times \Lambda_3)$ such that:*

$$\begin{aligned}\int_{\hat{\Omega}} c_{ijkl}(x)e_{ij}(\hat{u})e_{kl}(\hat{z}-\hat{u})dx \quad \geq \quad & \int_{\hat{\Omega}} \hat{f}_i(\hat{z}_i - \hat{u}_i)dx + \int_{\hat{\Gamma}_1} p_i(\hat{z}_i - \hat{u}_i)ds + \\ & \int_{\Gamma_2} \lambda_1(\hat{z}_2 - \hat{u}_2)ds + \int_{\Gamma_2} \lambda_2 \sigma_N(\hat{z}-\hat{u})ds + \\ & \int_{\Gamma_2} \lambda_3[\hat{u}_2\sigma_N(\hat{z}-\hat{u}) + (\hat{z}_2 - \hat{u}_2)\sigma_N(\hat{u})]ds \quad \forall \hat{z} \in \hat{K}\end{aligned} \tag{3.8}$$

$$\int_{\Gamma_2} \mu(\hat{u}_2 + v)ds = 0 \quad \forall \mu \in \Lambda_1 \tag{3.9}$$

$$\int_{\Gamma_2} \mu\sigma_N(\hat{u})ds = 0 \quad \forall \mu \in \Lambda_2 \tag{3.10}$$

$$\int_{\Gamma_2} \mu(\hat{u}_2 + v)\sigma_N(\hat{u})ds = 0 \quad \forall \mu \in \Lambda_3 \tag{3.11}$$

THEOREM 3.1. *For any $v \in U_{ad}$ problem (3.9)-(3.11) has a unique solution $(\hat{u}, \lambda) \in \hat{K} \times \Lambda$.*

Proof. The energy functional associated with problem (3.1) attains minimum $\hat{u}$ on the convex and closed set $\hat{K}$ [4, 10]. Strict convexity of the energy functional assures uniqueness of $\hat{u}$. There exists Lagrange multiplier $\lambda \in \Lambda$ rendering the Lagrangian to system (3.1) - (3.6) stationary at $\hat{u}$. □

THEOREM 3.2. *For any $v \in U_{ad}$ let $\hat{u} \in \hat{K}$ be a solution of the system (3.9)-(3.11). Then $u = \hat{u}\mid_{\Omega(v)}$ solves the problem (2.16).*

Proof. Let $u_1 \in K$ be the solution to (3.9) - (3.11) with test functions having compact support in $\Omega(v)$. Since the family $\{\hat{\Omega} \setminus \Omega(v), v \in U_{ad}\}$ has a uniform extension property [1, 7, 9, 16] there exists a uniform extension mapping from $\hat{\Omega} \setminus \Omega(v)$ on $\Omega(v)$ the norm of which does not depend on $v \in U_{ad}$. It implies the existence of bounded element $\lambda \in \Lambda$ defining the linear, continuous functional on $\Gamma_2(v)$ as well as assures the existence of solution u_2 to the system (3.9) - (3.11) in the domain $\hat{\Omega} \setminus \Omega(v)$ with test functions having a compact support in $\hat{\Omega} \setminus \Omega(v)$. Setting $\hat{u} = u_1$ on $\Omega(v)$ and $\hat{u} = u_2$ on $\hat{\Omega} \setminus \Omega(v)$ proves the Theorem. □

Let M be the set defined by (2.17) and $\hat{\Omega} \subset R^2$ be defined as in Section 2. We shall consider the following shape optimization problem: *For a given function $\phi \in M$, find function $v \in U_{ad}$ minimizing the cost functional*

$$J_\phi(v) = \int_{\Gamma_2(v)} \hat{\sigma}_N \phi_N ds \tag{3.12}$$

over the set U_{ad} given by (2.2). $\hat{\sigma}_N$ is a normal component of the stress field $\hat{\sigma} = \sigma(\hat{u})$ given by (3.9)-(3.11) calculated in domain $\hat{\Omega}$.

THEOREM 3.3. *For a given function $\phi \in M$ the problem (3.12) has at least one solution $v^\star \in U_{ad}$.*

Proof. From Theorems 3.1 and 3.2 and compactness of the set U_{ad} as well as the continuity of the cost functional (3.12) follows the existence of optimal solution $v^\star$ to problem (2.18) as well as the equivalence of problems (2.18) and (3.12). □

4 Finite-Dimensional Approximation

In order to solve numerically the optimization problem (3.12) we discretize it employing conforming finite element method [2]. By $\mathcal{T}_h$ we denote a regular (see [2]) family of partitions of domain $\hat{\Omega}$ depending on the parameter h given by division of the domain $\hat{\Omega}$ into triangular elements O_i, $i = 1, \dots, I$ such that:

$$\hat{\Omega} = \bigcup_{i=1}^{I} O_i \tag{4.1}$$

Let us introduce the finite dimensional space $\hat{V}_h$ approximating the space $\hat{V}$:

$$\hat{V}_h = \{z \in [C(\hat{\Omega})]^2 \cap \hat{V} \ : \ z_{|O_i} \in [P_1(O_i)]^2, \ \ \forall O_i \in \mathcal{T}_h\} \tag{4.2}$$

where $P_k(O_i)$, $k = 0, 1$ denotes the set of all polynomials of degree less then or equal to k on the element $O_i \in \mathcal{T}_h$. By $\hat{K}_h$ and $\hat{M}_h$ we denote the sets approximating the sets $\hat{K}$ and $\hat{M}$ respectively,

$$\hat{K}_h = \{z \in \hat{V}_h \ : \ z_{2h} \geq 0 \text{ on } \hat{\Gamma}_{2h} \} \tag{4.3}$$

$$\hat{M}_h = \{\phi \in [C(\Omega)]^2 \cap \hat{M} \ : \ \phi_{|O_i} \in [P_1(O_i)]^2, \forall O_i \in \mathcal{T}_h\} \tag{4.4}$$

Let us denote by $0 = s_0 < s_1 \cdots < s_\kappa = g_1$ the partition of the segment $[0, g_1]$, $H_i = s_i - s_{i-1}$, $i = 1, \dots, \kappa$, $H = \min\{H_i \mid i = 1, \dots, \kappa\}$. H is a discretization parameter. Using this partition we can defined the sets U_{Had} and Λ_{jH} approximating the sets U_{ad} and Λ_j respectively,

$$U_{Had} = \{v \in C([0, g_1]) \ : \ v_{|[s_{i-1}, s_i]} \in P_1([s_{i-1}, s_i])\} \cap U_{ad} \tag{4.5}$$

$$\Lambda_{jH} = \{\mu_H \in L^2([0, g_1]) \ : \mu_{H_{|[s_{i-1}, s_i]}} \in P_0([s_{i-1}, s_i])\} \cap \Lambda_j, \quad j = 1, 2, 3 \tag{4.6}$$

Mesh sizes H and h are assumed to satisfy the condition $b_1 \leq h/H \leq b_2$, b_1, b_2 positive constants. The discrete model can be characterized by the parameter h. The state system (3.9)-(3.11) is approximated by the following one:

Find an element $\hat{u}_h \in \hat{K}_h$ and $\lambda_H = (\lambda_{1H}, \lambda_{2H}, \lambda_{3H}) \in \Lambda_{1H} \times \Lambda_{2H} \times \Lambda_{3H}$ such that:

$$\begin{aligned}\int_{\hat{\Omega}_h} c_{ijkl}(x) e_{ij}(\hat{u}_h) e_{kl}(\hat{z} - \hat{u}_h) dx \geq \int_{\hat{\Omega}_h} \hat{f}_i(\hat{z}_i - \hat{u}_{hi}) dx + \int_{\hat{\Gamma}_{1h}} p_i(\hat{z}_i - \hat{u}_{hi}) ds + \\ \int_{\Gamma_{2h}} \lambda_{1H}(\hat{z}_2 - \hat{u}_{h2}) ds + \int_{\Gamma_{2h}} \lambda_{2H} \sigma_N(\hat{z} - \hat{u}_h) ds + \\ \int_{\Gamma_{2h}} \lambda_{3H}[\hat{u}_2 \sigma_N(\hat{z} - \hat{u}_h) + (\hat{z}_2 - \hat{u}_{h2}) \sigma_N(\hat{u})] ds \quad \forall \hat{z} \in \hat{K}_h\end{aligned} \tag{4.7}$$

$$\int_{\Gamma_{2h}} \mu_H(\hat{u}_{2h} + v_H)ds = 0 \quad \forall \mu_H \in \Lambda_{1H} \tag{4.8}$$

$$\int_{\Gamma_{2h}} \mu_H \sigma_N(\hat{u}_h)ds = 0 \quad \forall \mu_H \in \Lambda_{2H} \tag{4.9}$$

$$\int_{\Gamma_{2h}} \mu_H(\hat{u}_{2h} + v_H)\sigma_N(\hat{u}_h)ds = 0 \quad \forall \mu_H \in \Lambda_{3H} \tag{4.10}$$

THEOREM 4.1. *The system (4.7)-(4.10) has a unique solution* $(\hat{u}_h, \lambda_H) \in \hat{K}_h \times \Lambda_H$.

Proof. It is parallel to the proof of Theorem 3.1. □

The shape optimal design problem (3.12) is approximated by the following problem: *For a given function* $\phi_h \in M_h$, *find function* $v_H \in U_{adH}$ *minimizing the cost functional*

$$J_{\phi_h}(v_H) = \int_{\Gamma_2(v_H)} \hat{\sigma}_{Nh}\phi_{Nh}ds \tag{4.11}$$

over the set U_{adH}. $\hat{\sigma}_{Nh}$ *is a normal component of the stress field* $\hat{\sigma}_h = \sigma(\hat{u}_h)$.

THEOREM 4.2. *For a given function* $\phi_h \in M_h$ *and for any* $h > 0$ *the problem (4.11) has at least one solution.*

Proof. It consists in application of Weierstrass Theorem and is similar to the proof of Theorem 3.3. □

THEOREM 4.3. *Let* $v_H^* \in U_{adH}$ *be a solution of the optimization problem (4.11) and*

$$(\hat{u}_h(v_H^*), \lambda_H)$$

be the solution to the state system (4.7) - (4.10). Then there exist subsequences $\{v_{H_k}^*\} \subset \{v_H^*\}$, $\{\hat{u}_h(v_{H_k}^*)\} \subset \{\hat{u}_h(v_H^*)\}$ *and* $\{\lambda_{H_k}\} \subset \{\lambda_H\}$ *as well as elements* $v^* \in U_{ad}$, $\hat{u} \in \hat{K}$ *and* $\lambda \in \Lambda$ *such that*

$$v_{H_k}^* \to v^* \quad in\ [0, g_1] \tag{4.12}$$

$$\hat{u}_{h_k}(v_{H_k}^*) \to \hat{u}(v^*) \quad in\ H^1(\hat{\Omega}) \tag{4.13}$$

$$\lambda_{H_k} \to \lambda \quad in\ \Lambda \tag{4.14}$$

Moreover v^* *is a solution of problem (2.18) and* $\hat{u}(v^*)_{|\Omega(v^*)}$ *is a solution to (2.16).*

Proof. For construction of sequences $\{v_{H_k}^*\}$, $\{\hat{u}_h(v_{H_k}^*)\}$, $\{\lambda_{H_k}\}$ see [10, 12]. From Theorems 3.1, 3.2, 4.1 and compactness of the set U_{Had} as well as the continuity of the cost functional (4.11) follows the existence of optimal solution v^* to problem (2.18). For idea of the proof see [10, 12]. □

5 Matrix Formulation and Solution Methods

Let us formulate problem (4.11) in coefficients. We start with the formulation of the state system (4.7), (4.8). Let us denote,

$$\hat{M}_h = lin\{\hat{\varphi}_i\}_{i=1}^n, \quad \hat{V}_h = lin\{\hat{\varphi}_i\}_{i=1}^n, \quad \Lambda_{iH} = lin\{\hat{\lambda}_j^i\}_{j=1}^{\kappa}, i = 1,2,3, \quad U_{Had} = lin\{\zeta_i\}_{i=1}^{\kappa} \tag{5.1}$$

and

$$\hat{\phi}_h = \sum_{j=1}^n \gamma_j \hat{\varphi}_j, \quad \hat{u}_h = \sum_{j=1}^n q_j^i \hat{\varphi}_j, \; i = 1,2, \quad \lambda_{iH} = \sum_{j=1}^{\kappa} \beta_j^i \hat{\lambda}_j^i, \; i = 1,2,3, \quad v_H(x_1) = \sum_{i=1}^{\kappa} \alpha_i \zeta_i \tag{5.2}$$

By U_1 and U_2 we denote the sets corresponding to M_h and U_{Had} respectively,

$$U_1 = \{\gamma \in R^n \; : \; 0 \le \gamma_i \le 1, \quad i = 1,...n\} \tag{5.3}$$

$$U_2 = \{ \; \alpha \in R^{\kappa} \; : 0 \le \alpha_i \le c_0, \;\; \alpha_i - \alpha_{i-1} \le c_1 H_i, \;\; i = 1,...,\kappa, \\ \sum_{i=1}^{\kappa} [g_1 g_2 - \frac{1}{2}(\alpha_{i-1} + \alpha_i)] H_i = c_2 \; \} \tag{5.4}$$

The sets $\hat{K}_h$ and Λ_{iH} are described by:

$$U_3 = \{ \; q = \{q^i\}_{i=1}^2 \in R^{2n}, \; : \; q_j^2 \ge 0, \; j = 1,...\kappa, \;\; \text{on } \hat{\Gamma}_{2h} \; \} \tag{5.5}$$

$$U_4 = \{\beta = \{\beta^i\} \in R^{3\kappa}, \; i = 1,2,3 \; : \; \beta_j^i \ge 0, \;\; i = 1,2, \; j = 1,...,\kappa \; \} \tag{5.6}$$

respectively. Moreover let us denote:

$$A = \{\int_{\hat{\Omega}} c_{rskl} e_{rs}(\hat{\varphi}_i) e_{kl}(\hat{\varphi}_j)\}_{i,j=1}^n, \;\; k,l,r,s = 1,2, \quad F = \{\int_{\hat{\Omega}} f \hat{\varphi}_j dx + \int_{\hat{\Gamma}_{1h}} p_i \hat{\varphi}_j ds\}_{j=1}^n, \tag{5.7}$$

$$Q = \{\int_{\Gamma_{2h}} c_{rskl} e_{kl}(\hat{\varphi}_i) n_k n_l \hat{\varphi}_j ds\}_{i,j=1}^n, \;\; k,l,r,s = 1,2, \quad B = \{\int_{\Gamma_{2h}} \hat{\lambda}_i \hat{\varphi}_j ds\}_{i,j=1}^n \tag{5.8}$$

The problem (4.11) is equivalent to the following one: *For given $\gamma \in U_1$ find $\alpha \in U_2$ minimizing the function:*

$$J(\alpha) = \gamma^T Q q(\alpha) \tag{5.9}$$

where the pair $(q, \beta) \in U_3 \times U_4$ is the solution of the following state system,

$$Aq - F - B\beta \ge 0 \tag{5.10}$$

$$B(q+\alpha)=0. \tag{5.11}$$

Note that [12]

$$Qq = Aq - F - B\beta \tag{5.12}$$

To solve numerically problem (5.9) with constraints (5.10) and (5.11) we apply the Augmented Lagrangian approach [10]. We shall treat (5.11) as a constraint and we shall add it to the cost function (5.9) as a penalty term as well as we shall introduce the multiplier z associated with this constraint. We shall assume that $q = q(\alpha, \beta)$, $B = B(\alpha)$, i.e. depend on the control variables α and β. The augmented Lagrangian for the system (5.9) - (5.12) has the form:

$$L(\alpha,\beta,z) = J(\alpha,\beta) + \frac{1}{2}\varepsilon(q+\alpha)^T B^T B(q+\alpha) + z^T B(q+\alpha) \tag{5.13}$$

where $\varepsilon > 0$ is a penalty parameter and $z \in R^n, z \geq 0$. The problem (5.9)-(5.12) is equivalent to the following one:

$$\min_{(\alpha,\beta)\in U_2\times U_4} \max_{z\in U_4} L(\alpha,\beta,z) \tag{5.14}$$

where $q \in U_3$ is a solution to the inequality (5.10). To solve numerically problem (5.14) we can use Uzawa type algorithm [2, 10]. This algorithm has the form,
Step 0: Choose: $\alpha^{-1} \in U_2$, $\beta^{-1} \in U_4$, $z^0 \in U_4$, $\rho > 0$, $\bar{\varepsilon} > 0$, $\varepsilon > 0$. Set $n = 0$.
Step 1: Find $\alpha^n \in U_2$ and $\beta^n \in U_4$ satisfying:

$$L(\alpha^n,\beta^n,z^n) \leq L(\alpha,\beta,z^n) \quad \forall(\alpha,\beta) \in U_2 \times U_4 \tag{5.15}$$

Step 2: Set

$$z^{n+1} = Proj_{U_4}(z^n + \rho(B(q^n + \alpha^n))) \tag{5.16}$$

If $\mid z^{n+1} - z^n \mid \leq \bar{\varepsilon}$, Stop,

Otherwise: n = n+1, go to **Step 1**.

$Proj_U$ denotes the projection on the set U. For convergence analysis of the algorithm (5.15) - (5.16) see [2, 10].

5.1 Sensitivity Analysis

Let us calculate the directional derivatives of the Lagrangian (5.13) with respect to α and β. The directional derivative of the cost functional (3.12) has been calculated in [12, 14]. Assume the nodal points of the contact boundary Γ_2 are allowed to move vertically only [10]. In general the mapping $\alpha \to q(\alpha, \beta)$ is nondifferentiable (see [9]). Therefore we calculate the directional derivatives of the cost functional assuming the mapping is regular enough. For detailed discussion concerning differentiability see [10, 19].

By $w \in U_4$ we denote an adjoint variable satisfying the equation:

$$Aw = -[\gamma^T Q + zB + \varepsilon B^T B(q+\alpha)] \tag{5.17}$$

The directional derivative $dL(\alpha, \sigma\alpha)$ at a point $\alpha \in U_2$ in direction $\sigma\alpha \in U_2$ is given by:

$$dL(\alpha,\sigma\alpha) = [\gamma \frac{\partial Q}{\partial \alpha} q + \varepsilon I B^T B(q+\alpha) + z\frac{\partial B}{\partial \alpha}(q+\alpha) + zBI + \frac{1}{2}(q+\alpha)\frac{\partial}{\partial \alpha} B^T B(q+\alpha) - w^T \frac{\partial B}{\partial \alpha}\beta]\sigma\alpha \tag{5.18}$$

The directional derivative $dL(\beta, \sigma\beta)$ at a point $\beta \in U_4$ in direction $\sigma\beta \in U_4$ is given by:

$$dL(\beta, \sigma\beta) = -wB\sigma\beta \tag{5.19}$$

Using formulae for directional derivative of the cost functional (5.13) we can write the derivative of the cost functional (5.13) for given $\gamma \in U_1$, at a point $(\alpha, \beta) \in U_2 \times U_4$ in the direction $(\sigma\alpha, \sigma\beta) \in U_2 \times U_4$, in the following form,

$$dL_\gamma(\alpha, \sigma\alpha) = K_1\sigma\alpha \tag{5.20}$$

$$dL_\gamma(\beta, \sigma\beta) = K_2\sigma\beta \tag{5.21}$$

where the coefficients K_1 and K_2 are determined by (5.18) and (5.19).

Numerical algorithm for solving the optimization problem (5.15) has the following form [10]:
Step 0: Choose $\bar{\varepsilon} > 0$. Set $i = 0$.
Step 1:

a. For given $\alpha^i \in U_2$, and $\beta^i \in U_4$ find $q^i \in U_2$ satisfying (5.10).

b. For given $\alpha^i \in U_2$, $q^i \in U_1$, $z^i \in U^4$ find $w^i \in U_4$ satisfying (5.17).

c. Calculate K_1, K_2 using (5.18), (5.19). Set: $(\sigma\alpha)^i = -K_1$ and $\sigma\beta)^i = -K_2$ as well as $d_1^i = (\sigma\alpha)^i + e^i d_1^{i-1}$, $d_2^i = (\sigma\beta)^i + e^i d_2^{i-1}$ where e^i is Polak-Ribiera coefficient [10]. If $| d_j^i | \leq \bar{\varepsilon}$, $j = 1, 2$ Stop, otherwise go to **Step 2**.

Step 2: Find $t \in R$ such that:

$$\min_{t>0} L(\alpha^i + td_1^i, \beta^i + td_2^i, z^n) = L(\alpha^i + t^i d_1^i, \beta^i + t^i d_2^i, z^n) \tag{5.22}$$

Step 3: Set: $\alpha^{i+1} = Proj_{U_2}(\alpha^i + t^i d_1^i)$ and $\beta^{i+1} = Proj_{U_4}(\beta^i + t^i d_2^i)$, go to **Step 1**.

The Armijo line search method was employed in (5.22). Note, the SOR method can be used to solve problem (5.15) or the system (5.10) [12, 14]. The convergence of this algorithm is investigated in [2, 10].

6 Numerical Results

The discretized shape optimization problem (5.9), (5.10), (5.11) was solved numerically. The numerical algorithms described in previous section were employed. The algorithms were programmed in Matlab environment.

Fig 1. Example 6.1 – Initial and optimal shapes of the contact boundary

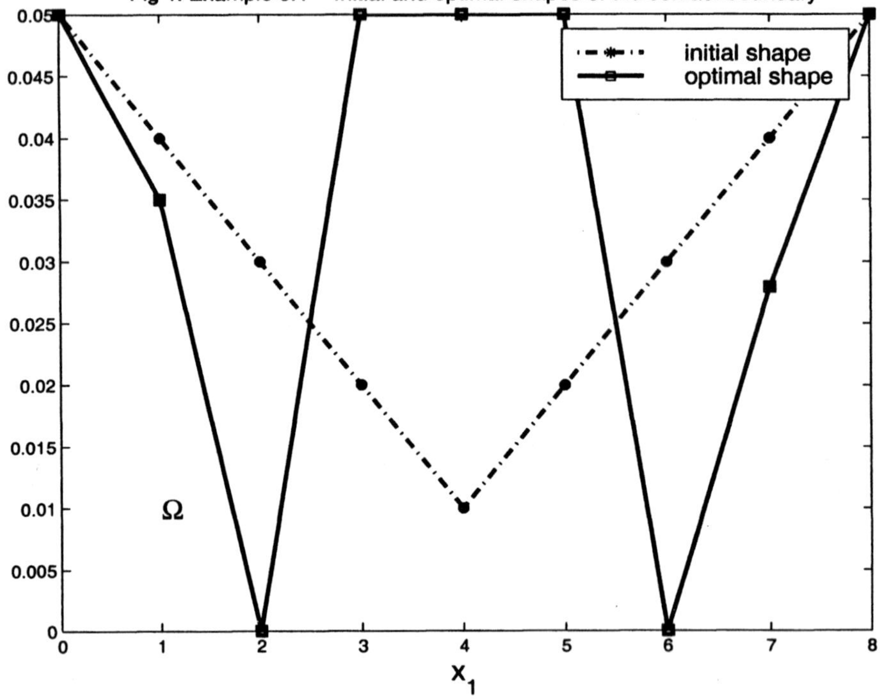

The computations were carried out for the elastic body consisting of the material characterized by the constant Poisson's ratio $\nu = 0.29$ and the Young modulus $E = 2.110^{11} Nm^{-2}$. The domain $\hat{\Omega}_h = (0, g_1) \times (0, g_2)$ is divided into 64 triangular elements. The boundary $\hat{\Gamma}_h$ of the domain $\hat{\Omega}_h$ was divided into three pieces,

$$\Gamma_0 = \{x \in R^2 \; x_1 = 0, g_1 \; x_2 \in (0, g_2)\}, \; \Gamma_1 = \{x \in R^2 \; x_1 = (0, g_1) \; x_2 = g_2\}, \tag{6.1}$$

$$\Gamma_2 = \{x \in R^2 \; x_2 = v_h(x_1)\}$$

Function $\phi_h \in M_h$ defined in (2.17) is selected as $\phi_h = 1$ on $\hat{\Omega} \supset \Omega(v_H)$ for all v_H. The other numerical data are as follows: $g_1 = 8$, $g_2 = 4$, $f_i = 0$ $i = 1, 2$, $p_1 = 0$, $\varepsilon = 10^{-5}$. We have obtained the following results:

EXAMPLE 6.1. *The computations were performed for the domain with initial contact boundary shape S_i given in Fig 1. Moreover: $p_2 = -5.610^6$ for $x_1 \in (3, 5)$, $c_0 = 0.05$, $c_1 = 0.1$, $c_2 = 3.01$. The obtained optimal shape S_o of the contact boundary is given in Fig 1. The cost functional value was reduced from 12.94 to -22.36. The error in satisfying constant volume constraint is 10%.*

EXAMPLE 6.2. *p_2, c_0, c_1 are the same as in Example 6.1. $c_2 = 3.04$ and the initial shape of the contact boundary S_i is given in Fig 2. The obtained optimal shape S_o of the contact boundary*

Fig 2. Example 6.2 – Initial and optimal shapes of the contact boundary

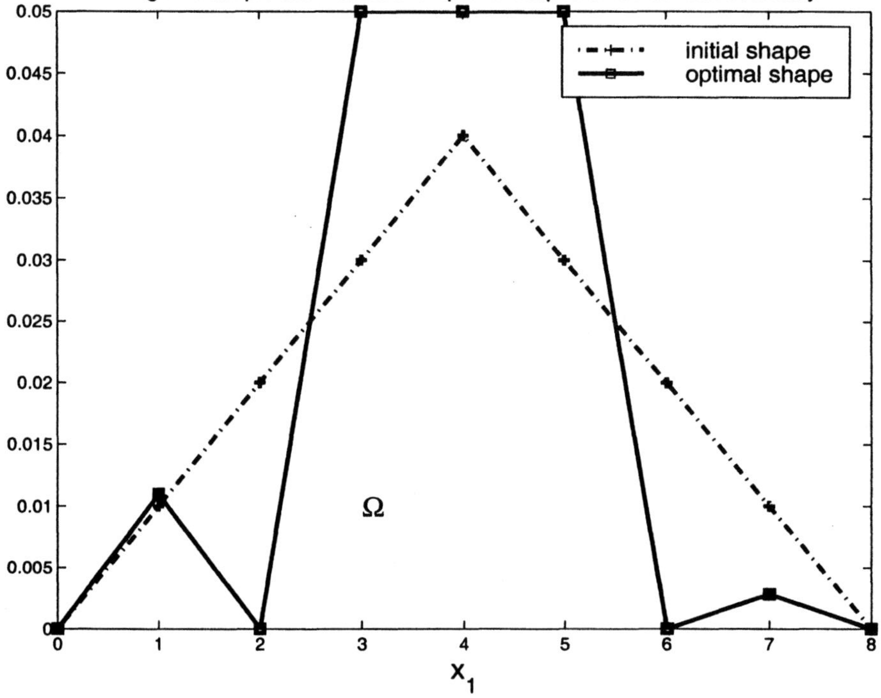

is given in Fig 2. The cost functional value was decreased from 13.79 to -6.26. The constant volume constraint was exceeded by 6%.

EXAMPLE 6.3. *In this case* $p_2 = -5.610^7$ *for* $x_1 \in (5,8)$, $c_0 = 0.06$, $c_1 = 0.1$, $c_2 = 3.055$, *the initial contact boundary shape* S_i *is given in Fig 3. The obtained optimal shape* S_o *is presented in Fig 3. The cost functional value was decreased from 122.33 to 46.35 while the constant volume constraint was exceeded by 16.6 %.*

Note [10, 11], that in Examples 6.1 and 6.2, for the body with boundary S_i the normal contact stress has its peak in the middle of the contact boundary. For the body with optimal shape boundary S_o normal contact stress, in all cases, is almost constant and shows very mild increase approaching the middle point of the contact boundary.

The speed of the convergence of the proposed algorithm is strongly dependent on the proper choice of parameter values in Uzawa and SOR algorithms as well as on the accuracy of the computing of the Lagrange multipliers. Most of the computational time is spent on solving the state system (5.10). Comparing with [12] we can note that the use of the augmented Lagrangian technique accelerates the computational process in particular, in the neighborhood of the starting point. The fictitious domain technique accelerates the solution of the state system.

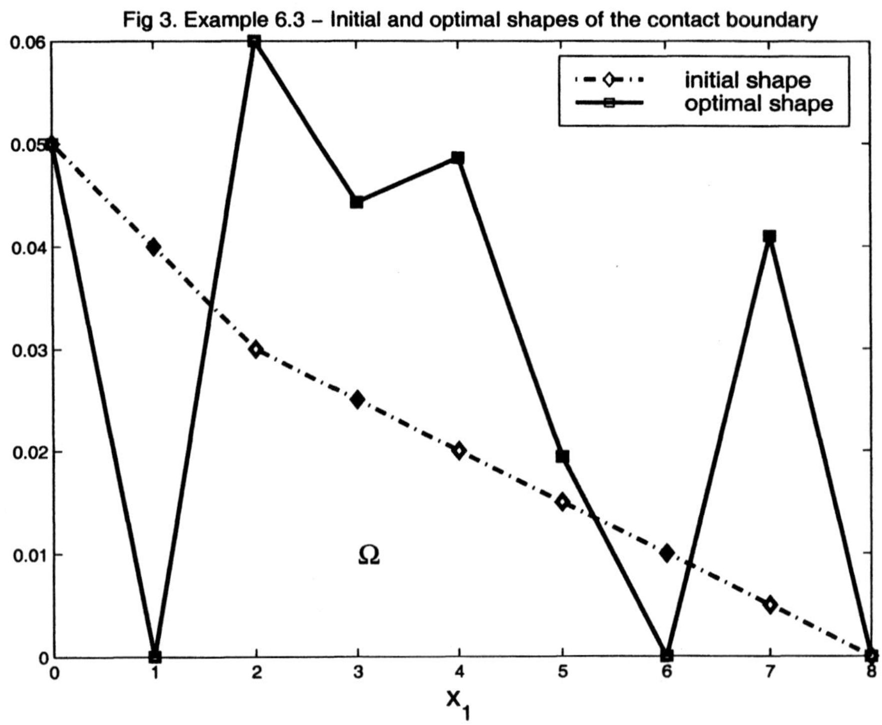
Fig 3. Example 6.3 – Initial and optimal shapes of the contact boundary
initial shape
optimal shape
0.06
0.05
0.04
0.03
0.02
0.01
0
0
1
2
3
4
5
6
7
8
Ω
x1

7 Conclusions

In the paper the fictitious domain formulation of the optimal shape design problem for unilateral contact systems was introduced. The problem was numerically solved. The proposed method seems to provide an efficient alternative to classical approach to numerical solving of shape optimal design problems. In order to increase the accuracy of obtained solutions the higher order approximation is being investigated [7, 18].

Bibliography

[1] R.A. ADAMS, *Sobolev Spaces*, Academic Press, New York, 1975.

[2] Ph.CIARLET, *Finite Element Method for Elliptic Systems*, North–Holland, Amsterdam, 1978.

[3] J. DANKOVA, J. HASLINGER, "Numerical Realization of a Fictitious Domain Approach used in Shape Optimization. Part I. Distributed Controls", *Applications of Mathematics*, **41**(1996), pp. 123 – 147.

[4] G. DUVAUT, J.L. LIONS, *Les inequations en mecanique et en physique*, Dunod, Paris, 1972.

[5] R. GLOWINSKI, T. PAN, J. PERIAUX, "A Fictitious Domain Method for Dirichlet Problem and Applications", *Computer Methods in Applied Mechanics and Engineering*, **111**(1994), pp. 283-303.

[6] R. GLOWINSKI, T. PAN, J. PERIAUX, "Combined Domain Decomposition/Fictitious Domain Methods with Lagrange Multipliers for the Unsteady Navier-Stokes Equations", in *Advances in Parallel and Vector Processing for Structural Mechanics* B.H.V. Topping and M. Papadrakakis eds.,Civil-Comp Press, Edinbourgh, 1994, pp. 119-126.

[7] R. GLOWINSKI, T.W. PAN, "Wavlet and Finite Element Solutions for the Neumann Problem using Fictitious Domain", *Journal of Computational Physics*, **126**(1996), pp. 40 - 51.

[8] J. HASLINGER, "Imbedding/Control Approach for Solving Optimal Shape Design Problems", *East - West Journal on Numerical Mathematics*, **1**(1993), pp. 111 - 119.

[9] J. HASLINGER, K.H. HOFFMANN, M. KOCVARA, "Control/Fictitious Domain Method for Solving Optimal Shape Design Problems", *Mathematical Modelling and Numerical Analysis*, **27**, No 2(1993), pp. 157-182.

[10] J. HASLINGER, P. NEITTAANMAKI, *Finite Element Approximation for Optimal Shape Design. Theory and Application*, John Wiley& Sons, 1988.

[11] J. HASLINGER, A. KLABRING, "Fictitious Domain Mixed Finite Element Approach for a Class of Optimal Shape Design Problems", *Mathematical Modelling and Numerical Analysis*, **29**(1995), pp. 435 - 450.

[12] A. MYŚLIŃSKI, "Mixed Finite Element Approximation of a Shape Optimization Problem for Systems Described by Elliptic Variational Inequalities", *Archives of Control Sciences*, **3**, No 3-4(1994), pp. 243 - 257.

[13] A. MYŚLIŃSKI, "Fictitious Domain Method with Lagrange Multipliers for Solving Optimal Shape Design Problems", In *Numerical Methods in Engineering '96* (J.A. Desideri, P. LeTallec, E. Onate, J. Periaux, E. Stein Eds), Proceedings of the second ECCOMAS Conference on "Numerical Methods in Engineering", John Wiley and Sons, Chichester, England, 1996, pp. 231 - 237.

[14] A. MYŚLIŃSKI, "Mixed Finite Element Approach for Shape Optimal Design of Large Displacement Contact Problems", *Computer & Structures*, **64**(1997), pp. 595 - 602.

[15] P. NEITTANMAKI, D. TIBA, "An Embedding of Domains Approach in Free Boundary Problems and in Optimal Design", *SIAM Journal on Control and Optimization*, **33**, No 5(1995), pp. 1587 - 1602.

[16] J. NECAS, *Les Methodes Directes en Theorie des Equations Elliptiques*, Masson, Paris, 1967.

[17] G. PEICHL, K. KUNISCH, "Shape Optimization for Mixed Boundary Value Problems based on an Embedding Domain Approach", *Preprint # 41 Graz University* (1995).

[18] A. RIEDER, "A Domain Embedding Method for Dirichlet Problems in Arbitrary Space Dimensions", *Mathematical Modelling and Numerical Analysis*, **32**(1998), pp. 405 - 432.

[19] J.P. ZOLESIO, J. SOKOLOWSKI, *Introduction to Shape Optimization. Shape Sensitivity Analysis*. Springer, Berlin, 1992.

Andrzej Myśliński. Warsaw University of Technology, Institute of Transport, 00-662 Warsaw, ul. Koszykowa 75, Poland
E-mail: e-mail: myslinsk@ibspan.waw.pl

Feedback Laws for the Optimal Control of Parabolic Variational Inequalities

Cătălin Popa

Abstract. For the optimal control of a certain class of parabolic variational inequalities, one proves that every optimal control is given by a feedback law (expressed by means of the optimal value function). The obtained result improves and generalizes previous results of the author (established for the optimal control of parabolic obstacle problem and semilinear parabolic equations).

1 Introduction

It is well known that heuristic considerations based on dynamic programming lead to optimal feedback laws expressed by means of the optimal value function. But, can these laws be rigorously justified? In many situations the answer is positive. In the paper [8], the author obtained such laws for the optimal control of semilinear parabolic equations and parabolic obstacle problem. In the present paper, we shall improve and generalize our previous results in the more general framework of the optimal control of parabolic variational inequalities.

2 The Framework and the Main Results

Let Ω be an open and bounded subset of $\mathbb{R}^n$ having a sufficiently smooth boundary and set $\mathcal{H} = L^2(\Omega)$. Let $\mathcal{U}$ be a real Hilbert space. (In what follows, $\mathcal{H}$ and $\mathcal{U}$ will be the state and the control spaces, respectively.) Denote the scalar products and norms of $\mathcal{H}$ and $\mathcal{U}$ by the same symbols: $(\cdot,\cdot)$ and $|\cdot|$.

The control system we deal with is described by the following nonlinear abstract differential equation (with initial condition):

$$\begin{cases} y'(t) + Ay(t) + \beta(y(t) - \psi) \ni Bu(t) + f(t), \ t \in (0,T), \\ y(0) = y_0. \end{cases} \tag{2.1}$$

We impose the following hypotheses on the data of equation (2.1):

(H1) $A : \mathcal{V} \to \mathcal{V}'$ is a linear continuous and symmetric operator, where $\mathcal{V}$ is a real Hilbert space continuously, densely and *compactly* embedded in $\mathcal{H}$ with $\mathcal{V}'$ its dual space. (Identifying

$\mathcal{H}$ with its own dual, we have $\mathcal{V} \subset \mathcal{H} \subset \mathcal{V}'$.) Denoting by $(\cdot,\cdot)$ the pairing between $\mathcal{V}$ and $\mathcal{V}'$, and by $|\cdot|_{\mathcal{V}}$ the norm of $\mathcal{V}$, we also assume that, for some $\omega > 0$ and $\alpha \in \mathbb{R}$,

$$(Ay, y) + \alpha|y|^2 \geq \omega|y|_{\mathcal{V}}^2 \text{ for all } y \in \mathcal{V}.$$

(H2) β is a maximal monotone graph in $\mathbb{R}^2$ and ψ is a given function in $\mathcal{H}$ such that, for some $c \geq 0$,

$$(Ay, \beta_\eta(y-\psi)) \geq -c(1+|\beta_\eta(y-\psi)|)(1+|y|) \text{ for all}$$
$$y \in D(A) = \{z \in \mathcal{V} : Az \in \mathcal{H}\} \text{ and } \eta > 0,$$

where $\beta_\eta(r) = \eta^{-1}(r - (I+\eta\beta)^{-1}r)$ for all $r \in \mathbb{R}$, $\eta > 0$. (The above inequality ensures the maximal monotonicity of the operator $A + \beta$ in $\mathcal{H} \times \mathcal{H}$.)

(H3) $B : \mathcal{U} \to \mathcal{H}$ is a linear continuous operator.

(H4) $f \in L^2(0, T; \mathcal{H})$.

In most situations $-A$ is a strictly elliptic differential operator in Ω having its principal part in divergence form, and $\mathcal{V}$ is a suitable Sobolev space.

If we define the convex function ϕ by

$$\phi(y) = \int_\Omega j(y(x) - \psi(x))dx \text{ for } y \in \mathcal{H},$$

where $j : \mathbb{R} \to (-\infty, +\infty]$ is a convex function whose subdifferential is β, then we may write equation (2.1) as

$$y' + Ay + \partial\phi(y) \ni Bu + f \text{ in } (0, T).$$

As regards the initial datum y_0, we assume that

(H5) $y_0 \in \mathcal{V} \cap D(\phi)$.

For the later use, set $K = \overline{D(\phi)}$.

Consider the following optimal control problem:

(P) Minimize

$$\int_0^T (h(u(t)) + g(y(t)))dt + \ell(y(T)) \tag{2.2}$$

over all $u \in L^2(0,T;\mathcal{U})$, where $y \in W^{1,2}([0,T];\mathcal{H})$ satisfies equation (2.1).

Here h, g and ℓ satisfy the following hypotheses:

(H6) $h : \mathcal{U} \to (-\infty, +\infty]$ is convex, lower semicontinuous and not identically $+\infty$.

(H7) $g, \ell : \mathcal{H} \to \mathbb{R}$ are Lipschitz continuous on bounded subsets and bounded from below by affine functions.

We associate with problem (P) the corresponding optimal value function $V : [0,T] \times K \to \mathbb{R}$, i.e.,

$$V(t,y) = \inf\{ \int_t^T (h(u(s)) + g(z(s)))ds + \ell(z(T)) :$$
$$z' + Az + \beta(z - \psi) \ni Bu + f \text{ a.e. } s \in (t,T),$$
$$z(t) = y, \; u \in L^2(t,T;\mathcal{U})\}.$$

One knows (see Proposition 2.1 in [8]), the function $y \mapsto V(t,y)$ is Lipschitz continuous on bounded subsets, uniformly with respect to $t \in [0,T]$. On the other hand, there are situations in which no point of K is interior for K (see Remark 2.1 in [8]). To give a sense to the gradient of V with respect to y also at those points of K which are not interior for K, we extend V to the whole space $\mathcal{H}$ in the following way:

$$\widetilde{V}(t,y) = V(t, P_K y) \text{ for } (t,y) \in [0,T] \times \mathcal{H},$$

where P_K is the projection operator of $\mathcal{H}$ onto K. Clearly, the function $\mathcal{H} \ni y \mapsto \widetilde{V}(t,y)$ is Lipschitz continuous on bounded subsets uniformly with respect to $t \in [0,T]$ because V has the same property and P_K is a contraction. So we may define

$$\partial_y V(t,y) = \partial_y \widetilde{V}(t,y) \text{ for } (t,y) \in [0,T] \times K,$$

where $\partial_y \widetilde{V}(t,y)$ is the generalized gradient (in Clarke's sense) of $y \mapsto \widetilde{V}(t,y)$.

Now, we are prepared to state the main result of our paper.

THEOREM 2.1. *Under hypotheses (H1)–(H7), if (u^*, y^*) is an optimal pair for problem (P), then the following feedback law holds:*

$$u^*(t) \in \partial h^*(-B^* \partial_y V(t, y^*(t))) \text{ a.e. } t \in (0,T). \tag{2.3}$$

Here h^* is the Legendre transform of h, which, in this case, is just the convex conjugate of h.

As we have said before, Theorem 2.1 improves and generalizes two previous results. The first (Theorem 2.3 in [8]) refers to the optimal control of semilinear parabolic equations. In this case β is a monotone function from $\mathbb{R}$ to $\mathbb{R}$ which is Lipschitz continuous on bounded subsets, and the obtained feedback law is just (2.3). The second result (Theorem 2.4 from [8]) concerns the optimal control of parabolic obstacle problem, where β is the maximal monotone graph in the plane defined by

$$\beta(r) = \begin{cases} 0 & \text{if } r > 0, \\ (-\infty, 0] & \text{if } r = 0, \\ \emptyset & \text{if } r < 0. \end{cases}$$

In this case the feedback law we established is less precise than (2.3), namely, inclusion (2.3) must be adjusted by adding the normal cone to K at ψ :

$$u^*(t) \in \partial h^*(-B^*(\partial_y V(t, y^*(t)) + N_K(\psi))) \text{ a.e. } t \in (0,T). \tag{2.4}$$

Here $K = \{y \in L^2(\Omega) : y \geq \psi$ a.e. in $\Omega\}$ (this is easy to see) and $N_K(\psi) = \{w \in L^2(\Omega) : \int_\Omega w(x)(y(x) - \psi(x))dx \leq 0$ for all $y \in K\}$. (If $\psi = 0$, then $N_K(0) = \{w \in L^2(\Omega) : w \leq 0$ a.e. in $\Omega\}$ and, by Theorem 2.4 from [8], the expected feedback law (2.3) must be corrected by adding an unspecified nonpositive term w. Theorem 2.1 says that this term may be chosen to be 0). In other words, Theorem 2.1 asserts that the presence of $N_K(\psi)$ in the expression of the feedback law (2.4) can be removed, and the feedback law (2.3) is valid for any *multi-valued* maximal monotone graph β in the plane.

The most natural way to prove Theorem 2.1 is to combine the reduced form of Pontryagin's maximum principle with the well-known relationship between the maximum principle and dynamic programming (that is, the relationship between the dual extremal arc in the maximum principle and the optimal value function). For problem (P), Pontryagin's maximum principle says that there exists a dual extremal arc $p \in L^\infty(0,T;\mathcal{H}) \cap L^2(0,T;\mathcal{V})$ which satisfies the following first order condition:

$$B^*p(t) \in \partial h(u^*(t)) \text{ a.e. } t \in (0,T), \tag{2.5}$$

or, in other words,

$$u^*(t) \in \partial h^*(B^*p(t)) \text{ a.e. } t \in (0,T). \tag{2.6}$$

The relationship between the maximum principle and dynamic programming is expressed in the following theorem:

THEOREM 2.2. *Under hypotheses (H1)-(H7), if (u^*, y^*) is an optimal pair for problem (P), then there exists $p \in L^\infty(0,T;\mathcal{H}) \cap L^2(0,T;\mathcal{V})$ which satisfies the reduced form (2.5) of the maximum principle and the inclusion*

$$-p(t) \in \partial_y V(t, y^*(t)) \text{ a.e. } t \in (0,T). \tag{2.7}$$

It is clear that (2.7) together with (2.6) gives the feedback relation (2.3); so, Theorem 2.1 is simply a consequence of Theorem 2.2.

Let us emphasize that Theorem 2.2 does not assure that the dual extremal arc p satisfies also an adjoint equation as Theorem 2.1 and 2.2 in [8] do, but in stronger hypotheses. This fact does not cause us any difficulty because the only statement of the maximum principle involving p we need here is the maximum relation (2.5).

3 Proof of Theorem 2.2

There are several ways to approach inclusion (2.7). One of these which leads to the most general results is based on an idea of F.H. Clarke and R.B. Vinter (used by them in the finite-dimensional case in [5]). We can express this idea in a few words like that: inclusion (2.7) can be interpreted as a (first order) maximum condition of type (2.5), but with respect to an auxiliary control variable v which arises in the state equation (2.1) as a perturbation. But the transformation of the preceding idea into a complete and rigorous proof is not an easy task in infinite dimensions. However, the heuristic considerations which lead to the basic point of the proof are quite simple and easy to present. To set forth these considerations, let us accept that the optimal value function V is continuously differentiable. Of course, this is not a realistic hypothesis, but accepting it in the beginning, the machinery of dynamic programming can run. Then, we shall see what changes must be done in our arguments in order to obtain a rigorous proof. For the sake of simplicity, we also suppose that both ψ and f are identically 0.

So, being continuously differentiable, the optimal value function V is a classical solution of the following Hamilton–Jacobi equation (called, in this context, the dynamic programming equation):

$$\partial_t V(t,y) - h^*(-B^*\nabla_y V(t,y)) - (Ay + \beta(y), \nabla_y V(t,y)) = -g(y). \tag{3.1}$$

Obviously, V also satisfies the final condition

$$V(T,y) = \ell(y). \tag{3.2}$$

As we have said, h^* is the Legendre transform of h, so one has

$$-h^*(-B^*p) = \inf\{h(u) + (B^*p, u) : u \in \mathcal{U}\} \text{ for } p \in \mathcal{H}. \tag{3.3}$$

Now, for any control $u \in L^2(0,T;\mathcal{U})$ and additional control $v \in L^2(0,T;\mathcal{H})$, let $y(\cdot)$ be the solution of the following perturbed version of the state equation (2.1) (with the same initial condition):

$$\begin{cases} y' + Ay + \beta(y) \ni Bu + v & \text{in } (0,T), \\ y(0) = y_0. \end{cases} \tag{3.4}$$

Replacing the variable y in the Hamilton–Jacobi equation (3.1) by $y(t)$ (that is, by the perturbed state at the moment t) and using (3.3), we easily obtain the following differential inequality:

$$\begin{aligned} &\partial_t V(t,y(t)) + h(u(t)) + (\nabla_y V(t,y(t)), Bu(t)) \\ &-(Ay(t) + \beta(y(t)), \nabla_y V(t,y(t))) \geq -g(y(t)). \end{aligned}$$

Looking at the last two terms of the left–hand side of the above inequality and looking at the perturbed equation (3.4) too, we see one can reconstitute $y'(t)$ as a factor in the scalar product by adding (and subtracting) the term $(\nabla_y V(t,y(t)), v(t))$:

$$\begin{aligned} \partial_t V(t,y(t)) + (\nabla_y V(t,y(t)), y'(t)) - (\nabla_y V(t,y(t)), v(t)) \\ +h(u(t)) + g(y(t)) \geq 0. \end{aligned}$$

But we can write the first two terms of the last inequality as the derivative with respect to t of $V(t,y(t))$:

$$\frac{d}{dt}V(t,y(t)) + h(u(t)) + g(y(t)) - (\nabla_y V(t,y(t)), v(t)) \geq 0.$$

Integrating then from 0 to T and taking (3.2) into account, we get the following inequality:

$$\begin{aligned} \int_0^T (h(u(t)) + g(y(t)))dt + \int_0^T (\nabla_y V(t,y(t)), -v(t))dt + \ell(y(T)) \\ \geq V(0,y_0). \end{aligned} \tag{3.5}$$

We see that the left–hand side of (3.5) is just the performance index (2.2) of problem (P) plus an additional term containing v. So, if we replace the functions u, v and y in (3.5) by $u^*, v^* \equiv 0$ and y^*, respectively, then the inequality clearly becomes equality. But this means that the triple $u^*, v^* \equiv 0$ and y^* solves the following auxiliary optimal control problem:

Minimize

$$\int_0^T (h(u(t)) + g(y(t)) + (\nabla_y V(t, y(t)), -v(t)))dt + \ell(y(T))$$

over all $u \in L^2(0, T; \mathcal{U})$, $v \in L^2(0, T; \mathcal{H})$, where y satisfies the perturbed state equation (3.4).

Now, it is easy to see that the part of the maximum principle which refers to the additional control variable v is just the relation (2.7), that is,

$$-p(t) = \nabla_y V(t, y^*(t)) \text{ a.e. } t \in (0, T).$$

But a problem arises here. Since the term in (3.5) containing v also contains the state variable y, the corresponding adjoint equation does not coincide with the original one. To avoid this situation, we would have to replace the term $(\nabla_y V(t, y), v)$ by something which does not depend on y. The most natural and suitable substitute of $(\nabla_y V(t, y), v)$ which does not depend on y is the function $k_\delta : [0, T] \times \mathcal{H} \to \mathbb{R}$ defined by

$$k_\delta(t, v) = \sup\{(\nabla_y V(t, y), v) : |y - y^*(t)| \leq \delta\} \text{ for } \delta > 0.$$

(Notice that k_δ is convex and Lipschitz continuous in v.) Looking at inequality (3.5) once again, it is clear (by the definition of k_δ) that the following lemma is true.

LEMMA 3.1. *Under hypotheses of Theorem 2.2, for any $u \in L^2(0, T; \mathcal{U})$ and $v \in L^2(0, T; \mathcal{H})$ such that the solution y of the perturbed equation (3.4) satisfies $|y(t) - y^*(t)| \leq \delta$ for all $t \in [0, T]$, we have the following version of inequality (3.5):*

$$\int_0^T (h(u(t)) + g(y(t)))dt + \ell(y(T)) + \int_0^T k_\delta(t, -v(t))dt \geq V(0, y_0). \tag{3.6}$$

As before, inequality (3.6) means that the triple $u^*, v^* \equiv 0$ and y^* solves the following optimal control problem:

(P_δ) Minimize the left–hand side of (3.6), i.e.,

$$\int_0^T (h(u(t)) + g(y(t)) + k_\delta(t, -v(t)))dt + \ell(y(T))$$

over all $u \in L^2(0, T; \mathcal{U})$, $v \in L^2(0, T; \mathcal{H})$, where y satisfies equation (3.4) and

$$|y(t) - y^*(t)| \leq \delta \text{ for all } t \in [0, T].$$

We apply the maximum principle to problem (P_δ). Its part which refers to the control variable v is the following inclusion (of type (2.5)):

$$-p_\delta(t) \in \partial_v k_\delta(t, 0) \text{ a.e. } t \in (0, T).$$

By the definition of k_δ it is not difficult to show (see [5] or [8]) that

$$-p_\delta(t) \in \overline{\mathrm{co}}\{\nabla_y V(t, y) : |y - y^*(t)| \leq \delta\} \text{ a.e. } t \in (0, T) \tag{3.7}$$

(where $\overline{\mathrm{co}}$ denotes the weak*-closed convex hull). Letting $\delta \to 0$, since V is continuously differentiable, we obtain:

$$-p(t) = \nabla_y V(t, y^*(t)) \text{ a.e. } t \in (0, T),$$

which is just the relation (2.7). Moreover, because k_δ does not depend on y, the adjoint equation satisfied by p is just the original one (which is an important detail when we take limits in (3.7)).

Now let us see how we can argue to prove Lemma 3.1 when the optimal value function V is not continuously differentiable. There are two places in which the continuous differentiability of V is used, namely: where we take limits in inclusion (3.7) as $\delta \to 0$, and where we apply the dynamic programming Hamilton–Jacobi equation (in its classical form). In the first case the continuous differentiability of V is not indispensable. Indeed, it suffices to use weaker continuity properties of the generalized gradient of V such as the upper semicontinuity of the generalized directional derivative of V. In the other case, the situation is more serious. Since the dynamic programming Hamilton–Jacobi equation is the infinitesimal expression of the dynamic programming principle, it is natural to apply this principle when V is nonsmooth. But the direct use of the dynamic programming principle is not convenient to our purpose because of the following reason: In the statement of Lemma 3.1 (in fact, in inequality (3.6)), the optimal value function V arises (via the function k_δ) together with the solution of the perturbed equation (3.4). But in the dynamic programming principle, the optimal value function V is indestructibly connected with the solution of the *original* state equation (2.1). How could we surpass this difficulty? We need the optimal value function V together with the solution of (3.4), but we can relate V only to the solution of (2.1). One would have to eliminate the perturbation term v in equation (3.4) in order to be able to apply the dynamic programming principle to the nonperturbed state equation (2.1). But how could we eliminate v?

The key idea is simple: on small time intervals, we decouple the perturbed equation (3.4) into two differential equations, the first of these being just the original state equation (2.1) and the other equation containing only the perturbation term v:

$$\begin{cases} y_j' + Ay_j + \beta(y_j - \psi) \ni Bu + f & \text{in } (\tau + iT/N_j, \tau + (i+1)T/N_j), \\ y_j(\tau + iT/N_j) = P_K z_j(\tau + (i+1)T/N_j - 0), \end{cases}$$

$$\begin{cases} z_j' = v(\tau + iT/N_j) & \text{in } (\tau + iT/N_j, \tau + (i+1)T/N_j), \\ z_j(\tau + iT/N_j) = y_j(\tau + iT/N_j - 0), \end{cases}$$

for all integers i from $m_j = \min\{i' \in \mathbb{Z} : \tau + i'T/N_j > 0\}$ to $n_j = \max\{i' \in \mathbb{Z} : \tau + i'T/N_j < T\}$. Here τ is a certain moment in $(0, T)$ and $\{N_j\}$ is a subsequence of the positive integers such that T/τ is irrational and

$$\sum_{i=m_j}^{n_j} \frac{T}{N_j} k_\delta \left(\tau + i\frac{T}{N_j}, -v\left(\tau + i\frac{T}{N_j}\right)\right) \to \int_0^T k_\delta(t, -v(t))dt \text{ as } j \to \infty.$$

(On the interval $[0, \tau + m_jT/N_j)$, y_j satisfies the equation $y_j' + Ay_j + \beta(y_j - \psi) \ni Bu + f$ and the initial condition $y_j(0) = y_0$.) Solving the second equation and eliminating its solution in the first equation, we obtain an initial value problem for the state equation (2.1) whose initial condition contains the term v:

$$\begin{cases} y_j' + Ay_j + \beta(y_j - \psi) \ni Bu + f \quad \text{in } (\tau + iT/N_j, \tau + (i+1)T/N_j), \\ y_j(\tau + iT/N_j) = P_K\left(y_j(\tau + iT/N_j - 0) + \dfrac{T}{N_j} v(\tau + iT/N_j)\right). \end{cases} \tag{3.8}$$

Now, on each interval $[\tau + iT/N_j, \tau + (i+1)T/N_j)$ (and also on $[0, \tau + m_jT/N_j)$) we may apply the dynamic programming principle, and adding the obtained inequalities, we get a discrete version of inequality (3.6) in Lemma 3.1 (see [8] for details). The only problem we still have to clear up is the convergence of y_j to the solution y of the perturbed equation (3.4). The following observation suggests us an approach to the convergence which is practicable also in infinite dimensions: The iterative scheme (3.8) can be visualized as a Lie–Trotter product formula approximation scheme for two appropriate contraction semigroups (see Lemma 3.1 in [8]). A suitable convergence result for the involved Lie–Trotter product formula yields:

$$y_j(t) \to y(t) \text{ strongly in } \mathcal{H} \text{ for all } t \in [0, T].$$

Moreover, an uniform estimate holds at the points $t = \tau + iT/N_j$. (For more details we refer to [9]; see also Lemma 3.1 in [8] for a somewhat simpler approximation scheme.)

Arriving at this point, let us emphasize that the removal of the normal cone $N_K(\psi)$ in the expression of the feedback law (2.4) is the effect of the presence of the projector operator P_K in the approximation scheme (3.8).

Now let us explain why the feedback law (2.3) is valid for more general control systems than those treated in [8] (that is, for any multi–valued maximal monotone graph β in the plane). In order to obtain first order necessary conditions of optimality for problem (P$_\delta$), in the same way as in §2, Ch.5 from [1], we associate with (P$_\delta$) a family (P$_{\delta,\eta}$) of smooth and penalized problems. (The new parameter η arises in the regularization process.) So, we have to take limits when both δ and η tend to zero. As in [8], we firstly obtain the optimality conditions for problem (P$_{\delta,\eta}$). These are satisfied by the corresponding optimal triple $u_{\delta,\eta}, v_{\delta,\eta}, y_{\delta,\eta}$ and by the dual extremal arc $p_{\delta,\eta}$. As an effect of penalization terms, we have as $\eta \to 0$ (see [8])

$$\begin{aligned} u_{\delta,\eta} &\longrightarrow u^* && \text{strongly in } L^2(0,T;\mathcal{U}), \\ v_{\delta,\eta} &\longrightarrow v^* \equiv 0 && \text{strongly in } L^2(0,T;\mathcal{H}) \end{aligned}$$

(so, $y_{\delta,\eta} \to y^*$ strongly in $C([0,T];\mathcal{H})$). But this shows us that we can choose η_δ such that $\eta_\delta \searrow 0$ as $\delta \searrow 0$ and

$$\int_0^T |u_{\delta,\eta_\delta}(t) - u^*(t)|^2 dt \le \delta \text{ and } \int_0^T |v_{\delta,\eta_\delta}(t)|^2 dt < \delta.$$

For simplicity, set $u_{\delta,\eta_\delta} = u_\delta$, $v_{\delta,\eta_\delta} = v_\delta$ and $p_{\delta,\eta_\delta} = p_\delta$. The key observation is the following result (whose proof can be found in Lemma 5.3 from [1]).

LEMMA 3.2. *Under hypotheses (H1)-(H7), for all $\delta > 0$, we have*

$$|p_\delta(t)| \leq const.\ for\ all\ t \in [0,T],$$
$$\int_0^T |p_\delta(t)|_{\mathcal{V}}^2 dt \leq const.,$$

where the constants are independent of δ.

This means that we can find $p \in L^\infty(0,T;\mathcal{H}) \cap L^2(0,T;\mathcal{V})$ such that, on a subsequence of $\{\delta\}$,

$$p_\delta \to p \text{ weak star in } L^\infty(0,T;\mathcal{H}) \text{ and weakly in } L^2(0,T;\mathcal{V}).$$

(Let us point out that we need nothing more than Lemma 3.2 in the process of passing to the limit as $\delta \to 0$.) So, the statement of Theorem 2.2 follows by passing to the limit as $\delta \to 0$ in the optimality conditions for problem $(\mathrm{P}_{\delta,\eta_\delta})$. (For more details, we refer to [8] once again).

4 Final Remarks

The approach before seems to be adequate also for other nonlinear distributed parameter systems such as that described by Navier–Stokes equations.

Another way to prove the relationship between the maximum principle and dynamic programming consists in the direct use of the dynamic programming Hamilton–Jacobi equations but, in this case, we must appeal to the infinite–dimensional theory of viscosity solutions. This can be done in the manner in which V. Barbu, E.N. Barron and R. Jensen establish Pontryagin's maximum principle in Hilbert spaces in [4].

Finally, we point out one more problem in the same circle of ideas. We have established the optimal feedback law (2.3). But to use this law in constructing optimal (or suboptimal) feedback controls (so that the feedback law should become effective), we must be able to compute (or to approximate) the optimal value function in a reasonable manner. Certainly, this function satisfies a Hamilton–Jacobi equation, but such an equation is a very complicated mathematical object. A promising way to compute solutions of Hamilton–Jacobi equations (which satisfy certain initial or final conditions) consists in splitting these equations into two or several simpler equations on small time intervals (by decomposing the associated Cauchy problem into two or several such problems). In this way, one obtains Lie–Trotter product formula approximation for Hamilton–Jacobi equations. Such an approach to the calculus (or approximation) of the optimal value function was initiated by V. Barbu in [2] and developed by V. Barbu and C. Popa in a series of papers (see, for example, [3], [6] and [7]).

Bibliography

[1] V. BARBU, *Optimal Control of Variational Inequalities,* Research Notes in Mathematics 100, Pitman, London, 1984.

[2] V. BARBU, *A product formula approach to nonlinear optimal control problems*, SIAM J. Control Optim., 26(1988), pp. 496–520.

[3] V. BARBU, *Approximation of the Hamilton–Jacobi equations via Lie–Trotter product formula*, Control Theory Adv. Tech., 4(1988), pp. 189–208.

[4] V. BARBU, E.N. BARRON and R. JENSEN, *The necessary conditions for optimal control in Hilbert spaces,* J. Math. Anal. Appl. 133(1988), pp. 151–162.

[5] F.H. CLARKE and R.B. VINTER, *The relationship between the maximum principle and dynamic programming*, SIAM J. Control Optim., 25(1987), pp. 1291–1311.

[6] C. POPA, *Trotter product formulae for Hamilton–Jacobi equations in infinite dimensions,* Differential Integral Equations, 4(1991), pp. 1251–1268.

[7] C. POPA, *Feedback laws for nonlinear distributed control problems via Trotter–type product formulae,* SIAM J. Control Optim., 33 (1995), pp. 971–999.

[8] C. POPA, *The relationship between the maximum principle and dynamic programming for the control of parabolic variational inequalities,* SIAM J. Control Optim., 35(1997), pp. 1711–1738.

[9] C. POPA, *The relationship between the maximum principle and dynamic programming for the control of parabolic obstacle problem*, to appear.

Cătălin Popa. Facultatea de Matematică, Universitatea "Al.I. Cuza", Bdul. Carol I 11, 6600 Iaşi, Romania
E-mail: cpopa@uaic.ro

Application of Special Smoothing Procedure to Numerical Solutions of Inverse Problems for Real 2-D Systems

Edward Rydygier and Zdzislaw Trzaska

Abstract. Novel and effective numerical procedure is introduced to stabilize the results of calculations in the instance of solving real $2-D$ inverse problems. For these problems a stabilization of results consists in an elaboration of regularization procedure. In presented work the special regularization procedure is connected with the combinatorial method for numerical solving inverse problems of $2-D$ systems. In order to construct a suitable regularization procedure many approximation procedures are investigated, e.g. a smoothing of scattered data, a two-dimensional interpolation, and an approximation of field sources function. The correct results were obtained for the approximation procedure elaborated on the basis of an inverse distance method of smoothing the scattered data for $2-D$ systems. This method named also the Shepard's method was adapted and modified to construct the special self-regularization method. The examples of inverse problem solutions for different $2-D$ systems taken from the practice are shown. On the basis of these examples the detailed problems were investigated like an identification of field sources, a finding steady field sources' distributions, a obtaining the solutions for complicated shapes of considered domain.

1 Introduction

Recently, it can be observed that the inverse problems are of increasing interest both in scientific centers and industry. The studies on inverse problems are carried out in two directions: development of the theory and numerical methods, and improvement of measurement technology, Kaczorek, 1985, Groetsch, 1993, Kurpisz, 1995, Tikhonov, 1995, Neittaanmaki, 1996. Problem of modeling the physical reality with suitable differential equations systems is relatively uncomplicated in the finite dimensional setting but becomes very difficult for various partial differential equations such as wave, heat, and electromagnetic equations. When it is impossible, or difficult, to obtain an exact solution of the partial differential equations governing a continuous system, the system is reduced to discrete form, John, 1978, Anger, 1990, Tikhonov, 1995.

Inverse problems are the ill-posed problems according to Hadamards definition of correctly posed problems, Hadamard, 1952. A solution of correctly posed problem must be unique and stable. The ill-posed nature of inverse problems causes that various methods for direct problems are inapplicable for wide range of inverse problems. Therefore, the special numerical procedures must be employed to stabilize the results of calculations, Tikhonov, 1995, Engl, 1996. In this paper a new numerical procedure is introduced to stabilize the results of calculations in the instance of a solving inverse problems. The considered problem of special regularization procedure is connected with the construction of a combinatorial method for solving inverse problems of $2-D$ systems. This method is an effective numerical approach applied a modern combinatorial analysis. This combinatorial approach consists in a use of new computational tools developed by authors of this paper during previous investigations in a field of combinatorics and the Fibonacci trigonometry.

2 Combinatorial Method for Numerical Solving Inverse Problems

Presently, a growing interest is observed in development of methods using a combinatorial analysis based on conceptions and objects from modern combinatorics, Vajda, 1989, Bergum, 1994, Ross, 1996. In a field of engineering problems various structures of the so-called numerical triangles and hyperbolic Fibonacci functions can be used for modeling and numerical analysis of distributed parameter systems, Trzaska, 1997, Rydygier and Trzaska, 1999. Monic non-zero polynomials which generate the first modified numerical triangle, FMNT, are defined by the following recurrence, Trzaska, 1993a, 1993b

$$T_{n+2}(x) = (2+x)T_{n+1}(x) - T_n(x), \quad n = 0,\ 1,\ 2,\ \dots, \tag{2.1}$$

with $T_0(x) = 1$ and $T_1(x) = 1 + x$ as initial elements. From the above recurrence, the following polynomials can be calculated

$$\begin{aligned}
T_0(x) &= 1 \\
T_1(x) &= 1 + x \\
T_2(x) &= 1 + 3x + x^2 \\
T_3(x) &= 1 + 6x + 5x^2 + x^3 \\
T_4(x) &= 1 + 10x + 15x^2 + 7x^3 + x^4 \\
T_5(x) &= 1 + 15x + 35x^2 + 28x^3 + 9x^4 + x^5 \\
\dots\dots\dots & \dots\dots\dots\dots\dots\dots\dots\dots\dots\dots\dots
\end{aligned}$$

Thus, the polynomial $T_n(x)$ can be written in the form

$$T_n(x) = \sum_{k=0}^{n} a_{n,k} x^k, \quad n = 0,\ 1,\ 2,\ \dots \tag{2.2}$$

where the coefficients $a_{n,k}, n = 0,\ 1,\ 2,\ \dots,\ 0 \le k \le n$, fulfill the relation

$$a_{n,k} = 2a_{n-1,k} + a_{n-1,k-1} - a_{n-2,k} \tag{2.3}$$

with $a_{0,0} = 1$ and $a_{1,0} = 1$ as initial values. Based on (3) the FMNT can be constructed. It is presented in Table 1.

Table 1. First modified numerical triangle, (**FMNT**)

$n \backslash k$	0	1	2	3	4	5	6	...	*Sum of Coeffs.*
0	1								1
1	1	1							2
2	1	3	1						5
3	1	6	5	1					13
4	1	10	15	7	1				34
5	1	15	35	28	9	1			89
6	1	21	70	84	45	11	1		233
...	...	...	...	...	...	...	...	...	...

To establish the second numerical triangle SMNT, the monic non-zero power polynomials are defined by the recurrence

$$P_{n+2}(x) = (2 + x)P_{n+1}(x) - P_n(x), \quad n = 0,\ 1,\ 2,\ ... \tag{2.4}$$

with $P_0(x) = 0$ and $P_1(x) = 1$ as initial elements. From (4) the following polynomials can be obtained

$$\begin{aligned}
P_0(x) &= 0 \\
P_1(x) &= 1 \\
P_2(x) &= 2 + x \\
P_3(x) &= 3 + 4x + x^2 \\
P_4(x) &= 4 + 10x + 6x^2 + x^3 \\
P_5(x) &= 5 + 20x + 21x^2 + 8x^3 + x^4 \\
\cdots\cdots\cdots & \cdots\cdots\cdots\cdots\cdots\cdots\cdots\cdots\cdots\cdots
\end{aligned}$$

From the above expressions the polynomial $P_n(x)$ can be written in the form

$$P_n(x) = \sum_{r=0}^{n-1} b_{n,r}x^r, \quad n = 0,\ 1,\ 2,\ ... \tag{2.5}$$

where the coefficients $b_{n,r}$, $n = 0,\ 1,\ 2,\ ...\ , 0 \le r \le n$ are defined by the recurrence

$$b_{n,r} = 2b_{n-1,r} + b_{n-1,r-1} - b_{n-2,r} \tag{2.6}$$

with $b_{0,0} = 0$ and $b_{1,0} = 1$ as initial values. Then, based on (6) the SMNT can be constructed. It is shown in Table 2.

Table 2. Second modified numerical triangle, **(SMNT)**

$n \backslash r$	0	1	2	3	4	5	6	...	*Sum of coeffs.*
0	0								0
1	1								1
2	2	1							3
3	3	4	1						8
4	4	10	6	1					21
5	5	20	21	8	1				55
6	6	35	56	36	10	1			144
...	...	...	...	...	...	...		...	...

Formally, both the FMNT and the SMNT are apparently similar to the classical Pascal triangle, Ross, 1996, but their elements cannot be evaluated directly by applying the rule corresponding to the classical Pascal triangle, Trzaska, 1993a. They must be computed in accordance with recurrence (3) and (6), respectively. The sum of all elements values in a row of FMNT or SMNT equals to f_{2n}, $n = 0, 1, 2, \ldots$, or f_{2n-1}, $n = 0, 1, 2, \ldots$, respectively, i. e. they are equal to successive elements of the Fibonacci sequence with even or odd indices, respectively Bergum, 1994

$$f_{n+2} = f_{n+1} + f_n, \quad n = 0,\ 1,\ 2,\ \ldots \tag{2.7}$$

with $f_0 = 1$ and $f_1 = 1$ as initial values. More properties of monic polynomials were described in details in Trzaska, 1993a, 1993b, Trzaska, 1997. Hyperbolic Fibonacci functions $sFh(x)$ and $cFh(x)$ are defined as follows

$$sFh(x) = \frac{\phi^{2x} - \phi^{-2x}}{\sqrt{5}}, \quad cFh(x) = \frac{\phi^{2x+1} + \phi^{-(2x+1)}}{\sqrt{5}} \tag{2.8}$$

where ϕ denotes the golden ratio, Trzaska, 1993a, Vajda, 1989.

It is easy to demonstrate that when a discrete variable $k \in I$ is used then the functions $sFh(k)$ and $cFh(k)$ in terms of corresponding elements of the Fibonacci sequence

$$f(p+1) = f(p) + f(p-1), \quad p = \ \ldots \ -3,\ -2,\ -1,\ 0,\ 1,\ 2,\ 3,\ \ldots\ , \tag{2.9}$$

with $f(0) = 0$ and $f(1) = 1$ can be written in the formulas

$$sFh(k) = f(2k), \qquad cFh(k) = f(2k+1) \tag{2.10}$$

Moreover, it is evident that Fibonacci hyperbolic functions and modified numerical triangles above presented can be very useful for practical problem studies, Trzaska, 1997, Rydygier, 1998, Rydygier and Trzaska, 1999.

On the basis of the stationary $2 - D$ space-continuous system described by the Poisson equations with specified boundary conditions, the computational algorithms are elaborated for solving direct and inverse problems with special regard to identification problems. The investigated system is described by the second order partial differential equation

$$\frac{\partial^2 u(x,y)}{\partial x^2} + \frac{\partial^2 u(x,y)}{\partial y^2} = f(x,y) \tag{2.11}$$

where $u(x, y)$ is the potential function and $f(x, y)$ is the field sources' function.

At first a direct problem will be solved. This problem consists in finding a solution of equation (11). This solution is the potential function u for known function f and for given boundary conditions. To solve numerically this problem a discretization using the finite difference method, Dahlquist, 1974, and an expanding the function f with a use of the Fourier series, Potter, 1973, for parameter $n = 1, 2, ..., N-1$ are done. The boundary conditions for function f are in the form

$$f_{m,0} = f_{m,N} = 0 \tag{2.12}$$

The discrete values of a field sources function can be calculated from following formula

$$f_{m,n} = \sqrt{2}\sum_{k=1}^{M-1} F_m(k) sin\frac{k\pi n}{M}, \quad m = 1, 2, ..., M-1 \tag{2.13}$$

In the same way a solution for the potential function u can be represented as follows

$$u_{m,n} = \sqrt{2}\sum_{k=1}^{M-1} U_m(k) sin\frac{k\pi n}{M}. \tag{2.14}$$

After modification with use some trigonometric identities, the second order difference equation can be established

$$[\frac{1}{h^2}(U_{m+1}(k) - 2U_m(k) + U_{m-1}(k) - (4sin^2\frac{k\pi}{2N})U_m(k))] = F_m(k), \tag{2.15}$$

where $m = 1, 2, ..., M-1$, values M and N define the limits of the space. To complete equation (15), the boundary conditions are formulated as follows

$$U_0(k) = 0 \text{ and } U_M(k) = C_k \tag{2.16}$$

where constants C_k (values of $U_M(k)$) are calculated from condition

$$u_{M,n} = 0, \quad n = 1, 2, ..., N$$

For the new parameter $q_k = 4sin^2\frac{k\pi}{2N}$ and on the basis of equation (4) generating $P_n(q)$ polynomials, the solution of equation (15) is obtained in the form

$$U_m(k) = P_m q_k)U_1(k) + \sum_{l=1}^{m-1} P_{m-1}(q_k)h^2 F_l(k), \quad m = 2, 3, ..., M-1 \tag{2.17}$$

From (17) the values $U_m(k)$ can be found in all nodes of discretization. The values $U_1(k)$ in the equation (21) are calculated from the boundary conditions

$$u_{0,n} = 0 \;\; and \;\; u_{M,n} = 0, \quad n = 0, 1, ..., N. \tag{2.18}$$

When $N = M$ and $n = 1$, the second equation of (18) is appeared as follows

$$0 = u_{M,1} = \sqrt{2}\sum_{k=1}^{M-1} U_M(k) sin\frac{k\pi}{M} \tag{2.19}$$

Doing similarly for $n = 2, \ ..., \ M-1$, the system of $M-1$ equations can be obtained to calculate a set of coefficients $U_M(k), k = 1, \ 2, \ ..., \ M-1$.

Then after a substitution of these coefficients to the equation (17) for $m = M$, a set of coefficients $U_1(k), k = 1, \ 2, \ ..., \ M-1$, can be found. Next, on the basis of equation (14), (17) the solution of equation (11) can be calculated as a set of potential function values at nodes of discretization. The above algorithm of calculation was implemented in the form of a computer program to analyze the system described by Poisson equation (11). The results of calculations are presented as graphs of potential function for different steps of discretization. Also the analytical solution if exists can be given on the input. It is used to calculate an error's distribution served for the comparison between the calculated and exact solution. After some modifications of the above algorithm, another algorithm was constructed to solve inverse problem of the system described by equation (11). The task consists in the calculations of unknown field sources function f for known the potential function u and given boundary conditions described by equations (12), (18). The solution is calculated using elaborated algorithm in two stages. Within the first stage, the task consists in calculation of the matrix of Fourier series coefficients for discrete values of potential function $u_{m,n}$ in accordance with equation (14). For example, when $m = 1$ the connection between coefficients $U_1(k)$ in demand and values $u_{1,n}$ can be presented in the matrix equation

$$\mathbf{U}_1 = \frac{1}{\sqrt{2}}\mathbf{S}^{-1}\mathbf{T}_1 \tag{2.20}$$

where $\mathbf{U}_1$ is a vector which consists of coefficients $U_1(k)$ for $k = 1, 2, ..., N-1$, the vector $\mathbf{T}_1$ consists of values of potential function $u_{1,n}$ for $n = 1, \ 2, \ ..., \ N-1$, and the matrix $\mathbf{S}$ is defined by suitable values of the *sine* function.

During the second stage of calculations, the matrix $\mathbf{F}$ of Fourier series coefficients is determined for the field sources function development. Elements values in rows 1 to $M-2$ can be calculated from following formula

$$F_l(k) = \frac{U_{l+1} - P_{l+1}(q_k))U_1(k) - \sum_{i=1}^{l-1} P_{l+1-i}(q(k))h^2 F_i(k)}{P_l(q(k))h^2} \tag{2.21}$$

Whereas the calculations M-1 row of matrix $\mathbf{F}$ are done on the basis of boundary conditions (18). The last operation is a determination the matrix **fp** which corresponds with the matrix **f** in the algorithm for a direct problem. For calculated elements of matrix $\mathbf{F}$, the elements of matrix **fp** can be constructed on the basis of following equation

$$f_p(m,n) = \sqrt{2}\sum_{k=1}^{M_1} F_m(k) sin\frac{kn\pi}{M} \tag{2.22}$$

where $m, \ n$ means succeeding row and column respectively.

The elaborated algorithms were established in the MATLAB language for PC computer. In order to test an accuracy of calculations, the computer simulations were carried out with special functions called the benchmark functions, Rydygier and Trzaska, 1999. The analytical solutions agree quite well with ones obtained by using the elaborated algorithms of numerical calculations. This proves the efficiency of the established method in the practical use. The detailed calculations are done in Rydygier, 1998.

3 Smoothing Procedure for Self-Regularization

The elaborated algorithms were tested using the experimental data to verify numerical calculations. During experimental verification, some detailed problems have been solved. These problems are connected with a data treatment like a two-dimensional interpolation and a smoothing of scattered data as well as an approximation of function circumscribed the field sources distribution, Dahlquist, 1974, Schumaker, 1976, Foley, 1984. These smoothing methods were used in order to stabilize the results, Rydygier, 1998. As a result of these investigations the special approximation procedure was elaborated which consists a kind of numerical regularization method, Lavrentiev, 1970, Tikhonov, 1995. The term self-regularization means that the inverse problems must be solved in their original formulation but some quantities that appear in solutions are determined in a way which ensures the stability of results of calculations. These quantities may refer to length of steps or degree of a polynomial if an unknown function is approximated. When solving inverse problems the upper limit of the time step exits as well as the lower limit. Usually, the lower limit should correspond with the values of measurement errors. In the instance of one-dimensional case for the heat flux, the number K of the time discretization sections can be derived from the equation, Kurpisz, 1995

$$\frac{1}{I}\sum_{k=1}^{K}\sum_{i=1}^{I}(u_{k,i}-T_{k,i})^2 \approx \delta^2 \tag{3.1}$$

where I stands for a number of interior locations, $T_{k,i}$ denotes the measured temperature at nodes (k,i), $u_{k,i}$- temperature resulting from a numerical method, the δ number is the estimation of the cumulative error

$$\delta = \sum_{k=1}^{K}\sigma_k^2 \tag{3.2}$$

where σ_k stands for standard deviation of measurements.

Many different approximation procedures for numerical obtained field sources function were investigated to construct a special numerical method leading to the self-regularization. There were a polynomial approximation method adapted to $2-D$ systems, a spline interpolation method for square grid and with the use of the cube spline function, Rydygier, 1998. The correct regularization results were obtained for approximation procedure elaborated on the basis of an inverse distance method named also the Shepard's method, Allasia, 1992, 1995, 1996, Gordon, 1978. This interpolation method applied to fitting surfaces to scattered data has regularization properties, which are useful to construct the approximation procedures helpful in solving inverse problems. In effect of a use this procedure the calculated field sources function has a distinct shape, the disturbances are eliminated and the background is reduced. These results are good for identification process because the maxima of sources' function are brought out, as it is shown in Fig. 2. An inverse distance method in order to interpolate arbitrarily spaced data was developed by D. Shepard in 1968 for interpolating irregularly spaced data in the context of geographic and demographic fitting. In the bivariate case for the interpolating function formula is defined by, Schumaker, 1976

$$f(x,y) = \begin{cases} (\sum_{i=1}^{N} \frac{F_i}{r_i^{\mu}})/(\sum_{i=1}^{N} \frac{1}{r_i^{\mu}}), & when\ r_i \neq 0\ for\ all\ i \\ F_i & when\ r_i = 0. \end{cases} \tag{3.3}$$

where $r_i = [(x - x_i)^2 + (y - y_i)^2]^{1/2}$, is a distance in Euclidean metric, F_i are the values at the points (x_i, y_i), $i = 1, 2, ..., N$ and $0 < \mu < \infty$.

The value of $f(x, y)$ at nondata points is obtained as a weighted average of all the data values, where the ith measurement is weighted according to the distance of (x, y) from the point (x_i, y_i). By converting all of the terms to a common denominator, the above formula may be transformed to the weighted mean form

$$f(x,y) = \sum_{i=1}^{N} F_i A_i(x,y) \tag{3.4}$$

where

$$A_i(x,y) = \frac{\prod_{j=1, j\neq i}^{N} [r_j(x,y)]^{\mu}}{\sum_{k=1}^{N} \sum_{l=1, l\neq k}^{N} [r_i(x,y)]^{\mu}}, \quad i = 1,\ 2,\ ...,\ N \tag{3.5}$$

which satisfy

$$A_i(x_j, y_j) = \delta_{i,j}, \quad i,\ j = 1,\ 2,\ 3,\ ...,\ N. \tag{3.6}$$

The representation (26) is numerically more stable than the formula (25). The function $f(x, y)$ is analytic every where in the plane except the points (x_i, y_i). Its behavior in the vicinity of the data points (x_i, y_i) depends on the size of μ. For $0 < \mu < 1$, function $f(x, y)$ has cusps at these points, for $1 < \mu$, $f(x, y)$ has flat spots at the data points, therefore the partial derivatives vanish there. To get smooth surfaces without cusps, the condition $1 < \mu$ will be better. If μ is relatively large, then the surface tends to became very flat near the data points and consequently quite steep at points in between. On the basis of expirements it was indicated that a choice of $\mu = 2$ is perhaps a good tradeoff, Gordon, 1978.

Exemplary results of self-regularization method are shown as the graph of approximated field sources function presented in Fig 2. This graph corresponds to the set of experimental data presented in Fig. 1. The experimental data were obtained from measurements of electrical potential distribution on the thin conductive plate and thin conductive layer placed on a plate of perfect insulator using a square net of 11 × 11 nodes of discretization. The experimental data presented in Fig. 1 were obtained for the plate of thin (0.07 *mm* thickness) sheet nickel alloy which consists a material for the cores of high frequency converters. The point current constraints are the inner sources of field. The program treated the input data (electric potential values in the grid nodes) as given discrete values of potential function within the investigated domain. It should be noted that the known locations and intensities regarding internal sources were used only to the verification. The accuracy of determination both localization and intensities is not worse than one percent. The full particulars are in Rydygier, 1998.

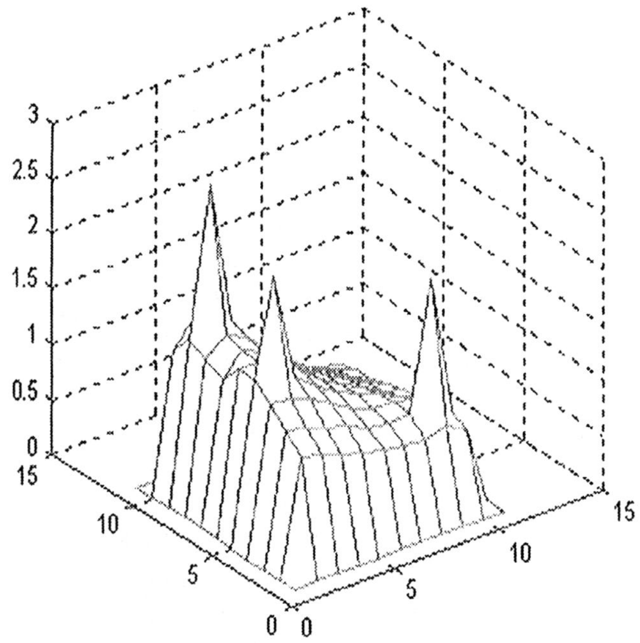

Figure 1: Experimental data for three sources

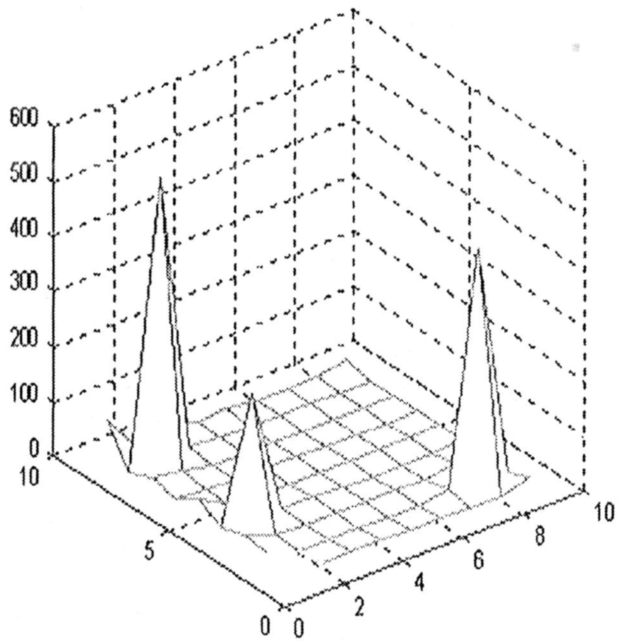

Figure 2: Identification of three sources

4 Examples

The elaborated combinatorial method was tested for different detailed problems. Especially, the heat transfer problem was examined for a resistance sintering of a tungsten rod. It corresponds to a tungsten rod manufacturing that is widely used in the practice. The investigated system is described by the Poisson equation

$$\frac{\partial^2 T(x,y)}{\partial x^2} + \frac{\partial^2 T(x,y)}{\partial y^2} = q \tag{4.1}$$

where $q = const.$, $T(x,y)$ means temperature distribution on a rectangular plane of cross-section of a rod. The problem described by the equation (29) is completed with zero-value boundary conditions. In a resistance sintering, heat is generated by an electrical current. The exemplary temperature data is shown in Fig. 3 for a rod of $0.011 \times 0.011 \times 0.446\ m$ sizes. This is a distribution of temperature refer to a plane of cross-section at $z = 0.2355\ m$ (the z axis is putted along the length of a rod). The inverse problem for the system described by equation (29) consists in a calculation a field sources function q. Calculations were done using square net of 15×15 nodes for discretization. As a result of calculations, the constant field sources function was obtained. Calculated function q after the regularization using the approximated procedure is shown in Fig. 4.

The conclusion on this investigated problem is that an estimation of field sources distribution cant be done only on the basis of experimental data of potential function. The reliable identification of field sources can be realized on the basis of field sources function obtained as a suitable inverse problem solution.

Another problem was considered for a torsion process of a metal I-bar. In is a problem from elastostatics where the $2-D$ system taken from the practice is described by the Poisson equation

$$\frac{\partial^2 \psi(x,y)}{\partial x^2} + \frac{\partial^2 \psi(x,y)}{\partial y^2} = q \tag{4.2}$$

where $\psi(x,y)$ means an auxiliary function connected with a torsion angle on a cross-section plane of a bar, q is a constant function.

The problem described by the equation (30) is completed with zero-value boundary conditions. This is a detailed inverse problem because combinatorial method together with the regularization procedure is used for $2-D$ systems that have a complicated shape. The considered problem is a torsion process of a metal I-bar so the investigated domain has a size shown in Fig. 5. In this Figure the input data on a cross-section plane of the I-bar of $16'' \times 6''$ sizes are shown too. The calculated field sources function is shown in Fig. 6.

It should be noted that the proposed approach of calculations for complicated shapes is simple and effective. In the event of disturbances, the additional calculations with a use of smoothing and approximation procedures must be done. Also there were done calculations for the heat distribution on a radiator plate. This problem is from electronics. The problem of distribution of electronic elements consists in its optimal localization. Presented example concerns only the identification of heat sources on a radiator plate of size $46 \times 76\ mm^2$ for 14×14 nodes of discretization. Illustrations are presented in Fig.7 and Fig.8. Detailed calculations are in Rydygier, 1998.

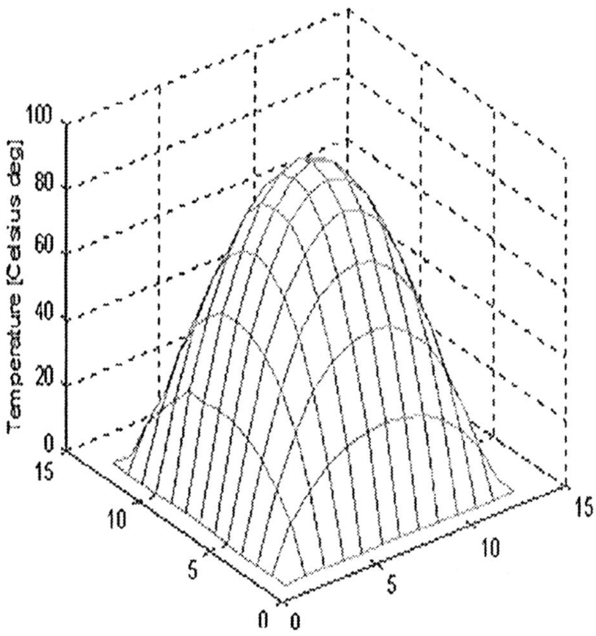

Figure 3: Temperature data for a tungsten rod

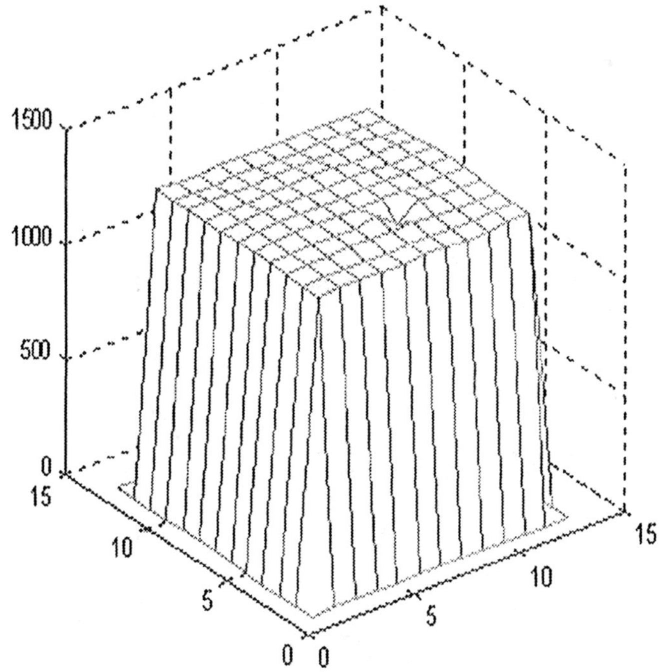

Figure 4: calculated sources' function

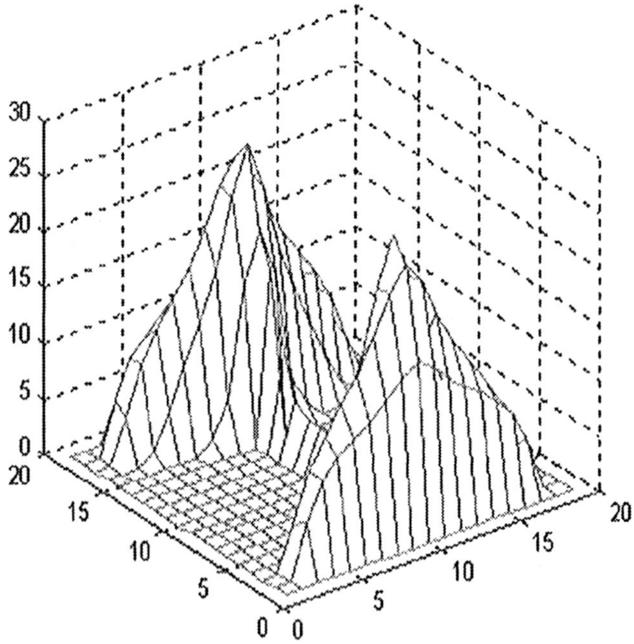

Figure 5: Input data for the I-bar

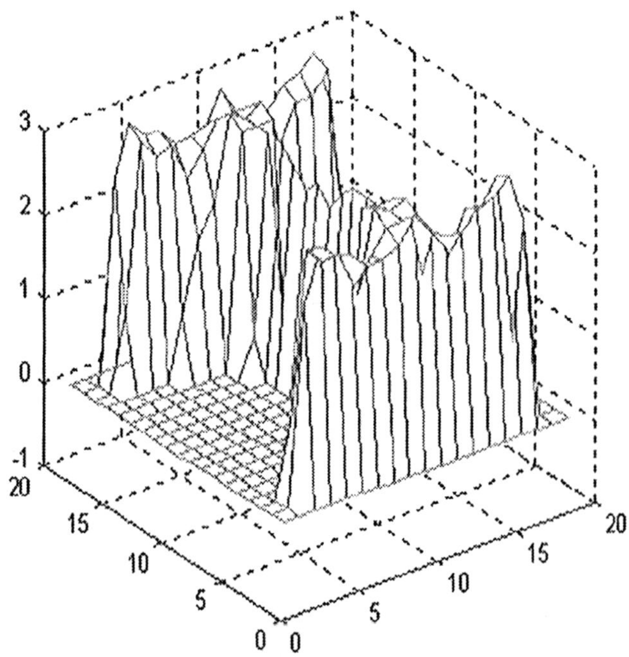

Figure 6: Sources function for the I-bar

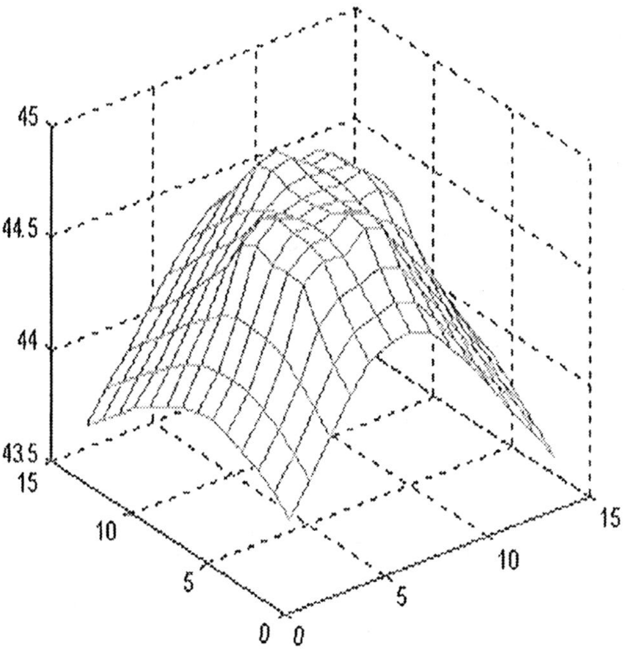

Figure 7: Temperature data for radiator plate

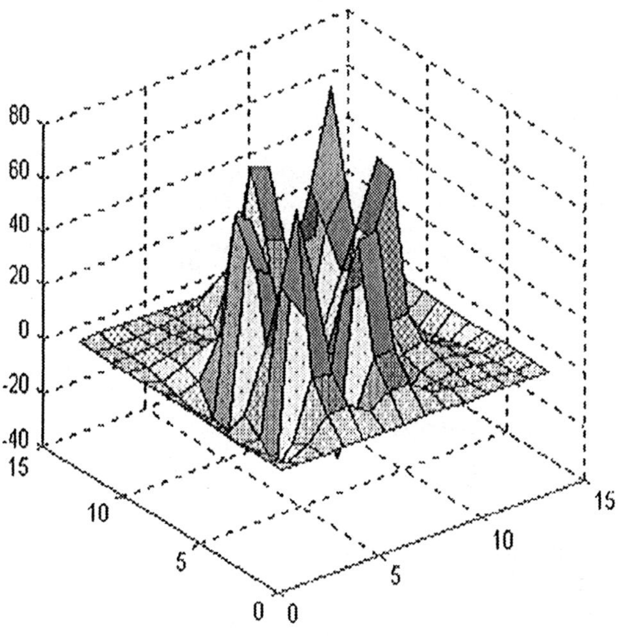

Figure 8: Sources function for radiator plate

5 Conclusions

The new approach to solve inverse problems is named the combinatorial method. Numerical algorithms are constructed using monic power polynomials generated by modified numerical triangles. After a comparison of this combinatorial method with another numerical methods used to solve different inverse problems, Anger, 1990, Engl et al, 1996, Naittaanmaki et al, 1996, Rydygier, 1998, it can be found that this new method is effective and easy to use. Advantages of presented method consist in a use combinatorics and application the recurrence formulas that are easy to implement in a form of computer programs. Using the power polynomials $P_n(x)$ and $T_n(x)$ with coefficients made by integer numbers increases the speed of computer calculations and allows an application procedures that are in the MATLAB package. In the future the authors will use the results of presented investigations to construct a special toolbox in MATLAB language to solve inverse problems.

The elaborated combinatorial method can be applied to determine inner heat sources, finding them is a solution of different practical heat transfer problems, and to locate corrosion domains on iron and carbon steel surfaces. Results of presented research can be utilized to improve usable properties of metal plates in production process in the industry. Also the results of this work can be used to build integrated computer systems for identification of thin layers properties in particular the heterogeneous spots in their structures.

Bibliography

[1] Allasia, G., 1992, Some Physical and Mathematical Properties of Inverse Distance Weighted Methods for Scattered Data Interpolation, Calcolo, Vol. 29, pp. 97-109.

[2] Allasia, G, 1995, A class of interpolating positive linear operators: theoretical and computational aspects, Approximation Theory, Wevelets, and Applications, Kluwer Academic Publishers, pp. 1-36.

[3] Allasia, G, Giolito, P., 1996, Fast evaluation algorithms for cardinal radial basis interpolants, Quaderni del Dipartimento di Matematica, No. 37, Univ. Torino.

[4] Anger, G., 1990, Inverse Problems in Differential Equations, Berlin, Akademie-Verlag.

[5] Bergum, G. E., Horadam, A. F., Philippou, A. N., 1994, Applications of Fibonacci Numbers, Vol. 1-6, Santa Clara, Fib. Assoc. Press.

[6] Dahlquist, G., Bjorck, A., 1974, Numerical Methods, New Jersey, Prentice Hall.

[7] Davis, P. J., 1963, Interpolation and Approximation, New York, Blaisdell Publ. Co.

[8] Engl, H. W., Hanke, M., Neubaer , A. ,1996, Regularization of Inverse Problems, Dordrecht, Kluwer.

[9] Foley, T. A., 1984, Three-stage interpolation to scattered data, Rocky Mount. J. of Math., Vol. 14,pp.141-149.

[10] Gordon, W. J., Wixon, J. A., 1974, Shepard Method of Metric Interpolation to Bivariate and Multivariate Interpolation, Math. Comp., Vol. 32, pp. 253-264.

[11] Groetsch, Ch. W., 1993, Inverse Problems in the Mathematical Sciences, Braunschweig, Vieweg.

[12] Hadamard, J, 1952,Lectures on Cauchy's Problem in Linear Partial Differential Equations, N.Y, Dover Publ.

[13] John, F., 1978, Partial Differential Equations, New York, Springer-Verlag.

[14] Kaczorek, T., 1985, Two-dimensional Linear Systems, Berlin, Springer-Verlag.

[15] Kurpisz, K., Nowak, A. J., 1995, Inverse Thermal Problems, Southampton, Com. Mech. Publ.

[16] Lavrentiev, M. M., Romanov, V. G., 1970, Multidimentional Inverse Problems for Differential Equations, Berlin, Springer-Verlag.

[17] MATLAB Reference Guide, 1994, New York, MathWorks Inc.

[18] Neittaanmaki, P., Rudnicki, M., Sawini, A., 1996, Inverse Problems and Optimal Design in Electricity and Magnetism, Oxford, Clarendon Press.

[19] Potter, D., 1973, Computational Physics, N.Y., J. Wieley.

[20] Ross, K .A., Wright, CH. R. B., 1992, Discrete Mathematics, New Jersey, Prentice Hall.

[21] Rydygier, E, 1998, Field Sources Identification Using a Combinatorial Method, Doctors Thesis, Warsaw University of Technology, Warsaw, Poland.

[22] Rydygier E., Trzaska Z., 1999, Inverse problems for distributed parameter systems solved with combinatorial method, Control and Cybernetics, Vol. 28, no. 2, pp.237-258

[23] Schumaker, L. L., 1976, Fitting surfaces to scattered data, Approximation Theory II, New York, Academic Press, pp. 203-268.

[24] Tikhonov, A. N., Goncharsky A. V., Stepanov, V. V., Yagola, A. G., 1995, Numerical Methods for the Solution of Ill-posed Problem, Dordrecht, Kluwer.

[25] Trzaska, Z., 1993a, Modified Numerical Triangles and the Fibonacci Sequence, The Fibonacci Quarterly, Vol. 32, pp. 124-129.

[26] Trzaska, Z., 1993b, Numerical Triangles, Fibonacci Sequence and Ladder Networks: Some Further Results, Appl. Math. Let., Vol. 6, pp. 55-61.

[27] Trzaska, Z., 1997, New Approach to Continuous-Discrete Systems, Bulletin of the Polish Academy of Sciences, Vol. 45, 3, pp. 433-443.

[28] Trzaska, Z., Rydygier, E., 1998, Identification of Field Sources in 2-D Systems Using a Combinatorial Method, Proc. 5th Int. Symp. Methods and Models in Autom. and Robotics, Miedzyzdroje, pp. 129-134.

[29] Vajda, S., 1989, Fibonacci and Lucas Numbers, and the Golden Section, Chichester, Ellis Horwood Ltd.

Edward Rydygier. The Andrzej Soltan Institute for Nuclear Studies, PL-05-400 Otwock-Swierk, Poland
E-mail: Edward.Rydygier@mst.gov.pl

Zdzislaw W. Trzaska. Department of Electrical Enginering, Warsaw University of Technology, PL-00-661 Warsaw, Poland
E-mail: trz@nov.iem.pw.edu.pl

Asymptotic Analysis of Aircraft Wing Model in Subsonic Airflow

Marianna A. Shubov

Abstract. In this paper we announce a series of results on the asymptotic and spectral analysis of an aircraft wing in a subsonic air flow and provide a brief outline of the proofs of these results. This model has been developed in the Flight Systems Research Center of UCLA and is presented in the works by A. V. Balakrishnan. The model is governed by a system of two coupled integro-differential equations and a two parameter family of boundary conditions modeling the action of the self-straining actuators. The unknown functions (the bending and torsion angle) depend on time and one spatial variable. The differential parts of the above equations form a coupled linear hyperbolic system; the integral parts are of convolution type. The system of equations of motion is equivalent to a single operator evolution-convolution type equation in the state space of the system equipped with the so-called energy metric. The Laplace transform of the solution of this equation can be represented in terms of the so-called generalized resolvent operator. The generalized resolvent operator is a finite-meromorphic function on the complex plane having the branch cut along the negative real semi-axis. The poles of the generalized resolvent are precisely the aeroelastic modes and the residues at these poles are the projectors on the generalized eigenspaces. In this paper, our main object of interest is the dynamics generator of the differential parts of the system, which is a nonselfadjoint operator in the state space with a purely discrete spectrum. We show that this operator has two branches of discrete spectrum, present precise asymptotic formulas for both branches, and outline the main steps of their derivation. The full proof will be given in another work. Based on these results in the subsequent papers, we will derive the asymptotics of the aeroelastic modes and approximations for the mode shapes.

1 Introduction

In this paper, we formulate a series of results on the asymptotic and spectral analysis of an aircraft wing model. The model has been developed in the Flight Systems Research Center of the University of California at LosAngeles. The mathematical formulation of the problem can be found in the work by A. V. Balakrishnan [5]. This model has been designed to find an approach to control of the flutter phenomenon in an aircraft wing in a surrounding airflow by using the so-called self-straining actuators.

The model, which is used in [5], is the 2-D strip model which applies to bare wings of high aspect ratio [11]. The structure is modeled by a uniform cantilever beam which bends and twists. The aerodynamics is simplified to the extreme: subsonic, incompressible, and inviscid. In addition, the author of [5] has added the self-straining actuators using a currently accepted model (see, e.g., [6-8, 12, 33, 35]).

Flutter, which is known as a very dangerous aeroelastic development, is the onset, beyond some speed-altitude combinations, of unstable and destructive vibrations of a lifting surface in an airstream. Flutter most commonly encountered on bodies subjected to large lateral aerodynamic loads of the lift type, such as aircraft wings, tails, and control surfaces. It is known to be an aeroelastic problem. The only air forces necessary to produce it are those due to the deflections of the elastic structures from the undeformed state.

The flutter or critical speed u_f and frequency ω_f are defined as the lowest airspeed and corresponding circular frequency at which a given structure flying at given atmospheric density and temperature will exhibit sustained, simple harmonic oscillations. Flight at u_f represents a borderline condition or neutral stability boundary, because all small motions must be stable at speeds below u_f, whereas divergent oscillations can ordinarily occur in a range of speeds (or at all speeds) above u_f. Because it is easier mathematically to describe the aerodynamic loads due to simple harmonic motion, theoretical flutter analysis often consists of assuming in advance that all dependent variables are proportional to $e^{i\omega t}$ (ω is real), and then finding such combinations of u and ω for which this actually occur. One is thus led to a complex or multiple eigenvalue problems involving eigenfunctions and associate functions. This is in contrast to free vibrations of a linear structure in vacuum, which is a real eigenvalue problem involving only eigenfunctions.

Due to its extreme importance, flutter prediction has been variously done by purely theoretical means, by analog computation, by wind-tunnel or rocket experiments on scaled dynamic models, and by flight testing full-scale aircraft.

Probably, the most important type of aircraft flutter results from coupling between the bending and torsional motions of a relatively large aspect-ratio wing and tail. A great deal of qualitative information can be obtained about the influence of various system parameters on this kind of flutter by studying the stability of the simple airfoil. An illustrative picture of such an airfoil can be found in [11], p.533. Based on aforementioned model, precise mathematical formulations of the continuous model dynamics can be found in [5].

Our main objective is to find the time-domain solution of the initial-boundary value problem from [5]. This objective requires very detailed mathematical analysis of the properties of the system. We expect that to present mathematically rigorous solution of this problem, we have to complete a series of at least three papers. In the current paper, we present the asymptotic representations for the so-called aeroelastic modes which are associated with the discrete spectrum of the problem. We are not going to give the detailed derivation of our formulas. Rigorous proofs can be found in our first paper from the aforementioned series. In the second, paper, we will discuss the so-called geometric properties of the mode shapes (or eigenfunctions) corresponding to the discrete spectrum. The final paper is expected to be devoted to the properties of the so-called continuous spectrum of the problem.

Now, we describe the content of the present paper. In Section 2, we give precise mathematical formulation of the problem. The model is governed by a system of two coupled partial integro-differential equations subject to two parameter family of boundary conditions. The parameters are introduced to model the action of the self-straining actuators as is accustomed in current engineering and mathematical literature. Secondly, we reformulate the problem and set it into the operator format. We also show that the dynamics is defined by two matrix operators in the energy space. One of the aforementioned operators is a matrix differential operator and the second one is a matrix integral convolution-type operator. That is why in the operator setting, we have the so-called evolution-convolution type problem. As it will be shown in the second paper of the aforementioned series, *the aeroelastic modes* (or the discrete spectrum of the problem) are asymptotically close to the discrete spectrum of the matrix differential operator while the continuous spectrum is completely defined by the matrix integral operator. *It is exactly the spectral properties of the differential operator which are of interest in the present paper.* We note that if the speed of an airstream $u = 0$, then the integral operators vanish and the appropriate purely structural problem will have only discrete spectrum. In Section 2, we also formulate the main result on the spectral asymptotics - Theorem 2.2.

In Section 3, we use the boundary conditions to construct two special matrices which we call *the left reflection matrix* and *the right reflection matrix* respectively. We would like to mention here that the approach developed in Section 3 of the present paper has already been used in our papers devoted to the spectral properties of the Timoshenko beam model with spatially nonhomogeneous coefficients and subject to two-parameter family of nonconservative boundary conditions (see [20-23]).

In the last section, we outline the main steps in the proof of the main result, i.e., we discuss the derivation of spectral asymptotics. In particular, we show that the aeroelastic modes from a countable set of complex numbers. This set is located in a strip parallel to the imaginary axis with points of accumulation only at infinity. There may be multiple modes of a finite multiplicity each. (For precise formulations see Theorem 2.2).

In the conclusion of the Introduction, we describe what kind of a control problem will be considered in connection with the flutter suppression. In the specific wing model considered in the current paper, both the matrix differential operator and the matrix integral operator contain entries depending on the speed u of the surrounding air flow. Therefore, the aeroelastic modes are functions of $u : \lambda_k = \lambda_k(u)$, $(k \in \mathbb{Z})$. The wing is stable if $Re\,\lambda_k(u) < 0$ for all k. However, if u is increasing, some of the modes move to the right half-plane. The flutter speed u_k^f for the $k - th$ mode is defined by the relation $Re\,\lambda_k(u_k^f) = 0$. To understand the flutter phenomenon, it is not sufficient to trace the motion of aeroelastic modes as functions of a speed of airflow. It is necessary to have efficient representations for the solutions of our boundary-value problem, containing the contributions from both the discrete and continuous parts of the spectrum. Such a representation will provide a precise description of the solution behavior. It is known, that flutter cannot be eliminated completely. To successfully suppress flutter, one should design self-straining actuators (i.e., in the mathematical language, to select parameters in the boundary conditions which are the control gains β and δ in formulas (2.10) and (2.11) of Section 2) in such a way that flutter does not occur in the desired speed range. This is highly nontrivial boundary control problem.

2 Statement of Problem. Operator Reformulation

In this section, we give a precise formulation of the initial-boundary value problem. Namely, we describe the system of two coupled damped integro-differential equations, which governs the dynamics of the model.

Following [5], let us introduce the following dynamical variables:

$$X(x,t) = \begin{pmatrix} h(x,t) \\ \alpha(x,t) \end{pmatrix}, \qquad -L \le x \le 0,\ t \ge 0, \tag{2.1}$$

where $h(x,t)$ - the bending and $\alpha(x,t)$- the torsion angle. The model, which we will consider, can be described by the following linear system:

$$(M_s - M_a)\ddot{X}(x,t) + (D_s - uD_a)\dot{X}(x,t) + (K_s - u^2K_a)X = \begin{bmatrix} f_1(x,t) \\ f_2(x,t) \end{bmatrix}. \tag{2.2}$$

From now on, we will use the notation "."(dot) to denote the differentiation with respect to t. We use the subscripts "s" and "a" to distinguish the structural and aerodynamical parameters respectively. All 2 x 2 matrices in Eq.(2.2) are given by the following formulas:

$$M_s = \begin{bmatrix} m & S \\ S & I \end{bmatrix}, \qquad M_a = (-\pi\rho)\begin{bmatrix} 1 & -a \\ -a & (a^2 + 1/8) \end{bmatrix}, \tag{2.3}$$

where m - the density of the flexible structure (mass per unit length), S - the mass moment, I - the moment of inertia, ρ - the density of air, a - linear parameter of the structure $(-1 \le a \le 1)$.

$$D_s = \begin{bmatrix} 0 & 0 \\ 0 & 0 \end{bmatrix} \qquad D_a = (-\pi\rho)\begin{bmatrix} 0 & 1 \\ -1 & 0 \end{bmatrix}, \tag{2.4}$$

$$K_s = \begin{bmatrix} E\frac{\partial^4}{\partial x^4} & 0 \\ 0 & -G\frac{\partial^2}{\partial x^2} \end{bmatrix} \qquad K_a = (-\pi\rho)\begin{bmatrix} 0 & 0 \\ 0 & -1 \end{bmatrix}, \tag{2.5}$$

where E - the bending stiffness, G - the torsion stiffness. The parameter u in Eq.(2.2) denotes the stream velocity. The right hand side of system (2.2) can be represented as the following system of two convolution-type integral operations:

$$f_1(x,t) = -2\pi\rho\int_0^t \left[uC_2(t-\sigma) - \dot{C}_3(t-\sigma)\right] g(x,\sigma)d\sigma, \tag{2.6}$$

$$f_2(x,t) = -2\pi\rho \int_0^t \Big[1/2C_1(t-\sigma) - auC_2(t-\sigma) + a\dot{C}_3(t-\sigma) + uC_4(t-\sigma) + 1/2\dot{C}_5(t-\sigma)\Big]\, g(x,\sigma)d\sigma, \tag{2.7}$$

$$g(x,t) = u\dot{\alpha}(x,t) + \ddot{h}(x,t) + (1/2-a)\ddot{\alpha}(x,t). \tag{2.8}$$

The aerodynamical functions C_i, $i = 1\ldots 5$, are defined in the following ways:

$$\begin{aligned}
\hat{C}_1(\lambda) &= \int_0^\infty e^{-\lambda t}C_1(t)dt = \frac{u}{\lambda}\frac{e^{-\lambda/u}}{K_0(\lambda/u)+K_1(\lambda/u)}, \quad Re\lambda > 0,\\
C_2(t) &= \int_0^t C_1(\sigma)d\sigma,\\
C_3(t) &= \int_0^t C_1(t-\sigma)(u\sigma - \sqrt{u^2\sigma^2+2u\sigma})d\sigma,\\
C_4(t) &= C_2(t) + C_3(t),\\
C_5(t) &= \int_0^t C_1(t-\sigma)((1+u\sigma)\sqrt{u^2\sigma^2+2u\sigma} - (1+u\sigma)^2)d\sigma.
\end{aligned} \tag{2.9}$$

where K_0 and K_1 are the modified Bessel functions of the zero and first order respectively [1, 17]. These formulas for the aerodynamical functions have been derived in [9]. It is known that the self-straining control actuator action can be modelled by the following boundary conditions:

$$Eh''(0,t) + \beta\dot{h}'(0,t) = 0, \qquad h'''(0,t) = 0, \tag{2.10}$$

$$G\alpha'(0,t) + \delta\dot{\alpha}(0,t) = 0, \qquad \beta,\delta \in \mathfrak{C}^+ \cup \{\infty\}, \tag{2.11}$$

where $\mathfrak{C}^+$ is the closed right half plane. Note that we essentially have tip "rate" controllers of the kind studied in [7, 8]. We consider the following boundary conditions at $x = -L$:

$$h(-L,t) = h'(-L,t) = \alpha(-L,t) = 0. \tag{2.12}$$

Let the initial state of the system be given as follows

$$h(x,0) = h_0(x), \qquad \dot{h}(x,0) = h_1(x), \qquad \alpha(x,0) = \alpha_0(x), \qquad \dot{\alpha}(x,0) = \alpha_1(x). \tag{2.13}$$

We will consider the solution of the problem given by Eqs.(2.2) and conditions (2.10)-(2.13) in the energy space $\mathcal{H}$. To introduce the metric of $\mathcal{H}$, we assume that the parameters satisfy the following two conditions:

$$det\begin{bmatrix} m & S \\ S & I \end{bmatrix} > 0, \tag{2.14}$$

$$0 < u \le \frac{\sqrt{2G}}{L\sqrt{\pi\rho}}. \tag{2.15}$$

Let $\{\tilde{C}_i\}_{i=1}^2$ be the kernels in the convolution operations in (2.6), (2.7), i.e.,

$$\tilde{C}_1(t) = -2\pi\rho(uC_2(t) - \dot{C}_3(t)), \tag{2.16}$$

$$\tilde{C}_2(t) = -2\pi\rho\,(1/2C_1(t) - auC_2(t) + a\dot{C}_3(t) + uC_4(t) + 1/2\dot{C}_5(t)), \tag{2.17}$$

and let M, D, K be the following matrices:

$$M = M_s - M_a, \quad D = D_s - uD_a, \quad K = K_s - u^2 K_a\,. \tag{2.18}$$

Then Eq.(2.2) can be written in the form

$$M\ddot{X}(x,t) + D\dot{X}(x,t) + KX(x,t) = \;(\mathcal{F}\dot{X})(x,t)), \quad t \geq 0, \tag{2.19}$$

where the matrix integral operator $\mathcal{F}$ is given by the formula

$$\mathcal{F} \;=\; \begin{bmatrix} \int_0^t \;\tilde{C}_1(t-\sigma)\left(\frac{d}{d\sigma}\cdot\right)\,d\sigma & \int_0^t \;\tilde{C}_1(t-\sigma)[u\cdot+(1/2-a)\left(\frac{d}{d\sigma}\cdot\right)]d\sigma \\ \int_0^t \;\tilde{C}_2(t-\sigma)\left(\frac{d}{d\sigma}\cdot\right)\,d\sigma & \int_0^t \;\tilde{C}_2(t-\sigma)[u\cdot+(1/2-a)\left(\frac{d}{d\sigma}\cdot\right)]d\sigma \end{bmatrix}$$

$$= \begin{bmatrix} \tilde{C}_1 * \left(\frac{d}{d\sigma}\cdot\right) & \tilde{C}_2 * \left(u\cdot+(1/2-a)\frac{d}{d\sigma}\cdot\right) \\ \tilde{C}_2 * \left(\frac{d}{d\sigma}\cdot\right) & \tilde{C}_1 * \left(u\cdot+(1/2-a)\frac{d}{d\sigma}\cdot\right) \end{bmatrix}, \tag{2.20}$$

where the notation " $*$ " has been used for the convolution.

REMARK 2.1. *We would like to mention that the model described by Eqs.(2.19) in the case* $u = 0$ *occurs actually in aeroelastic problems (see the classic textbook [11]) if one ignores aeroelastic forces. However, the boundary conditions in [11] which complemented the system of equations (2.19) are totally different. To the best of our knowledge, the whole problem consisting of Eqs.(2.19) (*$u = 0$*) and boundary conditions (2.10) and (2.11) has been considered only in one paper by Balakrishnan [10].*

Let $\mathcal{H}$ be the set of 4-component vector valued functions

$$\Psi = (h, \dot{h}, \alpha, \dot{\alpha})^T \equiv (\psi_0, \psi_1, \psi_2, \psi_3)^T$$

(the superscript "T" means the transposition) obtained as a closure of smooth functions satisfying the conditions

$$\psi_0(-L) = \psi_0'(-L) = \psi_2(-L) = 0 \tag{2.21}$$

in the following energy norm:

$$||\Psi||^2_{\mathcal{H}} = 1/2 \int_{-L}^{0} \Big[E|\psi_0''(x)|^2 + G|\psi_2'(x)|^2 + \tilde{m}|\psi_1(x)|^2 + \tilde{I}|\psi_3(x)|^2$$
$$+\tilde{S}(\psi_3(x)\bar{\psi}_1(x) + \bar{\psi}_3(x)\psi_1(x)) \;-\pi\rho u^2|\psi_2(x)|^2\Big]\,dx, \tag{2.22}$$

where

$$\tilde{m} = m + \pi\rho, \ \tilde{S} = S - a\pi\rho, \ \tilde{I} = I + \pi\rho(a^2 + 1/8). \tag{2.23}$$

Note that under conditions (2.14) and (2.15), formula (2.22) defines a positively definitive metric. Our goal is to rewrite Eq.(2.19) as the first order in time evolution-convolution equation in the energy space. As the first step, we will represent Eq.(2.19) in the form

$$\ddot{X} + M^{-1}D\dot{X} + M^{-1}KX = M^{-1}\mathcal{F}\dot{X}. \tag{2.24}$$

Note, due to condition (2.14), M^{-1} exists.

Using formulas (2.21), one can easily see that the initial-boundary value problem defined by Eq.(2.24) and conditions (2.10)-(2.13) can be represented in the form

$$\dot{\Psi} = i\mathcal{L}_{\beta\delta}\Psi + \tilde{\mathcal{F}}\dot{\Psi}, \quad \Psi = (\psi_0, \psi_1\psi_2, \psi_3)^T, \quad \Psi|_{t=0} = \Psi_0. \tag{2.25}$$

$\mathcal{L}_{\beta\delta}$ is the following matrix differential operator in $\mathcal{H}$:

$$\mathcal{L}_{\beta\delta} = -i\begin{bmatrix} 0 & 1 & 0 & 0 \\ -\frac{E\tilde{I}}{\Delta}\frac{d^4}{dx^4} & -\frac{\pi\rho u\tilde{S}}{\Delta} & -\frac{\tilde{S}}{\Delta}\left(G\frac{d^2}{dx^2} + \pi\rho u^2\right) & -\frac{\pi\rho u\tilde{I}}{\Delta} \\ 0 & 0 & 0 & 1 \\ \frac{E\tilde{S}}{\Delta}\frac{d^4}{dx^4} & \frac{\pi\rho u\tilde{m}}{\Delta} & \frac{\tilde{m}}{\Delta}\left(G\frac{d^2}{dx^2} + \pi\rho u^2\right) & \frac{\pi\rho u\tilde{S}}{\Delta} \end{bmatrix} \tag{2.26}$$

defined on the domain

$$\begin{aligned} \mathcal{D}(\mathcal{L}_{\beta\delta}) = \{\Psi \in \mathcal{H}: \ \psi_0 \ \in \ H^4(-L,0), \ \psi_1 \ \in H^2(-L,0), \ \psi_2 \ \in \ H^4(-L,0), \\ \psi_3 \ \in H^1(-L,0); \psi_1(-L) = \psi_1'(-L) = \psi_3(-L) = 0; \ \psi_0'''(0) = 0; \\ E\psi_0''(0) + \beta\psi_1'(0) = 0, \ G\psi_2'(0) + \delta\psi_3(0) = 0\}, \end{aligned} \tag{2.27}$$

where $H^i, i = 1, 2, 4$, are the standard Sobolev spaces [2]. $\tilde{\mathcal{F}}$ is a linear integral operator in $\mathcal{H}$ given by the formula

$$\tilde{\mathcal{F}} = \begin{bmatrix} 1 & 0 & 0 & 0 \\ 0 & [\tilde{I}(\tilde{C}_1*) - \tilde{S}(\tilde{C}_2*)] & 0 & 0 \\ 0 & 0 & 1 & 0 \\ 0 & 0 & 0 & [-\tilde{S}(\tilde{C}_1*) + \tilde{m}(\tilde{C}_2*)] \end{bmatrix} \begin{bmatrix} 0 & 0 & 0 & 0 \\ 0 & 1 & u & (1/2 - a) \\ 0 & 0 & 0 & 0 \\ 0 & 1 & u & (1/2 - a) \end{bmatrix}. \tag{2.28}$$

The main goal of the present paper is to derive the asymptotics of the spectrum of the operator $\mathcal{L}_{\beta\delta}$. It turns out that the spectral properties of both the diffential operator $\mathcal{L}_{\beta\delta}$ and the integral operator $\tilde{\mathcal{F}}$ are of crucial importance for the representation of the solution. Namely, as it will be shown in our second paper from the aforementioned series, the discrete spectrum of the entire problem is asymptotically close to the discrete spectrum of the operator $\mathcal{L}_{\beta\delta}$ and the continuous spectrum of it is completely determined by the operator $\tilde{\mathcal{F}}$. It will be the subsequent paper in which we will prove the geometric properties of the eigenfunctions and give complete characterization of the continuous spectrum.

We formulate now the lemma describing the properties of the operator $\mathcal{L}_{\beta\delta}$. As follows from the general theory [15, 34], $\mathcal{L}_{\beta\delta}$ is a closed linear operator in $\mathcal{H}$ whose resolvent is compact, and, therefore, the spectrum is discrete.

LEMMA 2.1. *a) Operator $\mathcal{L}_{\beta\delta}$ is nonselfadjoint unless both β and δ are purely imaginary. If Re $\beta \geq 0$ and Re $\delta \geq 0$, then this operator is dissipative, i.e., Im $(\mathcal{L}_{\beta\delta}\Psi, \Psi) \geq 0$ for all $\Psi \in \mathcal{D}(\mathcal{L}_{\beta\delta})$. b) When $\mathcal{L}_{\beta\delta}$ is dissipative, then it is maximal, i.e., it does not admit any more dissipative extensions.*

Important Remark. At this moment, we would like to emphasize the difference between the aircraft wing model considered in the present paper and models related to other flexible structures that have been studied by the author in recent years (see [19-23]). Each of the aforementioned models (a spatially nonhomogeneous damped string, a 3-dimensional damped wave equation with spatially nonhomogeneous coefficients, Timoshenko beam model, and coupled Euler-Bernoulli and Timoshenko beams) can be described by an abstract evolution equation in a Hilbert space $\mathcal{H}$

$$\dot{\Psi}(t) = iA\Psi(t), \quad \Psi(t) \in \mathcal{H}, \; t \geq 0. \tag{2.29}$$

The dynamics generator A is an unbounded nonselfadjoint operator in $\mathcal{H}$. This operator is our main object of interest in the aforementioned works. In each of the above examples, A is a specific differential or matrix differential operator in $\mathcal{H}$. The domain of A is defined by the differential expression and the corresponding boundary conditions.

For all of the above systems, A has a compact resolvent and, therefore, has purely discrete spectrum. The main results concerning the dynamics generator A established in the author's aforementioned works can be split into the following three parts:
(i) the explicit asymptotic formulas for the spectrum;
(ii) asymptotic representations for the generalized eigenvectors;
(iii) the Riesz basis property of the generalized eigenvectors in $\mathcal{H}$.
The author's approach to the most difficult results (iii) and the applications of these results to control problems can be found in the works [19,20,22,23,25,27-30].

In contrast with the evolution equation (2.29), the aircraft wing model is described by an evolution-convolution type equations of the form

$$\dot{\Psi}(t) = i\,A\,\Psi(t) + \int_0^t F(t-\tau)\dot{\Psi}(\tau)d\tau. \tag{2.30}$$

Here $\Psi(\cdot) \in \mathcal{H}$ - the energy space of the system, Ψ is a 4-component vector function, $A(A = \mathcal{L}_{\beta\delta})$ is a matrix differential operator, and $F(t)$ is a matrix valued function.

Eq.(2.30) does not define an evolution semigroup and does not have a dynamics generator. So, it is necessary to explain what is understood as the spectral analysis of Eq.(2.30).

Let us take the Laplace transformation of both parts of Eq.(2.24). Formal solution in the Laplace representation can be given by the formula

$$\hat{\Psi}(\lambda) = \left(\lambda I - iA - \lambda\hat{F}(\lambda)\right)^{-1}\left(I - \hat{F}(\lambda)\right)\Psi_0, \tag{2.31}$$

where Ψ_0 is the initial state, i.e., $\Psi(0) = \Psi_0$, and the symbol "^" is used to denote the Laplace transform. It is an extremely nontrivial problem to understand the precise meaning of Eq.(2.31) and, most importantly, to calculate the inverse Laplace transform of Eq.(2.31) in order to have the representation of the solution in the space-time domain. To do this, it is necessary to investigate the generalized resolvent operator

$$R(\lambda) = \left(\lambda I - iA - \lambda \hat{F}(\lambda)\right)^{-1}. \tag{2.32}$$

In the case of the 1-dimensional wing model, $R(\lambda)$ is an operator-valued meromorphic function on the complex plane with a branch cut along the negative real semi-axis. The poles of $R(\lambda)$ are called the eigenvalues, or *the aeroelastic modes.* The residues of $R(\lambda)$ at the poles are precisely the projectors on the corresponding generalized eigenspaces. The branch cut corresponds to the continuous spectrum.

As has already been mentioned, in the current paper, we present the asymptotic formulas for the aeroelastic modes. However, to find the space-time domain solution, we have to solve the following problems.

1. To prove that the generalized eigenvectors of the discrete spectrum form a Riesz basis in their closed linear span.
2. To obtain asymptotic formulas for the eigenfunctions of the continuous spectrum. (These eigenfunctions can be expressed in terms of the jump of the kernel of the generalized resolvent $R(\lambda)$ across the branch cut.)
3. To obtain an expansion theorem with respect to the eigenfunctions of the continuous spectrum.

Combination of 1 - 3 will allow us to "calculate" the inverse Laplace transform in Eq.(2.31) and, thus, to obtain the desired solution of the initial-boundary value problem.

In the conclusion of this section, we formulate the main result concerning the spectral asymptotics of the operator $\mathcal{L}_{\beta\delta}$.

THEOREM 2.2. *(a) The operator $\mathcal{L}_{\beta\delta}$ has a countable set of complex eigenvalues. Under the assumption*

$$\delta \neq \sqrt{G\tilde{I}} \tag{2.33}$$

the set of eigenvalues is located in a strip parallel to the real axis.
(b) The entire set of eigenvalues asymptotically splits into two disjoint subsets. We call them the β-branch and the δ-branch and denote by $\{\lambda_n^\beta\}_{n\in\mathbb{Z}}$ and $\{\lambda_n^\delta\}_{n\in\mathbb{Z}}$ respectively. If Re $\beta \geq 0$ and Re $\delta > 0$, then each branch is asymptotically close to its own horizontal line in the closed upper half plane. If Re $\beta > 0$ and Re $\delta = 0$, then both horizontal asymptotes coincide with the real axis. If Re β = Re $\delta = 0$, then the operator $\mathcal{L}_{\beta\delta}$ is selfadjoint and, thus, its spectrum is real. The entire set of eigenvalues may have only two points of accumulation: $+\infty$ and $-\infty$ in the sense that Re $\lambda_n^{\beta(\delta)} \longrightarrow \pm\infty$ and $Im \lambda_n^{\beta(\delta)} \longrightarrow const$ as $n \longrightarrow \pm\infty$ (see formulas (2.34) and (2.35) below).
(c) The following asymptotics is valid for the β-branch of the spectrum as $|n| \to +\infty$:

$$\lambda_n^\beta = \pm\pi\sqrt{E\tilde{I}/\Delta}\ (n + 1/4)^2 + \kappa_n(\omega), \quad \omega = |\delta|^{-1} + |\beta|^{-1} \tag{2.34}$$

with Δ being defined as $\Delta = m\tilde{I} - \tilde{S}^2$. A complex-valued sequence $\{\kappa_n(\omega)\}$ is bounded above in the following sense

$$\sup_{n\in\mathbf{Z}}\{|\kappa_n(\omega)|\} = C(\omega), \ \ C(\omega) \to 0 \ \textit{as} \ \omega \to +\infty$$

This branch may have a finite number of multiple eigenvalues of a finite algebraic multiplicity each. For such an eigenvalue, the geometric multiplicity may be less than the corresponding algebraic multiplicity, i.e., in addition to the eigenvector or eigenvectors, there may be the associate vectors. (Recall that "ϕ" is an associate vector of an operator A of the order m corresponding to the eigenvalue λ if $\phi \neq 0$, $(A - \lambda I)^m \phi = 0$, and $(A - \lambda I)^{m+1}\ \phi = 0$. If $m = 0$, then ϕ is an eigenvector).

(d) The following asymptotics is valid for the δ-branch of the spectrum:

$$\lambda_n^\delta = \frac{\pi n}{L\sqrt{\tilde{I}/G}} + \frac{i}{2L\sqrt{\tilde{I}/G}}\ ln\ \frac{\delta + \sqrt{G\tilde{I}}}{\delta - \sqrt{G\tilde{I}}} + O(|n|^{-1/2}),\ |n| \longrightarrow \infty. \tag{2.35}$$

In this branch, there may be only a finite number of multiple eigenvalues of a finite multiplicity each. Therefore, only a finite number of the associate vectors may exist.

3 Reflection Matrices

In this section, we will outline the main steps in the proof of Theorem 2.2. We start with the spectral equation for the operator $\mathcal{L}_{\beta\delta}$

$$\mathcal{L}_{\beta\delta}\Psi = \lambda\Psi \tag{3.1}$$

where $\Psi = (\psi_0, \psi_1, \psi_2, \psi_3)^T \in \mathcal{H}$. Eq.(3.1) written in a component-wise fashion is equivalent to a system of 4 coupled differential equations. Excluding ψ_j, $j = 0, 1, 3$, from this system, we arrive at the following 6th order ordinary differential equation with respect to the component ψ_2:

$$EG\psi_2^{VI} + E(\lambda^2\tilde{I} + \pi\rho u^2)\psi_2^{IV} - \lambda^2\tilde{m}G\psi_2'' - \psi_2\left[\lambda^2\pi\rho u^2\tilde{m} - \lambda^2(\pi\rho u)^2 + \lambda^4\Delta\right] = 0. \tag{3.2}$$

Rewriting Eq.(3.2) in the asymptotical form with respect to $|\lambda| \longrightarrow \infty$, we obtain

$$\psi_2^{VI} + \lambda^2(\tilde{I}/G)(1 + O(\lambda^{-2}))\psi_2^{IV} - \lambda^2(\tilde{m}/E)\psi_2'' - \lambda^4(\Delta/EG)(1 + O(\lambda^{-2}))\psi_2 = 0. \tag{3.3}$$

The characteristic equation corresponding to Eq.(3.3) has the form

$$z^6 + \lambda^2(\tilde{I}/G)(1 + O(\lambda^{-2}))z^4 - \lambda^2(\tilde{m}/E)z^2 - \lambda^4(\Delta/EG)(1 + O(\lambda^{-2})) = 0. \tag{3.4}$$

Setting t $= z^2$, we obtain the cubic equation of the form

$$t^3 + \lambda^2(\tilde{I}/G)(1 + O(\lambda^{-2}))t^2 - \lambda^2(\tilde{m}/E)t - \lambda^4(\Delta/EG)(1 + O(\lambda^{-2})) = 0. \tag{3.5}$$

Using the Cardano formulas, we obtain he following approximations for the roots of Eq.(3.5):

$$t_1 = \frac{\lambda\sqrt{T}}{\sqrt{R}} + \frac{RV - T}{2R^2} + O(\lambda^{-1}), \tag{3.6}$$

$$t_2 = -\lambda^2 R - \frac{RV - T}{R^2} + O(\lambda^{-2}), \tag{3.7}$$

$$t_3 = -\frac{\lambda\sqrt{T}}{\sqrt{R}} + \frac{RV - T}{2R^2} + O(\lambda^{-1}). \tag{3.8}$$

where R, T, and V are given by the formulas

$$R = R_0(1 + O(\lambda^{-2})),\ R_0 = \tilde{I}/G,\ V = \tilde{m}/E,\ T = \Delta/EG. \tag{3.9}$$

From (3.6) - (3.8), we obtain the following approximations for 6 roots of Eq.(3.4)

$$Z_{1,2} = \pm\left(C\sqrt{\lambda} + \frac{C_1}{\sqrt{\lambda}} + O(\lambda^{-1.5})\right) \equiv \pm\gamma(\lambda), \tag{3.10}$$

where

$$c = \left(\frac{T}{R_0}\right)^{1/4},\ c_1 = \frac{1}{4}\sqrt{\frac{R_0}{T}}\frac{R_0V - T}{2R_0^2},\ R_0 = \tilde{I}/G. \tag{3.11}$$

$$Z_{3,4} = \pm i\left(\lambda\sqrt{R_0} + \frac{d}{\lambda} + O(\lambda^{-3})\right) \equiv \pm\tilde{\gamma}(\lambda), \tag{3.12}$$

where

$$d = \frac{1}{2\sqrt{R_0}}\frac{R_0V - T}{2R_0} + \frac{\pi\rho u^2}{2G}\sqrt{R_0}. \tag{3.13}$$

$$Z_{5,6} = \pm i\left(c\sqrt{\lambda} - \frac{c_1}{\sqrt{\lambda}} + O(\lambda^{-1.5})\right) \equiv \Gamma(\lambda). \tag{3.14}$$

To obtain the spectral asymptotics, we proceed in the following way. First, having 6 roots (3.10), (3.12), and (3.14) of the characteristic equation (3.4), we can write the general solution of the differential equation (3.2). The general solution of Eq.(3.2) contains 6 different coefficients which are functions (still unknown) of the complex parameter λ. The eigenfunctions must satisfy the boundary conditions. So, we have to substitute the general solution of Eq.(3.2) into the boundary conditions. We will derive the constrains on the aforementioned six coefficients which have to be imposed in order to satisfy both the left - end and the right - end boundary conditions. As the result, we will obtain the left and the right reflection matrices.

The general solution of Eq.(3.2) can be represented in the form

$$\begin{aligned}\Psi(\lambda, x) &= \mathcal{A}(\lambda)\, e^{\gamma(\lambda)(x+L)} + \mathcal{B}(\lambda)\, e^{i\tilde{\gamma}(\lambda)(x+L)} + \mathcal{C}(\lambda)\, e^{i\,\Gamma(\lambda)(x+L)} \\ &+ \mathcal{D}(\lambda)\, e^{-\gamma(\lambda)(x+L)} + \mathcal{E}(\lambda)\, e^{-i\tilde{\gamma}(\lambda)(x+L)} + \mathcal{F}(\lambda) e^{-i\Gamma(\lambda)(x+\lambda)}.\end{aligned} \tag{3.15}$$

It is clear that in what follows, the function $\Psi(\lambda, \cdot)$ will be identified with the component ψ_2 of the 4-component eigenfunction of the operator $\mathcal{L}_{\beta\delta}$.

To find the left reflection matrix, we substitute the function $\Psi(\lambda,\cdot)$ (which has been identified with $\psi_2(\cdot)$) into the boundary conditions from (2.27) at $x=-L$. Using the equations connecting the functions $\psi_j, j=0,1,2,3$, we arrive at the following set of the boundary conditions for $\Psi(\lambda,\cdot)$ at $x=-L$:

$$\Psi(\lambda,-L)=\Psi''(\lambda,-L)=(G/\tilde{I})\Psi'''(\lambda,-L)+\lambda^2\left(1+O(\lambda^{-2})\right)\Psi'(\lambda,-L)=0. \tag{3.16}$$

Substituting (3.15) into conditions (3.16), we obtain the following equations for six unknown coefficients:

$$\mathcal{A}(\lambda)+\mathcal{B}(\lambda)+\mathcal{C}(\lambda)+\mathcal{D}(\lambda)+\mathcal{E}(\lambda)+\mathcal{F}(\lambda)=0, \tag{3.17}$$

$$\mathcal{A}(\lambda)\gamma^2(\lambda)-\mathcal{B}(\lambda)\hat{\gamma}^2(\lambda)-\mathcal{C}(\lambda)\Gamma^2(\lambda)+\mathcal{D}(\lambda)\gamma^2(\lambda)-\mathcal{E}(\lambda)\hat{\gamma}^2(\lambda)-\mathcal{F}(\lambda)\Gamma^2(\lambda)=0, \tag{3.18}$$

$$\begin{aligned}G/\tilde{I}\,&\left[\mathcal{A}(\lambda)\gamma^3(\lambda)-i\mathcal{B}(\lambda)\hat{\gamma}^3(\lambda)-i\mathcal{C}(\lambda)\Gamma^3(\lambda)-\mathcal{D}(\lambda)\gamma^3(\lambda)+i\mathcal{E}(\lambda)\hat{\gamma}^3(\lambda)+\right.\\ &\left.i\mathcal{F}(\lambda)\Gamma^3(\lambda)\right]+\lambda^2\left(1+O(\lambda^{-2})\right)\left[\mathcal{A}(\lambda)\gamma(\lambda)+i\mathcal{B}(\lambda)\hat{\gamma}(\lambda)+\right.\\ &\left.i\mathcal{C}(\lambda)\Gamma(\lambda)-\mathcal{D}(\lambda)\gamma(\lambda)-i\mathcal{E}(\lambda)\hat{\gamma}(\lambda)-i\mathcal{F}(\lambda)\Gamma(\lambda)\right]=0.\end{aligned} \tag{3.19}$$

Using the definitions of the functions $\gamma,\hat{\gamma}$, and Γ from (3.10), (3.12), and (3.14), we obtain the following asymptotic representation:

$$\gamma(\lambda)/\lambda=O(\lambda^{-1/2}),\quad \hat{\gamma}(\lambda)/\lambda=O(\lambda^{-1/2}),\quad \Gamma(\lambda)/\lambda=\sqrt{R_0}+O(\lambda^{-2}). \tag{3.20}$$

$$\frac{\hat{\gamma}(\lambda)}{\gamma(\lambda)}=1+O(\lambda^{-1}),\quad \frac{\Gamma(\lambda)}{\gamma(\lambda)}=\sqrt{\lambda}\frac{\sqrt{R_0}}{(T/R_0)^{1/4}}(1+O(\lambda^{-1}))\equiv\sqrt{\lambda}d_1(1+O(\lambda^{-1})). \tag{3.21}$$

Counting (3.10), (3.12), (3.14), (3.20), and (3.21), we can easily see that the system of 3 equations (3.17) - (3.19) can be written in the form of one matrix equation

$$\begin{bmatrix}1 & 1 & 1\\ 1 & -(1+O(\lambda^{-1})) & -\lambda d_1^2(1+O(\lambda^{-1}))\\ 1 & i(1+O(\lambda^{-1})) & O(\lambda^{-3/2})\end{bmatrix}\begin{pmatrix}\mathcal{A}(\lambda)\\ \mathcal{B}(\lambda)\\ \mathcal{C}(\lambda)\end{pmatrix}$$

$$=\begin{bmatrix}-1 & -1 & -1\\ -1 & (1+O(\lambda^{-1})) & \lambda d_1^2(1+O(\lambda^{-1}))\\ 1 & i(1+O(\lambda^{-1})) & O(\lambda^{-3/2})\end{bmatrix}\begin{pmatrix}\mathcal{D}(\lambda)\\ \mathcal{E}(\lambda)\\ \mathcal{F}(\lambda)\end{pmatrix}. \tag{3.22}$$

It is convenient to introduce the following notations:

$$\begin{pmatrix}\mathcal{A}(\lambda)\\ \mathcal{B}(\lambda)\\ \mathcal{C}(\lambda)\end{pmatrix}=X(\lambda),\quad \begin{pmatrix}\mathcal{D}(\lambda)\\ \mathcal{E}(\lambda)\\ \mathcal{F}(\lambda)\end{pmatrix}=Y(\lambda), \tag{3.23}$$

$$\mathcal{K}(\lambda)\ =\ \begin{bmatrix}1 & 1 & 1\\ \\ 1 & -(1+O(\lambda^{-1})) & -\lambda d_1^2(1+O(\lambda^{-1}))\\ \\ 1 & i(1+O(\lambda^{-1})) & O(\lambda^{-3/2})\end{bmatrix}. \tag{3.24}$$

It can be verified that the matrix equation (3.22) is equivalent to the following one:

$$X(\lambda) = \left\{ -I + 2\mathcal{K}^{-1}(\lambda) \begin{bmatrix} 0 & 0 & 0 \\ 0 & O(\lambda^{-1}) & O(1) \\ 1 & i(1+O(\lambda^{-1})) & O(\lambda^{-3/2}) \end{bmatrix} \right\} Y(\lambda). \tag{3.25}$$

Calculating the inverse matrix $\mathcal{K}^{-1}$ and completing the matrix multiplication in (3.25), we arrive at the following relation between the vectors X and Y:

$$X(\lambda) = \begin{bmatrix} i(1+O(\lambda^{-1})) & (i-1)(1+O(\lambda^{-1})) & O(\lambda^{-1}) \\ -(i+1)(1+O(\lambda^{-1})) & -i(1+O(\lambda^{-1})) & O(\lambda^{-1}) \\ O(\lambda^{-1}) & O(\lambda^{-1}) & -1+O(\lambda^{-3/2}) \end{bmatrix} Y(\lambda). \tag{3.26}$$

The matrix at the right hand side of Eq.(3.26) will be called the *left reflection matrix* and denoted by $\mathbb{R}_l(\lambda)$. Hence, Eq.(3.26) can be written in the form

$$X(\lambda) = \mathbb{R}_l(\lambda) Y(\lambda). \tag{3.27}$$

Now, we derive the second matrix equation for the vectors $X(\cdot)$ and $Y(\cdot)$. To this end, we will use the second set of the boundary conditions, i.e., the conditions at $x = 0$. If we rewrite the boundary conditions (2.27) at the end $x = 0$ in terms of Ψ, we will have

$$G\Psi'(0) + i\lambda\delta\Psi(\lambda), \quad G\Psi''''(0) + \lambda^2 \tilde{I}(1+O(\lambda^{-2}))\Psi'''(0) = 0, \tag{3.28}$$

$$GE\Psi''''(0) + i\lambda\beta G\Psi'''(0) + \lambda^2 \tilde{I} E(1+O(\lambda^2))\Psi''(0) + i\lambda^3\beta\tilde{I}(1+O(\lambda^{-2}))\Psi'(0) = 0. \tag{3.29}$$

In what follows, it is convenient to introduce the following notations:

$$e(\lambda) = e^{\gamma(\lambda)L}, \quad \hat{e}(\lambda) = e^{i\hat{\gamma}(\lambda)L}, \quad e_+(\lambda) = e^{i\Gamma(\lambda)L}. \tag{3.30}$$

Substituting the function $\Psi(\lambda, \cdot)$ in the form of (3.15) in each of the condition from (3.28) and (3.29), using the approximations (3.20) and (3.21) and conditions (2.33), we obtain the following relation between the vectors X and Y:

$$\mathbb{A}(\lambda)\mathbb{E}(\lambda)X(\lambda) = \mathbb{B}(\lambda)\mathbb{E}^{-1}(\lambda)Y(\lambda), \tag{3.31}$$

where the matrices $\mathbb{A}, \mathbb{B}$, and $\mathbb{E}$ are given by the formulas

$$\mathbb{E}(\lambda) = diag\left\{e(\lambda), \hat{e}(\lambda), e_+(\lambda)\right\}, \tag{3.32}$$

$$\mathbb{A}(\lambda) = \begin{bmatrix} 1 & -i(1+O(\lambda^{-1})) & O(\lambda^{-1/2}) \\ 1 & i(1+O(\lambda^{-1})) & O(\lambda^{-1}) \\ (1+O(\lambda^{-1/2})) & (1+O(\lambda^{-1/2})) & (1+\delta^{-1}G\sqrt{R})(1+O(\lambda^{-2})) \end{bmatrix}, \tag{3.33}$$

$$\mathbb{B}(\lambda) = \begin{bmatrix} (1+O(\lambda^{-1})) & -i(1+O(\lambda)^{-1}) & O(\lambda^{-1/2}) \\ (1+O(\lambda^{-1})) & i(1+O(\lambda^{-1})) & O(\lambda^{-1}) \\ -(1+O(\lambda^{-1/2})) & -(1+O(\lambda^{-1/2})) & -(1-\delta^{-1}G\sqrt{R})(1+O(\lambda^{-2})) \end{bmatrix}. \tag{3.34}$$

Using the standard methods of the matrix theory and asymptotic analysis, we calculate that $\mathbb{A}^{-1}$ can be asymptotically represented as

$$(\mathbb{A}(\lambda))^{-1} = \frac{1}{2}\begin{bmatrix} 1+O(\lambda^{-1/2}) & (1+O(\lambda^{-1/2})) & O(\lambda^{-1/2}) \\ i(1+O(\lambda^{-1/2})) & -i(1+O(\lambda^{-1/2})) & O(\lambda^{-1/2}) \\ \frac{-(1+i)(1+O(\lambda^{-1/2}))}{1+\delta^{-1}G\sqrt{R_0}} & \frac{(i-1)(1+O(\lambda^{-1/2}))}{1+\delta^{-1}G\sqrt{R_0}} & \frac{2(1+O(\lambda^{-1/2}))}{1+\delta^{-1}G\sqrt{R_0}} \end{bmatrix}. \tag{3.35}$$

Applying the matrix $\mathbb{A}^{-1}$ to the both sides of Eq.(3.31), we obtain the following relation:

$$X(\lambda) = \mathbb{R}_r(\lambda)Y(\lambda), \tag{3.36}$$

with $\mathbb{R}_r$ being given by the formula

$$\mathbb{R}_r(\lambda) =$$

$$(\mathbb{E}(\lambda))^{-1}\begin{bmatrix} (1+O(\lambda^{-1/2})) & O(\lambda^{-1/2}) & O(\lambda^{-1/2}) \\ O(\lambda^{-1/2}) & (1+O(\lambda^{-1/2})) & O(\lambda^{-1/2}) \\ \frac{-2(1+O(\lambda^{-1/2}))}{1+\delta^{-1}G\sqrt{R_0}} & \frac{-2(1+O(\lambda^{-1/2}))}{1+\delta^{-1}G\sqrt{R_0}} & \frac{\delta^{-1}G\sqrt{R_0}-1}{\delta^{-1}G\sqrt{R_0}+1}(1+O(\lambda^{-1/2})) \end{bmatrix}(\mathbb{E}(\lambda))^{-1}. \tag{3.37}$$

We call $\mathbb{R}_r$ *-the right reflection matrix.*

4 Spectral Asymptotics

In this section, we outline the derivation of the spectral asymptotics for the problem and the proof of the main results formulated in Theorem 2.2.

As was shown in the previous section, there exist two matrix equations for two unknown vectors $X(\cdot)$ and $Y(\cdot)$ defined in (3.23). Recall that these equations have the following forms:

$$X(\lambda) = \mathbb{R}_l(\lambda)Y(\lambda), \quad X(\lambda) = \mathbb{R}_r(\lambda)Y(\lambda), \tag{4.1}$$

where the left and right reflection matrices $\mathbb{R}_l$ and $\mathbb{R}_r$ are given in (3.26) and (3.37) respectively. If we exclude the vector $X(\cdot)$ from system (4.1), we obtain

$$Y(\lambda) = (\mathbb{R}_l(\lambda))^{-1}\mathbb{R}_r(\lambda)Y(\lambda). \tag{4.2}$$

Eq.(4.2) has a nontrivial solution if and only if the following equation is satisfied:

$$det\left(\mathbb{I} - (\mathbb{R}_l(\lambda))^{-1}\mathbb{R}_r(\lambda)\right) = 0. \tag{4.3}$$

Taking into account (3.26) and (3.37), we can rewrite Eq.(4.3) in the asymptotical form. It can be shown that the aforementioned asymptotical form of Eq.(4.3) is equivalent to the following two equations:

$$\frac{1-\delta^{-1}G\sqrt{R_0}}{1+\delta^{-1}G\sqrt{R_0}}(e_+(\lambda))^{-2} - 1 = O(\lambda^{-1/2}). \tag{4.4}$$

and

$$det\begin{bmatrix} (i(e(\lambda))^{-2}-1) & (i-1)(e(\lambda)\hat{e}(\lambda))^{-1} \\ -(i+1)(e(\lambda)\hat{e}(\lambda))^{-1} & (-(\hat{e}(\lambda))^{-2}-i) \end{bmatrix} = O(\lambda^{-1/2}). \tag{4.5}$$

To solve Eq.(4.5), we first consider an auxiliary equation in which the right-hand side has been replaced with zero. Then by applying the Rouchet Theorem argument (see, e.g., our works [22,23,26,31,32]), we arrive at the formula (2.35) for the δ-branch of the spectrum.
To derive the asymptotics for the β-spectral branch, we consider Eq.(4.5)

$$\left(\hat{e}(\lambda))^{-2} - i\right)\left((e(\lambda))^{-2} + i\right) - 2(e(\lambda)\hat{e}(\lambda))^{-2} = 0. \tag{4.6}$$

Using formulas (3.10)-(3.14) and (3.30), we rewrite Eq.(4.6) and have

$$i(-e^{2c\sqrt{\lambda}L} - e^{2ic\sqrt{\lambda}L}(1+O(\lambda^{-1/2}))) + 1 - e^{2c\sqrt{\lambda}L(1+i)}(1+O(\lambda^{-1/2})) = 0, \tag{4.7}$$

where the constant c is given in (3.11). Note that when λ belongs to the upper half-plane, then $\mu \equiv \sqrt{\lambda}$ changes in the first quadrant. To derive the asymptotics of the roots of Eq.(4.7), we divide the first quadrant into three regions. The first region is defined by the conditions $Im\ \mu > Re\ \mu \geq 0$, the second one by $Re\ \mu > Im\ \mu \geq 0$, and the third one by $Re\ \mu = Im\ \mu$.

Investigating Eq.(4.7) in each of the aforementioned regions separately, we arrive at the formula (2.34) for the β - branch.

Finally, we have to address the question of multiple eigenvalues. The possible existence of a finite number of multiple eigenvalues which produce a set of the associate vectors in the δ-branch follows from the analyticity of the function, $det\left(\mathbb{I} - (\mathbb{R}_l(\lambda))^{-1}\mathbb{R}_r(\lambda)\right)$ on the complex plane and the spectral asymptotics (2.35) for this branch. Indeed, formula (2.35) means that all sufficiently distant eigenvalues are simple, i.e., there are no associate vectors related to these eigenvalues. Similar argument is valid for the β-branch as well. With this we complete the discussion of the proof of Theorem 2.2.

Acknowledgement

Partial support by the National Science Foundation Grants DMS-9972748 and DMS-9706882 and Advanced Research Program -97 of Texas Grant #0036-44-045 is highly appreciated.

Bibliography

[1] Abramowitz M., Stegun I., Ed., *Handbook of Mathematical Functions*, Dover, New York, 1972.

[2] Adams R. A., *Sobolev Spaces*, Academic Press, New York, 1975.

[3] Ashley H., *Engineering Analysis of Flight Vehicles*, Dover Publ. Inc., New York, 1992.

[4] Ashley H. and Landahl M., *Aerodynamics of Wings and Bodies*, Dover Publ. Inc., New York, 1985.

[5] Balakrishnan A. V., Aeroelastic control with self-straining actuators: continous models, to appear in *Proceedings of SPIE's 5th Annual International Symposium on Smart Structures and Materials, March 1998, San Diego, California.*

[6] Balakrishnan A. V., Vibrating systems with singular mass-inertia matrices, *First International Conference on Nonlinear problems in Aviation and Aerospace, S. Sivasundaram, Ed., p.23-32, Dayton Beach, Florida, Embry-Riddle Aeronautical University Press.*

[7] Balakrishnan A. V., Theoretical limits of damping attainable by smart beams with rate feedback". In: *Smart Structures and Materials, 1997; Mathematics and Control in Smart Structures, Vasundara V. Varadan, Jagdish Chandra, Eds., Proceedings of SPIE Vol. 3039, (1997) p. 204-215, SPIE.*

[8] Balakrishnan A. V., Damping performance of strain actuated beams, *Comput. Appl. Math.*, **18**, No.1, (1999), p.31-86.

[9] Balakrishnan A. V. and Edwards J. W., Calculation of the transient motion of elastic airfoils forced by control surface motion and gusts, NASA TM 81351, 1980.

[10] Balakrishnan A. V., Control of structures with self-straining actuators: coupled Euler/ Timoshenko model, *Nonlinear Problems in Aviation and Aerospace. Gordon and Breach Science Publishers, Reading, United Kingdom*, 1998.

[11] Bisplinghoff R. L., Ashley H., and Halfman R. L., *Aeroelasticity*, Dover Publ. Inc., New York, 1996.

[12] Chen G., Krantz S. G., Ma D. W., Wayne C. E., and West H. H, The Euler -Bernoulli beam equations with boundary energy dissipation. *Operator Methods for optimal Control Problems, Lecture Notes in Mathematics* , Vol 108, p.67-96, (1987) *Marcel Dekker.*

[13] Evgrafov M. A., *Analytic Functions*, Dover Publ. Inc., 1978.

[14] Foias C., Sz-Nagy B., *Harmonic Analysis of Operators in Hilbert Spaces*, Rev. Ed., North Holland, Amsterdam, 1970.

[15] Istratescu V. I., *Introduction to Linear Operator Theory*, Pure Appl. Math Series of Monog., Marcel Dekker Inc., New York, 1981.

[16] Lee C. K., Chiang W. W., and O'Sullivan T. C., Piezoelectric modal sensor /actuator pairs for critical active damping vibration control, *J.Acoust. Soc .Am.*, **90** , p.384-394, (1991).

[17] Magnus W., Oberhettinger F., and Soni R. P., *Formulas and Theorems for the Special Functions of Mathematical Physics*, 3rd Ed. Springer VErlag, New York, 1966.

[18] Shubov M. A., Spectral operators generated by Timoshenko bean model, to appear in *Systems and Control Letters.*

[19] Shubov M. A., Timoshenko beam model: Spectral properties and control, *Proceedings of the 10th Int'l Workshop on Dynamics and Control "Complex Dynamics Process Incomplete Information"*, E. Reithmeier and G. Leitmsnn, Springer-Verlag, 1999.

[20] Shubov M. A., Exact controllability of Timoshenko beam, to appear in *IMA Journal of Cont. and Inform.*

[21] Shubov M. A., Asymptotics and spectral analysis of Timoshenko beam model, *Preprint of Texas Tech University*, 1999.

[22] Shubov M. A., Asymptotics of resonances and geometry of resonance states in the problem of scattering of acoustical waves by a spherically symmetric inhomogeneity of the density, *Dif. Int. Eq.*, **8**, (5), (1995), p.1073-1115.

[23] Shubov M. A., Asymptotics of spectrum and eigenfunctions for nonselfadjoint operators generated by radial nonhomogeneous damped wave equations, *Asymp. Anal.*, **16**, (1998), p.245-272.

[24] Shubov M. A., Nonselfadjoint operators generated by the equation of nonhomogeneous damped string, *Trans. of Amer. Math. Soc.*, Vol.**349**, (1997), p.4481-4499.

[25] Shubov M. A., Basis property of eigenfunctions of nonselfadjoint operator pencils generated by equation of nonhomogeneous damped string, *Integ. Eqs. Oper. Theory*, **25** (1996), p.289-328.

[26] Shubov M. A., Asymptotics of resonances and eigenvalues for nonhomogeneous damped string, *Asymptotic Analysis*, **13** (1996), p.31-78.

[27] Shubov M. A., Spectral operators generated by 3-dimensional damped wave equation and application to control theory, *Spectral and Scattering Theory (Ed. A.G. Ramm), Plenum Press , New York*, (1998), p.177-188.

[28] Shubov M. A., Spectral operators generated by damped hyperbolic equations, *Integ. Eqs. Oper. Theory*, **28** (1997), p.358-372.

[29] Shubov M. A., Unique controllability of damped wave equation, (jointly with C. Martin, J. Dauer, B. Belinskiy), *SIAM J. on Cont. Optim.*, Vol.**35** (5), (1997), p.1773-1789.

[30] Shubov M. A., Exact boundary and distributed controllability of radial damped wave equation, *J. de Math. Pures Appl.*, **77**, (1998), p.415-437.

[31] Shubov M. A., Asymptotics of the discrete spectrum for a radial Schrödinger equation with nearly Coulomb potential, *Integ. Eqs. Oper. Theory*, **14** (1991), p.586-608.

[32] Shubov M. A., Stark quantum defect for high Rydberg states of three-dimensional Schrödinger operator with screened Coulomb potential, *Il Nuovo Cimento*, **110B** (1995), p.1057-1092.

[33] Tzou H. S., and Gadre M., Theoretical analysis of a multi-layered thin shell coupled with piezoelectric shell actuators for distributed vibration controls, *J. Sound Vibr.*, 132, p.433-450, (1989).

[34] Weidmann J., *Spectral Theory of Ordinary Differential Operators*, Lecture notes in Math., 1258, Springer-Verlog, (1987).

[35] Yang S. M., and Lee Y. J., Modal analysis of stepped beams with piezoelectric materials, *J. Sound Vibr.*, **176**, p.289-300.

Marianna A. Shubov. Department of Mathematics and Statistics, Texas Tech University, Lubbock, TX 79409, USA
E-mail: mshubov@math.ttu.edu

Weak Set Evolution and Variational Applications

Jean-Paul Zolésio

Abstract. We consider a variational formulation associated with the usual incompressible Euler flow with a free bounday. The energy functional is minimized under three kinds of viscous regularizations. Basically, we introduce a "surface tension energy" at the interface between the two fluids and derive existence results. We show that smooth enough extremal point are indeed solution of the Euler equation. The surface energy uses a parabolic version of the usual perimeter and density perimeter. Application of several compactness and continuity resuluts associate with the weak convection of measurable sets is made to derive several topolgies on family of subsets of a given boounded domain D.

1 Introduction

The weak convection of measurable sets ([12], [11]) has been introduced in relation with the shape differential equation and related topics. We enlarge that approach to a variational problem related to the Euler equation for incompressible flows and we give several applications concerning the shape analysis. We focus on the compacity results and we propose three "families" of "viscosity" constraints on the vector fields V leading to existence results. The first one is the usual H^1 norm which leads to an "Eulerian" version for the Navier Stokes (dynamical) equation. The second is connected to the "parabolic version" of the compactness of the inclusion mapping from BV in L^1. The last one is based on the use of the "Density Perimeter" properties ([1], [2]).

The so-called "Speed Method" has been developed in relation with the shape optimization governed by Partial Differential Equations ([8],[9],[10],[12]). In the strong version we considered the flow mapping $T_t(V)$ of a smooth vector field

$V \in C^0([0,\infty[, C^k(R^N) \cap L^\infty(R^N), R^N)$ (V is smoothly globally defined over R^N at all times t).

For any set Ω_0 the set $\Omega_t(V) = T_t(V)(\Omega_0)$ is then defined at any time. In [12] the shape differential equation (introduced in [8] and [9]) is studied for shape functional governed by several classical boundary value problems (with use of the extractor estimate for the shape gradient [3]), then in [5] for non linear viscous flow.

The characteristic function of the evolution domain, $\chi_t = \chi_{\Omega_t(V)}$, is convected as $\chi_t = \chi_0 o T_t(V)^{-1}$ solves the convection problem

$$\frac{\partial}{\partial t}\chi + \nabla\chi.V = 0, \quad \chi(0) = \chi_{\Omega_0}$$

That this problem has a solution when the vector field $V \in L^2$ with $divV \in L^2$ and some growth assumption on the positive part $(divV)^+$. The incompressible situation was already introduced in [12], [11]. We give several continuity and compacity results. Applications are the existence results for variational principle related to the Euler incompressible flow equation (see also [11]) and metric on the family of subsets of a bounded domain D

2 Weak Convection of Characteristic Functions

2.1 A Uniqueness Result

Let $V \in L^2(0,\tau,L^2(D,R^N))$ be a vector field with $divV(t,.) = 0$ in D and $V(t,).n_{\partial D} = 0$ in $H^{-1/2}(\partial D)$. For any smooth element $\phi \in C^\infty([0,\tau] \times D)$ we consider the first order operator $D_V.\phi = \frac{\partial}{\partial t}\phi + \nabla_x\phi.V \in L^2(0,\tau,L^2(D))$. We introduce the Hilbert (resp. Banach) space $\mathcal{H}_V$ (resp. $\mathcal{W}_V$) defined as the completion of $C^\infty([0,\tau] \times D)$ for the following norm:

$$|\phi|_V = \left(|\phi|^2_{L^2(0,\tau,L^2(D))} + |D_V\phi|^2_{L^2(0,\tau,L^2(D))} \right)^{1/2}$$

$$(resp.\ |\phi|_V = \left(|\phi|_{L^\infty(0,\tau,L^2(D))} + |D_V\phi|_{L^1(0,\tau,L^2(D))} \right)$$

For any smooth ϕ, ψ we have:

$$\int_t^\tau \int_D D_V\phi\ \psi\, dxdt = \int_D \{\phi\psi(\tau) - \phi\psi(t)\}dx - \int_t^\tau \int_D \phi\ D_V\psi\, dxdt$$

That identity gives a weak sense to pointwise values $\phi(t)$ as follows: we consider elements $\psi \in \mathcal{R} = \{\psi \in H^1(0,\tau,L^2(D)) \cap L^\infty(0,\tau,W^{1,\infty}(D)),$ with $\psi(\tau) = 0\}$, so that

$$\nabla\psi \in L^\infty(0,\tau,L^\infty(D,R^N)),\ \nabla\psi.V \in L^2(0,\tau,L^2(D,R^N))$$

then $D_V\psi \in L^2(0,\tau \times D)$ and for any t, $0 \le t < \tau$, the element $\phi(t)$ is weakly defined by:

$$\int_D \phi(t)\psi(t)\, dx = -\int_t^\tau \int_D D_V\phi\ \psi\, dxdt - \int_t^\tau \int_D \phi\ D_V\psi\, dxdt$$

More precisely (as we will not characterize the dual space of $\mathcal{R}$), let $\theta \in W^{1,\infty}(D)$ we consider the element of $\mathcal{R}$ defined by $\psi(t,x) = (\tau - t)\theta(x)$. Then we get:

$$\int_D \phi(t)\theta dx = (\tau - t)^{-1} \int_0^t \int_D \{\, t\, D_V\phi\ \theta\ +\ \phi\,(\theta + t\nabla\theta.V)\,\}dxdt$$

Obviously we get $\mathcal{H}_V \subset C([0,\tau[, W^{1,\infty}(D)')$.

Also from the first identity we get (taking $t = 0$), for all smooth ϕ with $\phi(0) = 0$:

$$\int_0^\tau \int_D \phi^2\ dtdx\ =\ 2\int_0^\tau \int_D (\tau - t)D_V\phi\ \phi\, dtdx$$

(resp.

$$\forall t,\ 0 \le t \le \tau,\ \int_D \phi^2(t)dx = 2\int_0^\tau \int_D D_V\phi\ \phi\, dtdx$$
$$\le\ |\phi|_{L^\infty(0,\tau,L^2(D))}\ |D_V\phi|_{L^1(0,\tau,L^2(D))}\ ).$$

So by density we derive

$$\forall \phi \in \mathcal{H}_V \text{ with } \phi(0) = 0,\ |\phi|_{L^2(0,\tau,L^2(D))} \le 2\tau\ |D_V\phi|_{L^2(0,\tau,L^2(D))} \tag{2.1}$$

(resp.

$$\forall \phi \in \mathcal{W}_V\ ,\ \text{with}\ ,\phi(0) = 0, |\phi|_{L^\infty(0,\tau,L^2(D))} \le 2\ \ |D_V\phi|_{L^1(0,\tau,L^2(D))}).$$

We consider the dynamical system

$$u(0) = u_0,\ \ \frac{\partial}{\partial t}u(t) + < V(t), \nabla u(t) >= f \tag{2.2}$$

with initial condition $u_0 \in L^2(D)$ and right hand side

$$f \in L^2(0,\tau,\ L^2(D))\ (resp. f \in L^1(0,\tau,\ L^2(D))$$

The derive the following uniqueness result:

PROPOSITION 2.1. *For V given,*

$$V \in L^2(0,\tau, L^2(D, R^N))$$

with $divV(t,.) = 0$ in D and $V(t,).n_{\partial D} = 0$ in $H^{-1/2}(\partial D)$, let f be given in $L^2(0,\tau,L^2(D))$ (resp. in $L^1(0,\tau,L^2(D))$.) Then the problem (2) has at most one solution in $\mathcal{H}_V$ (resp. in $\mathcal{W}_V$.)

Proof. Let u^i be two such solutions, then $u = u^2 - u^1$ solves the homogeneous problem. From the previous estimate we get $u = 0$. □

2.2 The Galerkin Approximation

PROPOSITION 2.2. *Let $V \in L^1(0,\tau,L^2(D,R^3))$ with $divV \in L^2(0,\tau,L^2(D,R^3))$ verifying the following uniform integrability condition:*

$$\text{There exist } T_0 > 0, \rho < 1, s.t. \forall a \ge 0,\ \int_a^{a+T_0} ||V(t)||_{L^2(D,R^3)}dt\ \le \rho < 1$$

(for shortness we consider $V \in L^p(0,\tau,L^2(D,R^3))$ with $p > 1$). We assume that the positive part of the divergence $divV = (divV)^+ - (divV)^-$, verifies

$$||(divV(t))^+||_{L^\infty(D,R^3)} \in L^1(0,\tau).$$

Then if $< V(t,.), n >= 0$ (as an element of $L^1(0,\tau,H^{-\frac{1}{2}}(\partial D))$, $f \in L^1(0,\tau,L^2(D))$ and initial condition $\phi \in L^2(D)$, there exists solutions $u \in \mathcal{H}_V$ verifying also

$$u \in L^\infty(0,\tau,L^2(D)) \cap W^{1,p^*}(0,\tau,\ W^{-1,3}(D)) \subset C^0([0,\tau], W^{-\frac{1}{2},\frac{3}{2}}(D))$$

to the problem (2), where $\frac{1}{p}+\frac{1}{p^*}=1$. *Moreover there exists a constant* M *such that:*

$$\forall \tau,\ \|u\|_{L^\infty(0,\tau,L^2(D,R^N))} \leq M\{\ \|\phi\|_{L^2(D)} + \|f\|_{L^1(0,\tau,L^2(D))}\ \} \tag{2.3}$$

$$\{\ 1+\int_0^\tau (\|(divV(s))^+\|_{L^\infty(D,R^3)} + \|f(s)\|_{L^2(D,R^3)})$$

$$\int_s^\tau (\|(divV(\sigma))^+\|_{L^\infty(D,R^3)} + \|f(\sigma)\|_{L^2(D,R^3)})d\sigma)ds\ \}$$

When the initial condition is a characteristic function:

$$\phi = \chi_{\Omega_0} \in L^2(D)$$

and if $f=0$, *then the unique solution* $u \in \mathcal{H}_V$ *is itself a characteristic function:*

$$a.e.(t,x),\quad u(t,x)\ (1-u(t,x)\ ) = 0 \quad \text{that is } u = \chi_{Q_V}$$

Where Q_V *is a non cylindrical measurable set in* $]0,\tau[\times D$. *For a.e.t, we set*

$$\Omega_t(V) = \{x \in D|\quad (t,x) \in Q_V\ \}.$$

If V *is a free divergence field,* $divV(t,x)=0$ *for a.e.* $t,\in]0,\tau[$, *then the set* $\Omega_t(V)$ *verifies a.e.t,* $meas(\Omega_t(V)) = meas(\Omega_0)$.

Proof. Let us consider $V \in L^2(0,\tau,\ H_0^1(D))$ and a dense family $e_1,...e_m,...$ in $H_0^1(D)$ with each $e_i \in C^\infty_{comp}(D,R^3)$. Consider the approximated solution

$$u^m(t,x) = \Sigma_{i=1,\dots,m}\quad u_i^m(t)\, e_i(x)$$

with $U^m = (u_1^m,...,u_m^m)$ solution of the following linear ordinary differential system:

$$\forall t,\ \int_D (\frac{\partial}{\partial t}u^m(t) + < V(t), \nabla u^m(t) >)\, e_j(x)\, dx = \int_D f(t,x)e_j(x)\ dx,\ \ j=1,...,m$$

That is

$$\frac{\partial}{\partial t}U^m(t) + M^{-1}.A(t).U^m(t)\ =\ F(t) \tag{2.4}$$

where

$$M_{i,j} = \int_\Omega e_i(x)\, e_j(x)\, dx$$

$$A_{i,j}(t) = \int_D\ < V(t), \nabla e_i(x) >\ e_j(x)\, dx$$

That is an ordinary linear differential systems possessing a global solution when

$$V \in L^p(0,\tau,\ L^2(D,R^N))$$

for some $p,\ p>1$. By classical energy estimate, as

$$\int_D < V(t), \nabla u^m(t) > u^m(t)\, dx = -\frac{1}{2}\int_D < u^m(t), u^m(t) > divV(t)\, dx\ ,\ \ a.e.t$$

we get:

$$\forall \tau,\ \tau \leq T,\ \|u^m(\tau)\|^2_{L^2(D)} \leq \|u^m(0)\|^2_{L^2(\Omega)}$$

$$+\int_0^T \int_D < u^m(t,x), u^m(t,x) > (divV(t,x))^+ \, dtdx$$

$$+2\int_0^T \int_D f(t,x)u(t,x)dtdx$$

Setting

$$\psi(t) = \|(divV(t,.))^+\|_{L^\infty(D,R^3)}$$

When $f = 0$,

$$\frac{1}{2}\int_D u^m(t,x)^2 dx \leq \frac{1}{2}\int_D u^m(0,x)^2 dx$$

$$+\frac{1}{2}\int_0^t \psi(s) \int_D \|u^m(t,x)\|^2 \, dx$$

by use of the Gronwall's lemma we get:

$$\int_D u^m(t,x)^2 dx \leq \int_D u^m(0,x)^2 dx \ (1 + \int_0^t \psi(s) exp\{ \int_s^t \psi(\sigma) d\sigma \} ds \ )$$

By the choice of the initial conditions in the ordinary differential system we get

$$M > 0, s.t. \forall \tau, \leq T,\ \|u^m(\tau)\|_{L^2(D)}$$

$$\leq M\ \|\phi\|_{L^2(D)}\ (1 + \int_0^t \psi(s) exp\{ \int_s^t \psi(\sigma) d\sigma \} ds \ )$$

When $\psi = 0$, we get

$$\forall \tau,\ \tau \leq T,\ \|u^m(\tau)\|^2_{L^2(D)} \leq \|u^m(0)\|^2_{L^2(\Omega)}$$

$$+2\int_0^T \int_D f(t,x)u(t,x)dtdx$$

In the general case, we use

$$\|u^m\| \leq 1 + \|u^m\|^2$$

and we derive the following estimate:

$$\forall \tau,\ \tau \leq T,\ \|u^m(\tau)\|^2_{L^2(D)} \leq \|u^m(0)\|^2_{L^2(D,R^3)}$$

$$+\int_0^T \int_D < u^m(t,x), u^m(t,x) > (divV(t,x))^+ \, dtdx$$

$$+2\int_0^T \int_D f(t,x)u(t,x)dtdx$$

$$\leq \|u^m(0)\|^2_{L^2(D,R^3)} + \int_0^T \|f(t)\|_{L^2(D,R^3)} \, dt$$

$$+\int_0^T x < u^m(t,x), u^m(t,x) > (\psi(t) + \|f(t)\|_{L^2(D,R^3)}) \, dtdx$$

$$\leq M\,(||u_0||^2_{L^2(D,R^3)} + ||f||_{L^1(0,\tau,L^2(D,R^3))}\,)$$

$$+\int_0^t\ (\psi(t) + ||f(t)||_{L^2(D,R^3)})||u^m(s)||^2_{L^2(D)}ds$$

From Gronwall's inequality we derive:

$$||u^m(\tau)||^2_{L^2(D)}\ \leq M(||u_0||^2_{L^2(D,R^3)} + ||f||_{L^1(0,\tau,L^2(D,R^3))}\,)$$

$$\{\ 1+\int_0^t\ [\ (\psi(s) + ||f(s)||_{L^2(D,R^3)})\int_s^t(\psi(\sigma) + ||f(\sigma)||_{L^2(D,R^3)})d\sigma\ ]ds\ \}$$

In all cases n u^m remains bounded in $L^\infty(0,\tau,L^2(D,R^N))$ and there exists an element u in that space and a subsequence still denoted u^m which weakly-* converges to u. In the limit u itself verifies the previous estimate. It can be verified that u solves the problem in distribution sense. That is

$$\forall\phi\in H^1_0(0,\tau,L^2(D,R^3))\cap L^2(0,\tau,H^1_0(D,R^3)),\ \ \phi(\tau)=0,$$

$$-\int_0^\tau\int_D u(\frac{\partial}{\partial t}\phi + div(\ \phi V)\ )dxdt = \int_D\phi(0)\,u_0\,dx\ + \int_0^\tau\int_D < f,\phi > dxdt$$

When $V(t)\in L^2(D,R^3)$ the duality brackets $<\frac{\partial}{\partial t}u,\phi>$ are defined. If $\nabla\phi$ belongs to $L^\infty(D,R^3)$, this is verified, for example, when $\phi\in H^3_0(D)$ so that u_t is identified to an element of the dual space $H^{-3}(D)$. When $V(t)\in H^1(D,R^3)$ we get *a.e.t*, $u(t,.)\in L^2(D)$, $V(t,.)\in L^6(D)$, then $\nabla\phi(t)$ should be in $L^3(D,R^3)$,that is $\phi(t)\in W^{1,3}_0(D)$ and then, for *a.e.t*, the element u_t is in the dual space $W^{-1,\frac{3}{2}}(D)$ while $u_t\in L^{p^*}(0,\tau,\,W^{-1,\frac{3}{2}}(D))$ and

$$u\in W^{1,p^*}(0,\tau,\ W^{-1,\frac{3}{2}}(D))$$

Then we have

$$u\in L^\infty(0,\tau,L^2(D))\cap\cap W^{1,p^*}(0,\tau\,W^{-1,\frac{3}{2}}(D))$$
$$\subset L^2(0,\tau,W^{0,\frac{3}{2}}(D))\cap W^{1,p^*}(0,\tau,\ W^{-1,\frac{3}{2}}(D))$$
$$\subset C^0([0,\tau],W^{-\frac{1}{2},\frac{3}{2}}(D))$$

When the initial data is a characteristic function and $f=0$, $u_0=\chi_{\Omega_0}$, we shall verify that u^2 is also a solution which by uniqueness implies $u^2=u\ \ a.e.$.

Let $u_n\in C^\infty([0,\tau]\times\bar{D})$, $\longrightarrow u$ in $L^4(]0,\tau[\times D)$, then

$$\forall\phi\in H^1_0(0,\tau,L^2(D,R^3))\cap L^2(0,\tau,H^1_0(D,R^3)),$$

$$-\int_0^\tau\int_D u_n^2(\frac{\partial}{\partial t}\phi + <\nabla\phi,V>)dxdt =$$

$$2\int_0^\tau\int_D u_n\,(\frac{\partial}{\partial t}u_n + <\nabla u_n,V>)\phi\ dxdt\ + \int_D(u^n_0)^2\,\phi(0)\,dx$$

But

$$\frac{\partial}{\partial t}u_n + <\nabla u_n,V>\longrightarrow\frac{\partial}{\partial t}u + <\nabla u.V> = 0\ \text{ in } L^2(]0,\tau[,L^2(D)),$$

as $u_n \to u$ in L^2 and $u_n(0) \to u_0$ in the limit we get

$$-\int_0^\tau \int_D u^2(\frac{\partial}{\partial t}\phi + < \nabla\phi, V >)dxdt = \int_D \phi(0)\,(u_0)^2\,dx$$

then u^2 is a solution, we derive that $u^2 = u$. That is: u is a characteristic function. We show now that, as the divergence of the field is zero and the right hand side is zero too, the solution u can be written in the form

$$u = \chi_Q, \quad Q = \cup_{0<t<\tau} \{t\} \times \Omega_t, \text{ with } meas(\Omega_t) = meas(\Omega_0), \ a.e.t$$

it

We introduce $V^n \to V$ in $L^2(0,\tau,L^2(D,R^3))$ with $V^n \in C^\infty$ and $div(V^n)^- \in L^\infty([0,\tau] \times \bar{D})$. The solution u^n associated to these data is obtained through the flow of V^n as follows:

$$u^n(t,x) = \chi_{\Omega_0} o T_t(V^n)^{-1}(x)$$

as a consequence u^n is a characteristic function, $(u^n)^2 = u^n$, it is obviously solution to the equation associated to the initial condition u_0^2 and the field V^n. Now we can find a subsequence such that $u^n = (u^n)^2$ is $L^2(]0,\tau[\times D,\ R^n)$-weakly converging to some element v.

$$\forall\phi \in H_0^1(0,\tau,L^2(D,R^3)) \cap L^2(0,\tau,H_0^1(D,R^3)), \ \ \phi(\tau) = 0,$$

$$-\int_0^\tau \int_D < u^n, \frac{\partial}{\partial t}\phi + \nabla\phi.V^n > dxdt = \int_D \phi(0)\,u_0^n\,dx$$

And

$$-\int_0^\tau \int_D < (u^n)^2, \frac{\partial}{\partial t}\phi + \nabla\phi.V^n > dxdt = \int_D \phi(0)\,(u_0^n)^2\,dx$$

can be verified that the weak L^2 convergence of u^n to $u = \chi_{\Omega_0}$ is strong: we verify the behavior of the norm

$$lim_{n\to\infty} \int_0^\tau \int_D (u^n)^2\,dxdt \le lim_{n\to\infty}, \int_0^\tau \int_D (u^n)\,dxdt = \int_0^\tau \int_D (u^2)\,dxdt$$

(As $u^2 = u = \chi_{\Omega_0}$). Then $(u^n)^2$ itself is strongly converging to w, as a consequence we get $w = u^2 = \chi_{\Omega_0}$. □

PROPOSITION 2.3. *Let $V^n \to V$ in $L^2(0,\tau,L^2(D,R^3))$ with $divV^n(t,.) = divV(t,.) = 0$,*

$$\chi_{Q_{V^n}} \to \chi_{Q_{V^n}} \ \ in\ L^2(0,\tau,L^2(D))$$

2.3 Smooth Solutions

Bounded Speed Field

Let W be a given element in $L^\infty(D,R^N)$ with $divW = 0$. We consider the unbounded operator A_W in the Hilbert space $H = L^2(D)$,with dense domain $D_W = H_0^1(D)$ and defined by

$$A_W.\phi = W.\nabla\phi \tag{2.5}$$

It can easily be verified that the adjoint unbounded operator verifies

$$A_W^* = -A_W$$

PROPOSITION 2.4. *The unbounded operator A_W is the infinitesimal generator of a semi group of contraction in $H = L^2(D)$.*

Proof. As $H_0^1(D) \subset L^5(D)$, with $N = 3$, and as $\frac{1}{2} + \frac{1}{5} + \frac{3}{10} = 1$, let W_n be a sequence in $W^{1,\infty}(D, R^N)$ with $div\, W_n \to 0$ in $L^{\frac{5}{3}}(D)$ and converging to W in $L^{\frac{10}{3}}(D, R^N)$. We get:

$$\int_D A_W.\phi \;\phi dx = \int_D W.\nabla\phi \;\phi\, dx$$

$$= lim\{ \int_D W_n.\nabla\phi\phi\, dx \} = -lim\{ \int_D (divW_n \;\; (\phi)^2 \;\; +(W_n.\nabla\phi)\;\phi\,)dx \}$$

from which in the limit we deduce firstly that

$$\int_D (W.\nabla\phi)\;\; \phi\, dx = 0$$

and then that the operator A_W is dissipative: $\int_D (A_W.\phi)\; \phi\, dx = 0$.

We consider now the evolution hyperbolic problem associated to any element

$$V \in W^{1,\infty}(0, \tau, L^\infty(D, R^N))$$

with $div\, V(t,.) = 0$ *a.e.t.*We consider the unbounded operator $A(t)$ in the Hilbert space $H = L^2(D)$ defined by $A(t).\phi = V(t).\nabla\phi$ with dense domain $D = H_0^1(D)$ which is independent on t. The Triplet{ A(.),H, D } is then a CD-system in the sense of [6] (page 9)as we shall verify the following stability condition: for any times $t_1 < .... < t_k$,

$$\| R(t_k, \lambda)...R(t_2, \lambda).R(t_1, \lambda) \| \leq M(\lambda - \beta)^{-1}$$

where the resolvant $R(t, \lambda) = (\lambda\; I_d \; + \; A(t)\,)^{-1}$ exists for any $t > \beta$. That stability condition is obviously verified from the previous contraction property of each operator $A(t_k)$. we obtain the following result ([6], thm1.2 page 11) □

PROPOSITION 2.5. *Let $V \in W^{1,\infty}(0, \tau, L^\infty(D, R^N))$, $f \in Lip(0, \tau, L^2(D))$, and $\phi \in H_0^1(D)$, there exists a unique solution*

$$u \in C([0, \tau], H_0^1(D)) \cap C^1([0, \tau], L^2(D))$$

to the evolution problem (2).

Unbounded Speed Field

We directly applied the theory in the Hilbert space $L^2(D)$ and we derived that $V(t,.) \in L^\infty(D)$ was enough to describe the semigroup. From Sobolev embedding inequalities we have $H^1(D) \subset L^p(D)$ for any $p \leq \frac{2N}{N-1}$. Let V be a given element in $L^q(D, R^N)$ with $q > 2 + 2\frac{N-1}{N+1}$ and $divV = 0$.

Let us observe that in dimension 3 the following inclusion holds: $H^1(D) \subset L^6(D)$ then as soon as $V(t,.) \in L^3(D)$ we get $\forall\, \phi \in H^1(D)$, $\phi V(t,.).\nabla\phi \in L^1(D)$.

The semi group is also dissipative as well as his adjoint taking the Banach space $H = L^3(D)$ while the dense domain is $D = H_o^1(D)$. Then we get the

PROPOSITION 2.6. *Let $V \in W^{1,\infty}([0,\tau], L^3(D,R^N))$ with $divV(t,.) = 0$. i)Let $f = 0$ and initial data $\phi \in H_o^1(D)$ there exists a unique solution in $C^0([0,\tau], H_o^1(D)) \cap C^1([0,\tau], L^{\frac{6}{5}}(D))$ to the problem (2). ii) For $f \in W^{1,\infty}([0,\tau], L^{\frac{6}{5}}(D))$ the dynamical system (2) has a unique solution in*

$$C^0([0,\tau], H_o^1(D)) \cap C^1([0,\tau], L^{\frac{6}{5}}(D)).$$

3 Compacity Results

3.1 Bounded Vector Fields in $H_0^1(D)$

PROPOSITION 3.1. *Let V_n be a bounded sequence in E verying also:*

$$V_n \in L^2(0,\tau, H_0^1(D)), \quad ||V_n||_{L^2(0,\tau,H_0^1(D))} \leq M \tag{3.1}$$

Then if V_n weakly converges in $L^2(0,\tau, H_0^1(D))$ to V, the limiting element V verifies the previous regularity and $\chi_{V_n} \longrightarrow \chi_V$ strongly in $L^2(0,\tau.L^2(D))$.

LEMMA 3.2. *Let the sequence V_n be bounded in E, then the sequence $\frac{\partial}{\partial t}\chi_{V_n}$ is bounded in $L^2(0,\tau, H^{-1}(D))$. More precisely, for all $V \in E$ we have*

$$||\frac{\partial}{\partial t}\chi_V||_{L^2(0,\tau,H^{-1}(D))} \leq ||V||_{L^2(0,\tau,L^2(D))} \tag{3.2}$$

Proof. $\forall \phi \in L^2(0,\tau, H_0^1(D))$ we have

$$\int_0^\tau < \frac{\partial}{\partial t}\chi, \phi >_{H^{-1}(D)\times H_0^1(D)} dt = \int_0^\tau \int_D \chi\ V.\nabla\phi\ dxdt$$

Then

$$|\int_0^\tau < \frac{\partial}{\partial t}\chi, \phi >_{H^{-1}(D)\times H_0^1(D)} dt| \leq ||\chi V||_{L^2(I\times D)}\ ||\phi||_{L^2(I,H^{-1}(D))}$$

□

Proof. (of proposition 3.1) The sequence χ_{V_n} is itself obviously bounded in $L^2(0,\tau, L^2(D))$ and as the continuous inclusion mapping $L^2(D) \to H^{-\epsilon}(D)$ is compact,$\forall \epsilon > 0$, from a classical result [7] we get the strong convergence of the sequence:

$$\chi_{V_n} \to \chi \text{ strongly in } L^2(0,\tau, H^{-\epsilon}(D))$$

but also χ_{V_n} is $L^2(0,\tau, L^2(D))$-weakly convergent to the same element χ which verifies:

$$0 \leq \chi \leq 1 \ a.e.$$

From the $L^2(0,\tau, H_0^1(D))$ boundedness of the sequence V_n we derive the weak-$L^2(0,\tau, H_0^1(D))$ convergence of V_n to V. This enables us to get in the limit in the weak version of the equation of χ_{V_n}:

$$\forall n, \int_0^\tau \int_D \chi_{V_n}\phi\, dxdt = \int_{\Omega_0} \phi(0,x)dx + \int_0^\tau \int_D (\tau - t)\chi_{V_n} V_n \nabla\phi\ dxdt$$

which gives in the limit:

$$\int_0^\tau \int_D \chi\phi\, dxdt = \int_{\Omega_0} \phi(0,x)dx + \int_0^\tau \int_D (\tau - t)\chi_V \nabla\phi\, dxdt$$

That is:

$$\chi(0) = \chi_{\Omega_0}, \quad \frac{\partial}{\partial t}\chi + \nabla\chi.V = 0$$

That problem is a unique solution, then $\chi = \chi_V = \chi^2$. Then χ_{V_N} converges to χ_V strongly in $L^2(0,\tau,L^2(D))$. □

3.2 Boundedness of the Perimeter

PROPOSITION 3.3. *Let f_n be a bounded sequence in $L^2(0,\tau,BV(D))$ such that $\frac{\partial}{\partial t}f_n$ is bounded in $L^2(0,\tau,H^{-2}(D))$. Then there exists a subsequence strongly convergent in $L^2(0,\tau,L^1(D))$.*

We adapt to the present situation the proof of [7] theorem 5.1 page 58 (in the R.Temam's version given in foot notes).

LEMMA 3.4.

$$\forall\eta > 0, \textit{there exists a constant } c_\eta \quad \textit{with} \ \forall\phi \in BV(D),$$

$$||\phi||_{L^1(D)} \le \eta||\phi||_{BV(D)} + c_\eta||\phi||_{H^{-2}(D)}$$

Proof. (of the lemma) Assume that it is wrong. Then, $\forall\eta > 0$, there exists $\phi_n \in BV(D)$ and $c_n \to \infty$ such that

$$||\phi_n||_{L^1(D)} \ge \eta||\phi_n||_{BV(D)} + c_n||\phi_n||_{H^{-2}(D)}$$

We introduce $\psi_n = \phi_n/||\phi_n||_{BV(D)}$, and we derive:

$$||\psi_n||_{L^1(D)} \ge \eta + c_n||\psi_n||_{H^{-2}(D)} \ge \eta$$

But also $||\psi_n||_{L^1(D)} \le c||\psi_n||_{BV(D)} = c$, for some constant c. Then: $||\psi_n||_{H^{-2}(D)} \to 0$. But as $||\psi_n||_{BV(D)} = 1$, there exists a subsequence strongly convergent in $L^1(D) \subset H^{-2}(D)$, which turns to be strongly convergent to zero. This is a contradiction with $||\psi_n||_{L^1(D))} \ge \eta$. □

Proof. (of the proposition) From the lemma, $\forall\eta > 0$, there exists a constant d_n such that

$$\forall f \in L^2(0,\tau,BV(D)),$$

$$||f||_{L^2(0,\tau,L^1(D))} \le \eta||f||_{L^2(0,\tau,BV(D))} + d_n||f||_{L^2(0,\tau,H^{-2}(D))}$$

Given $\epsilon > 0$, as $||f_n||_{l^2(0,\tau,BV(D))} \le M$, we shall get ;

$$||f_n||_{L^2(0,\tau,L^1(D))} \le 1/2\epsilon + d_n\,||f_n||_{L^2(0,\tau,H^{-2}(D))}$$

if we chose η such that $\eta M \le 1/2\,\epsilon$. At that point the conclusion will derive if we establish strong convergence to zero of f_n in $L^2(0,\tau,H^{-2}(D))$. Now, as $L^1(D) \subset H^{-2}(D)$, we get $f_n \in H^1(0,\tau,H^{-2}(D)) \subset C^0([0,\tau],H^{-2}(D))$ so that by use of Lebesgue dominated convergence theorem it will be sufficient to prove the pointwise convergence of $f_n(t)$ strongly to zero in $H^{-2}(D)$. We shall prove it for $t = 0$. We have $f_n(0) = a_n + b_n$, with

$$a_n = 1/s \int_0^s f_n(t)dt, \quad b_n = -1/s \int_0^s (s-t) f_n'(t)dt$$

If $\epsilon > 0$ is given we chose s such that

$$||b_n||_{H^{-2}(D)} \leq \int_0^s ||f_n'(t)||_{H^{-2}(D)}\, dt$$

Finally we observe that $a_n \to 0$ weakly in $BV(D)$, then strongly in $H^{-2}(D)$. □

THEOREM 3.5. *Let V_n be a bounded sequence in E such that*

$$t \longrightarrow ||\nabla \chi_{V_n}(t)||_{M^1(D,R^N)}$$

is bounded in $L^2(0,\tau)$ and V_n weakly converges to V in $L^2(0,\tau,L^2(D))$. Then χ_{V_n} strongly converges to χ_V in $L^2(0,\tau,L^2(D))$.

The idea is to apply the previous proposition to $f_n = \chi_{V_n}$. From lemma 3.2 we get the boundedness in $L^2(0,\tau,H^{-2}(D))$, then we get $\chi_{V_n} \to \chi$ strongly in $L^2(0,\tau,L^2(D))$ to a characteristic function χ. Then the conclusion derives.

3.3 Boundedness of the Density Perimeter

Density Perimeter

Following [1], [2], we consider for any closed set A in D the density perimeter associated to any $\gamma > 0$ by the following.

$$P_\gamma(A) = sup_{\epsilon \in (0,\gamma)} \left[\frac{meas(A^\epsilon)}{2\,\epsilon}\right] \tag{3.3}$$

Where A^ϵ is the dilation $A^\epsilon = \cup_{x \in A} B(x,\epsilon)$. We recall some main properties:

The mapping $\Omega \to P_\gamma(\partial\Omega)$ is lower-semi continuous in the H^c-topology

The property $P_\gamma(\partial\Omega) < \infty$ implies that $\text{meas}(\partial\Omega) = 0$ and $\Omega - \partial\Omega$ is open in D.

If $P_\gamma(\partial\Omega_n) \leq m$ and Ω_n converges in the H^c – topology to some open subset $\Omega \subset D$, then the convergence holds in the $L^2(D)$topology of the characteristic functions

Clean Open Tube

A Clean open tube is a set $\tilde{Q}$ in $]0,\tau[\times D$ such that for $a.e.t$, $\tilde{\Omega}_t = \{x \in D \mid (t,x) \in \tilde{Q}\}$ is an open set in D verifying for almost every t in $(0,\tau)$ the following cleanness property:

$$meas(\partial\tilde{\Omega}_t) = 0, \quad \tilde{\Omega}_t = \text{ interior of } cl(\tilde{\Omega}_t). \tag{3.4}$$

Notice that as the previous openness condition holds at almost every time t, the set Q is not necessarelly itself an open subset in $]0,\tau[\times D$. Nethertheless when the field V is smooth the tube $\bigcup_{0<t<\tau}\{t\}\times T_t(V)(\Omega_0)$ is open (resp. open and clean open) when the initial set Ω_0 is open (resp. clean open) in D.

We say that two tubes Q and Q' are equivalent if the characteristic functions are equal as elements in $L^2(0,\tau,L^2(D))$ (i.e. $\chi_Q = \chi_{Q'}$), that is to say that at almost every time t the two sets $\{x \in D \mid (t,x) \in Q\}$ and $\{x \in D \mid (t,x) \in Q'\}$ are the same up to a measurable subset E of D verifying $meas(E) = 0$.

LEMMA 3.6. *Let Q be a measurable set in $]0,\tau[\times D$, if there exists a clean open tube $\tilde{Q}$ such that $\chi_Q = \chi_{\tilde{Q}}$, then that clean tube is unique.(There exists at most one equivalent clean open tube)*

Proof. assume two such clean tubes $\tilde{Q}$ and $\tilde{Q}'$. Then at a.e. t we have $\tilde{\Omega}_t = \tilde{\Omega}'_t$ up to a measurable set E_t verifying $meas(E_t) = 0$. As those two open set verify (3.4) they are equals. □

When Ω_0 is a clean open in D and Q is a clean open tube in $]0,\tau[\times D$, with

$$\chi_Q \in C^0([0,\tau], H^{-1/2}(D))$$

and such that there exists a divergence free field V in E such that:

$$\frac{\partial}{\partial t}\chi_Q + \nabla\chi_Q.V = 0, \quad \chi_Q(0) = \chi_{\Omega_0} \tag{3.5}$$

we say that V builds the tube and we note $Q = Q_V$. Now such field when it exists is not unique. The set of fields that built the clean open tube Q is closed and convex:

$$\mathcal{V}_Q = \{V \in E \mid Q_V = Q\} \tag{3.6}$$

LEMMA 3.7. *If the set $\mathcal{V}_Q$ is non empty, then it is closed and convex in E so it contains a unique element V_Q which minimizes the $L^2(0,\tau,L^2(D))$ norm in that class.*

When the tube Q is built by a smooth field $V \in L^1(0,\tau,W_0^{1,\infty}(D,R^N))$, that is $Q = Q_V$ (i.e. $\Omega_t = \{x \in D \mid (t,x) \in Q\} = T_t(V)(\Omega_0)$), obviously the convex set $\mathcal{V}_{Q_V}$ is non empty as $V \in \mathcal{V}_{Q_V}$. But in general V_{Q_V}, the minimum L^2-norm element in $\mathcal{V}_{Q_V}$, is different from V.

We describe now a construction of clean open tubes for which the set $\mathcal{V}_Q$ is non empty.

The "Parabolic" Situation

We turn to the situation of dynamical domain. One could think to use the time-space perimeter, see [11]. For any smooth free divergence vector field, $V \in C^0([0,\tau],W_0^{1,\infty}(D,R^N))$, we consider,

$$\Theta_\gamma(V,\Omega_0) = Min \left\{ \int_0^\tau (\frac{\partial}{\partial t}\mu)^2 \, dt \mid \mu \in \mathcal{M}_\gamma(V,\Omega_0) \right\} \tag{3.7}$$

Where

$$\mathcal{M}_\gamma(V,\Omega_0) = \{ \mu \in H^1(0,\tau), \, P_\gamma(\partial\Omega_t(V)) \le \mu(t) \; a.e.t, \quad \mu(0) \le (1+\gamma)P_\gamma(\partial\Omega_0) \}$$

Where are many examples in which that set is non empty. When that set is empty we put $\Theta_\gamma(V,\Omega_0) = +\infty$. Notice that even when the mapping $p = (t \longrightarrow P_\gamma(\Omega_t(V))\ )$ is an element of $H^1(0,\tau)$ (then $p \in \mathcal{M}_\gamma(V,\Omega_0)\ \})$, we may have: $\Theta(V,\Omega_0) < ||p'||^2_{L^2(0,\tau)}$ as the minimizer will escape to possible variation of the function p.

PROPOSITION 3.8. *For any smooth free divergence field $V \in C^0([0,\tau], W_0^{1,\infty}(D,R^N))$, we have:*

$$P_\gamma(\partial\Omega_t(V)) \leq 2P_\gamma(\partial\Omega_0) + \sqrt{\tau}\ \Theta(V,\Omega_0)^{1/2} \tag{3.8}$$

Proof. As

$$P_\gamma(\partial\Omega_t(V)) \leq \mu(t) \leq 2P_\gamma(\partial\Omega_0) + \int_0^\tau \frac{\partial}{\partial t}\mu(t)dt$$

Then

$$P_\gamma(\partial\Omega_t(V_n)) \leq \mu(t) \leq 2P_\gamma(\partial\Omega_0) + \sqrt{\tau}(\int_0^\tau (\frac{\partial}{\partial t}\mu(t))^{1/2}dt$$

As μ is chosen being the minimizer element associated with V, the estimation is proved. □

PROPOSITION 3.9. *Let $V_n \in C^0([0,\tau], W_0^{1,\infty}(D,R^N))$, with the following convergence:*

$$V_n \longrightarrow V \ in\ \ L^2((0,\tau)\times D, R^N)$$

and the uniform boundedness:

$$\exists M > 0,\ \ \Theta(V_n,\Omega_0) \leq M$$

Then

$$\Theta(V,\Omega_0) \leq liminf\ \Theta(V_n,\Omega_0)$$

Proof. With the boundedness assumption:

$$P_\gamma(\partial\Omega_t(V_n)) \leq C = 2P_\gamma(\partial\Omega_0) + \sqrt{\tau}\ M^{1/2}$$

Let μ_n be the unique minimizer in $H^1(0,\tau)$ associated with $\Theta(V_n,\Omega_0)$. There exists a subsequence, still denoted μ_n, which weakly converges to an element $\mu \in H^1(0,\tau)$. That convergence holds strongly in $L^2(0,\tau)$, then almost every where. By definition we have

$$P_\gamma(\partial\Omega_t(V_n)) \leq\ \mu_n\ a.e.$$

Then in the limit:

$$P_\gamma(\partial\Omega_t(V)) \leq\ liminf P_\gamma(\partial\Omega_t(V_n)) \leq\ \mu(t),\ \ a.e.t$$

Also the square of the norm being weakly lower semi continuous we have

$$\int_0^\tau (\frac{\partial}{\partial t}\mu(t))^2 dt\ \leq\ liminf\ \int_0^\tau (\frac{\partial}{\partial t}\mu_n(t))^2 dt$$

that leads to

$$\Theta(V,\Omega_0) \leq \int_0^\tau (\frac{\partial}{\partial t}\mu(t))^2 dt \leq liminf\ \Theta(V_n,\Omega_0)$$

□

PROPOSITION 3.10. *Let $V_n \in C^0([0,\tau], W_0^{1,\infty}(D,R^N))$, with the following convergence:*

$$V_n \longrightarrow V \text{ in } L^2((0,\tau)\times D, R^N)$$

and the uniform boundedness:

$$\exists M > 0, \quad \Theta(V_n, \Omega_0) \le M$$

We assume that Ω_0 is an open subset in D verifying

$$\Omega_0 = \text{interior of } \bar{\Omega}_0$$

Then there exists a clean open tube $\tilde{Q} = \bigcup_{0<t<\tau} \{t\} \times \tilde{\Omega}_t$ such that:

$$\text{at a.e.} t \in (0,\tau), \ \chi_{\Omega_t(V_n)} \to \chi_{\tilde{\Omega}_t} \text{ in } L^2(D), \ \Omega_t(V_n) \to \tilde{\Omega}_t \text{ in} H^c \text{ topology.}$$

a.e.$t \in (0,\tau)$, the set $\tilde{\Omega}_t$ is an open set verifying cleanness property (3.4). Moreover $\chi_{\tilde{\Omega}_t}$ is the single a.e.t open set verifying those conditions (3.4) and whose characteristic function solves the convection problem:

$$\frac{\partial}{\partial t}\chi_{\tilde{\Omega}_t} + \nabla\chi_{\tilde{\Omega}_t}.V(t) = 0, \quad \chi_{\tilde{\Omega}_0} = \chi_{\Omega_0}$$

That is: $\tilde{\Omega}_t$ is the unique open family in D verifying the previous cleanness property and such that $\chi_{\tilde{\Omega}_t} = \chi_{\Omega_t(V)}$a.e.t.

Proof. We have $\chi_{Q_{V_n}} \to \chi_{Q_V}$ in $L^2(I \times D)$. Then for almost every t we have $\chi_{\Omega_t(V_n)} \to \chi_{\Omega_t(V)}$ in $L^2(D)$. At each t there exist a subsequence (depending on t) which converges in H^c-topology to an open set: $\Omega_t(V_{n_k}) \to \omega_t$. Now for a.e.$t$ we know that $\Omega_t(V_n) \to \Omega_t(V)$ in measure (i.e. for the $L^2(D)$-norm of the characteristic functions), then at a.e.t, $\chi_{\omega_t} = \chi_{\Omega_t(V)}$. From the boundedness of $P_\gamma(\Omega_t(V_n))$ we derive that for almost every t, $\omega_t(V)$ is an open set in D and $meas(\partial\omega_t(V)) = 0$. Then we set $\tilde{\Omega}_t = cl(\omega_t) - \partial\omega_t$. □

We consider now for any smooth field

DEFINITION 3.11.

$$p_\gamma(V,\Omega_0) = \int_0^\tau P_D(\Omega_t(V))dt \ + \gamma\,\Theta_\gamma(V,\Omega_0)$$

THEOREM 3.12. *Let $V_n \in C^0([0,\tau], W_0^{1,\infty}(D,R^N))$ which weakly converges in E to V with the boundedness condition:*

$$p_\gamma(V_n,\Omega_0) \le M. \tag{3.9}$$

Then there exists a clean open tube $\tilde{Q}$ built by V with the following convergence

$$\text{a.e.} t \in (0,\tau), \ \chi_{V_n}(t) \to \chi_V(t), \quad \Omega_t(V_n) \to \tilde{\Omega}_t \text{ in } H^c \text{ topology}$$

and $p_\gamma(V) \le liminf\ p_\gamma(V_n)$.

4 Variational Principle in Euler Problem

The two fluids functional: in the moving domain Ω_t we assume the density $\rho_i = 1 + a$ while in the exterior domain $\Omega_t^c = \bar{D} - \Omega_t$, the density is $\rho_e = a$. The one fluid configuration will correspond to $a = 0$. Then we assume $a \geq 0$. The Euler functional is

$$J(V) = \int_I \int_D [(a + \chi_V) \ 1/2 \ |V(t,x)|^2 - \chi_V g(t,x) \ - \ f(t,x).V(t,x) \ ] \, dxdt \qquad (4.1)$$

where χ_V (that we shall denote by χ when no confusion is possible) is the solution to the following problem:

$$\chi(0) = \chi_{\Omega_0}, \ \frac{\partial}{\partial t}\chi + \nabla\chi.V = 0, \ \ \chi = \chi^2 \qquad (4.2)$$

In that section we assume that V is a smooth divergence free vector field so that we get:

THEOREM 4.1.

i) The functional J is Gateaux differentiable on $E^\infty = E \cap C^0([0,\tau], C^\infty(\bar{D}, R^N))$.

ii) Assume that $V \in E^\infty$ is such that $\forall W \in E^\infty$, $J'(V,W) = 0$, then there exists $P \in \mathcal{D}'(D, R^N)$ such that:

$$\frac{\partial}{\partial t}(\,(a + \chi_V)\,V\,) + D(\,(a + \chi_V)\,V\,).V \ + \ \nabla P = g\ \nabla\chi_V \qquad (4.3)$$

Proof. The proof follows the classical control theory approach: the field V can be considered as a control parameter while χ solution of (4.2) is the state variable. In that approach the derivative of equation (4.2) leads to the derivative χ' solution of (4.4) bellow. That equation is not completely studied here, but as the field V is assumed to be smooth, χ is obtained by the flow mapping $T_t(V)$ so that it is easily verified that the mapping $V \longrightarrow \chi$ is Gateaux differentiable in $L^2(0,\tau, H^2(D) \cap H_0^1(D))$. By direct calculus, as V is smooth we have:

$$J'(V,W) = \int_I \int_D (\ \chi'\,(1/2|V|^2 \ - g) \ + \ [(a+\chi)V - f]W \ )dxdt$$

Where χ' is the Gateaux derivative of χ at V in the direction W, given by:

$$\chi'(0) = 0, \ \frac{\partial}{\partial t}\chi' + \nabla\chi'.V = -\nabla\chi.W \qquad (4.4)$$

We introduce the adjoint problem:

$$\lambda(\tau) = 0, \ -\frac{\partial}{\partial t}\lambda \ - \ \nabla\lambda.V = 1/2\ |V|^2 \ - g \qquad (4.5)$$

$$J'(V,W) = \int_I \int_D ((-\frac{\partial}{\partial t}\lambda - \ \nabla\lambda.V)\ \chi' \ \ + [\ (a+\chi)V - f\,]W)dxdt$$

$$= \int_I \int_D ((\frac{\partial}{\partial t}\chi' + \ \nabla\chi'.V)\ \lambda \ \ + [\ (a+\chi)V - f\,]W)dxdt$$

$$= \int_I \int_D ((-\nabla\chi.W)\ \lambda\ + [\ (a+\chi)V - f\,]W)dxdt$$

But as $\int_D ((-\nabla\chi.W)\ \lambda\, dx = \int_D\ \chi.W\ (\nabla\lambda)dx$ we get

$$J'(V,W) = \int_I \int_D [\chi\,\nabla\lambda\ +\ (a+\chi)V\ -\ f\,]\,W dxdt \tag{4.6}$$

□

4.1 The Linear Decoupled Problem

We set $\Lambda = \chi_V\ \nabla\lambda$ and Λ solves the following problem:

PROPOSITION 4.2. *The variable* Λ *solves the backward problem*

$$\Lambda(\tau) = 0, \quad -\frac{\partial}{\partial t}\Lambda - D\Lambda.V\ -\ D^*V.\Lambda = \chi\,\nabla(\ 1/2|V|^2\ ) \tag{4.7}$$

Proof: with (3.5) we get

$$-\frac{\partial}{\partial t}\chi\ \ \nabla\lambda\ -\ \nabla\chi.V\ \ \nabla\lambda = 0 \tag{4.8}$$

Also from (4.5) we get

$$\frac{\partial}{\partial t}\Lambda = \frac{\partial}{\partial t}\chi\,\nabla\lambda + \chi\,\nabla(\frac{\partial}{\partial t}\lambda)$$

But from the equation verified by λ we get:

$$-\nabla(\frac{\partial}{\partial t}\lambda) - \nabla(\nabla\lambda.V) = 1/2\ \nabla(|V|^2\,) - \nabla g$$

After multiplying by χ and with $\nabla(A.B) = D^*A.B + D^*B.A$ we get:

$$-\chi\ \nabla(\frac{\partial}{\partial t}\lambda)\ -\ \chi(D^*(\nabla\lambda).V\ +\ D^*V.\nabla\lambda) = 1/2\,\chi\,\nabla(\,|V|^2\,) - \chi\,\nabla g \tag{4.9}$$

By adding (4.8) and (4.9) we get:

$$-\chi\ \nabla(\frac{\partial}{\partial t}\lambda)\ -\ \frac{\partial}{\partial t}\chi\ \ \nabla\lambda\ -\ \nabla\chi.V\ \ \nabla\lambda - \ \chi D^*(\nabla\lambda).V\ -\ D^*V.(\chi\nabla\lambda)$$

$$= 1/2\,\chi\,\nabla(\,|V|^2\,) - \chi\,\nabla g$$

Notice that

$$[\,\nabla\chi.V\ \ \nabla\lambda + \chi D^*(\nabla\lambda).V\,]_i\ =\ \frac{\partial}{\partial x_j}\chi\,V_j\ \frac{\partial}{\partial x_i}\lambda\ +\ \chi\,\frac{\partial}{\partial x_i}(\ \frac{\partial}{\partial x_j}\lambda)V_j$$

$$= (\ \frac{\partial}{\partial x_j}\chi\,\frac{\partial}{\partial x_i}\lambda\ +\ \chi\,\frac{\partial}{\partial x_j}(\ \frac{\partial}{\partial x_i}\lambda)\,)\,V_j$$

$$=\ \frac{\partial}{\partial x_j}(\chi\,\frac{\partial}{\partial x_i}\lambda\,)\,V_j\ =\ \frac{\partial}{\partial x_j}(\Lambda_i\,)\,V_j\ =\ (D\Lambda.V)_i$$

and then $\nabla\chi.V\ \ \nabla\lambda + \chi D^*(\nabla\lambda).V\ = D\Lambda\,.V$

And finally the proposition is proved. Notice that as far as the field V is smooth the sensitivity analysis of the element χ' could have been traited by the transverse field approached developed in [12],[4].

4.2 Necessary Optimality Condition

From (4.6) the necessary condition for extremality of J at a smooth vector field V would be: there exists a distribution π such that:

$$\chi_V \, \nabla\lambda + (a+\chi_V)\, V \; - \; f \; = \; -\nabla\pi \tag{4.10}$$

That is condition:

$$\Lambda = f - (a+\chi_V)V \; - \; \nabla\pi \tag{4.11}$$

Moreover Plugging the necessary condition (4.11) in (4.7), we get:

$$-\frac{\partial}{\partial t}(f - (a+\chi_V)V \; - \; \nabla\pi) \; - D(f - (a+\chi_V)V \; - \; \nabla\pi).V$$

$$- D^*V.(f - (a+\chi_V)V \; - \; \nabla\pi) = \chi\,\nabla(\,1/2|V|^2 \; - g\,)$$

Which can be rewritten as follows:

$$\begin{aligned} \frac{\partial}{\partial t}(\,(a+\chi_V)V\,) + D(\,(a+\chi_V)V\,).V \; + \; \frac{\partial}{\partial t}(\nabla\pi - f) \\ +D(\,\nabla\pi).V \; + \; D^*V.(\,\nabla\pi) \\ -(\,D(f).V \; + \; D^*V.f) \\ = \; - D^*V.(\,(a+\chi_V)V\,) \; + \chi\,\nabla(\,1/2|V|^2 \; - g\,) \\ = \; -a\,D^*V.V \; - \chi_V\,D^*V.V + 1/2\,\chi_V\,\nabla(|V|^2) - \chi\,\nabla g \end{aligned}$$

Here we have the interesting simplification through the classical

LEMMA 4.3.

$$D^*V.V = 1/2\;\nabla(|V|^2) \tag{4.12}$$

Then we get

$$= -a/2\;\nabla(\,|V|^2\,) - \chi\,\nabla g$$

That simplication must be underline here. The previous lemma applies as the intgrand function, in the very definition of the functional J is chosen as the kinetic energy itself. If the functional J was in the form

$$J(V) = \int_0^T \int_D (a+\chi_V)\mathcal{R}(V) + \ldots.$$

We would face here to the term

$$\chi_V(\; -D^*V.V \; + 1/2\;\nabla(\mathcal{R}(V)\,)\;)$$

Which would not cancel.... Finally, as $D(\,\nabla\pi).V \; + \; D^*V.(\,\nabla\pi) \; = \nabla(\,\nabla\pi.V\,)$, We get:

$$\frac{\partial}{\partial t}(\,(a+\chi_V)V\,) + D(\,(a+\chi_V)V\,).V\;)$$

$$\nabla\left(\frac{\partial}{\partial t}(\pi) + \nabla\pi.V + a/2\,|V|^2 \right)$$

$$= \frac{\partial}{\partial t}f + \left(D(f).V + D^*V.f \right) - \chi\nabla g$$

$$= \left(D^*(f).V + D^*V.f \right) + \left(D(f) - D^*f \right).V\chi\nabla g$$

$$= \nabla(f.V) + \left(D(f) - D^*f \right).V - \chi\nabla g$$

then we get V as a solution to the problem

$$\frac{\partial}{\partial t}(\rho_V\, V) + D(\rho_V\, V).V + \nabla P \tag{4.13}$$

$$= \frac{\partial}{\partial t}f + \left(D(f) - D^*f\right).V - \chi\nabla g$$

Where the density is

$$\rho_V = (a + \chi_V)$$

And the pressure is given by

$$P = \frac{\partial}{\partial t}\pi + \nabla\pi.V + a/2\,|V|^2 - f.V$$

If we assume that $\sigma(f) = Df - D^*f$ is zero, then $curlf = 0$ (as $\sigma..\sigma = |curlf|^2$) and f derives from a potential, for example in the following form:

$$f(t,x) = \int_0^t \nabla F(\sigma, x)\, d\sigma$$

Then V solves the following Euler equation

$$\frac{\partial}{\partial t}(\rho_V\, V) + D(\rho_V\, V).V + \nabla P$$

$$= \nabla F - \chi\nabla g$$

In fact "F is of no use in the functional" as any additive gradient term can be "absorbed" by the pressure term P as follows:

$$\frac{\partial}{\partial t}(\rho_V\, V) + D(\rho_V\, V).V + \nabla P = g\,\nabla\chi \tag{4.14}$$

With

$$P = \frac{\partial}{\partial t}\pi + \nabla\pi.V + a/2\,|V|^2 - f.V + \chi\, g - F \tag{4.15}$$

5 Existence Results

for any given positive constants $\sigma \geq 0$, $\mu \geq 0$ and $\nu \geq 0$ we shall consider the minimization associated to the following functional

$$J_{\sigma,\mu,\nu}(V) = J(V) + \sigma \int_0^{\tau} ||\nabla(\chi_V(t))||_{M^1(D)} dt + \mu\Theta(V, \Omega_0) + \nu \int_0^{\tau} \int_D DV..DV\, dxdt \qquad (5.1)$$

In the sequel, with $a > 0$ we shall consider the three situations associated with $\sigma+\mu+\nu > 0$ and $\sigma\,\mu\,\nu = 0$. When ν is zero the terms σ and μ will play a surface tension role at the dynamical interface while the case $\sigma + \mu = 0$ should be consider as a mathematical regularisation as in the non usual variational interpretation developed in the previous section $\nu > 0$ does not lead to the usual viscosity term (i.e. does not lead to the Navier Stoke equations) Let

$$E = \{V \in L^2(0,\tau, L^2(D, R^N)) \,:\, divV(t) = 0 \; a.e.t,\; V(t).n_{\partial D} = 0 \text{in} \;\; H^{-1/2}(\partial D)\}$$

THEOREM 5.1. *Assuming $\sigma.\eta.\nu > 0$, there exists $V \in E$ such that, $\forall W \in E$:*

$$J_{\sigma,\eta,\nu}(V) \leq J_{\sigma,\eta,\nu}(W)$$

Let us consider a minimizing sequence V_n of $J_{\sigma,\eta,\nu}$ in E. Then V_n is bounded in

$$L^2(0, \tau, L^2(D, R^N))$$

we consider a subsequence weakly converging to V. By the compacity results (3.1),(3.5) or (3.12), we get the strong convergence of the characteristic functions:χ_{V_n} strongly converges to χ_{V_n}. For any $\Phi \in L^2(0,\tau, L^2(D))$ we get $\chi_{V_n}\Phi$ which strongly converges to $\chi_{V_n}\Phi$ (as χ_{V_n} converges almost every where and is dominated by 1). Then $\chi_{V_n} V_n$ weakly converges to $\chi_V V$ in $L^2(0,\tau, L^2(D))$:

$$\forall \Phi \in L^2(0,\tau, L^2(D)),\; \int_0^{\tau}\int_D \chi_{V_n} V_n\, \Phi dxdt = \int_0^{\tau}\int_D (\chi_{V_n}\Phi)\; V_n\, dxdt$$

$$\longrightarrow \int_0^{\tau}\int_D (\chi_V \Phi)\; V\, dxdt$$

We set:

$$\chi_{V_n}(V_n)^2 = (\chi_{V_n} V_n)^2$$

Then

$$\int_0^{\tau}\int_D (\chi_V\, V)^2\, dxdt \leq \; liminf \int_0^{\tau}\int_D (\chi_{V_n}\, V_n)^2\, dxdt$$

It derives classiquely that the limiting element V realizes the minimum of the functional $J_{\sigma,\eta,\nu}$ over the linear space E.

6 Differentiability

6.1 Field in $L^2(0,\tau,L^4(D,R^N))$

We consider a divergence free field V in $L^2(0,\tau,L^4(D,R^N))$. Notice that $V.V = |V|^2 \in L^1(0,\tau,L^2(D))$. We consider

$$J(V) = MIN_{\zeta\in L^\infty(0,\tau,L^\infty(D))}\ MAX_{\phi\in\mathcal{H}_V}\ \ \mathcal{L}(\zeta,\phi)$$

with

$$\mathcal{L}(\zeta,\phi) = \int_0^\tau \int_D \{\ 1/2(a+\zeta)\,|V|^2\ -\ g\zeta\ +\ \zeta\,(\frac{\partial}{\partial t}\phi + \nabla\phi.V)\,\}dxdt - \int_{\Omega_0}\phi(0)dx$$

or, for any integer $m \geq 1$:

$$\mathcal{L}^m(\zeta,\phi) = \int_0^\tau \int_D \{\ 1/2\zeta^m\,|V|^2\ -\ g\zeta\ +\ \zeta\,(\frac{\partial}{\partial t}\phi + \nabla\phi.V)\,\}dxdt - \int_{\Omega_0}\phi(0)dx$$

The Lagrangian $\mathcal{L}$ is concave-convex on $L^\infty(0,\tau,L^\infty(D)) \times \mathcal{H}_V$.

Saddle points (χ,λ) are characterized by:

$$\frac{\partial}{\partial t}\chi + \nabla\chi.V = 0,\ \ \chi(0) = \chi_{\Omega_0}$$

$$\frac{\partial}{\partial t}\lambda + \nabla\lambda.V\ = 1/2\,|V|^2\ -\ g\,,\ \ \ \lambda(\tau) = 0$$

or

$$\frac{\partial}{\partial t}\lambda^m + \nabla\lambda^m.V\ = m/2\ \ \chi_V\,|V|^2\ -\ g\,,\ \ \ \lambda^m(\tau) = 0$$

If $V \in L^2(0,\tau,L^4(D,R^3))$ we get $f = 1/2\,\chi_V\,|V|^2\ -\ g \in L^1(0,\tau,L^2(D))$, then that uncoupled system possesses a unique solution

$$(\chi_V,\lambda)\ \text{in}\ \ L^2(0,\tau,L^2(D)) \times \mathcal{W}_V.$$

In general such functional J in form of min max has a well known Gateaux derivative.Formally, assuming also that $V \in L^2(0,\tau,H^1_0(D,R^N))$, we have

$$J'(V,W) = \int_0^\tau \int_D \chi_V < (V + \nabla\lambda), W >\ dxdt$$

Notice that if V was a minimum of J (with the previous regularities) we would get the necessary optimality condition in the form $\chi_V(V + \nabla\lambda) = \nabla\pi$. That calculus cannot be justified as the differentiability of a functional in Min Max form requires that the sets are not parameter dependant, while here the space $\mathcal{H}_V$ (or $\mathcal{W}_V$) does depends on the parameter V. (It does not depends on the field V when V is smooth enough, then that MinMax derivative is justified). Netherveless the necessary condition is the same one we discovered through the direct calculus.This let us think that the differentiability of J at any divergence free vector field $V \in L^2(0,\tau,L^2(D,R^N))$ depends on the well posedness of the equation in λ with right hand side in $L^1(0,\tau,L^1(D))$.

7 Eulerian Metric on Measurable Sets

Besides the Courant Metric we show how we can easily define with the previous material a "Eulerian metric" with analogous compacity properties. For convenience here (in view of the Euler variational formulation) we considered vector fields V in Hilbert space in the form $L^2(0,\tau,H)$. In that section we could have taken V in $L^1(0,\tau,H)$ which would have eliminated the infimum in the definition of the metric bellow. Let D be a bounded domain in R^N and $\mathcal{O}_a$ the family of measurable subsets Ω in D with $meas(\Omega) = a$. For any vector field V in $\mathcal{E}_0 = \{\ V \in L^2([0,\tau], L^2(D,R^N))$ with $divV(t) = 0\ a.e.t\ ,\ \ V(t,.).n_{\partial D} = 0$ in $H^{-1/2}(\partial D)\}$ and any $\Omega_0 \in \mathcal{O}_a$ we associate the solution $\chi \in L^2([0,\tau], Char(D)) \cap C^0([0,\tau], H^{-\frac{1}{2}}(D))$, to the convection problem

$$\chi(0) = \chi_{\Omega_0},\ \ \frac{\partial}{\partial t}\chi + \nabla\chi.V = 0 \tag{7.1}$$

For convenience, some times, we shall denote that characteristic function as χ_V. We consider the family of elements in $\mathcal{O}_a$ which can be reached from Ω_0 in the following sense:

$$\mathcal{O}_{a,\Omega_0} = \{\,\Omega \in \mathcal{O}_a\ s.t.\ \text{there exits } V \in \mathcal{E}_0 \text{ with } \chi_\Omega = \chi(1)\ \} \tag{7.2}$$

$$d(\Omega_0,\Omega) = \ INF_{\{\ V\in\mathcal{E}_0\ s.t.\ \chi_\Omega=\chi(1)\ \}} \int_0^1 ||V(t)||_{L^2(D,R^N)}\ dt \tag{7.3}$$

If $d(\Omega_0,\Omega) = 0$, then for any ϵ there exist a field V_ϵ with $\chi_\Omega = \chi_{V_\epsilon}(1)$ and

$$\int_0^1 ||V_\epsilon(t)||_{L^2(D,R^N)}\ dt\ \le \epsilon$$

Then we haave a sequence V_{ϵ_n} converging to zero in $L^1(0,1,\ L^2(D,R^N))$

$$\forall\phi,\ \int_0^\tau \int_D \chi_n\ (-\frac{\partial}{\partial t}\phi - \nabla\phi.V_n)dtdx\ =\ \int_D \chi_{\Omega_0}\ \phi(0)\ dx$$

Let χ_n weakly$-*$ converges to χ^* in $L^\infty(0,1,L^2(D,R^N))$. So that in the limit we get:

$$\forall\phi,\ \int_0^\tau \int_D \chi^*\ (-\frac{\partial}{\partial t}\phi)dtdx\ =\ \int_D \chi_{\Omega_0}\ \phi(0)\ dx$$

That is, in distribution sense, $\frac{\partial}{\partial t}\chi^* = 0$, and we derive $\chi^*(t) = \chi_{\Omega_0}$ in $H^{-1/2}(D)$ for all t. The symmetry of d is obvious as $\Omega \in \mathcal{O}_{a,\Omega_0}$ if and only if $\Omega_0 \in \mathcal{O}_{a,\Omega}$. The triangle property derives for the L^1-norm property in the definition of d: let Ω^1 and Ω^2 belonging to $\mathcal{O}_{a,\Omega_0}$, let V^1 and V^2 vector fields associated with those sets in the definition of $\mathcal{O}_{a,\Omega_0}$. We have $\chi^i(1) = \chi_{\Omega^i}$ while $\chi^i(0) = \chi_{\Omega_0}$. Consider the field defined by

$V(t,x) = -2V^1(1-2t,x)$ for $0 \le t < 1/2$ and $V(t,x) = 2V^2(2t-1,x)$ for $1/2 \le t \le 1$. If χ is the solution to (1) with that field V we get $\chi(0) = \chi_{\Omega^1}$, $\chi(1/2) = \chi_{\Omega_0}$ and $\chi(1) = \chi_{\Omega^2}$. So that this field V is associated to Ω^2 in $\mathcal{O}_{a,\Omega^1}$. That is

$$d(\Omega^1,\Omega^2) \le \int_0^1 ||V(t)||_{L^2(D,R^N)}\ dt$$

$$= 2\int_0^{1/2} ||V^1(1-2t)||_{L^2(D,R^N)}\, dt + 2\int_{1/2}^1 ||V(2t-1)||_{L^2(D,R^N)}\, dt$$

$$= \int_0^1 ||V^1(s)||_{L^2(D,R^N)}\, ds + \int_0^1 ||V(s)||_{L^2(D,R^N)}\, dt$$

We consider in that inequality minimizing sequences V_n^i in the definitions of $d(\Omega_0, \Omega^i)$. For any positive ϵ we get

$$d(\Omega^1, \Omega^2) \leq d(\Omega^1, \Omega_0) + d(\Omega^2, \Omega_0) + 2\epsilon$$

In that majorization Ω_0 can be replaced by any element in $\mathcal{O}_{a,\Omega_0}$.

7.1 Pseudo Metrics

We consider, for each p, $1 < p \leq 2$, $s \geq 0$ the following:

$$\delta_{p,s}(\Omega_0, \Omega) = INF_{\{\ V\in\mathcal{E}_0\ s.t.\ \chi_\Omega = \chi_V(1)\}}\ (\int_0^1 ||V(t)||^p_{H^s(D,R^N)}\, dt)^{1/p} \tag{7.4}$$

Notice that the Infimum is reached for some field in $L^p(0, \tau, L^2(D, R^N))$. We have $d(\Omega_0, \Omega) = \delta_{1,0}(\Omega_0, \Omega)$.

By the some arguments we derive the following properties

PROPOSITION 7.1.

$$\delta_{p,s}(\Omega_1, \Omega_2) = 0 \text{ impliess } \chi_{\Omega_1} = \chi_{\Omega_2}$$

$$\delta_{p,s}(\Omega_1, \Omega_2) = \delta_{p,s}(\Omega_2, \Omega_1)$$

$$\delta_{p,s}(\Omega_1, \Omega_2) \leq 2^{1-1/p}\, [\delta_{p,s}(\Omega_1, \Omega_0) + \delta_{p,s}(\Omega_0, \Omega_2)]$$

7.2 Compact Subsets in $\mathcal{O}_{a,\Omega_0}$

Let be given a positive real number $s > 0$. We consider

$$\mathcal{O}^s_{a,\Omega_0} = \{\ \Omega \in \mathcal{O}_{a,\Omega_0}\ \ s.t.\ \text{ there exist } V \in L^2(0,1,H^s(D,R^N)),\ \chi_V(1) = \chi_\Omega\,\}$$

We define the metric d_s by replacing $L^2(D, R^N) = H^0(D, R^N)$ in the definition of D by the Sobolev space $H^s(D, R^N)$. And for any given constant $M > 0$, we consider the subset

$$\mathcal{K}^{s,M}_{a,\Omega_0} = \{\ \Omega \in \mathcal{O}^s_{a,\Omega_0}\ \text{ with } ||V||_{L^2(0,1,H^s(D,R^N))} \leq\ M\ \}$$

PROPOSITION 7.2. *The set $\mathcal{K}^{s,M}_{a,\Omega_0}$ is compact in the metric space $\mathcal{O}^s_{a,\Omega_0}$ (equipped with d_s).*

The prof is based on the continuity property:

PROPOSITION 7.3. *Let V_n be a weakly convergent sequence in*

$$L^2(0, \tau, H^s(D, R^N))$$

with $divV_n(t,.) = 0$ and $V_n(t).n_{\partial D} = 0$, then we have the strong convergence of the Characteristic functions associated to the χ_{Ω_0}-convections: $\chi_n \longrightarrow \chi$ strongly in $L^2(0, \tau, L^2(D))$.

We make use of the following boundedness:

LEMMA 7.4. *(similar to 3.2.) For all*

$$V \in L^2(0,\tau,\ L^2(D,R^N))$$

with $divV(t)=0$ and $V(t).n_{\partial D}=0$, we have:

$$||\frac{\partial}{\partial t}\chi_V||_{L^2(0,\tau,\ H^{-1}(D))} \leq ||V||_{L^2(0,\tau,L^2(D))} \tag{7.5}$$

$$||\frac{\partial}{\partial t}\chi_V||_{L^1(0,\tau,\ H^{-1}(D))} \leq ||V||_{L^1(0,\tau,L^2(D))} \tag{7.6}$$

$$||\frac{\partial}{\partial t}\chi_V||_{L^2(0,\tau,\ W^{-1,1}(D))} \leq ||V||_{L^2(0,\tau,L^\infty(D))} \tag{7.7}$$

For each n, $\phi \in \mathcal{D}(]0,\tau[\times D)$, we have:

$$\int_0^\tau \int_D \chi_n\ (-\phi_t\ - V_n.\nabla\phi\ )dxdt = \int_{\Omega_0} \phi(0)\,dx$$

We conclude as χ_n weakly converges in $L^2(0,\tau,\ L^2(D))$ (up to a subsequence) to an element λ (with $0 \leq \lambda \leq 1$ *a.e.*),then as $\frac{\partial}{\partial t}\chi_n$ remains bounded in $L^2(0,\tau,\ H^{-1}(D))$ (then in $L^2(0,\tau,\ H^{-1}(D))$), and as the continuous embedding $L^(D)$ in $H^{-1/2}(D)$ is compact, we conclude that the sequence of characteristic functions χ_n strongly converges to λ in

$$L^2(0,\tau,\ H^{-1/2}(D))$$

Then we pass to the limit in previous weak formulation as $V_n.\nabla\phi$ weakly converges to $V.\nabla\phi$ in $L^2(0,\tau,\ H^{1/2}(D))$. In the limit we get λ as the solution of the convection to the problem associated to the field V with initial domain Ω_0. Then $\lambda=\lambda^2$ is the characteristic function $\lambda = \chi_V$. It derives that the norms are also convergent and we have then the strong $L^2(0,\tau,\ L^2(D))$ convergence of χ_n to χ_V.

Compactness

THEOREM 7.5. *Let Ω_n be a sequence in $\mathcal{O}_{a,\Omega_0}$ such that $d_s(\Omega_n,\Omega_0) \leq M$. With $s>0,\ M>0$. Then for any σ, $0 \leq \sigma < s$, there exist a subsequence (still denoted Ω_n) and a set $\Omega \in \mathcal{O}^\sigma_{a,\Omega_0}$ such that Ω_n converges to Ω in $\mathcal{O}^\sigma_{a,\Omega_0}$, i.e. $d_\sigma(\Omega_n,\Omega) \to 0$ as $n\to\infty$ and $d_\sigma(\Omega,\Omega_0) \leq liminf_{n\to\infty} d_\sigma(\Omega_n,\Omega_0)$.*

Proof. Let V_n be an associated sequence such that

$$d_s(\Omega_n,\Omega_0) \leq \int_0^1 ||V_n(t)||_{H^s(D,R^N)}\,dt +\ 1/n$$

With $\chi_{\Omega_n} = \chi_{V_n}(1)$. Then V_n remains bounded in $L^1(0,1,H^s(D,R^N))$ then we consider a subsequence, still denoted V_n, which weakly converges

$$\sigma(\ L^1(0,1,\ H^s(D,R^N)),\ L^\infty(0,1,\ H^{-s}(D,R^N))\ )\text{ to an element}$$

$V \in L^1(0,1,H^s(D,R^N))$. We adapt to that situation the previous proof of continuity. We get $\frac{\partial}{\partial t}\chi_{V_n}$ bounded in $L^1(0,1,\ H^{-1}(D))$ then χ_{V_n} strongly converges in $L^1(0,1,\ H^{-1/2}(D))$ to some element $\chi = \chi^2 = \chi_V$. We set $\chi_\Omega = \chi(1)$ and the idea is to estimate $d_\sigma(\Omega,\Omega_n)$ For any $\phi \in \mathcal{D}(]0,1[\times D)$ we have

$$\int_0^1 \int_D \chi_{V_n}\ (-\phi_t - V_n.\nabla\phi)dxdt = \int_{\Omega_0} \phi(0)\ dx$$

The conclusion follows. □

Bibliography

[1] D. Bucur, J.P.; Zolésio. *Free Bondary Problems and Density Perimeter.* J. Differential. Equations 126(1996),224-243.

[2] D. Bucur and J.P. Zolésio. *Boundary Optimization under Pseudo Curvature Constraint.* Annali dela Scuola Normale Superiore di Pisa, IV, XXIII (4), 681-699, 1996.

[3] J. Cagnol and J.P. Zolésio. Shape Derivative in the Wave Equation with Dirichlet Boundary Conditions. J. Differential Equations 157, 1999.

[4] R. Dziri and J.P. Zolésio. Dynamical Shape Control in Non-cylindrical Navier-Stokes Equations. J. convex analysis, vol. 6, 2, 293-318, 1999.

[5] N. Gomez and J.P. Zolésio. Shape sensitivity and large deformation of the domain for norton-hoff flow. In G. Leugering, editor, *Proceedings of the IFIP-WG7.2 conference, Chemnitz,* volume 133 of *Int. Series of Num. Math.*, pages 167–176, 1999.

[6] T. Kato. Abstract Differential Equations and Non Linear Mixed Problems. Lezioni Fermiane. Scuola Norm. Sup. Pisa, 1985.

[7] J. L. Lions. Quelques méthodes de résolution des problèmes aux limites non linéaires. Dunod, Gauthier -Villars, Paris. 1969.

[8] J.P. Zolésio. Un résultat d'existence de vitesse convergente. C.R. Acad. Sc. Paris *Serie A*, volume 283, pp. 855, 1976.

[9] J.P. Zolésio. Identification de domaine par déformations. *Thèse de doctorat d'état.*, Univerité de Nice, 1979.

[10] J.P. Zolésio. In Optimization of Distributed Parameter structures, vol.II, (E. Haug and J. Céa eds.), Adv. Study Inst. Ser. E: Appl. Sci., 50, Sijthoff and Nordhoff, Alphen aan den Rijn, 1981:

i) The speed method for Shape Optimization, 1089-1151. ii) Domain Variational Formulation for Free Boundary Problems, 1152-1194. iii) Semiderivative of reapeted eigenvalues, 1457-1473.

[11] J.P. Zolésio. Variational Principle in the Euler Flow. In G. Leugering, editor, *Proceedings of the IFIP-WG7.2 conference, Chemnitz*, volume 133 of *Int. Series of Num. Math.*, 1999.

[12] J.P. Zolésio. Shape Differential Equation with Non Smooth Field. In Computational Methods for Optimal Design and Control. J. Borggard, j. Burns, E. Cliff and S. Schreck eds., volume 24 of *Progress in Systems and Control Theory*, pp.426-460, Birkhauser, 1998.

Jean-Paul Zolésio. Research Director at CNRS, Centre de Mathématiques Appliquées, Ecole des Mines de Paris, 2004 route des Lucioles, BP. 93, 06902 Sophia Antipolis Cedex, France
E-mail: Jean-Paul.Zolesio@sophia.inria.fr

Index